Managing Plant Production Under Changing Environment

Mirza Hasanuzzaman • Golam Jalal Ahammed •
Kamrun Nahar
Editors

Managing Plant Production Under Changing Environment

Springer

Editors
Mirza Hasanuzzaman
Department of Agronomy
Sher-e-Bangla Agricultural University
Dhaka, Bangladesh

Golam Jalal Ahammed
College of Horticulture and Plant Protection
Henan University of Science and Technology
Luoyang, China

Kamrun Nahar
Department of Agricultural Botany
Sher-e-Bangla Agricultural University
Dhaka, Bangladesh

ISBN 978-981-16-5058-1 ISBN 978-981-16-5059-8 (eBook)
https://doi.org/10.1007/978-981-16-5059-8

This Springer imprint is published by the registered company Springer Nature Singapore Pte Ltd.
The registered company address is: 152 Beach Road, #21-01/04 Gateway East, Singapore 189721, Singapore

Preface

The world population is expected to reach 9.8 billion in 2050 but agricultural lands will not expand the same speed. Already, one in nine people around the world suffer from hunger, and the only way to feed them is by doubling food production in a sustainable way. To meet this challenge, we must increase the food production by 70% to feed the increasing population by the year 2050. Global climate change is becoming a great concern for future agriculture that manipulates different complexes of nature and environmental components, thus restricting the utility of natural resources of the planet by disturbing the ecosystem. As water cycle is a vital component of this ecosystem, it has been severely hampered by the climate change in such an extent that it becomes unable to provide the required environment for plants. Moreover, the levels of environmental pollutants are increasing significantly due to climatic adversities and anthropogenic activities due to rapid urbanization and industrialization. All of these negative impacts are accelerating with the passage of time and ultimately link with agriculture. According to FAO, around 60% of the world population depends on agriculture and thus extensive agricultural activities provoke substantial negative impacts on the environment. Climate change affects agriculture and forestry globally, though its impacts may vary depending on the geographical region at different levels. But a minor change in the climate can affect the productivity of the crops to a great extent. One of the major effects of climate change is plant abiotic stress which includes salinity, drought, flooding, heat, cold, metal/metalloid toxicity, etc. These negatively affect plant growth, physiology, and yield to a great extent.

Agriculture is a significant factor in reducing poverty and developing the economy of the world. Agriculture and food production mainly depend on two natural resources, water and land. Advancements in technologies have contributed to more food production but it is not enough for current growing population. An increase in the production of growth that ultimately leads to an increase in economic growth put a huge cost into the natural environment. Therefore, enhancing plant productivity in changing climate is one of the major tasks for plant biologists.

In this book, we organized 19 chapters written by 139 experts in the field of agronomy, plant science, soil science, and environmental science. It provides comprehensive literature of recent advances on the aspects of decline in plant productivity under climate changes and how it affects food security globally. Emphasis has also been given to elaborate the recent advancement on the techniques of mitigating such negative consequences.

We would like to thank the authors for their time and effort in producing such chapters. We are grateful to Md. Rakib Hossain Raihan, Md. Mahabub Alam, Abdul Awal Chowdhury Masud, Khursheda Parvin, and Taufika Islam Anee of Sher-e-Bangla Agricultural University, for their help in reviewing and formatting of the manuscripts. We are highly thankful to Aakanksha Tyagi (Associate Editor, Springer India) and Jayesh Kalleri, Production Editor and Metilda Nancy Marie Rayan Thomas Albert Rayan, Project Manager of this book and all other editorial staffs for their editorial assistance and timely production of the book. The editors and contributing authors hope that this book will include a practical update on the knowledge for crop production in the era of climate change.

Dhaka, Bangladesh — Mirza Hasanuzzaman
Luoyang, China — Golam Jalal Ahammed
Dhaka, Bangladesh — Kamrun Nahar

Contents

Editors and Contributors

About the Editors

Mirza Hasanuzzaman is Professor of Agronomy at Sher-e-Bangla Agricultural University in Dhaka. He received his Ph.D. on "Plant Stress Physiology and Antioxidant Metabolism" from Ehime University, Japan, with a scholarship from the Japanese government (MEXT). Later, he completed his postdoctoral research at the Center of Molecular Biosciences, University of the Ryukyus, Japan, as a recipient of the Japan Society for the Promotion of Science (JSPS) postdoctoral fellowship. He was also the recipient of the Australian Government Endeavour Research Fellowship for postdoctoral research as an Adjunct Senior Researcher at the University of Tasmania, Australia. Dr. Hasanuzzaman's current work is focused on the physiological and molecular mechanisms of environmental stress tolerance. Dr. Hasanuzzaman has published over 150 articles in peer-reviewed journals. He has edited 10 books and written over 40 book chapters on important aspects of plant physiology, plant stress tolerance, and crop production. According to Scopus®, Dr. Hasanuzzaman's publications have received over 10000 citations with an *h*-index of 51. He has been named as a Highly Cited Researcher by Clarivate. He is an editor and reviewer for more than 50 peer-reviewed international journals and was a recipient of the "Publons Peer Review Award" 2017, 2018, and 2009. He has been honored by different authorities for his outstanding performance in different fields like research and education, and he has received the World Academy of Science Young Scientist Award (2014) and University Grants Commission (UGC) Award 2018. He is a fellow of Bangladesh Academy of Sciences (BAS) and Royal Society of Biology.

Golam Jalal Ahammed is Associate Professor of Horticulture at Henan University of Science and Technology, Luoyang, China. Dr. Ahammed received a Ph. D. in Olericulture from Zhejiang University, China. His major research interests include plant stress physiology, phytohormones, climate change effects on plants, and environmental pollution. He authored over a hundred research papers in peer-reviewed journals. He is a Senior Editorial Board Member of *Scientific Reports* and an Associate Editor of *AoB Plants* and *Journal of Plant Growth Regulation*. He also served as Flanders (FWO) internal expert panel member of Bio2: Functional Biology (Ph.D. fellowship call 2020). Dr. Ahammed was awarded two research grants for International Young Scientists by the National Natural Science Foundation of China and one National Foreign Expert Project (Foreign Youth Talent Program) by the Ministry of Science and Technology of China. He was also a recipient of several grants from the China Postdoctoral Science Foundation.

Kamrun Nahar is Professor of Agricultural Botany at Sher-e-Bangla Agricultural University, Dhaka, Bangladesh. She received her Ph.D. Degree in "Environmental Stress Physiology of Plants" from the United Graduate School of Agricultural Sciences, Ehime University, Japan, with Japanese Government (MEXT) Scholarship. Dr. Nahar has been involved in research with field crops emphasizing stress physiology since 2006. She has completed several research works and also continuing research project funded by Sher-e-Bangla Agricultural University Research System and Ministry of Science and Technology (Bangladesh). She is supervising M.S. students. Dr. Nahar published over 100 articles and chapters related to plant physiology and environmental stresses. Her publications reached about 9500 citations with h-index: 51 (according to Google Scholar). She is involved in editorial activities and reviewer of international journals. She is active member of about 20 professional societies. Dr. Nahar attended different international conferences and presented 20 papers, and posters in national and international conferences in different countries.

Contributors

Mohsin Abbas Department of Environmental Sciences, University of Gujrat, Hafiz Hayat Campus, Gujrat, Pakistan

Muhammad Abubakar Department of Environmental Sciences, University of Gujrat, Hafiz Hayat Campus, Gujrat, Pakistan
Institute of Biomedical Sciences ABEL SALAZAR, University of Porto, Porto, Portugal

Semih Açıkbaş Department of Field Crops, Faculty of Agriculture, Siirt University, Siirt, Turkey

Nisha Agrawal School of Biochemistry, Devi Ahilya University, Takshashila Campus, Indore, MP, India

Zahoor Ahmad University of Central Punjab, Bahawalpur Campus, Bahawalpur, Pakistan

Ehsan Ahmed Department of Environmental Sciences, University of Gujrat, Hafiz Hayat Campus, Gujrat, Pakistan

Muhammad Ahmed Department of Environmental Sciences, University of Gujrat, Hafiz Hayat Campus, Gujrat, Pakistan

Naznin Ahmed Department of Agronomy, Faculty of Agriculture, Sher-e-Bangla Agricultural University, Dhaka, Bangladesh

Sharif Ahmed International Rice Research Institute, Bangladesh office, Dhaka, Bangladesh

Temoor Ahmed National Key Laboratory for Rice Biology, Institute of Biotechnology, Zhejiang University, Hangzhou, P. R. China

Afsana Hoque Akhi Molecular Breeding Lab, Plant Breeding Division, Bangladesh Agricultural Research Institute, Gazipur, Bangladesh

Tahmina Akter Molecular Breeding Lab, Plant Breeding Division, Bangladesh Agricultural Research Institute, Gazipur, Bangladesh

Sadia Sabrina Alam Molecular Breeding Lab, Plant Breeding Division, Bangladesh Agricultural Research Institute, Gazipur, Bangladesh

Shafaqat Ali Department of Botany, Government College Women University, Sialkot, Pakistan
Department of Biological Sciences and Technology, China Medical University, Taichung, Taiwan
Department of Environmental Sciences and Engineering, Government College University, Faisalabad, Pakistan

Taufika Islam Anee Department of Agronomy, Faculty of Agriculture, Sher-e-Bangla Agricultural University, Dhaka, Bangladesh

Memona Arooj Department of Environmental Sciences, University of Gujrat, Hafiz Hayat Campus, Gujrat, Pakistan

Doğan Arslan Departmant of Field Crops, Faculty of Agriculture, Siirt University, Siirt, Turkey

Hüseyin Arslan Department of Field Crops, Faculty of Agruculure, Harran Üniversity, Osmanbey Campus, Şanliurfa, Turkey

Zaki Ul Zaman Asam Department of Environmental Sciences, University of Gujrat, Hafiz Hayat Campus, Gujrat, Pakistan

Habib-ur-Rehman Athar Institute of Pure and Applied Biology, Bahauddin Zakariya University, Multan, Pakistan

Nicolle Louise Ferreira Barros Instituto de Ciências Biológicas, Universidade Federal do Pará (UFPA), Belém, PA, Brazil

Martin Battaglia Department of Animal Sciences, Cornell University, Ithaca, NY, USA

Rajan Bhatt Department of Soil Science, Punjab Agricultural University, Ludhiana, Punjab, India

Rajani Bisen Jawahar Lal Nehru Krishi Vishwavidyalaya, Jabalpur, MP, India

Marian Brestic Department of Botany and Plant Physiology, Faculty of Agrobiology, Food and Natural Resources, Czech University of Life Sciences Prague, Prague, Czechia

Maham Asif Bukhari Department of Environmental Sciences, COMSATS University Islamabad, Vehari, Pakistan

Muhammad Adnan Bukhari Department of Agronomy, University College of Agriculture and Environmental Sciences, The Islamia University of Bahawalpur, Bahawalpur, Pakistan

Mustafa Ceritoğlu Department of Field Crops, Faculty of Agriculture, Siirt University, Siirt, Turkey

Fatih Cig Department of Field Crops, Faculty of Agriculture, Siirt University, Siirt, Turkey

Juel Datta State Key Laboratory of Rice Biology, China National Rice Research Institute, Hangzhou, China

Anamika Dubey Metagenomics and Secretomics Research Laboratory, Department of Botany, Dr. Harisingh Gour University (A Central University), Sagar, Madhya Pradesh, India

Murat Erman Department of Field Crops, Faculty of Agriculture, Siirt University, Siirt, Turkey

Muniba Farhad Department of Chemistry, Government College University, Faisalabad, Pakistan

Mujahid Farid Department of Environmental Sciences, University of Gujrat, Hafiz Hayat Campus, Gujrat, Pakistan

Sheharyaar Farid Institute of Biomedical Sciences ABEL SALAZAR, University of Porto, Porto, Portugal
Department of Biology, Ecology and Evolution, University of Liege, Liege, Belgium
University of the Basque Country, Bilbao, Spain

Ahmad Fayyaz Department of Environmental Sciences, University of Gujrat, Hafiz Hayat Campus, Gujrat, Pakistan

Masayuki Fujita Laboratory of Plant Stress Responses, Faculty of Agriculture, Kagawa University, Miki-cho, Kagawa, Japan

Rafaqat Ali Gill Oil Crops Research Institute, Chinese Academy of Agriculture Sciences, Wuhan, China

Ntsomboh-Ntsefong Godswill Department of Plant Biology, Faculty of Science, University of Yaounde, Yaounde, Cameroon

Vinod Kumar Goyal Jawahar Lal Nehru Krishi Vishwavidyalaya, Jabalpur, MP, India

Dipinte Gupta Plant Biotechnology Lab, Department of Botany, Faculty of Science, Dayalbagh Educational Institute (Deemed University), Agra, Uttar Pradesh, India

Adel Hadifa Rice Research and Training Center (RRTC), Field Crops Research Institute, Agricultural Research Center, Sakha, Kafr Elsheikh, Egypt

Khalid Rehman Hakeem Department of Biological Sciences, Faculty of Science, King Abdulaziz University, Jeddah, Saudi Arabia

Mirza Hasanuzzaman Department of Agronomy, Faculty of Agriculture, Sher-e-Bangla Agricultural University, Dhaka, Bangladesh
Department of Agronomy, Sher-e-Bangla Agricultural University, Dhaka, Bangladesh

Umer Hayat Department of Environmental Sciences, University of Gujrat, Hafiz Hayat Campus, Gujrat, Pakistan

Akbar Hossain Bangladesh Wheat and Maize Research Institute, Dinajpur, Bangladesh

Atif Hussain Department of Environmental Sciences, University of Gujrat, Hafiz Hayat Campus, Gujrat, Pakistan

Alia Idrees Department of Environmental Sciences, University of Gujrat, Hafiz Hayat Campus, Gujrat, Pakistan

Muhammad Zahid Ihsan Baghdad-ul-Jadeed Campus Cholistan Institute of Desert Studies (CIDS), The Islamia University of Bahawalpur (IUB), Bahawalpur, Pakistan

Asif Iqbal Department of Agronomy, Faculty of Agriculture, University of Agriculture Faisalabad, Faisalabad, Pakistan

Muhammad Iqbal Department of Environmental Sciences and Engineering, Government College University, Faisalabad, Pakistan

Muhammad Aamir Iqbal Department of Environmental Sciences, COMSATS University Islamabad, Vehari, Pakistan
Department of Agronomy, Faculty of Agriculture, University of Poonch Rawalakot (AJK), Rawalakot, Pakistan
Department of Agronomy, University of Poonch Rawalakot, Rawalakot, Azad Kashmir, Pakistan

Sameera Iqbal Department of Environmental Sciences, University of Gujrat, Hafiz Hayat Campus, Gujrat, Pakistan

Sanobar Iqbal Department of Environmental Sciences, University of Gujrat, Hafiz Hayat Campus, Gujrat, Pakistan

Muhammad Irfan SoMine Pakyürekil and Environmental Sciences Division, Nuclear Institute of Agriculture (NIA), Tandojam, Pakistan

Maria Islam Department of Environmental Sciences, University of Gujrat, Hafiz Hayat Campus, Gujrat, Pakistan

Md Sohidul Islam Department of Agronomy, Hajee Mohammad Danesh Science and Technology University, Dinajpur, Bangladesh

Mohammad Sohidul Islam Deparment of Agronomy, Hajee Mohammad Danesh Science and Technology University, Dinajpur, Bangladesh

Tasmiya Jabeen Department of Agronomy, Hajee Mohammad Danesh Science and Technology University, Dinajpur, Bangladesh

Meeta Jain School of Biochemistry, Devi Ahilya University, Takshashila Campus, Indore, MP, India

Kiran Javed Department of Environmental Sciences, University of Gujrat, Hafiz Hayat Campus, Gujrat, Pakistan

Muhammad Tariq Javed Department of Botany, Government College University, Faisalabad, Pakistan

Talha Javed College of Agriculture, Fujian Agriculture and Forestry University, Fuzhou, China

Gagandeep Kaur Yadavindra College of Engineering, Talwandi Sabo, Bathinda, Punjab, India

Yüksel Kaya Department of Field Crops, Faculty of Agriculture, Siirt University, Siirt, Turkey

Afifa Khizar Department of Environmental Sciences, University of Gujrat, Hafiz Hayat Campus, Gujrat, Pakistan

Rukiye Kılıç Department of Field Crops, Faculty of Agriculture, Siirt University, Siirt, Turkey

Ashwani Kumar Metagenomics and Secretomics Research Laboratory, Department of Botany, Dr. Harisingh Gour University (A Central University), Sagar, Madhya Pradesh, India

Arpna Kumari Department of Botanical and Environmental Sciences, Guru Nanak Dev University, Amritsar, Punjab, India

Liu Liyun Graduate School of Integrated Sciences for Life, Hiroshima University, Higashi-Hiroshima, Japan

Analía Llanes Plant Physiology Laboratory, Department of Natural Sciences, FCEFQyN, Universidad Nacional de Río Cuarto, Río Cuarto, Argentina

Jubayer Al Mahmud Department of Agroforestry and Environmental Science, Faculty of Agriculture, Sher-e-Bangla Agricultural University, Dhaka, Bangladesh

Deyvid Novaes Marques Departamento de Genética, Universidade de São Paulo (USP), Escola Superior de Agricultura "Luiz de Queiroz" (ESALQ), Piracicaba, SP, Brazil

Aiman Masood Department of Environmental Sciences, University of Gujrat, Hafiz Hayat Campus, Gujrat, Pakistan

Abdul Awal Chowdhury Masud Department of Agronomy, Faculty of Agriculture, Sher-e-Bangla Agricultural University, Dhaka, Bangladesh

Ram Swaroop Meena Department of Agronomy, Institute of Agricultural Sciences, Banaras Hindu University, Varanasi, Uttar Pradesh, India

Kinza Mehak Department of Environmental Sciences, University of Gujrat, Hafiz Hayat Campus, Gujrat, Pakistan

Sayed Mohammad Mohsin Laboratory of Plant Stress Responses, Faculty of Agriculture, Kagawa University, Miki-cho, Kagawa, Japan
Department of Plant Pathology, Faculty of Agriculture, Sher-e-Bangla Agricultural University, Dhaka, Bangladesh

Mahmuda Binte Monsur Department of Genetics and Plant Breeding, Bangabandhu Sheikh Mujibur Rahman Agricultural University, Gazipur, Bangladesh

Muhammad Mubeen Department of Environmental Sciences, COMSATS University Islamabad, Vehari, Pakistan

Sidra Tul Muntaha Department of Environmental Sciences, University of Gujrat, Hafiz Hayat Campus, Gujrat, Pakistan

Hina Mushtaq Department of Environmental Sciences, University of Gujrat, Hafiz Hayat Campus, Gujrat, Pakistan

Kamrun Nahar Department of Agricultural Botany, Faculty of Agriculture, Sher-e-Bangla Agricultural University, Dhaka, Bangladesh

Wajid Nasim Department of Agronomy, University College of Agriculture and Environmental Sciences, The Islamia University of Bahawalpur (IUB), Bahawalpur, Pakistan

Misbah Naz Institute of Environment and Ecology, School of The Environment and Safety Engineering, Jiangsu University, Zhenjiang, Jiangsu, China
State Key Laboratory of Crop Genetics and Germplasm Enhancement, MOA Key Laboratory of Plant Nutrition and Fertilization in Low-Middle Reaches of the Yangtze River, Nanjing Agricultural University, Nanjing, China

Muhammad Noman National Key Laboratory for Rice Biology, Institute of Biotechnology, Zhejiang University, Hangzhou, P. R. China

Sumaiya Haque Omy Molecular Breeding Lab, Plant Breeding Division, Bangladesh Agricultural Research Institute, Gazipur, Bangladesh

Aynur Bilmez Özçinar Department of Field Crops, Faculty of Agriculture, Siirt University, Siirt, Turkey

Gülen Özyazici Departmant of Field Crops, Faculty of Agriculture, Siirt University, Siirt, Turkey

Mehmet Arif Ozyazici Department of Field Crops, Faculty of Agriculture, Siirt University, Sıirt, Turkey

Mine Pakyürek Department of Horticulture, Siirt University Faculty of Agriculture, Siirt, Turkey

Khursheda Parvin Department of Horticulture, Sher-e-Bangla Agricultural University, Dhaka, Bangladesh

Laboratory of Plant Stress Responses, Faculty of Agriculture, Kagawa University, Miki-cho, Kagawa, Japan
Department of Horticulture, Faculty of Agriculture, Sher-e-Bangla Agricultural University, Dhaka, Bangladesh

Asma Qadri Department of Environmental Sciences, University of Gujrat, Hafiz Hayat Campus, Gujrat, Pakistan

Khussboo Rahman Department of Agronomy, Faculty of Agriculture, Sher-e-Bangla Agricultural University, Dhaka, Bangladesh

Hasan Uz Zaman Raihan Molecular Breeding Lab, Plant Breeding Division, Bangladesh Agricultural Research Institute, Gazipur, Bangladesh

Aman Raj Metagenomics and Secretomics Research Laboratory, Department of Botany, Dr. Harisingh Gour University (A Central University), Sagar, Madhya Pradesh, India

Muhammad Ramzan Department of Environmental Sciences, University of Gujrat, Hafiz Hayat Campus, Gujrat, Pakistan

Musarrat Ramzan Department of Botany, The Islamia University of Bahawalpur, Bahawalpur, Pakistan

Dua Nouman Rana Department of Environmental Sciences, University of Gujrat, Hafiz Hayat Campus, Gujrat, Pakistan

Rajiv Ranjan Plant Biotechnology Lab, Department of Botany, Faculty of Science, Dayalbagh Educational Institute (Deemed University), Agra, Uttar Pradesh, India

Zulakha Rasheed Department of Environmental Sciences, University of Gujrat, Hafiz Hayat Campus, Gujrat, Pakistan

Disna Ratnasekera Department of Agricultural Biology, Faculty of Agriculture, University of Ruhuna, Matara, Sri Lanka

Ali Raza Oil Crops Research Institute, Chinese Academy of Agricultural Sciences (CAAS), Wuhan, Hubei, China

Muhammad Ali Raza College of Agronomy, Sichuan Agricultural University, Chengdu, China

Nighat Raza Department of Food Science and Technology, Muhammad Nawaz Shareef University of Agriculture, Multan, Pakistan

Abdul Rehman Department of Crop Sciences and Biotechnology, Dankook University, Cheonan-si, South Korea

Muhammad Rizwan Department of Botany, Government College Women University, Sialkot, Pakistan

Department of Environmental Sciences and Engineering, Government College University, Faisalabad, Pakistan

Md. Motiar Rohman Molecular Breeding Lab, Plant Breeding Division, Bangladesh Agricultural Research Institute, Gazipur, Bangladesh

Ayman E. L. Sabagh Departmant of Field Crops, Faculty of Agriculture, Siirt University, Siirt, Turkey
Department of Agronomy, Faculty of Agriculture, Kafrelsheikh University, Kafrelsheikh, Egypt

Zahid Saqib Department of Environmental Sciences, University of Gujrat, Hafiz Hayat Campus, Gujrat, Pakistan

Pooja Saraswat Plant Biotechnology Lab, Department of Botany, Faculty of Science, Dayalbagh Educational Institute (Deemed University), Agra, Uttar Pradesh, India

Wajiha Sarfraz Department of Environmental Sciences and Engineering, Government College University, Faisalabad, Pakistan
Department of Botany, Government College Women University, Sialkot, Pakistan

Khadeja Sultana Sathi Department of Agronomy, Faculty of Agriculture, Sher-e-Bangla Agricultural University, Dhaka, Bangladesh

Seyithan Seydoşoğlu Department of Field Crops, Faculty of Agriculture, Siirt University, Sıirt, Turkey

Rubab Shabbir College of Agriculture, Fujian Agriculture and Forestry University, Fuzhou, China

Muhammad Shahid Department of Bioinformatics and Biotechnology, Government College University, Faisalabad, Pakistan

Tariq Shah INRAE, Paris, France College of Agriculture, West Virginia University, Morgantown, WV, USA
College of Tropical Crops, Hainan University, Haikou, China

Madiha Shahzadi Department of Environmental Sciences, University of Gujrat, Hafiz Hayat Campus, Gujrat, Pakistan

Muhammad Bilal Shakoor College of Earth and Environmental Sciences, University of Punjab, Lahore, Punjab, Pakistan

Kulvir Singh Punjab Agricultural University, Faridkot, Punjab, India

Pritpal Singh Department of Soil Science, Punjab Agricultural University, Ludhiana, Punjab, India

Milan Skalicky Department of Botany and Plant Physiology, Faculty of Agrobiology, Food and Natural resources, Czech University of Life Sciences Prague, Prague, Czechia

Cláudia Regina Batista de Souza Instituto de Ciências Biológicas, Universidade Federal do Pará (UFPA), Belém, PA, Brazil

Sipan Soysal Kurtalan Vocational High School, Siirt University, Siirt, Turkey

Oksana Sytar Department of Plant Biology, Kiev National University of Taras Shevchenko, Kiev, Ukraine

Md Tahjib-Ul-Arif Department of Biochemistry and Molecular Biology, Faculty of Agriculture, Bangladesh Agricultural University, Mymensingh, Bangladesh

Rana Muhammad Sabir Tariq Department of Agriculture & Agribusiness Management, University of Karachi, Karachi, Pakistan

Hafiz Muhammad Tauqeer Department of Environmental Sciences, University of Gujrat, Gujrat, Pakistan

Sadia Haque Tonny Molecular Breeding Lab, Plant Breeding Division, Bangladesh Agricultural Research Institute, Gazipur, Bangladesh

Rachana Tripathi School of Biochemistry, Devi Ahilya University, Takshashila Campus, Indore, MP, India

Nizamettin Turan Department of Field Crops, Faculty of Agriculture, Siirt University, Sıirt, Turkey

Veysel Turan Faculty of Agriculture, Institute of Soil Science and Plant Nutrition, Bingöl University, Bingöl, Turkey

Özge Uçar Department of Field Crops, Faculty of Agriculture, Siirt University, Siirt, Turkey

Akihiro Ueda Graduate School of Integrated Sciences for Life, Hiroshima University, Higashi-Hiroshima, Japan

Mubeen Umer Department of Environmental Sciences, University of Gujrat, Hafiz Hayat Campus, Gujrat, Pakistan

Allah Wasaya College of Agriculture, Bahauddin Zakariya University, Bahadur Sub-Campus, Layyah, Pakistan

Nimra Yaseen Department of Bioinformatics and Biotechnology, Government College University, Faisalabad, Pakistan

Tayyaba Younas Department of Environmental Sciences, University of Gujrat, Hafiz Hayat Campus, Gujrat, Pakistan

Saira Zaka Department of Environmental Sciences, University of Gujrat, Hafiz Hayat Campus, Gujrat, Pakistan

Muhammad Zubair Department of Chemistry, University of Gujrat, Hafiz Hayat Campus, Gujrat, Pakistan

Nanoremediation: An Innovative Approach for Environmental Safety

Muhammad Shahid, Nimra Yaseen, Muhammad Noman, Temoor Ahmed, and Muhammad Tariq Javed

Abstract Nanotechnology is an emerging field of science with a wide range of applications in various areas including environmental safety. Cleanup of environmental pollutants by the application of nanotechnology has tremendous prospects due to the superiority of nanomaterials (NMs) over micro-materials. This superiority lies in their small size and large surface area. Many types of metallic nanoparticles (NPs) such as the zero-valent iron, gold, silver, copper, and metal oxide NPs such as copper oxide (CuO/Cu_2O), zinc oxide (ZnO), and titanium oxide (TiO_2) have been extensively studied in the environmental remediation process. This chapter encompasses the significance of nanoscience and nanotechnology in the clean-up of environmental pollutants, types of NMs being used for environmental remediation, and their production and application in surface, ground, and brackish water remediation, soil remediation, and atmospheric pollution control. The effects of NMs on soil microbial communities with reference to their impact on bioremediation are also discussed. Finally, the current state of knowledge on the challenges faced by the use of NMs for environmental remediation and future prospects of nanotechnology to ensure environmental safety is briefly described.

Keywords Environment · Fabrication · Nanotechnology · Pollutants · Remediation

M. Shahid (✉) · N. Yaseen
Department of Bioinformatics and Biotechnology, Government College University, Faisalabad, Pakistan
e-mail: mshahid@gcuf.edu.pk

M. Noman · T. Ahmed
National Key Laboratory for Rice Biology, Institute of Biotechnology, Zhejiang University, Hangzhou, P. R. China

M. T. Javed
Department of Botany, Government College University, Faisalabad, Pakistan

M. Hasanuzzaman et al. (eds.), *Managing Plant Production Under Changing Environment*, https://doi.org/10.1007/978-981-16-5059-8_1

1 Introduction

The remediation of the environment by cleaning the pollution and its sources is imperative to protect ecological as well as general public health. Environmental cleanliness is extremely important for a safe and healthy life due to our direct reliance on the environment for numerous vital life-supporting resources such as air for breathing, water for drinking and agricultural/industrial use, and fertile land for agricultural practices. However, the current dilemma is the continuous contamination of vital environmental resources such as water, land and, air by a variety of natural and human-oriented pollutants (Brusseau et al. 2019). According to reports, there are about 900 sites in the United States (USEPA 2009) and 3,50,000 sites in Europe, contaminated with human and environmental health-threatening pollutants, which are required to be cleanup on a priority basis (EEA 2014). The contamination of environmental resources is the major source of environmental pollution. Environmental pollution is caused by various sources including industrial effluents, residual gases, sewage, oil spills, uncontrolled use of fertilizers, pesticides, and herbicides, and gases from burning fossil fuels (Fulekar and Pathak 2017).

A wide range of strategies has been employed for environmental remediation including the methods involving the removal or degradation of pollutants contaminating the aquatic, lithospheric, and atmospheric environment. The purpose of all these remediating techniques is to reduce/remove the potential risks to human and environmental health caused by contaminants. Some remediating techniques excavate the contaminated sites for the treatment of contaminants (ex situ), while, others degrade the contaminants within the polluted site (in situ) (Karn et al. 2011). Some factors such as cost and time required for environmental cleanup are the major constraints limiting the application of conventional remediating techniques on the commercial level. Moreover, the production of toxic intermediates during the remediation process requiring downstream processing for their complete degradation also pioneered the research activities to develop new technologies for environmental remediation in a better, cheaper, faster, and eco-friendly way (USEPA 2013).

Nanotechnology, in recent years, has emerged as an attractive and cost-efficient alternative to conventional remediation methods to sense, reduce, and degrade environmental contaminants. Currently, significant efforts have been made to implement nanotechnology-based solutions to environmental problems, which aim at the fabrication and use of NMs to enhance the overall effectiveness of environmental remediation techniques (El-Ramady et al. 2017; Patil et al. 2016). These nanoparticles (NPs) display unique properties due to their small size and large surface area as compared to the original substrate (Noman et al. 2019). Due to their unique characteristics, NPs have been engineered and utilized in various fields of science including environmental remediation. The term nanoremediation refers to remediation strategies involving the use of engineered NMs to reduce or degrade contaminants. It has been estimated that almost 45–70 contaminated sites around the world have utilized NPs-based remediation for environmental clean-up. The engineering of new and advanced NMs is currently required to remove toxic and harmful pollutants from the environment (Agarwal and Joshi 2010).

This chapter describes briefly nanoremediation potential of various types of NMs along with the benefits associated with their application. Finally, the status of knowledge and future challenges associated with these nanoremediation-based approaches are discussed.

2 Nanoremediation for Environmental Cleanup

During the past few years, an astonishing increase in the number of contaminated sites was observed worldwide. The water bodies have been reported to be contaminated by pesticides, oil and greases, heavy metals, and chlorinated toxic compounds (Jadhav and Singhal 2013). Moreover, it has been reported that the leachate from solid waste landfill sites increases the heavy metal (Al, Cd, Cr, Fe, Zn, Ni, and Pb) contents in the surrounding water bodies including groundwater (Kale et al. 2010). The recent stressful trend on conventional remediation technologies made nanotechnology a potential candidate for environmental cleanup (Fig. 1).

Nanoremediation due to its better efficacy and eco-friendly approach provides the best solution to control and treat environmental pollution by helping in pollution sensing, monitoring, and remediation (Rajan 2011). This NPs-based process of site remediation has the potential to solve obstacles of remediation processes in a cost-effective manner. It has been speculated that nanoremediation will outclass the conventional, partially effective, and tedious processes including thermal treatment, chemical oxidation, and surfactant co-solvent flushing in the next few years (Löffler and Edwards 2006; Patil et al. 2016). Further, the traditional methods of environmental clean-up demand high capital investment to remediate soil, water, and other environmental media, whereas, Karn et al. (2011) reported after a comprehensive review of nanoremediation approaches that NMs-based remediation is not only cost-efficient but also reduce the time required for the contaminant removal/degradation and drop the concentration of contaminants near to zero. Nanoremediation

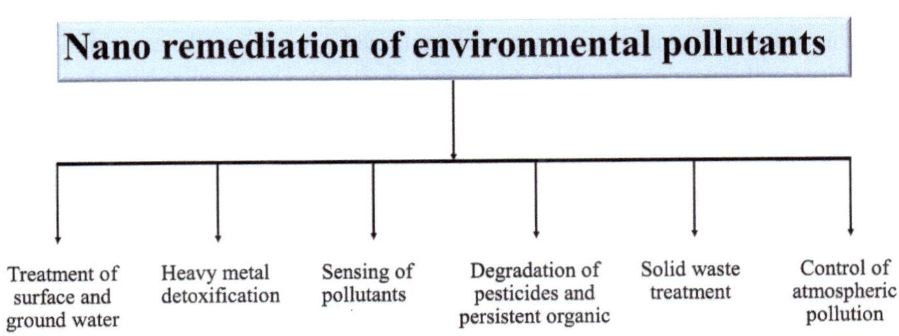

Fig. 1 Role of nanoremediation in environmental clean-up

approaches employ a wide variety of NMs such as zeolites, metal and metal oxides, carbon nanofibers and tubes, and bimetallic NPs for the mineralization and transformation of environmental contaminants. Amongst all, zero-valent iron NPs are most widely used for the remediation process due to their high surface area to volume ratio and reactivity (Garner and Keller 2014; Tosco et al. 2014). Other factors making zero-valent iron NPs excellent agents for the mineralization of pollutants in water and soil include favorable quantum size, less reduction potential, and better transport potential via an underground matrix of groundwater. The unique properties of NPs enable them to carry out both chemical reduction and catalysis to degrade a wide array of pollutants including chlorine-containing compounds, pesticides, heavy metals, and polychlorinated biphenyls (PCB), etc. Besides zero-valent iron NPs, other metallic NMs (such as titanium dioxide NPs, copper NPs, and zinc NPs, etc.), bimetallic NPs (such as Ag/Fe; Ni/Fe; Cu/Fe), and carbon nanotubes have also been successfully applied for the remediation of polluted soil and water, whereas, nano-scale filters have been developed for the purification of polluted air (Patil et al. 2016; Rajan 2011).

According to a report published by the USEPA 2004, nanoremediation techniques have been successfully implemented at 44 sites in seven countries including the United States, China, North and South Korea, Japan, Hong Kong, and Canada, which resulted in efficient reclamation of polluted sites (USEPA 2004). These studies revealed the high cost-efficiency of nanoremediation processes and saved almost 80–90% of the predicted budget as compared to the conventional methods of environmental clean-up. Moreover, the time required for the remediation process was also reduced from months to days. It has been reported that the amount of TCE reduced by 99% within a few days after the treatment with zero-valent iron NPs, for which conventional methods were operated for years (Zhang 2003).

3 Type of Nanomaterials Used for Nanoremediation

3.1 Metallic Nanoparticles

It has been well-established that metallic NPs lead a significant role in the transformation and mineralization of contaminants released directly into the environment by industrial units (Amde et al. 2017; Khurana et al. 2019). The metallic NPs along with their oxides were reported to clean up the environmental pollution effectively (Majedi and Lee 2016). Similarly, the NPs-based remediation technique includes the determination of potential benefits and associated drawbacks as well. The toxicity level of metallic NPs depends on environmental conditions as well as the type and concentration of a particular pollutant in the contaminated environmental media. However, the redox reactivity and catalytic activity of metallic NPs make them able to degrade and reduce the toxicity of the pollutants (Zhang and Fang 2010). Some widely used NPs in environmental nanoremediation have been described herein (Table 1).

Table 1 Role of different nanoparticles in nanoremediation

Nanoparticles	Pollutants	Applications	References
Metallic nanoparticles			
Silver nanoparticles	Heavy metals Fungus or bacterial strains	Act as nanocatalyst Carry efficient photolytic activity Play role in the purification of water	Khanna et al. (2019)
Iron nanoparticles	Chemicals Oil and grease from the water surface	Helpful in cleaning of toxic metals through magnetic separation	Nguyen et al. (2018)
Gold nanoparticles	Organic and inorganic dyes Heavy metal residues	Efficient in the degradation process Helpful in the neutralization of toxins	Santhoshkumar et al. (2017)
Copper nanoparticles	Textile effluent and azo dyes	Photocatalysts for azo dye degradation and treatment of textile effluent	Noman et al. (2019)
Metal oxide nanoparticles			
Iron oxide nanoparticles	Mercury Lead Dyes	Used for disinfection of water Help to improve water quality Efficient absorbents	Sharma et al. (2017)
Platinum Nanoparticle	Solid waste Carbon black Metabolites such as gallic acid and ascorbic acid	Effective for removal of organic ligands Serve as electrocatalyst for remediation	Zhao et al. (2020)
Titanium oxide Nanoparticles	Particulate matter Pieces of metals Solid pollutants especially	Effective in the treatment of wastewater Noncorrosive and more stable Powerful in photocatalytic activity of organic compounds	Lai et al. (2018); Saleh et al. (2019)
Copper oxide nanoparticles	Heavy metals like arsenic or lead Remove pest or herbs Remove cadmium and nickel from aqueous solution	Effective in removing by-products, produced through disinfection Efficient in killing bacterial cells present in water	Sharma et al. (2017)
Zinc oxide nanoparticles	Cadmium Nickel Chromium Manganese Silver Lead	Enhance dissolution Avoid aggregation of contaminants Most effective in the treatment of solid waste	Gagné et al. (2019)
Magnesium Oxide nanoparticles	Remove phosphate Bacterial growth Carbon deposition	Act as a catalyst for reforming Remediate phosphate from the aqueous phase	Li et al. (2018); Jung and Ahn (2016)
Nickel oxide Nanoparticles	Safranin Methylene blue	Perform antimicrobial activity in wastewater treatment Efficient in sorption nature	Sabouri et al. (2019); Yao et al. (2018)

(continued)

Table 1 (continued)

Nanoparticles	Pollutants	Applications	References
Polymer-based nanoparticles			
Silica-based nanoparticles	Reduce greenhouse gases Remediate hazardous waste	Remove dyes from water Act as photocatalyst for dye degradation	Kuznetsova et al. (2019)
Cadmium sulfide Nanoparticles	Removal of inorganic Solutes Removal of nitrates and heavy metals	Useful in water purification units Help in biosensing of pollutants	Nisha et al. (2015)
Aluminum-based Nanoparticles	Degrade heavy metals Remove viruses and biological contaminants	Helpful in degrading contaminated soil Effective for sedimentation and coagulation of water	Alinejad and Mahmoodi (2009)

3.1.1 Iron Nanoparticles

Iron nanoparticles (FeNPs) are highly efficient for the remediation of polluted environments such as groundwater, soil aquifers, and cyanobacterial-growing regions. The zero-valent FeNPs possess a high reducing potential and have been extensively used in the detoxification of harmful metal ions and chlorinated ethylene surfaces. The FeNPs have been successfully applied as nanocatalysts for the remediation of water bodies contaminated with phosphate and nitrate-containing compounds (Nguyen et al. 2018). The reason behind the extensive use of FeNPs is their eco-friendly nature with high pollutant removing efficiency (Lei et al. 2018). Moreover, FeNPs also lead to a significant role in the remediation of heavy metals contaminated soil. Various types of FeNPs including zero-valent FeNPs, ferric oxide (Fe_3O_4), and ferrous sulfide (FeS) have been emerged as key players to overcome environmental issues in an eco-friendly manner. The unique properties of these NPs, encouraging their large-scale application for the reclamation of metal-contaminated soil, include their small size, large surface area to volume ratio, and powerful reduction potential to detoxify the metal ions present in the soil medium (Peng et al. 2019). In order to achieve maximum remediation efficiency of FeNPs, it is necessary to determine the type and concentration of particular pollutants present in the soil, water, or other environmental media along with the conditions under which that particular compound can be degraded or transformed completely. Prabhakar et al. (2017) reported that FeNPs mineralized the contaminants using co-precipitation, absorption, photocatalytic, and other mechanisms.

3.1.2 Gold Nanoparticles

Gold nanoparticles (AuNPs) have shown powerful potential for nanoremediation due to their long-term stability in the contaminated medium, which is associated with their ability to resist the oxidation process. The AuNPs can be used to increase the efficiency of the remediation process for environmental clean-up (Qu et al. 2017). Green AuNPs with associated capping agents (such as phenols, flavonoids, alkaloids, reductases, and terpenes) were found very efficient in the neutralization of toxic compounds present in the environment. In an aquatic environment, AuNPs get dissolved in water and eliminate toxic metals and metal ions present in water by chelating them on their surface, thus resulting in the cleaning of water (Santhoshkumar et al. 2017).

AuNPs also possess good catalytic activity and can oxidize various organic pollutants present in the atmosphere (Barakat et al. 2013). AuNPs also have applications in biosensing and electro biocatalysis. They are used in fuel cells to improve the efficiency of enzymatic electrodes which are ultimately used for the removal of toxic compounds present in the atmosphere. (Kizling et al. 2018). Moreover, AuNPs are reported to have excellent heavy metals and chemical sensing potential. The biosensing ability of these NPs helps in the detection of the toxicity level of a particular chemical in soil, water, or atmosphere and their removal from the contaminated environmental medium (de Almeida et al. 2014). AuNPs were also found proficient in the biological control of fungus as they inhibited the fungal growth by inducing oxidative stress in fungi in soil and water (Andries et al. 2016).

3.1.3 Silver Nanoparticles

In nanoremediation approaches, silver (Ag) NPs are considered as one of the important components due to their efficient antimicrobial potential. These NPs have been tremendously used for the disinfection of environmental medium (viz., soil and water, etc.), textile, and food contaminated with infectious microorganisms (Marambio-Jones and Hoek 2010; Quadros and Marr 2010). The AgNPs are reported to bind with phosphorous and sulfur molecules of bacteria, thus blocking key metabolic functions occurring within bacterial cells resulted in the death of the bacteria (Deshmukh et al. 2018). However, the uncontrolled use of AgNPs resulted in their aggregation in the environment, which negatively affects the health of plants, animals, and other living beings including humans.

The toxicity level of AgNPs is associated with its method of production. The AgNPs synthesized through green routes were found less toxic as compared to those produced through physical and chemical methods. AgNPs-based nano-scale filters have been used to remove infectious particles present in the air. The use of AgNPs in the remediation of contaminated aquatic and soil environments has also been extensively reported. Kumar et al. (2013) reported photocatalytic degradation of methyl red using green AgNPs synthesized from *Ulva lactuca* and concluded that biogenic AgNPs possess excellent photocatalytic potential for the degradation of methyl red dye. Similarly, Selvam and Sivakumar (2015) reported biogenic AgNPs-mediated photocatalytic degradation of methyl orange dye. Further, AgNPs are reported extensively for the reclamation of salt-affected soil environments (Abou-Zeid and Ismail 2018). The nanoremediation potential of AgNPs can be enhanced by doping them with other types of NMs such as TiO_2 NPs. The knowledge of contaminant type and concentration further improves the efficiency of the AgNPs-based nanoremediation process.

3.1.4 Copper Nanoparticles

Copper nanoparticles (CuNPs) have shown tremendous potential for removing/degrading organic dyes from industrial wastewater. The excellent degradation potential of CuNPs lies in their long-term stability in the contaminated environment and the best catalytic potential. Moreover, they have been found effective in the recycling of wastewater due to their hydrolytic and oxidative properties (Noman et al. 2019; Usman et al. 2019). CuNPs are also equipped with a strong antimicrobial effect that is useful for remediation of pathogen-contaminated environmental mediums (such as water, soil, and air). It has been reported that CuNPs transformed almost all amounts of toxic chemicals into nontoxic forms along with the killing of pathogenic microorganisms when applied on contaminated soil thus helped in the soil reclamation in an eco-friendly manner (Chaudhary et al. 2019). Similarly, the role of CuNPs in nanofiltration has also been explored, where they extracted toxic chemicals from the air and transformed them into less toxic forms (Rajput et al. 2018).

3.1.5 Metal Oxide Nanoparticles

Metal oxide NPs (such as TiO_2 NPs, ZnO NPs, Fe_2O_3 NPs, Fe_3O_4 NPs, and CuO NPs) are one of the largest and major groups of NPs with vast applications in the environmental sector including environmental nanoremediation (Djurišić et al. 2015). These NPs are extensively used for the remediation of polluted aquatic environments. The properties that make metal oxide NPs excellent nanoremediators include their high surface area to volume ratio, catalytic potential, long-term stability under challenged environment, and their ability to prevent the production of toxic intermediates during the degradation/remediation process (Sharma et al. 2017).

Among metallic oxide NPs, TiO_2 NPs have been extensively investigated for the removal of contaminants from the environment. It has been reported that TiO_2 NPs have significant potential for the treatment of contaminated surface water, soil, and wastewater (Yang et al. 2018). The remediating potential of TiO_2 NPs was found to be associated with their unique physicochemical properties such as large surface area, stability, reducing potential, non-corrosive nature, and production of non-toxic degradation products during the remediation process (Lai et al. 2018). Further, CuO NPs have also been reported to screen and mineralize the toxic by-products formed during the disinfection of water. That's why Cu pipes are commonly used for the distribution of drinking water (Sharma et al. 2017).

3.2 Polymer-Based Nanomaterials

Polymers of carbon and silicon are most widely used in nanoremediation, as they are much efficient, faster, and cost-effective. Polymer-based NMs have gained importance due to their large surface area, stability, and remediation capability (Table 1). These polymer-based NMs have been developed to remove contaminants from soil and water bodies in a more eco-friendly way (Zhu et al. 2019). Polymers of carbonaceous material possess unique characteristics for absorption, catalysis, and bio-sensing (Kim et al. 2011). Carbon-based nanocomposites are of several types that include graphite, fullerene, carbon black fullerene, reduced graphene oxide, carbon nanotubes (CNTs), and graphene oxides. Depending upon their physico-chemical parameters, these polymers are considered efficient in removing pollutants from the environment (Madannejad et al. 2019). They have better stability under natural environmental conditions, which reduces the overall cost of the remediation process. However, their remediation performance can be enhanced by driving the remediation process (Siddiqui et al. 2018).

Carbon nanotubes are the polymers of carbon, which are either single-walled or multi-walled in nature. Carbon nanotubes combine to form carbon nano-fibers, which ultimately form complex nano-composites. Composites developed from such nanomaterials have proved themselves a powerful tool in remediating pollutants when used in conjunction with other treatment processes. Such polymers have outstanding physicochemical properties like flexibility, low density, and insulation which facilitate the overall remediation process (Ji et al. 2016). Moreover, carbon- and silicon-based nano-emitters have been developed that possess the ability to absorb harmful radiations emitted from nuclear reactors (Gu and Chen 2018). For the treatment of wastewater activated carbon-based nano-absorbents have been developed, that decontaminate water via adhesive action. These nano-absorbents are both cost-effective and environment-friendly as they remain stable for a long period without any change in their own physicochemical properties (Alaba et al. 2018).

4 Production and Fabrication of Nanomaterials

As far as NMs production is concerned, there are numerous physicochemical processes available for the fabrication of inorganic NPs, but certain limitations associated with these conventional methods (such as toxicity, cost, and complexity) hinder their implementation for NPs production. The modern scientific community has been attracted by a more reliable, clean, cost-effective, green, and eco-friendly approach for the synthesis of NPs by functional microorganisms and plant material. The green routes for the synthesis of numerous types of NPs have been reported extensively (Fathima et al. 2018; Noman et al. 2019; Thangamani and Bhuvaneshwari 2019). The biogenic NMs considered cheaper, safe, environment-friendly, and reactive than those produced through physical and chemical methods. The better remediation potential of biogenic NMs was due to the associated capping agents involved in the long-term stabilization and bioactivity of the NMs (Ahmed et al. 2015).

5 Nanoremediation for Aquatic Environment

The clean water resources are day by day due to the discharge of untreated wastes from industrial and municipal sectors worldwide. The contaminants present in water ultimately have negative effects on the health of plants, animals, and humans due to their long-range transport, persistence, and bioaccumulation up to toxic levels (Wenning and Martello 2014). Therefore, it is necessary to develop better remediation techniques to purify contaminated water resources. To address this alarming situation, nanotechnology provides the best solution with the lowest impact on the integrity of the environment. Nowadays, CNTs have been used extensively for environmental clean-up particularly for the remediation of wastewater. These nanotubes are elastic in nature and easily spread into the aquatic environments and adsorb contaminants on their surface thus removing pollutants from the wastewater. Multi-walled CNTs were preferred over single-walled CNTs for the cleaning of the aquatic environment as they provide more surface area for the absorption of contaminants (Kwadijk et al. 2013). Moreover, FeNPs have also been used for preventing the befouling of marine water by restricting the growth of algal blooms and cyanobacteria (Nguyen et al. 2018). Currently, the world is facing plastic pollution as a major issue due to the extensive use of plastic products. The waste microplastics in huge amounts have been discharged into the water bodies where they become accumulated and affect the aquatic life and food chains. The plastic due to their recalcitrant nature remains in water bodies for a long time without any degradation. The management of plastic waste is the key issue of the modern world. In this context, magnetic NPs serve the purpose as they have the capability to reduce plastic pollution in the water system. These NMs deteriorate the structure of the plastic by penetrating and breaking the strong bonding between the molecules maintaining the overall structure of plastic (Pico et al. 2019).

The contamination of water with heavy metals, such as arsenic (As), lead (Pb), cadmium (Cd), mercury (Hg), and nickel (Ni), is another major issue making the water unfit for consumption with long-term negative effects in the eco-system. An innovative technology called nano-filtration has been developed and found more effective in removing heavy metals and metal ions from the wastewater (Fig. 2). The nano-filtration process to purify water from metal contamination is more advantageous than previously available conventional methods due to the low utilization of energy and cost-efficient nature (Carolin et al. 2017). Besides nano-filtration, nanotechnology developed nano-absorbents for removing toxic metal ions and heavy metals from wastewater. The nano-absorbents perform their action rapidly as they have a better ability to soak toxic metals present in the wastewater. Among various

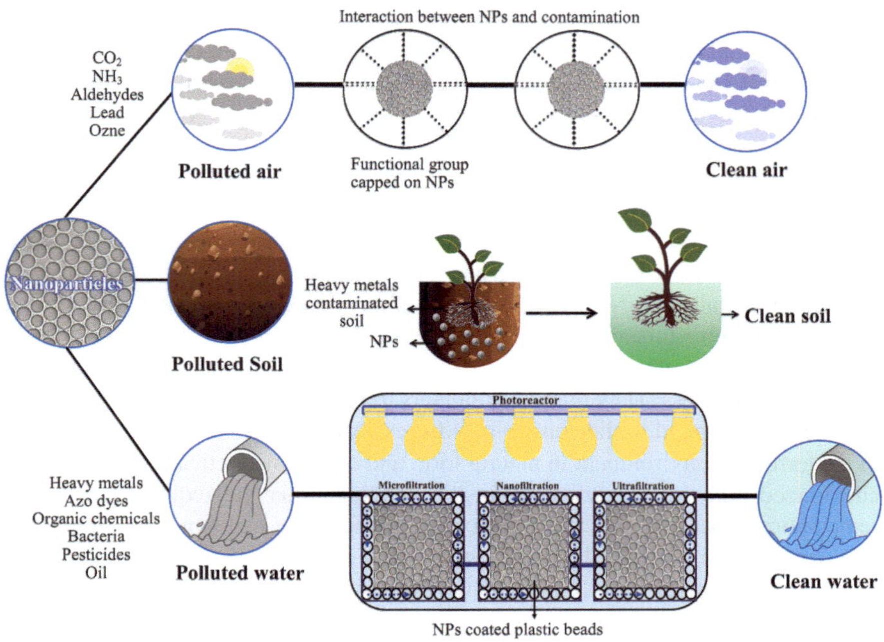

Fig. 2 Proposed mechanism of environmental nanoremediation

types of NPs, CNTs are most commonly used as nano-absorbents as they adhere to a variety of pollutants by hydrophobic interaction, and covalent and electrostatic bonding (Dubey et al. 2017). These nano-absorbents can also be used for the removal of emerging contaminants such as pharmaceuticals and drugs. These contaminants are released into water bodies without any treatment and have toxic effects on human health when entering into the body as they disturb hormonal, genetic, and enzymatic systems. The nano-absorbents have the ability to remove pollutants even when they are in a very low amount due to their unique high affinity for pollutant absorption (Basheer 2018). Remediation of aquifers is a challenging task as compared to the remediation of surface water as the groundwater contains both bioaccumulative and toxic contaminants. For groundwater remediation, CNTs are effective in remediating such contaminants in a cost-effective and eco-friendly manner as they are more stable and sustainable for the purification of water (Enfrin et al. 2019; Neto et al. 2019).

6 Nanoremediation for Terrestrial Environment

The remediation of contaminated terrestrial sites is the biggest challenge faced by the world during the past few years. Recently, nanotechnology engineered various types of NPs having the tremendous capability of remediating the terrestrial environment.

The NMs have been preferred over conventional remediation techniques for cleaning up the terrestrial environment because of their rapid action, high efficiency, stability, and low cost (Visentin et al. 2019b). AgNPs, due to their excellent antimicrobial potential, have been found effective in the treatment of soil contaminated with infectious biological entities (Zhou et al. 2017). The use of graphene-based NMs and CNTs has been successfully applied in the cleaning up of metals and organic pollutants from soil (Wang et al. 2013). Recently, FeS NPs have gained much attention for the decontamination of the polluted terrestrial environments due to their smaller size and better reactive potential (Chen et al. 2018).

Further, metal oxide NPs are the most commonly used NPs for the remediation of polluted environments due to their ability to perform a variety of eco-friendly activities. Their transformation activities are associated with the soil characteristics and its aging process. They remove pollutants from the soil by adhering them to their surface when applied to the contaminated sites (Amde et al. 2017). The NMs occur in the terrestrial environment in natural form and can be synthetically designed. The customized NPs have been developed depending upon the requirements of the remediation process in order to enhance the efficiency of nanoremediation process. It is necessary to evaluate the quality of soil and sediment, type of polluting agent, amount of pollutant, and optimum conditions under which pollutants can be degraded (Corsi et al. 2018).

7 Nanoremediation for Atmospheric Environment

The quality of air has been challenged by a variety of pollutants released from automobiles, refrigerators, and industrial units causing atmospheric pollution, one of the serious threats affecting the quality of life, especially in urban areas. Among air contaminants, heavy metals rank first that badly affect the atmospheric environment (Liu et al. 2018). A number of tools have been developed to evaluate the reliability of remediation techniques for atmospheric decontamination. Among all, nanotechnology was found much more sustainable than others for the remediation of atmospheric pollutants (Visentin et al. 2019a). Different types of engineered NPs have been synthesized with the inherent potential to clean the atmospheric environment. The nano-strips have been developed that can detect and remove particular chemical pollutants present in the atmosphere. These nano-strips are installed on roadsides for capturing the heavy metals and metal oxides present in the air and exhaust from automobiles and other machinery. After capturing the heavy metals, these nano-strips transform the pollutants into a less toxic form (Abdullayev et al. 2011).

Nuclear industries like nuclear energy production, uranium mining ores, nuclear weapon manufacture, and waste fuel treatment units have resulted in the release of toxic gases in the atmosphere. For the remediation of such gases, CNTs are preferred over others due to their unique characteristics such as electric and thermal stability. Such tubes are fixed on roadsides and separate pollutants by the hydrothermal method (Xie et al. 2019). Moreover, FeO NPs have also been proved themselves as efficient candidates for detoxification and the transformation of hazardous pollutants present in the atmosphere (Ingle et al. 2014). To cope with the atmospheric pollutants, nanotechnology has also developed nano-sensors for monitoring and assessing the toxicity level of pollutants present in the air (Brusseau et al. 2019; Nasrollahzadeh et al. 2019). The nanosensors are extremely sensitive to pollutants and can detect an extremely low concentration of air pollutants. They not only detect particular contaminants in the air but also adhere to pollutants on their surface and transform it into non-toxic form (da Silveira Petruci et al. 2019; Nasrollahzadeh et al. 2019).

8 Potential Risks of Using Nanomaterials as Environmental Remediating Agents

The potential use of NMs as an environmentally beneficial technology has gained significant attention from scientists, however; many uncertainties and ecological risks are associated with their extensive use in environmental sectors (Patil et al. 2016). The major environmental concern is that the nanoparticles could end up in environmental bodies infesting drinking water sources harming the health of humans and animals. The eco-toxicity ability of nanoparticles shows possible risks, which adversely affecting organisms in the environment. Although nanotechnology potentially shows a valuable substitute for present traditional environmental remediation approaches, research related to nanoparticle's effects on the environment and health required some attention. Oberdörster et al. (2005) stated that studies related to toxicity should not pay attention only to humans and wildlife but also focus on the lower organisms constituting the food chain. Previous studies have shown that nanoparticles have some opposing influence on pure cultures of bacteria like *Bacillus subtilis, Escherichia coli* due to the synthesis of reactive oxygen species in living cells (Diao and Yao 2009; Lee et al. 2008; Patil et al. 2016).

Many studies have shown that NMs have a toxics effect on soil microorganisms like CuO and Fe_3O_4 were found to alter the microbial community in soil by producing toxicity at 1% and 5% w/w dry soil (Ben-Moshe et al. 2013). Nanotechnology related to hazards on human health comprise direct contact or inhalation of nanoparticles at the place of work via air contamination, ingestion of food or water, and dermal in contact, mainly via application of personal care (Quadros and Marr 2010). Human health risks associated with nanotechnology include inhalation or exposure of nanoparticles at the workplace through contaminated air, ingestion of water or food, and dermal contact, primarily through the application of personal care (Yang et al. 2008). Nanoparticles have some hazardous effects on humans in case of direct contact like lung diseases, oxidative stress, genotoxicity, pulmonary pathological changes, lipid peroxidation, etc. It might be possible concerns related to safety that may limit the extensive application of NMs environmental remediation (Bouwmeester et al. 2009; Quadros and Marr 2010). Hence, to make this technology more beneficial than harmful, monitoring and intervention measures need to be implemented sooner than later.

9 Current Trends of Nanoremediation and Future Perspective

Nanoremediation has attracted the attention of the entire scientific community by using advanced materials bearing unique physicochemical properties for the removal of contaminants from the environment. Currently, the world is facing the problem of environmental deterioration, which needs to be resolved using effective and innovative technology with the lowest downstream impact on the environment. Nanotechnology has developed numerous innovative tools such as nano-sensors, nano-absorbents, and nano-catalysts for the nanoremediation of contaminated environmental mediums (such as soil, water, and air) (Vishwakarma 2012). NMs as nano-sensors and nano-absorbents helped in the detection, monitoring, and transformation of heavy metals and other organic pollutants present in any type of environmental medium (Vishwakarma 2012). Moreover, nano-catalysts have been used to enhance the efficiency of a remediation process used to treat heavy metals and residual dyes containing wastewater (Varma 2014). Nanoremediation has also proved its mark in the detection and degradation of extremely toxic emerging contaminants released by pharmaceutical and drug companies directly into the environment (Petrie et al. 2015). NPs-based remediation process has also been used to overcome the issue of plastic pollution. Plastic waste due to its recalcitrant nature is very difficult to degrade but numerous types of NMs (such as zero-valent FeNPs) have shown significant potential to degrade plastic material by destroying the structure (da Costa et al. 2016). Further, research insights are required to explore the mechanisms of NPs-mediated remediation of environmental pollutants. In addition, the investigations regarding NPs application to control pesticide and smog pollution also need research-based investigations.

In future studies, attention must be focused to evaluate the efficiency/efficacy of nanoremediation under field conditions along with the assessment of potential risks associated with the use of NMs on the health of the environment and life forms. Moreover, extensive research insights are required to evaluate the toxicity level of transformed products and their downstream impacts on the overall functioning of the ecosystem. Additionally, the impacts of intermediate products produced during nanoremediation process on environmental health need serious research focus.

10 Conclusions

Water, soil, and air are important environmental resources and their continuous contamination due to human intervention is a problem of serious concern. The dawn of nanotechnology presents the engineered NMs to remediate polluted environmental resources in a cost-effective, time-saving, and eco-friendly manner. Furthermore, nanoremediation of terrestrial, aquatic, and atmospheric environments may employ the fabricated metallic NPs, metal oxide NPs, polymer NMs, and other nanocomposites for cleaning up the polluted environments. However, investigations regarding the evaluation of potential environmental hazards of nanoremediation techniques are needed to decipher the ecological risks of nanoremediation.

References

Abdullayev E, Sakakibara K, Okamoto K, Wei W, Ariga K, Lvov Y (2011) Natural tubule clay template synthesis of silver nanorods for antibacterial composite coating. ACS Appl Mater Interfaces 3:4040–4046

Abou-Zeid H, Ismail G (2018) The role of priming with biosynthesized silver nanoparticles in the response of Triticum aestivum L to salt stress. Egypt J Bot 58(1):73–85

Agarwal A, Joshi H (2010) Application of nanotechnology in the remediation of contaminated groundwater: a short review. Recent Res Sci Technol 2:6

Ahmed MJ, Murtaza G, Mehmood A, Bhatti TM (2015) Green synthesis of silver nanoparticles using leaves extract of Skimmia laureola: characterization and antibacterial activity. Mater Lett 153:10–13

Alaba PA, Oladoja NA, Sani YM, Ayodele OB, Mohammed IY, Olupinla SF, Daud WMW (2018) Insight into wastewater decontamination using polymeric adsorbents. J Environ Chem Eng 6: 1651–1672

Alinejad B, Mahmoodi K (2009) A novel method for generating hydrogen by hydrolysis of highly activated aluminum nanoparticles in pure water. Int J Hydrogen Energy 34:7934–7938

Amde M, Liu JF, Tan ZQ, Bekana D (2017) Transformation and bioavailability of metal oxide nanoparticles in aquatic and terrestrial environments. A review. Environ Pollut 230:250–267

Andries M, Pricop D, Oprica L, Creanga DE, Iacomi F (2016) The effect of visible light on gold nanoparticles and some bioeffects on environmental fungi. Int J Pharm 505:255–261

Barakat T, Rooke JC, Genty E, Cousin R, Siffert S, Su BL (2013) Gold catalysts in environmental remediation and water-gas shift technologies. Energy Environ Sci 6:371–391

Basheer AA (2018) New generation nano-adsorbents for the removal of emerging contaminants in water. J Mol Liq 261:583–593

Ben-Moshe T, Frenk S, Dror I, Minz D, Berkowitz B (2013) Effects of metal oxide nanoparticles on soil properties. Chemosphere 90:640–646

Bouwmeester H, Dekkers S, Noordam MY, Hagens WI, Bulder AS, De Heer C, Ten Voorde SE, Wijnhoven SW, Marvin HJ, Sips AJ (2009) Review of health safety aspects of nanotechnologies in food production. Regul Toxicol Pharmacol 53:52–62

Brusseau ML, Pepper IL, Gerba C (2019) Environmental and pollution science. Academic Press

Carolin CF, Kumar PS, Saravanan A, Joshiba GJ, Naushad M (2017) Efficient techniques for the removal of toxic heavy metals from aquatic environment: a review. J Environ Chem Eng 5: 2782–2799

Chaudhary J, Tailor G, Yadav B, Michael O (2019) Synthesis and biological function of nickel and copper nanoparticles. Heliyon 5:e01878

Chen Y, Liang W, Li Y, Wu Y, Chen Y, Xiao W, Zhao L, Zhang J, Li H (2018) Modification, application and reaction mechanisms of nano-sized iron sulfide particles for pollutant removal from soil and water: a review. Chem Eng J 362:144

Corsi I, Winther-Nielsen M, Sethi R, Punta C, Della Torre C, Libralato G, Lofrano G, Sabatini L, Aiello M, Fiordi L (2018) Ecofriendly nanotechnologies and nanomaterials for environmental applications: key issue and consensus recommendations for sustainable and ecosafe nanoremediation. Ecotoxicol Environ Saf 154:237–244

da Costa JP, Santos PS, Duarte AC, Rocha-Santos T (2016) (Nano) plastics in the environment–sources, fates and effects. Sci Total Environ 566:15–26

da Silveira Petruci JF, Piccoli JP, Fortes PR, Cardoso AA (2019) Nanomaterials in air pollution trace detection. In: Nanomaterials applications for environ matrices. Elsevier, pp 427–447

de Almeida MP, Pereira E, Baptista P, Gomes I, Figueiredo S, Soares L, Franco R (2014) Gold nanoparticles as (bio) chemical sensors. In: Comprehensive analytical chemistry, vol 66. Elsevier, pp 529–567

Deshmukh S, Patil S, Mullani S, Delekar S (2018) Silver nanoparticles as an effective disinfectant: a review. Mater Sci Eng: C 97:954

Diao M, Yao M (2009) Use of zero-valent iron nanoparticles in inactivating microbes. Water Res 43:5243–5251

Djurišić AB, Leung YH, Ng AM, Xu XY, Lee PK, Degger N, Wu R (2015) Toxicity of metal oxide nanoparticles: mechanisms, characterization, and avoiding experimental artefacts. Small 11:26–44

Dubey S, Banerjee S, Upadhyay SN, Sharma YC (2017) Application of common nano-materials for removal of selected metallic species from water and wastewaters: a critical review. J Mol Liq 240:656–677

El-Ramady H, Alshaal T, El-Henawy A, Abdalla N, Taha HS, Elmahrouk M, Shalaby T, Elsakhawy T, Omara AED, El-Marsafawy S (2017) Environmental nanoremediation under changing climate. Environ Biodivers Soil Secur 1:109–128

Enfrin M, Dumée LF, Lee J (2019) Nano/microplastics in water and wastewater treatment processes–origin, impact and potential solutions. Water Res 161:621

European Environment Agency (EEA) (2014) Progress in management of contaminated sites. Report CSI 015. Copenhagen, Denmark. http://www.eea.europa.eu/data-and-maps/indicators/progress-in-anagement-of-contaminated-sites-3/assessment

Fathima JB, Pugazhendhi A, Oves M, Venis R (2018) Synthesis of eco-friendly copper nanoparticles for augmentation of catalytic degradation of organic dyes. J Mol Liq 260:1–8

Fulekar M, Pathak B (2017) Environmental nanotechnology. CRC Press

Gagné F, Auclair J, Turcotte P, Gagnon C, Peyrot C, Wilkinson K (2019) The influence of surface waters on the bioavailability and toxicity of zinc oxide nanoparticles in freshwater mussels. Comp Biochem Physiol Part C Toxicol Pharmacol 219:1–11

Garner KL, Keller AA (2014) Emerging patterns for engineered nanomaterials in the environment: a review of fate and toxicity studies. J Nanopart Res 16:2503

Gu Q, Chen J (2018) Carbon-nanotube-based nano-emitters: a review. J Lumin 200:181–188

Ingle AP, Seabra AB, Duran N, Rai M (2014) Nanoremediation: a new and emerging technology for the removal of toxic contaminant from environment. In: Microbial biodegradation and bioremediation. Elsevier, pp 233–250

Jadhav SB, Singhal RS (2013) Polysaccharide conjugated laccase for the dye decolorization and reusability of effluent in textile industry. Int Biodeterior Biodegrad 85:271–277

Ji T, Feng Y, Qin M, Feng W (2016) Thermal conducting properties of aligned carbon nanotubes and their polymer composites. Compos Part A: Appl Sci Manuf 91:351–369

Jung KW, Ahn KH (2016) Fabrication of porosity-enhanced MgO/biochar for removal of phosphate from aqueous solution: application of a novel combined electrochemical modification method. Biores Technol 200:1029–1032

Kale SS, Kadam AK, Kumar S, Pawar N (2010) Evaluating pollution potential of leachate from landfill site, from the Pune metropolitan city and its impact on shallow basaltic aquifers. Environ Monit Assess 162:327–346
Karn B, Kuiken T, Otto M (2011) Nanotechnology and in situ remediation: a review of the benefits and potential risks. Cien Saude Colet 16:165–178
Khanna P, Kaur A, Goyal D (2019) Algae-based metallic nanoparticles: synthesis, characterization and applications. J Microbiol Meth 163:105656
Khurana P, Thatai S, Kumar D (2019) Destruction of recalcitrant nanomaterials contaminants in industrial wastewater. In: Emerging and nanomaterial contaminants in wastewater. Elsevier, pp 137–158
Kim MI, Ye Y, Won BY, Shin S, Lee J, Park HG (2011) A highly efficient electrochemical biosensing platform by employing conductive nanocomposite entrapping magnetic nanoparticles and oxidase in mesoporous carbon foam. Adv Funct Mater 21:2868–2875
Kizling M, Dzwonek M, Wieckowska A, Bilewicz R (2018) Gold nanoparticles in bioelectrocatalysis–the role of nanoparticle size. Curr Opin Electrochem 12:113–120
Kumar P, Govindaraju M, Senthamilselvi S, Premkumar K (2013) Photocatalytic degradation of methyl orange dye using silver (ag) nanoparticles synthesized from Ulva lactuca. Colloid Surf B: Biointer 103:658–661
Kuznetsova YV, Letofsky-Papst I, Sochor B, Schummer B, Sergeev AA, Hofer F, Rempel AA (2019) Greatly enhanced luminescence efficiency of CdS nanoparticles in aqueous solution. Colloids Surfaces A: Physicochem Eng Aspect 581:123814
Kwadijk C, Velzeboer I, Koelmans A (2013) Sorption of perfluorooctane sulfonate to carbon nanotubes in aquatic sediments. Chemosphere 90:1631–1636
Lai C, Zhou X, Huang D, Zeng G, Cheng M, Qin L, Yi H, Zhang C, Xu P, Zhou C (2018) A review of titanium dioxide and its highlighted application in molecular imprinting technology in environment. J Taiwan Inst Chem Eng 91:517
Lee C, Kim JY, Lee WI, Nelson KL, Yoon J, Sedlak DL (2008) Bactericidal effect of zero-valent iron nanoparticles on Escherichia coli. Environ Sci Technol 42:4927–4933
Lei C, Sun Y, Tsang DC, Lin D (2018) Environmental transformations and ecological effects of iron-based nanoparticles. Environ Pollut 232:10–30
Li J, Li J, Zhu Q (2018) Carbon deposition and catalytic deactivation during CO2 reforming of CH4 over CO/MgO catalyst. Chinese J Chem Eng 26:2344–2350
Liu A, Ma Y, Gunawardena JM, Egodawatta P, Ayoko GA, Goonetilleke A (2018) Heavy metals transport pathways: the importance of atmospheric pollution contributing to stormwater pollution. Ecotoxicol Environ Saf 164:696–703
Löffler FE, Edwards EA (2006) Harnessing microbial activities for environmental cleanup. Curr Opin Biotechnol 17:274–284
Madannejad R, Shoaie N, Jahanpeyma F, Darvishi MH, Azimzadeh M, Javadi H (2019) Toxicity of carbon-based nanomaterials: Reviewing recent reports in medical and biological systems. Chemico-biological Interactions
Majedi SM, Lee HK (2016) Recent advances in the separation and quantification of metallic nanoparticles and ions in the environment. TrAC Trend Anal Chem 75:183–196
Marambio-Jones C, Hoek EM (2010) A review of the antibacterial effects of silver nanomaterials and potential implications for human health and the environment. J Nanopart Res 12:1531–1551
Nasrollahzadeh M, Sajadi SM, Sajjadi M, Issaabadi Z (2019) Applications of nanotechnology in daily life. In: Interface science and technology, vol 28. Elsevier, pp 113–143
Neto VDOS, Freire PDTC, do Nascimento RF (2019) Groundwater remediation using nanomaterials. In: Nanomaterials applications for environmental matrices. Elsevier, pp 381–402
Nguyen NH, Von Moos NR, Slaveykova VI, Mackenzie K, Meckenstock RU, Thűmmler S, Bosch J, Ševců A (2018) Biological effects of four iron-containing nanoremediation materials on the green alga *Chlamydomonas* sp. Ecotoxicol Environ Saf 154:36–44

Nisha K, Navaneethan M, Dhanalakshmi B, Murali KS, Hayakawa Y, Ponnusamy S, Muthamizhchelvan C, Gunasekaran P (2015) Effect of organic-ligands on the toxicity profiles of CdS nanoparticles and functional properties. Colloid Surf B: Biointer 126:407–413
Noman M, Shahid M, Ahmed T, Niazi MBK, Hussain S, Song F, Manzoor I (2019) Use of biogenic copper nanoparticles synthesized from a native Escherichia sp. as photocatalysts for azo dye degradation and treatment of textile effluents. Environ Pollut:113514
Oberdörster G, Oberdörster E, Oberdörster J (2005) Nanotoxicology: an emerging discipline evolving from studies of ultrafine particles. Environ Health Perspect 113:823–839
Patil SS, Shedbalkar UU, Truskewycz A, Chopade BA, Ball AS (2016) Nanoparticles for environmental clean-up: a review of potential risks and emerging solutions. Environ Technol Innovation 5:10–21
Peng D, Wu B, Tan H, Hou S, Liu M, Tang H, Yu J, Xu H (2019) Effect of multiple iron-based nanoparticles on availability of lead and iron, and micro-ecology in lead contaminated soil. Chemosphere 228:44–53
Petrie B, Barden R, Kasprzyk-Hordern B (2015) A review on emerging contaminants in wastewaters and the environment: current knowledge, understudied areas and recommendations for future monitoring. Water Res 72:3–27
Pico Y, Alfarhan A, Barcelo D (2019) Nano-and microplastic analysis: focus on their occurrence in freshwater ecosystems and remediation technologies. Trend Anal Chem 113:409
Prabhakar R, Samadder SR, Jyotsana (2017) Aquatic and terrestrial weed mediated synthesis of iron nanoparticles for possible application in wastewater remediation. J Clean Prod 168:1201–1210
Qu Y, Shen W, Pei X, Ma F, You S, Li S, Wang J, Zhou J (2017) Biosynthesis of gold nanoparticles by Trichoderma sp. WL-go for azo dyes decolorization. J Environ Sci 56:79–86
Quadros ME, Marr LC (2010) Environmental and human health risks of aerosolized silver nanoparticles. J Air Waste Manag Assoc 60:770–781
Rajan C (2011) Nanotechnology in groundwater remediation. Int J Environ Sci Dev 2:182
Rajput V, Minkina T, Fedorenko A, Sushkova S, Mandzhieva S, Lysenko V, Duplii N, Fedorenko G, Dvadnenko K, Ghazaryan K (2018) Toxicity of copper oxide nanoparticles on spring barley (Hordeum sativum distichum). Sci Total Environ 645:1103–1113
Sabouri Z, Akbari A, Hosseini HA, Hashemzadeh A, Darroudi M (2019) Bio-based synthesized NiO nanoparticles and evaluation of their cellular toxicity and wastewater treatment effects. J Mol Struct 1191:101–109
Saleh MG, Badawy AA, Ghanem AF (2019) Using of titanate nanowires in removal of lead ions from waste water and its biological activity. Inorg Chem Commun 107508
Santhoshkumar J, Rajeshkumar S, Kumar SV (2017) Phyto-assisted synthesis, characterization and applications of gold nanoparticles–a review. Biochem Biophys Rep 11:46–57
Selvam GG, Sivakumar K (2015) Phycosynthesis of silver nanoparticles and photocatalytic degradation of methyl orange dye using silver (ag) nanoparticles synthesized from Hypnea musciformis (Wulfen) JV Lamouroux. Appl Nanosci 5:617–622
Sharma VK, Yang X, Cizmas L, McDonald TJ, Luque R, Sayes CM, Yuan B, Dionysiou DD (2017) Impact of metal ions, metal oxides, and nanoparticles on the formation of disinfection byproducts during chlorination. Chem Eng J 317:777–792
Siddiqui M, Nizamuddin S, Baloch HA, Mubarak N, Al-Ali M, Mazari SA, Bhutto A, Abro R, Srinivasan M, Griffin G (2018) Fabrication of advance magnetic carbon nano-materials and their potential applications: a review. J Environ Chem Eng 7:1
Thangamani N, Bhuvaneshwari N (2019) Green synthesis of gold nanoparticles using Simarouba glauca leaf extract and their biological activity of micro-organism. Chem Phys Lett 732:136587
Tosco T, Papini MP, Viggi CC, Sethi R (2014) Nanoscale zerovalent iron particles for groundwater remediation: a review. J Cleaner Prod 77:10–21
USEPA (2004) Cleaning up the Nation's waste sites: markets and technology trends, EPA 542-R-04-015, 2004. US Environmental Protection Agency, Washington, DC
USEPA (2009) US Environmental Protection Agency (USEPA). National priorities list (NPL). http://www.epa.gov/superfund/sites/npl/

USEPA (2013) US Environmental Protection Agency (USEPA) Remediation Technologies. http://www.epa.gov/superfund/remedytech/remed.htm

Usman M, Ahmed A, Yu B, Peng Q, Shen Y, Cong H (2019) Photocatalytic potential of bio-engineered copper nanoparticles synthesized from Ficus carica extract for the degradation of toxic organic dye from waste water: growth mechanism and study of parameter affecting the degradation performance. Mater Res Bull 120:110583

Varma RS (2014) Journey on greener pathways: from the use of alternate energy inputs and benign reaction media to sustainable applications of nano-catalysts in synthesis and environmental remediation. Green Chem 16:2027–2041

Visentin C, da Silva Trentin AW, Braun AB, Thomé A (2019a) Application of life cycle assessment as a tool for evaluating the sustainability of contaminated sites remediation: a systematic and bibliographic analysis. Sci Total Environ 672:893–905

Visentin C, da Silva Trentin AW, Braun AB, Thomé A (2019b) Lifecycle assessment of environmental and economic impacts of nano-iron synthesis process for application in contaminated site remediation. J Cleaner Prod 231:307

Vishwakarma V (2012) The Role of Nanotechnology R&D Institutes to Enhance Competitiveness of Small and Medium Enterprises. Promoting Nanotechnology Applications, Special features

Wang S, Sun H, Ang HM, Tadé M (2013) Adsorptive remediation of environmental pollutants using novel graphene-based nanomaterials. Chem Eng J 226:336–347

Wenning R, Martello L (2014) POPs in marine and freshwater environments. Environmental forensics for persistent organic pollutants. In: Environmental forensics for persistent organic pollutants. Elsevier Newnes, p 424

Xie Y, Chen C, Ren X, Wang X, Wang H, Wang X (2019) Emerging natural and tailored materials for uranium-contaminated water treatment and environmental remediation. Prog Mater Sci

Yang W, Peters JI, Williams RO III (2008) Inhaled nanoparticles—a current review. Int J Pharm 356:239–247

Yang X, Chen Y, Liu X, Guo F, Su X, He Q (2018) Influence of titanium dioxide nanoparticles on functionalities of constructed wetlands for wastewater treatment. Chem Eng J 352:655–663

Yao Y, Zhang J, Gao M, Yu M, Hu Y, Cheng Z, Wang S (2018) Activation of persulfates by catalytic nickel nanoparticles supported on N-doped carbon nanofibers for degradation of organic pollutants in water. J Colloid Interface Sci 529:100–110

Zhang L, Fang M (2010) Nanomaterials in pollution trace detection and environmental improvement. Nano Today 5:128–142

Zhang WX (2003) Nanoscale iron particles for environmental remediation: an overview. J Nanopart Res 5:323–332

Zhao Q, Li H, Zhang X, Yu S, Wang S, Sun G (2020) Platinum in-situ catalytic oleylamine combustion removal process for carbon supported platinum nanoparticles. J Energy Chem 41: 120–125

Zhou D, Song X, Zhao F, Baohua G (2017) Soil environment and pollution remediation. Pedosphere 27(3):387–388

Zhu Y, Liu X, Hu Y, Wang R, Chen M, Wu J, Wang Y, Kang S, Sun Y, Zhu M (2019) Behavior, remediation effect and toxicity of nanomaterials in water environments. Environ Res 174:54

Sustainable Agriculture and Plant Production by Virtue of Biochar in the Era of Climate Change

Hafiz Muhammad Tauqeer, Veysel Turan, Muniba Farhad, and Muhammad Iqbal

Abstract In recent years, rapid increase in population growth, improper usage of synthetic fertilizers, organic matter depletion, nutrient imbalance, and land degradation owing to several anthropogenic activities have significantly exerted considerable pressure on agriculture which negatively influences sustainable plant production. Therefore, it is necessary to sustain the most appropriate levels of organic matter in degraded soils, which supports sustainable crop production and maintains nutrient cycling in them. Biochar has been broadly used for sustainable plant production among different organic matters due to its several advantages such as mitigating global warming, excellent soil conditioner, and as a potential amendment for various environmental applications over other soil additives. Moreover, biochar additions in agricultural soils also promoted the seed germination, growth, biomass, yield, and nutritional qualities of crops grown on biochar amended soils. In addition to these benefits, biochar also supports soil microorganisms by providing them habitat due to its porous structure and releases essential nutrients from its matrix, improving microbial communities. Thus, it is suggested that biochar could play a vital role in reducing the adverse impacts of climate change and threats to sustainable crop production.

Keywords Activated carbon · Abiotic stress · Soil amendment · Crop growth · Plant nutrition · Carbon sequestration

---

H. M. Tauqeer
Department of Environmental Sciences, University of Gujrat, Gujrat, Pakistan

V. Turan
Institute of Soil Science and Plant Nutrition, Faculty of Agriculture, Bingöl University, Bingöl, Turkey

M. Farhad
Department of Chemistry, Government College University, Faisalabad, Pakistan

M. Iqbal (✉)
Department of Environmental Sciences and Engineering, Government College University, Faisalabad, Pakistan

M. Hasanuzzaman et al. (eds.), *Managing Plant Production Under Changing Environment*, https://doi.org/10.1007/978-981-16-5059-8_2

1 Background

In recent years, land degradation, intense agriculture, soil fertility loss, environmental stresses (heat, drought, salinity, cold, metals), and nutrient imbalance significantly decreased sustainable agriculture and plant production owing to a decline in soil organic matter. Besides, the world's population dramatically increased during the last four to five decades which exerted stress on food production (Riaz et al. 2019). Soil nutrient depletion and fertility loss are the key concerns linked with sustainable food production and food uncertainty owing to extensive land use (Agegnehu et al. 2017). The applications of inorganic fertilizers have played a crucial role in enhancing crop and plant production during the last half-century. However, the use of synthetic fertilizers alone is not a wise solution in maintaining soil fertility and enhancing crop yield because the chemical fertilizers, particularly, nitrogen (N) may result in soil degradation and other associated environmental problems such as the rapid organic matter decomposition of organic matter resulted in the reduction of soil carbon stocks (Agegnehu et al. 2017). Thus, maintaining the suitable organic matter in degraded arable lands and ensuring effective biological nutrient cycling is critical for sustainable plant production and soil management. After understanding land degradation and environmental issues, research on numerous organic additives such as composts, mulches, manures, and other carbonaceous additives e.g. biochar has evolved extensively with vital findings on agronomic benefits, greenhouse gas emissions, carbon sequestration, soil quality, and fertility as well as a potential soil amendment (Bis et al. 2018). Biochar, a carbon rich porous material produced through the slow pyrolysis and/or by the combustion, thermolysis, or gasification of various feedstock such as plant residue (Knicker 2007; Naeem et al. 2021; Preston and Schmidt 2006), anthropogenic sources (Warnock et al. 2007), forest waste, biomass from energy crops, (Agegnehu et al. 2017) forage plant biomass (Husk and Major 2011), swine manure (Ren et al. 2020; Tsai et al. 2012), sewage biosolids (Gao et al. 2020; Li et al. 2018; Zhou et al. 2017), empty fruit bunches (Abdulrazzaq et al. 2015; Yavari et al. 2016, 2019), poultry litter and manure (Abd El-Mageed et al. 2021; Sehrish et al. 2019; Wang et al. 2015; Chan et al. 2008; Jin et al. 2016), human manure (Liu et al. 2014), goat manure (Touray et al. 2014; Tayyab et al. 2018), and paper-mill waste (Hmid et al. 2015), kitchen waste (Xu et al. 2020a) and rice husks (Islam et al. 2021; Wang et al. 2020a). The physical and chemical properties of biochar entirely depend upon the feedstock type, heating rate, pyrolysis conditions, residence time, pressure, design of reaction vessel, the flow rate of inert gas, and other treatments (sieving, crushing, activation) after pyrolysis (Joseph and Lehmann 2009; Qambrani et al. 2017). For instance, wood biomass-derived biochar was relatively more resistant to biodegradation due to higher lignin content (Windeatt et al. 2014) compared to biochar derived from crop residues and animal manures (El-Naggar et al. 2018; Singh et al. 2014). The biochar derived from manure feedstock, however, is thought to be nutrient (Mg, Ca, and P) rich (Bandara et al. 2020; Cao et al. 2011) accompanied by higher cation exchange capacity (CEC) and stability (Cely et al. 2015). Previous studies revealed that biochar obtained from the chicken manure at different pyrolysis conditions exhibited dissimilar characteristics of pH, electrical conductivity (EC), N and P concentrations (Chan et al. 2008; Meier et al. 2017). Additionally, biochar is drawing attention as potential input in agriculture to support sustainable crop production and increase yield via improving soil fertility, water holding capacity, providing essential nutrients, carbon capturing benefits, simultaneously alleviating the negative consequences of numerous biotic and abiotic stresses, reducing greenhouse gas emissions and pollution (Akhtar et al. 2015; Beesley and Dickinson 2011; Lehmann and Joseph 2015). Moreover, agricultural activities also deteriorate the soil organic carbon (SOC) day by day. The resilient carbon fraction of biochar enhanced the total carbon pool in the soil, resultantly improved soil fertility (Niar et al. 2017; Lorenz and Lal 2014). This SOC plays a key role via maintaining the nutrients (P, N, K) and water retention and by providing habitat for soil microorganisms that improve soil structure and support plant growth (Kolton et al. 2011; Lorenz et al. 2007). Land use practices and extreme weather conditions (especially high temperatures) are also known to reduce SOC and soil fertility. The addition of biochar as a soil conditioner is recommended to enhance both SOC and soil fertility. Apart from this, biochar also increases carbon sequestration and reduces greenhouse gas emissions released from biomass breakdown and thus reduces the global warming issue (Qambrani et al. 2017). Likewise, biochar may also be utilized as an excellent adsorbent to remove toxic environmental pollutants from the soil or wastewater (Yu et al. 2021). The occurrence of numerous functional groups onto the surface of biochar served as excellent binding sites for the adsorption of toxic heavy metals such as lead (Pb), cadmium (Cd), and nickel (Ni) consequently prevent their accumulation in plants (Tauqeer et al. 2021).

Thus, this chapter aims to collect information about the potential applications of biochar for sustainable plant production after its incorporation into agricultural soils.

2 Benefits of Biochar Additions in Soils

2.1 Soil Quality Improvement

The addition of biochar in soils has a remarkable influence on numerous physical characteristics of soil such as porosity, texture, depth and structure, surface area, particle and pore size distribution, and bulk density. This improvement in the physical traits of soil consequently has a positive influence on water availability at deeper depths and aeration in the root zones which support plant growth (Chan et al. 2008). Additionally, the different merits of biochar additions in agricultural soils significantly raised interest in its utilization as a soil conditioner due to an increase in physical and biological traits of soils such as water and nutrient retention which further improved plant growth (Riaz et al. 2019). For instance, among nine numerous sorts of biochar each produced from various feedstock (500 °C), miscanthus feedstock biochar significantly enhanced soil fine and medium pores, EC, available water content, CEC and reduced pH, bulk and particle density, and soil-wide pore (Khan et al. 2017) (Table 1).

2.2 Soil Physical Properties

The biochar addition in soils increases water-holding capacity (WHC) and decreases bulk density. This rise in WHC capacity could be attributed to the larger surface area as well as the highly porous structure of biochar which enhanced water uptake capacity and hence improved plant growth (Kinney et al. 2012; Laghari et al. 2016). For example, the biochar derived from pine sawdust feedstock at numerous pyrolysis conditions (400, 500, 600, 700, and 800 °C) were added in desert soil, a significant improvement in sorghum yield by 32% and 19% was observed at 700 and 400 °C, accordingly, over control. Additionally, WHC of the desert soil was improved by 16% and 59% which enhanced water use efficiency by 52% and 74% as well as total soil carbon stock, CEC, and plant nutrient content under 400 and 700 °C treatments (Laghari et al. 2016) (Table 1).

2.3 Soil Chemical Properties

Biochar application also improved the chemical traits of soil such as CEC, soil pH, soil fertility, and nutrient uptake by the plants (Lehmann and Joseph 2015). Likewise, the oxidation process occurring onto biochar surfaces and the abundance of different negative charge sites increased the CEC of the soil which increases nutrient retention and subsequently supports plant growth (Cheng et al. 2008; Laird et al. 2010). In contrast, biochar additions to agricultural soils also increased anion exchange capacity (AEC) of the soil owing to the presence of oxonium functional groups which reduced the leaching of anionic nutrients (NO_3^-, PO_4^{3-}) from the soil (Lawrinenko and Laird 2015).

2.4 Soil Biological Properties

Soil microorganisms such as fungi, bacteria, algae, nematodes, actinomycetes, archaea, protozoa, and bacteriophages perform a crucial role in maintaining soil functions such as soil structure formation and improvement, nutrient cycling, suppression of pathogens and diseases, organic matter decomposition, secretion of plant growth supporters, and mineralization of organic toxicants (Gorovtsov et al. 2019). The presence of biochar in agricultural soils elicits the diversity and functioning of these microorganisms owing to the overall improvement in physicochemical traits of

Table 1 The influence of biochar applications on different traits of soil and plant

Feedstock type	Results	References
Cow-bone derived biochar (application rate = 2.5, 5 and 10% w/w Pyrolysis conditions = 500 °C and 800 °C).	Increased total N, total dissolved organic carbon, and total P. The application of 2.5 and 5% biochar (500 °C) improved the activities of alkaline phosphatase and β-glucosidase, over control. Moreover, a significant improvement in maize growth, polyphenol oxidase (PPO), lipid peroxidase (POD), phenylalanine ammonia-lyase (PAL), chlorophyll, and carotene contents were observed, over control	Azeem et al. (2021)
Spartina alterniflora feedstock	The sole and combined application of biochar with effective micro-organisms promisingly improved seed germination rate, stem diameter, plant height, total biomass, and nutrient uptake by *Sesbania cannabina*. Moreover, a remarkable reduction in salt content and improvement in total carbon, available P, total N, and available K, soil NO_3^- and NH_4^+, microbial biomass carbon, soil enzymes, and soil fertility was recorded in the sole and combined treatments of biochar and effective micro-organisms. Overall, the integrated use of biochar at 3% and effective micro-organisms could be an effective approach for the management of coastal saline-alkali soil	Cui et al. (2021)
Peanut shells derived biochar	Biochar utilization in aluminum (Al) and acid-toxic soil improved nutrients availability, exchangeable cations (Mg^{2+}, Ca^{2+}, K^+), soil organic matter, N use efficiency, and overall soil quality. Further, an improvement in the root and shoot biomass of maize by 44% and 89%, respectively, were recorded over control. Results suggested that biochar may use to improve soil quality and support plant production through alleviating Al toxicity	Xia et al. (2020)
Woodchips derived biochar	The utilization of biochar (at 1%) promisingly improved *Arenosols* health, various microbial groups, and populations, maize biomass especially root biomass, and the activities of antioxidants grown on poorly humus sandy soil	Kocsis et al. (2020)
Cassava straw	Applications of N fertilizers coupled with biochar improved soil quality, morphological traits of roots and photosynthesis resultantly increased the yield and yield-related traits of noodle rice	Ali et al. (2020)

(continued)

Table 1 (continued)

Feedstock type	Results	References
Miscanthus and wheat straw biochar	The provision of wheat straw biochar improved bacterial abundance, actinomycetes, soil enzymes, soil fertility index, the geometric mean of enzyme activities index which resultant in an overall improvement in the soil quality	Mierzwa-Hersztek et al. (2017)
Cotton gin trash (pyrolyzed at 450 °C)	Biochar promisingly improved SOM, the contents of Ca, P, Mn and K, and EC in clay loam and sandy loam soils in comparison to the rest of the biochar treatment	Zhang et al. (2016)
Wood and manure-derived biochar treatments	Increased water content in the soil and plant water use efficiency. Additionally, improved CEC and total N while reduced NH_4-N leaching	Ajayi et al. (2016); Abel et al. (2013)
Wood, peanut shell −chicken manure −wheat chaff	Enhanced the availability of P up to 208% while reducing AMF abundance in the soil	Madiba et al. (2016); Warnock et al. (2007)
Wheat straw	Improved soil pH, the contents of SOC, N, and reduced N_2O release	Li et al. (2015)
Eucalyptus logs, maize Stover	Biochar applications significantly enhanced the contents of total N in the soil from the atmosphere	Güereña et al. (2015)
Acacia whole tree green waste	Improved porosity of the soil and aggregate stability	Hardie et al. (2014)
Different biochar prepared from various feedstock	Improved pH, microbial biomass, microbial habitat, and the contents of P, N, K, and total carbon.	Thies et al. (2015); Biederman and Harpole (2013)

soil as well as the porous structure of biochar which serves as habitat and also prevent them from predation (Khan et al. 2020; Palansooriya et al. 2019; Warnock et al. 2007). Moreover, biochar additions to soil also increased carbon-to-nitrogen (C:N) ratios, dissolved organic carbon (C), and K^+ concentrations which support numerous microbial community structures (Wong et al. 2019). Likewise, biochar also enhanced the activities of soil enzymes which increase microbial communities and improved overall soil health (Ramzani et al. 2017; Khan et al. 2020) (Table 1).

2.5 *Provision and Retention of Essential Nutrients*

Biochar also supports sustainable plant production by providing essential mineral nutrients to plants as well as microorganisms. Though, biochar increases soil pH which influences the availability of micronutrients. However, biochar slowly released micronutrients from its matrix and makes them available for plants

(Ahmed et al. 2016). Moreover, biochar additions remarkably promoted the grain quality and yield of *Zea mays* after Mg and Ca uptake (Major et al. 2010). Likewise, the application of acidified biochar produced from maize cob (350 °C) promisingly improved the growth, yield, physiological, chemical, and biochemical traits, antioxidants, and anti-nutrients in *Chenopodium quinoa* grown on drought, salt and Ni stressed soils. The results suggested that the acidified biochar effectively increased the bioavailability and aerial transport of nutrients and subsequent accumulation in quinoa seed (Ramzani et al. 2017).

2.6 CO_2 Sequestration and Reduction of Greenhouse Gas Emission

Agriculture contributes its share in releasing the substantial magnitudes of greenhouse gases which is an alarming and universal global warming and climate change issue (Burney et al. 2010). Usually, CO_2 is released into the atmosphere by the microbial decay or burning of agricultural by-products as well as through the breakdown of organic matter (Smith et al. 2010). Carbon emissions from the soil are considered as one of the prime signals of land degradation which is a challenging task for sustainable plant production, biodiversity conservation, and acclimatizing to climate change (Barrow 2012; Mchunu and Chaplot 2012). The presence of vegetation cover is a natural and effective method of CO_2 captured from the air via photosynthesis. The efficacy of this practice for carbon sequestration is inadequate owing to the instability of captured carbon which returned into the environment as CO_2 through respiration or decomposition (Semida et al. 2019).

As mentioned earlier, biochar additions to agricultural soils may mitigate greenhouse gas emissions and combat climate change through a range of mechanisms (Mohammadi et al. 2020). For instance, inhibition of CO_2 and CH_4 (particularly from rice fields), reduced nitrous oxide (N_2O) released from agricultural soils, consequently decreased the use of artificial fertilizers. The improvement in crop yield are the additional key benefits of biochar applications in agro-ecosystem due to the improved soil aeration (Mohammadi et al. 2020; Qambrani et al. 2017; Rogovska et al. 2011; Zhang et al. 2012).

Various microorganisms produced CH_4 under anoxic conditions via methanogenesis. Approximately CH_4 is considered 20 times more powerful than CO_2 in absorbing thermal radiation in the earth's lower troposphere and increased global warming (Watson et al. 2000). It was observed that after adding biochar in the soil, a remarkable reduction in CH_4 emission was observed (Rondon et al. 2005a). This reduction in CH_4 emission could be due to the porous characteristics of biochar which increased aeration and reduced the favorable anaerobic environments causing methanogenesis (Verheijen et al. 2010). In another study, biochar utilization also reduced CH_4 and CO_2 emissions from the rice field (Liu et al. 2011). Thus, biochar from animal manure may help in this context.

Nitrous oxide is also an important gas having over 300 times more potential than CO_2 in absorbing thermal radiation in the troposphere and causing global warming (Watson et al. 2000). Primarily, N_2O is produced in the soil by numerous microorganisms via denitrification and nitrification. The presence of moisture content in the soil significantly influenced the production of N_2O. For instance, higher moisture (>70%) levels support anoxic conditions, which promote denitrification, while reduced moisture (<50%) levels stimulate nitrification. It was reported that the higher moisture level (up to 80%) produced 8–23 times more N_2O in contrast to lower moisture levels (40%) (Bruun et al. 2011). Similarly, the findings of a study revealed that the utilization of biochar in the form of charcoal significantly declined N_2O release up to 89% (Yanai et al. 2007). Moreover, over 80% decrease in N_2O emissions from biochar amended soil was observed in the greenhouse, and field trials in Columbia (Renner 2007) consequently reduced the applications of synthetic fertilizer. This reduction in N_2O emission from the soil could be due to the adsorption of nitrate (NO_3^-) onto the large surfaces of biochar. Additionally, biochar applications also influence the N transfer and N dynamics which reduced N_2O release (DeLuca et al. 2006; Rondon et al. 2007; Yanai et al. 2007). The presence of biochar in agricultural soils also supports biological stabilization of inorganic N, resultantly reduced ammonia volatilization owing to the higher C: N ratios and lower N content in biochar (Taghizadeh-Toosi et al. 2011).

Likewise, a recent field study (24 months) was conducted in Moso bamboo forest to evaluate the effectiveness of various biochar application rates (0, B5, and B15 Mg ha^{-1}) on SOC stocks, greenhouse gas emissions, and vegetation carbon stocks. Results suggested that the maximum SOC stocks were increased up to 66%, while the greenhouse gas emissions increased by 21%, respectively, in B5 and B15 treatments over control. Moreover, the addition of biochar remarkably reduced N_2O release by 24% in B15, whereas increased CH_4 emission by 16% in B5, respectively, over control. Overall, biochar utilization improved the total ecosystem carbon stock of the moso bamboo forest by 486% and 252% for B5 and B15 treatments and is recommended as an excellent and effective approach for the management of forest soils (Xu et al. 2020b). A recent two-year field study also investigated the potential of biochar as a soil conditioner to combat climate change in sandy loam soil under the influence of drip irrigation with mulch. Biochar was prepared from the corn residue and applied in the soil at various rates (Bo, B15, B30, and B45 t ha^{-1}). The average CH_4 reduction by 124% and 132% was observed in B15 and B30 treatments, respectively over control. Likewise, B30 and B45 treatments improved SOC in the top upper layer (15 cm) by 19% and 37% during the first growing season and by 12% and 15% during the second growing season. Among all applied rates, B30 was efficient in reducing CH_4 and N_2O emissions and improved corn yields (Yang et al. 2020).

Moreover, the applications of rice straw, bamboo, and wood chip-derived biochar promisingly reduced CO_2 emissions from the paddy (Liu et al. 2011) as well as silt loam soil (Spokas et al. 2009). Previously, it was observed that the amending soybean cropland and *Brachiaria humidicola* grass stands with biochar (at 20 g kg^{-1}) eliminate CH_4 releases whereas reduced NO_2 emissions by 50% and 80% (Rondon et al. 2005b) (Table 2).

2.7 Heavy Metal Immobilization and Food Safety

In recent years, a lot of research work has been done so far on biochar and its numerous applications as a potential amendment especially for the removal of heavy metals and other environmental toxicants from the soil and water owing to its majestic properties such as alkaline nature, higher CEC, and porosity (Khan et al. 2020; Tauqeer et al. 2021). Results revealed that the combined application of lignin-derived biochar and arbuscular mycorrhizal fungi (AMF) significantly improved barley grain and was safer for human consumption grown on Pb contaminated soil (Khan et al. 2020). A recent study conducted by (Zubair et al. 2021) revealed that the textile waste biochar coated with chitosan remarkably reduced Cd distribution in roots and shoots of *Moringa oleifera* L while improving the overall growth, dietary parameters, antioxidants as well as soil enzymes over control (Table 3).

3 Sustainable Plant Production under the Influence of Biochar

This section provides selected studies on the usage of biochar as a potential soil additive and its influence on sustainable plant production.

3.1 Seed Germination and Plant Growth

Up till now, limited research work on the influence of biochar from different feedstocks either on improvement or inhabitation of seed germination has been conducted so far (Semida et al. 2019). For instance, among nine different biochars (poultry manure, rice straw, vegetable waste, neem leaves, cotton sticks, wheat straw, domestic waste, citrus leaves, and eucalyptus leaves), the addition of vegetable waste-derived biochar at 2% w/w significantly improved seed germination of maize (Qayyum et al. 2015). Amending soil with biochar (0.5, 2.5 kg m^{-2}) enhanced *Amaranthus palmeri,* seed sprouting but no influence on *Senna obtusifolia* and *Digitaria ciliaris* (Soni et al. 2014). Similarly, the addition of biochar in sandy soil increased maize growth by improving leaf osmotic potential and relative water content as well as photosynthesis. The possible mechanism for this enhanced seed germination and improved growth is due to the overall improvement in soil quality,

Table 2 The influence of biochar incorporation on carbon sequestration and greenhouse gas emissions

Biomass feedstock	Application rate	Results				References
		CO_2 emissions	Soil carbon	N_2O emission	CH_4 emission	
Fir sawdust (650 °C)	20 mg ha^{-1}	Significant decline in forest soil while no influence on orchard soil under 55%, 100% moisture levels.	No influence	No influence	Completely utilized in both orchard and forest soil at 100% moisture level	Walkiewicz et al. (2020)
Wheat straw (500 °C)	4% (w/w) in 100 g soil	Decreased by 7–9% over control.	Increased dissolved organic carbon and total dissolved nitrogen	Reduced up to 36–44%	No influence	Wang et al. (2020a, b)
Bamboo derived biochar (800 °C)	10, 30 t ha^{-1}	No noteworthy variation	No influence	No influence	No influence	Zhou et al. (2017)
Sawdust and chicken manure derived biochar (400 °C)	24 t ha^{-1}	No influence	No influence	No substantial modification	No substantial modification	Zhibin et al. (2017)
Douglas-fir derived biochar (420 °C)	1, 10% (w/w)	Significantly enhanced CO_2 emission	No influence	Addition of 1% biochar had no influence while the 10% addition enhanced N_2O emission by 191%.	Reduced	Hawthorne et al. (2017)
Douglas-fir slash (420 °C)	20 t ha^{-1}	Enhanced by 6%	No influence	No influence	Reduced by 8%	Johnson et al. (2017)
Bamboo leaf derived biochar (500 °C)	5 t ha^{-1}	No influence	No influence	Reduced emission by 20%	No influence	Xiao et al. (2016)
Combined spruce sawdust and maple (350–450 °C)	5 t ha^{-1}	No influence	No influence	No influence	No influence	Sackett et al. (2015)

Sugar maple wood as a feed-stock (500 °C)	5, 10, and 20 t ha^{-1}	Significantly enhanced	No influence	No influence	No influence	Mitchell et al. (2015)
Bamboo leaf derived biochar (500 °C)	5 t ha^{-1}	No influence	No influence	No influence	No influence	Wang et al. (2014)
Cornstalk	24 t ha^{-1}	No influence	Enhanced dissolved organic carbon	No influence	Reduced by 61% Reduced by 63%	Feng et al. (2012)
Bamboo	2.50%	No change	No influence	No influence	Reduced by 63% 51	Liu et al. (2011)

Table 3 Some selective studies on the immobilization of heavy metals by the virtue of biochar

Feedstock type	Pollutant type	Results	References
Cow-bone derived biochar (applied at 0%, 2.5%, 5% and 10%, w/w, pyrolysis temperature 500 °C and 800 °C)	Cd and Zn in mine–smelters contaminated soil.	The addition of biochar significantly reduced Zn and Cd concentrations in the roots and shoots of maize over control	Azeem et al. (2021)
Chitosan-coated textile waste biochar	Cd-polluted soil	Amending Cd polluted soil with the textile waste-derived biochar coated with chitosan resulted in the significant improvement in growth, biomass, nutritional quality, and soil enzymology while reducing Cd in roots, shoots, and in the soil over control	Zubair et al. (2021)
Lignin-derived biochar	Pb-acid batteries	The utilization of lignin-derived biochar coupled with arbuscular mycorrhizal fungi (AMF) reduced labile Pb concentrations over control. Additionally, Pb concentrations in barley grain were found below the critical limit and fit for human consumption	Khan et al. (2020)
Manure waste	Cu-mining	Promisingly reduced the accumulation and uptake of different heavy metals and support *Brassica napus* by producing excessive biomass	Gascó et al. (2019)
Cymbopogon flexuosus waste-derived biochar	Coal mining	Biochar treatment improved soil health and alleviate soil acidity which supports plant productivity	Jain et al. (2020)
Eucalyptus wood and sewage sludge biochar	Zn mining	Significantly reduced labile fractions of Zn, Pb, and Cd	Penido et al. (2019)
Miscanthus derived biochar and zeolite	Ni-polluted soil	A significant reduction in Ni bioavailability and its accumulation in wheat, sunflower, and maize were observed over control	Shahbaz et al. (2018a, b, 2019)
Eucalyptus wood biochar	Zn mining	Results revealed that biochar additions improved soil pH and support plant establishment via improving germination	Martins et al. (2018)

(continued)

Table 3 (continued)

Feedstock type	Pollutant type	Results	References
Biochar obtained from the quercus ilex wood	Cu-mining	Remarkably reduced the bioavailability of heavy metals and their uptake by the plants	Forján et al. (2018)
Dairy manure	E-waste recycling site	Reduction in the bioavailability of Zn, Cu, Pb, and Cd was observed due to the improvement in CEC, pH, and available P	Chen et al. (2018)
Pine needles, soybean Stover, wheat straw	Military shooting range soil	The addition of both biochar treatments reduced the labile fractions of Pb and Cu over control but the results were more prominent in soybean straw biochar treatment.	Ahmad et al. (2016)
Rice hull	Arable land within the surrounding of the abandoned mining area.	The increase in soil pH was observed subsequently decreased NH_4NO_3 fractions as well as accumulation in lettuce	Kim et al. (2015)
Wheat straw	Electroplating area	Biochar significantly immobilized heavy metals in the soil	Gan et al. (2012)
Rice husk, straw, and bran	Agricultural area-mining site	Promisingly reduced the concentrations of Zn, Cd, and Pb by 83%, 98%, and 72% in pore water	Zheng et al. (2012)

structure, moisture availability as well as the reduction in bulk density (Haider et al. 2015). Thus, biochar addition as an amendment may evoke poor emergence and crop establishment owing to poor soil conditions.

3.2 Improvement in Physiological Characteristics of Plants

Reportedly, the improvement in crop productivity and growth after biochar addition reflects the overall enhancement in the physiological traits of plants. For example, an increase in P availability and its uptake by maize was observed when biochar was applied with Arbuscular mycorrhiza fungi over other plants (Mau and Utami 2014). Likewise, an increase in stomatal conductance, chlorophyll fluorescence, and photosynthetic rate of *Abutilon theophrasti* was recorded when grown on soil amended with mixed biochar (Seehausen et al. 2017). The combined applications of biochar with zeolite (BC75% + ZE25%) considerably improved the physiology, grain yield, biochemistry, biomass, antioxidant activities in maize and sunflower (Shahbaz et al. 2018b). In another study, the up-gradation in plant water use efficiency, stomatal pore aperture, membrane stability index, stomatal density, relative water content, photosynthetic rate, and stomatal conductance were increased in tomato plants grown in sandy loam soil amended with biochar (Akhtar et al. 2014).

3.3 Crop Yield

Biochar applications in agricultural soils increased crop yield however, this increase mainly depends on several factors such as soil type, soil pH, fertilizer application, dosage, and feedstock of biochar and crop species (Jeffery et al. 2011). It was observed that the application of biochar at various rates (10, 15, 20 t ha^{-1}) not only improved the maize grain, water use efficiency, nutrient uptake, and yield when grown on arid sandy soil (Uzoma et al. 2011). Likewise, the yield components of sunflower were remarkably increased under the influence of biochar addition (Furtado et al. 2016). This improved yield could be due to several factors associated with biochar such as the increase in soil specific surface area, CEC, water and nutrient retention on to the large surfaces of the biochar, porosity as well as liming behavior which overall support plant growth (Zubair et al. 2021). Moreover, overall improvement in soil features resulted in the enhancement of spinach biomass, antioxidants enzymes in spinach leaves, soil enzymes, and sandy soil health (Khan et al. 2017).

3.4 Stress Alleviation by the Virtue of Biochar

During the last decades, numerous studies have proven that biochar not only enhances crop yield under ordinary circumstances but also supports plant establishment under adverse environments such as drought, salinity, heat, and pollution (Haider et al. 2015; Pressler et al. 2017; Shaaban et al. 2018). It has been reported that biochar addition alleviates drought stress and improved plant growth by increasing WHC consequently promote plant growth (Hafeez et al. 2017; Haider et al. 2015; Liu et al. 2016). Likewise, biochar can also nullify the adverse effects of salt stress by adsorbing Na^+ thereby promote crop production (Akhtar et al. 2015; Kim et al. 2016). Additionally, when biochar was used as an additive for decreasing the bioavailability of heavy metals and their accumulation by different plants, significant results were found (Shahbaz et al. 2018a, b; Shahbaz et al. 2019). Biochar additions remarkably reduced Pb, Ni, and Cd concentrations by adsorbing them onto its larger inner surfaces, or via ion exchange resultantly support plant production under heavy metal stress (Khan et al. 2020; Shahbaz et al. 2018a, b, 2019; Zubair et al. 2021). Apart from this, biochar is also known to improve and support plant establishment under heat stress by increasing WHC which increases plant water uptake and alleviates heat stress from the plants (Busscher et al. 2011; Karhu et al. 2011). Additionally, biochar also provides essential mineral nutrients by releasing them from its matrix and make available them for plant uptake which further improved nutritional quality and crop production (Taghizadeh-Toosi et al. 2012).

4 Conclusion and Way Forward

In recent years, biochar applications have myriad benefits such as increased soil pH, CEC, overall soil structure, and SOC, which significantly improved crop production, their nutritional quality under various biotic and abiotic stresses. Similarly, biochar additions to agricultural soils also reduced the transport of toxic pollutants via binding them onto its larger surfaces which support plant growth and enhance crop yield from degraded soils. Moreover, biochar also controls greenhouse gas emissions and increases carbon sequestration from the atmosphere, resultantly supports plant production under changing climatic conditions. Thus, biochar has a strong potential as an amendment and a soil conditioner that supports plant production from degraded soils. Besides these advantages, we provide some additional guidelines for future studies on exploring the potential of biochar in agricultural soils. Reportedly, biochar utilization promisingly influences the structure and diversity of microbial diversity in soil. However, scarce literature is available concerning the influence of biochar additions on particular functions performed by microorganisms and their gene functions linked with nitrogen and carbon cycling. Though, biochar addition significantly improved various traits of soil that support plant growth. However, it requires a lot of biomass to produce biochar for long-term field-scale experiments that potentially support plant production. Thus, it is necessary to study the interaction of biochar with other suitable mineral fertilizers to prevent the depletion of organic matter in the soil.

References

Abd El-Mageed TA, Abdelkhalik A, Abd El-Mageed SA, Semida WM (2021) Co-composted poultry litter biochar enhanced soil quality and eggplant productivity under different irrigation regimes. J Plant Nutr Soil Sci:1–17

Abdulrazzaq H, Jol H, Husni A, Abu-Bakr R (2015) Biochar from empty fruit bunches, wood, and rice husks: effects on soil physical properties and growth of sweet corn on acidic soil. J Agric Sci 7(1):192

Abel S, Peters A, Trinks S, Schonsky H, Facklam M, Wessolek G (2013) Impact of biochar and hydrochar addition on water retention and water repellency of sandy soil. Geoderma 202:183–191

Agegnehu G, Srivastava AK, Bird MI (2017) The role of biochar and biochar-compost in improving soil quality and crop performance: a review. Appl Soil Ecol 119:156–170

Ahmad M, Ok YS, Rajapaksha AU, Lim JE, Kim BY, Ahn JH, Lee SS (2016) Lead and copper immobilization in a shooting range soil using soybean Stover-and pine needle-derived biochars: chemical, microbial and spectroscopic assessments. J Hazard Mater 301:179–186

Ahmed MB, Zhou JL, Ngo HH, Guo W (2016) Insight into biochar properties and its cost analysis. Biomass Bioenergy 84:76–86

Ajayi AE, Holthusen D, Horn R (2016) Changes in microstructural behaviour and hydraulic functions of biochar amended soils. Soil Till Res 155:166–175

Akhtar SS, Andersen MN, Liu F (2015) Residual effects of biochar on improving growth, physiology and yield of wheat under salt stress. Agric Water Manag 158:61–68

Akhtar SS, Li G, Andersen MN, Liu F (2014) Biochar enhances yield and quality of tomato under reduced irrigation. Agric Water Manag 138:37–44

Ali I, He L, Ullah S, Quan Z, Wei S, Iqbal A, Ligeng J (2020) Biochar addition coupled with nitrogen fertilization impacts on soil quality, crop productivity, and nitrogen uptake under double-cropping system. Food Energy Secur 9(3):e208

Azeem M, Ali A, Jeyasundar PG, YSA L, Abdelrahman H, Latif A, Zhang Z (2021) Bone-derived biochar improved soil quality and reduced cd and Zn phytoavailability in a multi-metal contaminated mining soil. Environ Pollut 277:116800

Bandara T, Franks A, Xu J, Bolan N, Wang H, Tang C (2020) Chemical and biological immobilization mechanisms of potentially toxic elements in biochar-amended soils. Crit Rev Env Sci Technol 50(9):903–978

Barrow CJ (2012) Biochar: potential for countering land degradation and for improving agriculture. Appl Geogr 34:21–28

Beesley L, Dickinson N (2011) Carbon and trace element fluxes in the pore water of an urban soil following greenwaste compost, woody and biochar amendments, inoculated with the earthworm *Lumbricus terrestris*. Soil Biol Biochem 43(1):188–196

Biederman LA, Harpole WS (2013) Biochar and its effects on plant productivity and nutrient cycling: a meta-analysis. GCB Bioenergy 5(2):202–214

Bis Z, Kobyłecki R, Ścisłowska M, Zarzycki R (2018) Biochar–potential tool to combat climate change and drought. Ecohydrol Hydrobiol 18(4):441–453

Bruun EW, Müller-Stöver D, Ambus P, Hauggaard-Nielsen H (2011) Application of biochar to soil and N2O emissions: potential effects of blending fast-pyrolysis biochar with anaerobically digested slurry. Eur J Soil Sci 62(4):581–589

Burney JA, Davis SJ, Lobell DB (2010) Greenhouse gas mitigation by agricultural intensification. Proc Natl Acad Sci USA 107(26):12052–12057

Busscher WJ, Novak JM, Ahmedna M (2011) Physical effects of organic matter amendment of a southeastern US coastal loamy sand. Soil Sci 176(12):661–667

Cao X, Ma L, Liang Y, Gao B, Harris W (2011) Simultaneous immobilization of lead and atrazine in contaminated soils using dairy-manure biochar. Env Sci Technol 45(11):4884–4889

Cely P, Gascó G, Paz-Ferreiro J, Méndez A (2015) Agronomic properties of biochars from different manure wastes. J Anal Appl Pyrolysis 111:173–182

Chan KY, Van Zwieten L, Meszaros I, Downie A, Joseph S (2008) Using poultry litter biochars as soil amendments. Soil Res 46(5):437–444

Chen Z, Zhang J, Liu M, Wu Y, Yuan Z (2018) Immobilization of metals in contaminated soil from E-waste recycling site by dairy-manure-derived biochar. Environ Technol 39(21):2801–2809

Cheng CH, Lehmann J, Engelhard MH (2008) Natural oxidation of black carbon in soils: changes in molecular form and surface charge along a climosequence. Geochim Cosmochim Acta 72(6): 1598–1610

Cui Q, Xia J, Yang H, Liu J, Shao P (2021) Biochar and effective microorganisms promote *Sesbania cannabina* growth and soil quality in the coastal saline-alkali soil of the Yellow River Delta, China. Sci Total Environ 756:143801

DeLuca TH, MacKenzie MD, Gundale MJ, Holben WE (2006) Wildfire-produced charcoal directly influences nitrogen cycling in ponderosa pine forests. Soil Sci Soc Am J 70(2):448–453

El-Naggar A, Lee SS, Awad YM, Yang X, Ryu C, Rizwan M, Ok YS (2018) Influence of soil properties and feedstocks on biochar potential for carbon mineralization and improvement of infertile soils. Geoderma 332:100–108

Feng Y, Xu Y, Yu Y, Xie Z, Lin X (2012) Mechanisms of biochar decreasing methane emission from Chinese paddy soils. Soil Biol Biochem 46:80–88

Forján R, Rodríguez-Vila A, Cerqueira B, Covelo EF (2018) Comparison of compost with biochar versus technosol with biochar in the reduction of metal pore water concentrations in a mine soil. J Geochem Explor 192:103–111

Furtado GDF, Chaves LHG, de Sousa JRM, Arriel NHC, Xavier DA, de Lima GS (2016) Soil chemical properties, growth and production of sunflower under fertilization with biochar and NPK. Embrapa Algodão-Artigo em periódico indexado (ALICE)

Gan W, He Y, Zhang X, Zhang S, Lin Y (2012) Effects and mechanisms of straw biochar on remediation contaminated soil in electroplating factory. J Ecol Rural Env 28(3):305–309

Gao J, Zhao T, Tsang DC, Zhao N, Wei H, Feng M, Qiu R (2020) Effects of Zn in sludge-derived biochar on cd immobilization and biological uptake by lettuce. Sci Total Environ 714:136721

Gascó G, Álvarez ML, Paz-Ferreiro J, Méndez A (2019) Combining phytoextraction by *Brassica napus* and biochar amendment for the remediation of a mining soil in Riotinto (Spain). Chemosphere 231:562–570

Gorovtsov AV, Minkina TM, Mandzhieva SS, Perelomov LV, Soja G, Zamulina IV, Yao J (2019) The mechanisms of biochar interactions with microorganisms in soil. Environ Geochem Health:1–24

Güereña D, Lehmann J, Thies J, Enders A, Karanja N, Neufeldt H (2015) Partitioning the contributions of biochar properties to enhanced biological nitrogen fixation in common bean (*Phaseolus vulgaris*). Biol Fertil Soils 4:479–491

Hafeez Y, Iqbal S, Jabeen K, Shahzad S, Jahan S, Rasul F (2017) Effect of biochar application on seed germination and seedling growth of *Glycine max* (L.) Merr. Under drought stress. Pak J Bot 49(51):7–13

Haider G, Koyro HW, Azam F, Steffens D, Müller C, Kammann C (2015) Biochar but not humic acid product amendment affected maize yields via improving plant-soil moisture relations. Plant Soil 395(1):141–157

Hardie M, Clothier B, Bound S, Oliver G, Close D (2014) Does biochar influence soil physical properties and soil water availability? Plant Soil 376(1):347–361

Hawthorne I, Johnson MS, Jassal RS, Black TA, Grant NJ, Smukler SM (2017) Application of biochar and nitrogen influences fluxes of CO_2, CH_4 and N_2O in a forest soil. J Environ Manag 192:203–214

Hmid A, Al Chami Z, Sillen W, De Vocht A, Vangronsveld J (2015) Olive mill waste biochar: a promising soil amendment for metal immobilization in contaminated soils. Environ Sci Pollut Res 22(2):1444–1456

Husk B, Major J (2011) Biochar commercial agriculture field trial in Québec, Canada–year three: Effects of biochar on forage plant biomass quantity, quality and milk production. International Biochar Initiative

Islam MS, Song Z, Gao R, Fu Q, Hu H (2021) Cadmium, lead, and zinc immobilization in soil by rice husk biochar in the presence of low molecular weight organic acids. Environ Technol:1–14

Jain S, Khare P, Mishra D, Shanker K, Singh P, Singh RP, Baruah BP (2020) Biochar aided aromatic grass [*Cymbopogon martini* (Roxb.) Wats.] vegetation: a sustainable method for stabilization of highly acidic mine waste. J Hazard Mater 390:121799

Jeffery S, Verheijen FG, van der Velde M, Bastos AC (2011) A quantitative review of the effects of biochar application to soils on crop productivity using meta-analysis. Agric Ecosyst Environ 144(1):175–187

Jin Y, Liang X, He M, Liu Y, Tian G, Shi J (2016) Manure biochar influence upon soil properties, phosphorus distribution and phosphatase activities: a microcosm incubation study. Chemosphere 142:128–135

Johnson MS, Webster C, Jassal RS, Hawthorne I, Black TA (2017) Biochar influences on soil CO 2 and CH 4 fluxes in response to wetting and drying cycles for a forest soil. Sci Rep 7(1):1–9

Joseph S, Lehmann J (2009) Biochar for environmental management: science and technology (no. 631.422 B615bi). Earthscan, London, GB

Karhu K, Mattila T, Bergström I, Regina K (2011) Biochar addition to agricultural soil increased CH_4 uptake and water holding capacity–results from a short-term pilot field study. Agric Ecosyst Environ 140(1–2):309–313

Khan WUD, Ramzani PMA, Anjum S, Abbas F, Iqbal M, Yasar A, Ihsan MZ, Anwar MN, Baqar M, Tauqeer HM, Virk ZA, Khan SA (2017) Potential of miscanthus biochar to improve sandy soil health, in situ nickel immobilization in soil and nutritional quality of spinach. Chemosphere 185:1144–1156

Khan MA, Ramzani PMA, Zubair M, Rasool B, Khan MK, Ahmed A, Iqbal M (2020) Associative effects of lignin-derived biochar and arbuscular mycorrhizal fungi applied to soil polluted from Pb-acid batteries effluents on barley grain safety. Sci Total Environ 710:136–294

Kim HS, Kim KR, Kim HJ, Yoon JH, Yang JE, Ok YS, Kim KH (2015) Effect of biochar on heavy metal immobilization and uptake by lettuce (*Lactuca sativa* L.) in agricultural soil. Environ Earth Sci 74(2):1249–1259

Kim HS, Kim KR, Yang JE, Ok YS, Owens G, Nehls T, Kim KH (2016) Effect of biochar on reclaimed tidal land soil properties and maize (*Zea mays* L.) response. Chemosphere 142:153–159

Kinney TJ, Masiello CA, Dugan B, Hockaday WC, Dean MR, Zygourakis K, Barnes RT (2012) Hydrologic properties of biochars produced at different temperatures. Biomass Bioenergy 41: 34–43

Knicker H (2007) How does fire affect the nature and stability of soil organic nitrogen and carbon? A review. Biogeochemistry 85(1):91–118

Kocsis T, Kotroczó Z, Kardos L, Biró B (2020) Optimization of increasing biochar doses with soil–plant–microbial functioning and nutrient uptake of maize. Environ Technol Innovations 20: 101–191

Kolton M, Meller Harel Y, Pasternak Z, Graber ER, Elad Y, Cytryn E (2011) Impact of biochar application to soil on the root-associated bacterial community structure of fully developed greenhouse pepper plants. Appl Environ Microbiol 77(14):4924–4930

Laghari M, Naidu R, Xiao B, Hu Z, Mirjat MS, Hu M, Fazal S (2016) Recent developments in biochar as an effective tool for agricultural soil management: a review. J Sci Food Agric 96(15): 4840–4849

Laird DA, Fleming P, Davis DD, Horton R, Wang B, Karlen DL (2010) Impact of biochar amendments on the quality of a typical Midwestern agricultural soil. Geoderma 158(3–4): 443–449

Lawrinenko M, Laird DA (2015) Anion exchange capacity of biochar. Green Chem 17(9): 4628–4636

Lehmann J, Joseph S (eds) (2015) Biochar for environmental management: science, technology and implementation. Routledge

Li B, Fan CH, Xiong ZQ, Li QL, Zhang M (2015) The combined effects of nitrification inhibitor and biochar incorporation on yield-scaled N_2O emissions from an intensively managed vegetable field in southeastern China. Biogeosciences 12(6):2003–2017

Li J, Yu G, Xie S, Pan L, Li C, You F, Wang Y (2018) Immobilization of heavy metals in ceramsite produced from sewage sludge biochar. Sci Total Environ 628:131–140

Liu C, Wang H, Tang X, Guan Z, Reid BJ, Rajapaksha AU, Sun H (2016) Biochar increased water holding capacity but accelerated organic carbon leaching from a sloping farmland soil in China. Environ Sci Pollut Res 23(2):995–1006

Liu X, Li Z, Zhang Y, Feng R, Mahmood IB (2014) Characterization of human manure-derived biochar and energy-balance analysis of slow pyrolysis process. Waste Manag 34(9):1619–1626

Liu Y, Yang M, Wu Y, Wang H, Chen Y, Wu W (2011) Reducing CH_4 and CO_2 emissions from waterlogged paddy soil with biochar. J Soils Sediments 11(6):930–939

Lorenz K, Lal R, Preston CM, Nierop KG (2007) Strengthening the soil organic carbon pool by increasing contributions from recalcitrant aliphatic bio (macro) molecules. Geoderma 142(1–2):1–10

Lorenz K, Lal R (2014) Biochar application to soil for climate change mitigation by soil organic carbon sequestration. J Plant Nutr Soil Sci 177(5):651–670

Madiba OF, Solaiman ZM, Carson JK, Murphy DV (2016) Biochar increases availability and uptake of phosphorus to wheat under leaching conditions. Biol Fertil Soils 52(4):439–446

Major J, Rondon M, Molina D, Riha SJ, Lehmann J (2010) Maize yield and nutrition during 4 years after biochar application to a Colombian savanna oxisol. Plant Soil 333(1):117–128

Martins GC, Penido ES, Alvarenga IFS, Teodoro JC, Bianchi ML, Guilherme LRG (2018) Amending potential of organic and industrial by-products applied to heavy metal-rich mining soils. Ecotoxicol Environ Saf 162:581–590

Mau AE, Utami SR (2014) Effects of biochar amendment and arbuscular mycorrhizal fungi inoculation on availability of soil phosphorus and growth of maize. J Degrade Min Land Manag 1(2):69–74

Mchunu C, Chaplot V (2012) Land degradation impact on soil carbon losses through water erosion and CO_2 emissions. Geoderma 177:72–79

Meier S, Curaqueo G, Khan N, Bolan N, Cea M, Eugenia GM, Borie F (2017) Chicken-manure-derived biochar reduced bioavailability of copper in a contaminated soil. J Soils Sediments 17(3):741–750

Mierzwa-Hersztek M, Gondek K, Klimkowicz-Pawlas A, Baran A (2017) Effect of wheat and Miscanthus straw biochars on soil enzymatic activity, ecotoxicity, and plant yield. Int Agrophys 31(3):367

Mitchell PJ, Simpson AJ, Soong R, Simpson MJ (2015) Shifts in microbial community and water-extractable organic matter composition with biochar amendment in a temperate forest soil. Soil Biol Biochem 81:244–254

Mohammadi A, Khoshnevisan B, Venkatesh G, Eskandari S (2020) A critical review on advancement and challenges of biochar application in paddy fields: environmental and life cycle cost analysis. PRO 8(10):1275

Naeem I, Masood N, Turan V, Iqbal M (2021) Prospective usage of magnesium potassium phosphate cement combined with Bougainvillea alba derived biochar to reduce Pb bioavailability in soil and its uptake by Spinacia oleracea L. Ecotoxicol Environ Saf 208:111723

Nair VD, Nair PK, Dari B, Freitas AM, Chatterjee N, Pinheiro FM (2017) Biochar in the agroecosystem–climate-change–sustainability nexus. Front Plant Sci 8:2051

Palansooriya KN, Wong JTF, Hashimoto Y, Huang L, Rinklebe J, Chang SX, Ok YS (2019) Response of microbial communities to biochar-amended soils: a critical review. Biochar 1(1): 3–22

Penido ES, Martins GC, Mendes TBM, Melo LCA, do Rosário Guimarães I, Guilherme LRG (2019) Combining biochar and sewage sludge for immobilization of heavy metals in mining soils. Ecotoxicol Environ Saf 172:326–333

Pressler Y, Foster EJ, Moore JC, Cotrufo MF (2017) Coupled biochar amendment and limited irrigation strategies do not affect a degraded soil food web in a maize agroecosystem, compared to the native grassland. GCB Bioenergy 9(8):1344–1355

Preston CM, Schmidt MW (2006) Black (pyrogenic) carbon: a synthesis of current knowledge and uncertainties with special consideration of boreal regions. Biogeosciences 3(4):397–420

Qambrani NA, Rahman MM, Won S, Shim S, Ra C (2017) Biochar properties and eco-friendly applications for climate change mitigation, waste management, and wastewater treatment: a review. Renew Sust Energ Rev 79:255–273

Qayyum MF, Abid M, Danish S, Saeed MK, Ali MA (2015) Effects of various biochars on seed germination and carbon mineralization in an alkaline soil. Pak J Agric Sci 51:977–982

Ramzani PMA, Shan L, Anjum S, Ronggui H, Iqbal M, Virk ZA, Kausar S (2017) Improved quinoa growth, physiological response, and seed nutritional quality in three soils having different

stresses by the application of acidified biochar and compost. Plant Physiol Biochem 116:127–138
Ren J, Zhao Z, Ali A, Guan W, Xiao R, Wang JJ, Li R (2020) Characterization of phosphorus engineered biochar and its impact on immobilization of cd and Pb from smelting contaminated soils. J Soils Sediments 20(8):3041–3052
Renner R (2007) Rethinking biochar. Environ Sci Technol 41(17):5932–5933
Riaz M, Arif MS, Hussain Q, Khan SA, Tauqeer HM, Yasmeen T, Haider MS (2019) 18 Application of biochar for the mitigation of abiotic stress-induced damages in plants. Plant Tolerance to Environmental Stress: Role of Phytoprotectants
Rogovska N, Laird D, Cruse R, Fleming P, Parkin T, Meek D (2011) Impact of biochar on manure carbon stabilization and greenhouse gas emissions. Soil Sci Soc Am J 75(3):871–879
Rondon M, Ramirez JA, Lehmann J (2005a) Charcoal additions reduce net emissions of greenhouse gases to the atmosphere. In Proceedings of the 3rd USDA Symposium on Greenhouse Gases and Carbon Sequestration in Agriculture and Forestry (Vol. 208, pp. 21–24). USDA Baltimore
Rondon M, Ramirez JA, Lehmann J (2005b) Greenhouse gas emissions decrease with charcoal additions to tropical soils. In Proceedings of the 3rd USDA symposium on greenhouse gases and carbon sequestration, baltimore, USA (Vol. 208)
Rondon MA, Lehmann J, Ramírez J, Hurtado M (2007) Biological nitrogen fixation by common beans (*Phaseolus vulgaris* L.) increases with bio-char additions. Biol Fertil Soils 43(6):699–708
Sackett TE, Basiliko N, Noyce GL, Winsborough C, Schurman J, Ikeda C, Thomas SC (2015) Soil and greenhouse gas responses to biochar additions in a temperate hardwood forest. GCB Bioenergy 7(5):1062–1074
Seehausen ML, Gale NV, Dranga S, Hudson V, Liu N, Michener J, Thomas SC (2017) Is there a positive synergistic effect of biochar and compost soil amendments on plant growth and physiological performance? Agronomy 7(1):13
Sehrish AK, Aziz R, Hussain MM, Rafiq MT, Rizwan M, Muhammad N, Ali S (2019) Effect of poultry litter biochar on chromium (Cr) bioavailability and accumulation in spinach *(Spinacia oleracea)* grown in Cr-polluted soil. Arab J Geosci 12(2):57
Semida WM, Beheiry HR, Sétamou M, Simpson CR, Abd El-Mageed TA, Rady MM, Nelson SD (2019) Biochar implications for sustainable agriculture and environment: a review. S Afr J Bot 127:333–347
Shahbaz AK, Iqbal M, Jabbar A, Hussain S, Ibrahim M (2018a) Assessment of nickel bioavailability through chemical extractants and red clover (*Trifolium pratense* L.) in an amended soil: related changes in various parameters of red clover. Ecotoxicol Environ Saf 149:116–127
Shahbaz AK, Lewińska K, Iqbal J, Ali Q, Iqbal M, Abbas F, Ramzani PMA (2018b) Improvement in productivity, nutritional quality, and antioxidative defense mechanisms of sunflower (*Helianthus annuus* L.) and maize (*Zea mays* L.) in nickel contaminated soil amended with different biochar and zeolite ratios. J Environ Manag 218:256–270
Shahbaz AK, Ramzani PMA, Saeed R, Turan V, Iqbal M, Lewińska K, Rahman MU (2019) Effects of biochar and zeolite soil amendments with foliar proline spray on nickel immobilization, nutritional quality and nickel concentrations in wheat. Ecotoxicol Environ Saf 173:182–191
Shaaban M, Van Z, Bashir L, Younas S, Núñez-Delgado A, Chhajro MA, Hu R (2018) A concise review of biochar application to agricultural soils to improve soil conditions and fight pollution. J Environ Manag 228:429–440
Singh B, Macdonald LM, Kookana RS, van Zwieten L, Butler G, Joseph S, Weatherley A, Kaudal BB, Regan A, Cattle J, Dijkstra F, Boersma M, Kimber S, Keith A, Esfandbod M (2014) Opportunities and constraints for biochar technology in Australian agriculture: looking beyond carbon sequestration. Soil Res 52(8):739–750
Smith JL, Collins HP, Bailey VL (2010) The effect of young biochar on soil respiration. Soil Biol Biochem 42(12):2345–2347
Soni N, Leon RG, Erickson JE, Ferrell JA, Silveira ML, Giurcanu MC (2014) Vinasse and biochar effects on germination and growth of palmer amaranth (*Amaranthus palmeri*), sicklepod (*Senna obtusifolia*), and southern crabgrass (*Digitaria ciliaris*). Weed Technol 28(4):694–702

Spokas KA, Koskinen WC, Baker JM, Reicosky DC (2009) Impacts of woodchip biochar additions on greenhouse gas production and sorption/degradation of two herbicides in a Minnesota soil. Chemosphere 77(4):574–581

Taghizadeh-Toosi A, Clough TJ, Condron LM, Sherlock RR, Anderson CR, Craigie RA (2011) Biochar incorporation into pasture soil suppresses in situ nitrous oxide emissions from ruminant urine patches. J Environ Qual 40(2):468–476

Taghizadeh-Toosi A, Clough TJ, Sherlock RR, Condron LM (2012) Biochar adsorbed ammonia is bioavailable. Plant Soil 350(1):57–69

Tauqeer HM, Fatima M, Rashid A, Shahbaz AK, Ramzani PMA, Farhad M, Iqbal M (2021) The current scenario and prospects of immobilization remediation technique for the management of heavy metals contaminated soils. Appl Remediat Inorg Pollut:155

Tayyab M, Islam W, Arafat Y, Pang Z, Zhang C, Lin Y, Zhang H (2018) Effect of sugarcane straw and goat manure on soil nutrient transformation and bacterial communities. Sustainability 10(7): 2361

Thies JE, Rillig MC, Graber ER (2015) Biochar effects on the abundance, activity and diversity of the soil biota. Biochar Environ Manage Sci Technol Implementation 2:327–389

Touray N, Tsai WT, Chen HR, Liu SC (2014) Thermochemical and pore properties of goat-manure-derived biochars prepared from different pyrolysis temperatures. J Anal App Pyrolysis 10:116–122

Tsai WT, Liu SC, Chen HR, Chang YM, Tsai YL (2012) Textural and chemical properties of swine-manure-derived biochar pertinent to its potential use as a soil amendment. Chemosphere 89(2): 198–203

Uzoma KC, Inoue M, Andry H, Fujimaki H, Zahoor A, Nishihara E (2011) Effect of cow manure biochar on maize productivity under sandy soil condition. Soil Use Manag 27(2):205–212

Verheijen F, Jeffery S, Bastos AC, Van der Velde M, Diafas I (2010) Biochar application to soils. A critical scientific review of effects on soil properties, processes, and functions. EUR, 24099, 162

Walkiewicz A, Kalinichenko K, Kubaczyński A, Brzezińska M, Bieganowski A (2020) Usage of biochar for mitigation of CO_2 emission and enhancement of CH_4 consumption in forest and orchard Haplic Luvisol (Siltic) soils. Appl Soil Ecol 156:103711

Wang H, Yi H, Zhang X, Su W, Li X, Zhang Y, Gao X (2020a) Biochar mitigates greenhouse gas emissions from an acidic tea soil. Pol J Environ Stud 29:1

Wang L, Bolan NS, Tsang DC, Hou D (2020b) Green immobilization of toxic metals using alkaline enhanced rice husk biochar: effects of pyrolysis temperature and KOH concentration. Sci Total Environ 720:137584

Wang Y, Lin Y, Chiu PC, Imhoff PT, Guo M (2015) Phosphorus release behaviors of poultry litter biochar as a soil amendment. Sci Total Environ 512:454–463

Wang Z, Li Y, Chang SX, Zhang J, Jiang P, Zhou G, Shen Z (2014) Contrasting effects of bamboo leaf and its biochar on soil CO 2 efflux and labile organic carbon in an intensively managed Chinese chestnut plantation. Biol Fertil Soils 50(7):1109–1119

Warnock DD, Lehmann J, Kuyper TW, Rillig MC (2007) Mycorrhizal responses to biochar in soil–concepts and mechanisms. Plant Soil 300(1):9–20

Watson RT, Noble IR, Bolin B, Ravindranath NH, Verardo DJ, Dokken DJ (2000) Land use, land-use change and forestry: a special report of the intergovernmental panel on climate change. Cambridge University Press

Windeatt JH, Ross AB, Williams PT, Forster PM, Nahil MA, Singh S (2014) Characteristics of biochars from crop residues: potential for carbon sequestration and soil amendment. J Environ Manag 146:189–197

Wong JTF, Chen X, Deng W, Chai Y, Ng CWW, Wong MH (2019) Effects of biochar on bacterial communities in a newly established landfill cover topsoil. J Environ Manag 236:667–673

Xia H, Riaz M, Zhang M, Liu B, El-Desouki Z, Jiang C (2020) Biochar increases nitrogen use efficiency of maize by relieving aluminum toxicity and improving soil quality in acidic soil. Ecotoxicol Environ Saf 196:110531

Xiao YH, Li YF, Wang ZL, Jiang PK, Zhou GM, Liu J (2016) Effects of bamboo leaves and their biochar additions on soil N2O flux in a Chinese chestnut forest. J Plant Nutr Fertil 22(3): 697–706

Xu C, Zhao J, Yang W, He L, Wei W, Tan X, Lin A (2020a) Evaluation of biochar pyrolyzed from kitchen waste, corn straw, and peanut hulls on immobilization of Pb and cd in contaminated soil. Environ Pollut 261:114133

Xu L, Fang H, Deng X, Ying J, Lv W, Shi Y, Zhou Y (2020b) Biochar application increased ecosystem carbon sequestration capacity in a Moso bamboo forest. For Ecol Manage 475: 118447

Yanai Y, Toyota K, Okazaki M (2007) Effects of charcoal addition on N_2O emissions from soil resulting from rewetting air-dried soil in short-term laboratory experiments. Soil Sci Plant Nutr 53(2):181–188

Yang W, Feng G, Miles D, Gao L, Jia Y, Li C, Qu Z (2020) Impact of biochar on greenhouse gas emissions and soil carbon sequestration in corn grown under drip irrigation with mulching. Sci Total Environ 729:138752

Yavari S, Malakahmad A, Sapari NB (2016) Effects of production conditions on yield and physicochemical properties of biochars produced from rice husk and oil palm empty fruit bunches. Environ Sci Pollut Res 23(18):17928–17940

Yavari S, Sapari NB, Malakahmad A, Yavari S (2019) Degradation of imazapic and imazapyr herbicides in the presence of optimized oil palm empty fruit bunch and rice husk biochars in soil. J Hazard Mater 366:636–642

Yu W, Hu J, Yu Y, Ma D, Gong W, Qiu H, Gao HW (2021) Facile preparation of sulfonated biochar for highly efficient removal of toxic Pb (II) and Cd (II) from wastewater. Sci Total Environ 750:141545

Zhang A, Bian R, Pan G, Cui L, Hussain Q, Li L, Yu X (2012) Effects of biochar amendment on soil quality, crop yield and greenhouse gas emission in a Chinese rice paddy: a field study of 2 consecutive rice growing cycles. Field Crops Res 127:153–160

Zhang Y, Idowu OJ, Brewer CE (2016) Using agricultural residue biochar to improve soil quality of desert soils. Agriculture 6(1):10

Zheng RL, Cai C, Liang JH, Huang Q, Chen Z, Huang YZ, Sun GX (2012) The effects of biochars from rice residue on the formation of iron plaque and the accumulation of Cd, Zn, Pb, As in rice (*Oryza sativa* L.) seedlings. Chemosphere 89(7):856–862

Zhibin LIN, Qi LIU, Gang LIU, Cowie AL, Qicheng BEI, Benjuan LIU, Zubin XIE (2017) Effects of different biochars on *Pinus elliottii* growth, N use efficiency, soil N_2O and CH_4 emissions and C storage in a subtropical area of China. Pedosphere 27(2):248–261

Zhou D, Liu D, Gao F, Li M, Luo X (2017) Effects of biochar-derived sewage sludge on heavy metal adsorption and immobilization in soils. Int J Environ Res Public Health 14(7):681

Zubair M, Ramzani PMA, Rasool B, Khan MA, Akhtar I, Turan V, Iqbal M (2021) Efficacy of chitosan-coated textile waste biochar applied to Cd-polluted soil for reducing cd mobility in soil and its distribution in moringa (*Moringa oleifera* L.). J Environ Manag 284:112047

Soil Management Vis-à-Vis Carbon Sequestration in Relation to Land Use Cover/Change in Terrestrial Ecosystem—A Review

Rajan Bhatt, Pritpal Singh, and Gagandeep Kaur

Abstract Soil carbon (C) sequestration has gained worldwide attention for sustainable agricultural production while mitigating C footprints through reduced greenhouse gases (GHGs) emission. Soil organic C (SOC) directly impacts the soil health and long-term sustainability of terrestrial cropland ecosystems. SOC pool of surface soils (0–30 cm) across India and the world varied from 0.12–7.67 Pg (1 Pg = 10^{15} g) and 17–162 Pg, respectively, while 0.34–20.3 and 38–267 Pg in a soil profile (upto 100 cm depth). The restoration of historic losses of SOC levels (42–72 Pg) in soils requires an understanding of the ecological processes in relation to crop production and soil management practices. Among the best agricultural management practices (BMPs), conservation tillage including zero tillage (ZT), cropping systems, nutrient management, inclusion of legume in cropping systems, crop cultivars, crop residue management, crop diversification, agri- and industrial waste management, land use management, biochar application, and irrigation management are considered important with considerable potential of C sequestration in soil. Nutrient management had considerable importance in sequestering the higher fraction of atmospheric C back into the soil. Crop residue management within the ecosystem helps conserving soil C with multifaceted impacts on crop productivity in terms of increased plant-mediated C input and C sustainability with mitigated environmental implications. Irrigation water management had considerable impacts on C sequestration and GHGs emissions especially related to the methane (CH_4) emissions under anaerobic soil environment. Rice-based ecosystems had considerable capabilities to sequester C in recalcitrant pool. Agroforestry system has potential significance for ecosystems' C budget with high C input into soils.

Keywords Carbon sequestration · Conservation tillage · Crop production · Cropping sequences · Soil organic carbon

R. Bhatt · P. Singh (✉)
Department of Soil Science, Punjab Agricultural University, Ludhiana, Punjab, India

G. Kaur
Yadavindra College of Engineering, Talwandi Sabo, Bathinda, Punjab, India

M. Hasanuzzaman et al. (eds.), *Managing Plant Production Under Changing Environment*, https://doi.org/10.1007/978-981-16-5059-8_3

1 Introduction

Soils constitute the largest carbon (C) pool in the terrestrial ecosystem, and its stock in various organic forms varied from ~1500–2000 Pg (1 Pg = 10^{15} g) C upto a depth of 100 cm. Presently, worldwide C sink for the terrestrial regions is projected to include ~550–700 Pg C in vegetation and ~ 1200–1600 Pg C in total C (organic + inorganic) upto a depth of 100 cm (Paustian et al. 1997a). However, according to Eswaran et al. (2000), soil upto a depth of 100 cm may constitute ~2500 Pg of total C (organic + inorganic). Generally, C is divided into five major pools per Intergovernmental Panel on Climate Change (IPCC 2007), namely, oceanic (~38,000 Pg), geologic (~5000 Pg), pedologic (~2460 Pg), soil organic C (SOC; ~950 Pg), and atmospheric (~800 Pg). There is a globally growing concern on soil C sequestration through crop production and soil management practices (Russell et al. 2005; Wilson and Kaisi 2008; Singh and Benbi 2018a, b) not only to enhance soil C storage and to mitigate global greenhouse gases (GHGs) emission (Singh and Benbi 2020a, b) but also to improve soil health for long-term sustainability of a crop production system. Soil C sequestration is a technique in which atmospheric C is sequestered into the soil by improved technologies, which improved the soil health and yield potentials (Sundermeier et al. 2008). The oxidation of soil organic matter (SOM) in cultivated agricultural soils has been estimated to contribute ~50 Pg C to the atmosphere (Paustian et al. 2000). During the last decade, the carbon dioxide (CO_2) concentration in the atmosphere has increased by ~85 ppm (Raich and Potter 1995). For this, agriculture sector has been considered as major culprit as a sector contributed ~one-fourth of the historic anthropogenic CO_2 emissions during the past two centuries. On a global scale, agriculture sector has the potential to offset GHGs emission by 1.15–3.30 Pg C year^{-1} (Cole et al. 1997). Therefore, soil C sequestration by terrestrial vegetation has been the promising approach for mitigation of GHGs (Cole et al. 1997), as the terrestrial system associated with the soil and crop management has been the key player in global C budget (Lal 2004a). Bowen and Rovira (1999) reported that the buildup of 1.0 t SOM in soil removes 3.667 t CO_2 from the atmosphere. Schlesinger (1997) has reported that >2/3rd of organic C storage in the terrestrial ecosystem occurred in SOM with a net C flux of ~60 Pg C year^{-1} from the soil to the atmosphere.

Among the best agricultural management practices, tillage frequency (Campbell et al. 1995; Six et al. 2004; Calegari et al. 2008; Lopez-Garrido et al. 2009; Anikwe 2010), tillage depth (Jitareanu et al. 2009), type of tillage implements (Dong et al. 2008; Patra et al. 2010), nutrient management (Kaur et al. 2008; Banger et al. 2010; Singh and Benbi 2018a; Singh et al. 2021a), organic manure application (Raji and Ogunwole 2006; Singh and Benbi 2018a), crop residue incorporation (Wang et al. 2010b; Sharma et al. 2020a), green manuring of soil (Yadvinder-Singh et al. 2004; Ghosh et al. 2010), land use change (Singh and Singh 2017), land use cover (Benbi et al. 2015; Singh and Benbi 2018b), crop cover (Entz et al. 2002; Wang et al. 2010b), cropping intensity (Campbell et al. 2007; Singh and Benbi 2020a, b), cropping sequence/rotation (Campbell et al. 1995; West and Post 2002; Wilson and Kaisi 2008; Singh and Benbi 2020a, b), alley cropping (Okonkwo et al. 2009), and irrigation water management (Minamikawa and Sakai 2007) are important that contribute significantly toward soil C sequestration and consequent mitigation of GHGs emissions. Optimum management of row crop field coupled with reduced soil tillage, efficient nutrient management, and water conservation are the credible management practices that hold good promise as a means of soil C sequestration, although the photosynthetic CO_2 assimilation rates vary largely depending upon climatic conditions, soil fertility, and other soil and plant factors (Lal 2004b; Sharma et al. 2021a).

Soil C sequestration contributes significantly toward soil tilth (Lal et al. 1997), fertility (Sainju et al. 2008), water holding capacity (Singh and Benbi 2016), and ultimately leads to improvement in crop yield (Singh and Benbi 2018a; Sharma et al. 2020a). Widespread adoption of recommended crop production practices in the United States (U.S.) has the potential to sequester ~45–98 Tg (1 Tg = 10^{18} g) SOC in croplands (Lal 2003). It had a special significance for tropical and subtropical ecosystems, where the soils are inherently low in SOC, and the production systems are low in soil fertility owing to greater oxidation of SOM under high temperature conditions. The Council for Agricultural Science and Technology (CAST) suggested that adoption of improved agricultural practices in croplands can enhance C sequestration rates from 0.1 to 1.0 Mg ha^{-1} year^{-1}, although the assimilation rate decreased as soil approaches a steady equilibrium state (CAST 2004).

Soils do have the capacity to sequester atmospheric C that varied with the soil texture, quantity of organic inputs, and the initial SOC status (Ingram and Fernades 2001; Paustian et al. 2000; Singh and Benbi 2021). Soil organic C in swell-shrink soils in subhumid tropics attains quasi-equilibrium values between 5 and 30 years depending upon soil texture and native vegetation (Naitam and Bhattacharyya 2004). Soil management practices have the prospective to contribute greatly toward soil C sequestration since the C sink capacity of the world's agricultural and degraded soils is ~50–66% of historic C loss (~42–72 Pg) (Lal 2004b; Bangroo et al. 2011). The global historic losses of soil C due to intensive cultivation have been estimated to be around 25% (~55 Pg C) of original C in virgin and uncultivated soils (Cole et al. 1997). Soil organic C acts as a simultaneous source and sink for essential plant nutrients (Campbell et al. 1995; Bationo et al. 2007) and plays a vital role in maintaining soil fertility and productivity (Bronson et al. 2004) by improving the physical and chemical characteristics of soils (Lal et al. 1997). Crop production and soil management practices for C sequestration simultaneously improved the organic matter inputs to soil and decreased the loss of soil organic matter through decomposition (Sharma et al. 2020a). Nonetheless, the practices leading to increased soil C sequestration are both site and situation specific (Paustian et al. 1997a; Koul and Panwar 2008; Singh et al. 2020a), which inter alia depend on environmental and socioeconomic factors affecting soil C dynamics. Recent global estimates have revealed that the capacity of agricultural soils to sequester ~20–30 Pg C over the next 50–100 years (Paustian et al. 1997a). Soil C sequestration is necessary not only to enhance soil C storage for C trading and to mitigate GHG emission but also to enhance the farmer's livelihoods (Sainju et al. 2008).

Conventional rice–wheat cropping system in the entire Indo-Gangetic Plains (IGPs) depends much on the fertilizer–N application, but its higher doses lead to serious environmental implications, namely, emission of GHGs and underground water pollution (Bhatt et al. 2021). The lack of proper crop rotations with legumes, intensive tillage, open-field burning of rice residues, overexploitation of groundwater and overfertilization, malnutrition, biodiversity loss, and so on is considered as major reasons for the deterioration of soil health and sustainability of a rice–wheat cropping system (Sharma et al. 2021b). The faulty irrigation practices, namely, flood irrigation, as well as crop establishment methods such as intensive tilling in rice–wheat cropping systems, is considered responsible for large-scale C equivalent emissions (Singh et al. 2020a). Crop residue burning has been the most threatening sustainability issues in the IGPs, particularly under rice-based cropping systems affecting the environmental sustainability due to reduced C sustainability. Therefore, to sequester C back in soil for a longer period of time or to hinder C gases emitted to the atmosphere, researchers across the country invented, tested, and recommended a large number of techniques for producing more using less land, energy, or water after reducing the C footprints. In view of highly variable crop production and soil management practices in crop production (Singh et al. 2019a, b; Singh and Benbi 2020a, b; Sharma et al. 2020b), an attempt was made to gather information pertaining to C sequestration potential and dynamics of soil C associated with best crop production and soil management practices.

2 Soil Organic C in Different Soil Types

The SOC concentration in Indian soils ranges from 2.1 to14.8 g kg^{-1}, with a mean value typically <0.5%. The loss of C from historic levels due to higher C footprints urgently needs diligent attention to cut down C losses and increased organic C concentration to the desired levels 0.5% to 1.0% (Swarup et al. 2000). India has great diversity of soil types, namely, Alfisols (~81 Mha), Vertisols (~60.4 Mha), Inceptisols (~51.7 Mha), Ultisols (~36.6 Mha), Entisols (~24.8 Mha), Aridisols (~18.3 Mha), Mollisols (~1.8 Mha), and Gelisols (~0.8 Mha) (Velayutham et al. 2000). Soil organic C pool in different soil orders in India and the world.

3 Crop Production and Soil Management Practices for Increased C Sequestering

3.1 Conservation Tillage and C Sequestration

Tillage management has been the most credible option capable of mitigating CO_2 emissions due to its role in preventing organic matter oxidation (Singh et al. 2020a, b). Tillage accentuates soil organic matter disintegration processes through the physical disturbance by breaking bigger aggregates into the smaller ones (Benbi et al. 2016), which reduces the emission of CO_2 gas (Oades 1984; Beare et al. 1994). Tillage also affects soil temperature, aeration, and water relation by its impacts on surface residue cover and soil structure (Paustian et al. 1997b). Reducing tillage intensity with stubble mulching helps in improving the C sequestration and hence the organic matter content of soils (Hu et al. 2010; Sainju et al. 2010) through C accumulation (Garcia-Franco et al. 2015). In a rice–wheat cropping system, the conventional tillage (CT) coupled with open field residue burning prior to wheat establishment has high C footprints and low C sustainability (Singh et al. 2020a; Singh et al. 2021b). They have reported that compared to the CT, rice residue retention with happy seeder technology for wheat establishment leads to a significant reduction of C and energy footprints by 14.1% and 12.9%, respectively, which is a must for sustainable agriculture in the region (Fig. 1) (Singh et al. 2020a).

Liu et al. (2016) revealed significantly higher microbial biomass carbon (MBC) and microbial biomass nitrogen (MBN) concentrations in surface soil layers than in the subsurface layers which further linked with lesser labile C and N pools as the soil becomes deeper and deeper. Similar reports have already been reported by researchers under different ecosystems, namely, forestland (Agnelli et al. 2004), grassland (Fierer et al. 2003), and arable land (Taylor et al. 2002). The MBC concentration accounted for 6.79%, 3.90%, 2.84%, and 2.24% of the SOC concentration, while MBN concentration accounted for 3.13%, 3.09%, 2.29%, and 1.55% of TN concentration under grassland, forest, plough tillage (PT), and no-tillage (NT), respectively. At the 5–15 cm depth, the MBC:MBN ratio was higher under grassland and forestland than cultivation practices. At the 15–25 cm depth, the MBC–MBN ratios were generally lower under PT and NT than grassland and forestland. The MBC concentration accounted for 4.94%, 3.20%, 2.45%, and 1.50% of SOC concentration, while MBN concentration accounted for 2.44%, 1.75%, 1.74%, and 1.78% of TN concentration under grassland, forestland, PT, and NT, respectively. The ratio of MBC–MBN is generally not affected by variation in soil depths under different land-use systems. Under zero tillage crop establishment systems, MBC: MBN ratios significantly decreased with the increase in soil depth. However, retention of crop residues further promoted the soil health indicators and reduced the soil bulk density thus enhancing other soil properties.

Reduced tillage instead of intensive or zero tillage (ZT) is a win–win strategy for promoting C sequestration sustainably (Sainju et al. 2010; Garcia-Franco et al. 2015) through C accumulation within the small macroaggregates and microaggregates at

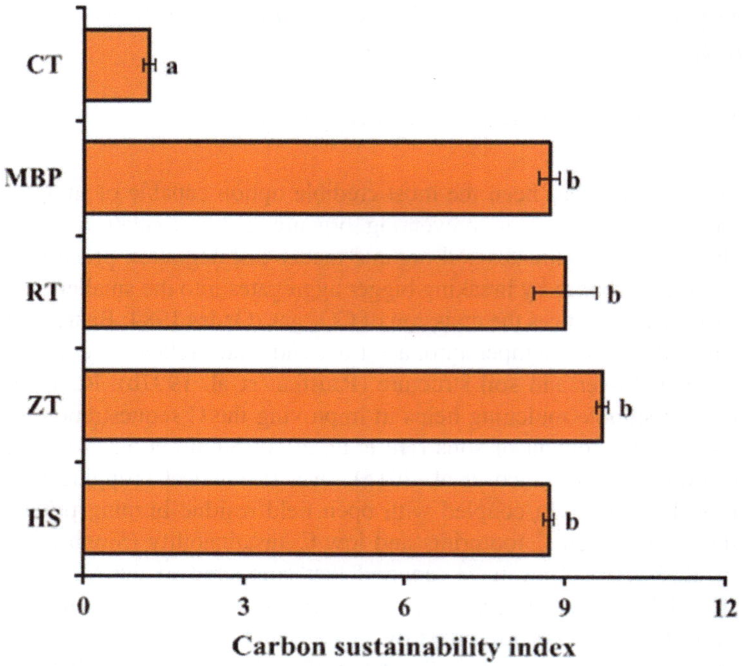

Fig. 1 Influence of tillage intensity for rice residue management and wheat sowing on carbon sustainability index (Acronyms: HS-happy seeder, ZT-zero tillage, RT-rotavator tillage, MBP-mold board plough tillage, and CT-conventional tillage (Source: Singh et al. 2020a)

the 5- to 15-cm depth (Garcia-Franco et al. 2015). Reduced tillage is considered important to promote the formation of new aggregates and soil C within them and therefore improved the soil properties because of enhanced SOC (Sainju et al. 2010). The soil management options that contribute to reduced decomposition or soil respiration, vis-à-vis reduced tillage or NT practices, mulch farming, and reducing bare fallow or increased cropping intensity, lead to C sequestration in soil (Halvorson et al. 2002). The adoption of appropriate tillage practices has received worldwide attention owing to the high C sequestration potential (Singh et al. 2020a). Campbell et al. (1995) reported that continuous wheat (*Triticum aestivum* L.) cultivation under NT system gained ~1.5 t ha^{-1} more C than continuous wheat in conventional tillage (CT) plots in comparison to NT sown fallow–wheat that gained ~0.5 t ha^{-1} more C than fallow–wheat in CT plots. An increase in SOC content with decreased fallow frequency in a cropping sequence shows the effect of added crop biomass (aerial and underground) and reduced soil disturbance during tillage operation (Campbell et al. 1995). The underground crop biomass has been reported to boost SOC buildup to a greater extent than aerial biomass, as the crop roots have a significant effect on SOC content (Kaisi and Grote 2007) because ~40% of the photosynthates synthesized in different plant parts are released in the rhizosphere through plant roots within an hour of their production (Kumar et al. 2006). Such an increase in SOC content with root biomass can be explained by the exudation of organic compounds that bind soil particles and stabilizes SOC (Chevallier et al. 2004). On the contrary, however, Halvorson and Reule (1999) reported that C sequestration efficiency, when measured on the basis of C in aerial biomass, was increased by ~30% compared to only ~11% when estimated by considering C in both aerial + root biomass with the application of N over no-N plots.

The accumulation of SOC to a greater extent in the surface layer might be attributed to the accumulation of crop residue in the top layer, and even the root biomass grows abundantly in the top soil layer (Kaisi and Grote 2007; Sharma et al. 2020a). Wright et al. (2007) reported the most influential effect of tillage on SOC restoration at 0–5 cm soil depth and extended impact to the 15–30 cm depth for wheat (*Triticum aestivum*) and sorghum (*Sorghum bicolor* L.). Wright and Hons (2005) reported significantly higher soil C sequestration in NT than CT treatment imposition under sorghum–wheat–soybean cropping sequence at 0–5 cm soil depth. In a *Rhodic Hapludox* of Southern Brazil, NT resulted in 6.84 Mg C ha^{-1} in the upper (0–10.0 cm) soil layer, which represents ~64.6% high C than CT under various winter crop cover treatments (Calegari et al. 2008). Upto ~20–40% loss of SOM is reported to occur after shifting the land for agricultural production from original untilled conditions (Davidson and Ackerman 1993). Over and above, the role of tillage management on soil C sequestration is considered time-dependent (Six et al. 2004). In humid and dry temperate climates of the United States, Six et al. (2004) reported that newly converted NT cropping systems rather increase net global warming potential (GWP) relative to CT practices. However, more than 10-year long-term adoption of CT practice in humid climatic conditions leads to a significant reduction in GWP. In comparison, >20-year long-term adoption of NT practice is required under dry climatic conditions for reducing GWP and that too with a high degree of uncertainty (Six et al. 2004). In Mollisols of Argentina, SOC in the upper 7.5 cm layer of degraded soils was greater (27 g kg^{-1}) in NT than in CT (24 g kg^{-1}) when compared to non-degraded soil, where there was no difference in total organic C content after 8 years of different tillage treatment (Fabrizzi et al. 2003). The effect of reduced tillage on soil C restoration relates to improved soil aggregation (Paustian et al. 1997b) and decreased soil respiration (Buyanovsky et al. 1987) that leads to the buildup of C in soil. Halvorson et al. (2002) reported an inverse relationship between tillage intensity and soil C sequestration and reported an increase in soil C sequestration potential as the tillage intensity decreased (NT > MT > CT). Six et al. (2002) reported that in tropical and temperate soils, an increase in C levels was ~325 $\pm$ 113 kg C ha^{-1} $year^{-1}$ under NT compared to CT. Tillage induced CO_2 emission and a sharp increase in CO_2 emission with a maximum value of 6.24 g CO_2 m^{-2} h^{-1} under long-term (15 years) CT plots (Lopez-Garrido et al. 2009). The losses of C through CO_2 emission were higher (801 and 905 g C m^{-2} $year^{-1}$ for short (3 years) and long term (15 years), respectively in CT treatment compared to 764 and 718 g C m^{-2} $year^{-1}$ for RT) and NT treatments practiced on long term (15 years), respectively (Lopez-Garrido et al. 2009). On the contrary, however, Minoshima et al. (2007) reported that despite low assimilation of newly added crop residue C in

NT soil, a similar amount of CO_2 was emitted from CT and NT, probably due to the high activity of microbes in the rhizosphere of the residue in NT soil.

Soil disturbance during tillage greatly influences soil C dynamics owing to accelerated erosion destructing soil aggregates and catalyzed microbial decomposition of soil organic matter (Singh and Benbi 2018b). The adoption of reduced tillage practices reduces C losses by offsetting CO_2 emission and checking soil losses. In low-fertility tropical soils, NT is considered efficient soil management practice that improves the physical and chemical properties of soils (Lal et al. 1997). Gama-Rodrigues et al. (2010) reported that the extent of soil C sequestration depends upon the physical protection of soil organic C, as C occlusion in soil aggregates is the major mechanism for C protection in soils. About 40% higher soil aggregate stability has been reported under the NT management system compared to the CT management system (Jung et al. 2008). Patra et al. (2010) compared the traditional plow tractor-mounted cultivator, and power tillers for puddling the rice field in IGPs of West Bengal (India) to investigate the effect on C buildup in soil and reported that light soil puddling using traditional plow exhibited the highest buildup of soil organic C and microbial biomass. Dong et al. (2008) reported that after 5 years of moldboard plowing (MBP) and rotary tillage (RT) of soil, there was higher annual CO_2 efflux from the soil because of less immobilization of soil organic C by microorganisms under long-term intensive tillage when compared to NT soils.

3.2 *Zero Tillage Wheat*

The number of sustainability issues pertaining to conventional rice–wheat cropping system reported in the region (Bhatt et al. 2019), namely, reduced yields and poor soil health (Bhandari et al. 2002), depleting ground water tables (Hira et al. 2004), and polluting air (Bijay-Singh et al. 2008). For irrigation purposes, mostly underground water is pumped up by submersible motors, which produce GHGs through hydrolysis of the applied N-fertilizers (Humphreys et al. 2010). For the harvesting of the field crops, especially rice and wheat, combines preferred in the region which left 0.3–0.6 m high anchored straw and lose straw in windrows. The management of crop residues especially rice is difficult due to its higher silica contents; hence, farmers generally used to burn it openly for timely sowing of the next upland wheat crop. Extensive tillage operations are required for straw management, coupled with the need to allow time for the straw to decompose sufficiently to avoid N immobilization (Gajri et al. 2002). Incorporation of the rice straw often delays wheat sowing beyond the optimum date (before 15th November) for maximum yield beyond which the wheat yields start decreasing. The farmers, therefore, opt for burning rice residues that result in air pollution and loss of organic C and nutrients, namely, ~35 kg N, ~21 kg K, and ~ 3 kg ha^{-1} each of P and S (Yadvinder-Singh et al. 2008). Residue burning exerts harmful effects on soil health due to the degradation of soil physical and biological properties (Yadvinder-Singh et al. 2005). The burning of rice residues is a major source of air pollution in the region, in the form of GHGs which led to global warming (Gupta et al. 2004). Zero tillage is the best option to date for the timely sowing of the wheat crop while ensuring successful and ecofriendly management of rice straw and saving presowing irrigation (Sidhu et al. 2007, 2008). Direct drilling of wheat seeds into the soil is considered a viable option to sow wheat seeds into the soil. Happy seeder—a modified zero tillage machine capability of seeding wheat seed in standing stubble—has been the most promising technology for eco-friendly rice residue management. The presence of the loose rice straw acts as mulch that further helps to regulate the soil temperature, reduces vapor pressure gradient and upflow of the vapor, reduces wind speed and its vapor lifting capacity, and finally, the soil evaporation that further encourages higher partitioning of the share of evaporation to transpiration (Balwinder-Singh et al. 2011; Balwinder-Singh et al. 2014). Reduced tillage benefits on land productivity (Paccard et al. 2015), water productivity (Guan et al. 2015), C intake in soil (Zhangliu et al. 2015), and finally on farmer's livelihoods (Tripathi et al. 2013) are well recognized. The fuel-related CE emissions were significantly lower in the HS method by 35.9–94.1 kg CO_{2e} ha^{-1} (25.5–47.3%) than the moldboard tillage, rotavator tillage, zero tillage (in residue removed fields), and the conventional tillage followed by residue burning (Singh et al. 2020a).

3.3 *Cropping Systems and C Sequestration*

Some interventions such as increasing cropping intensity than the fallow period (Hurisso et al. 2013; Lefèvre et al. 2014; Gan et al. 2012), including perennial forages, such as alfalfa (*Medicago sativa* L.) (Sainju and Lenssen 2011), planting high root biomass-to-aboveground biomass ratio plants (Negash and Kanninen 2015), and optimum and judicious fertilizer management (Zentner et al. 2011), helps in reducing C emissions in the atmosphere and thereby reduces the C footprints. For the optimization of C sequestration efficiency, cropping system (Wilson and Kaisi 2008; Singh and Benbi 2020b), intercropping (Makumba et al. 2007), cover cropping (Wang et al. 2010b), and ratoon cropping play a leading role in soil C sequestration (Wang et al. 2010a). In brown Chernozem soils of Saskatchewan, Canada, Campbell et al. (1995) have reported that during 12 years of continuous wheat (*Triticum aestivum*) cultivation, the soil has gained ~2.0 t ha^{-1} more C in 0–15 cm soil layer than in fallow–wheat cropping sequence, with most of the increase occurring during first 5 years of cropping. In dry land soils of North Dakota, Halvorson et al. (2002) reported that continuous use (12 years) of wheat–fallow system, even NT, results in loss of SOC, as fallow period represents the time of high microbial activity and decomposition of SOM with no input of crop residue. Kroodsma and Field (2006) reported that C sequestration was lowest in non-rice annual cropland that sequestered 9.0 g C m^{-2} year^{-1} of soil C and highest on land that switched from annual crops to vineyards sequestered 68 g C m^{-2} year^{-1} and land switched from annual crops to orchards sequestered 85 g C m^{-2} year^{-1} in comparison to rice field that sequestered 55 g C m^{-2} year^{-1} and referred this low C sequestration to the result of burning of rice residue after harvesting. Wang et al. (2010b) conducted a phytotron study to compare biomass accumulation and C sequestration by growing cover crops and evaluated that among winter cover crops, the highest and lowest amount of C assimilation occurred by bell bean (*Vicia faba* L.) 597 and 149 g m^{-2} by white clover (*Trifolium repens*), respectively, in fine sandy soils of United States. However, among the summer crops, sunhemp (*Crotalaria juncea* L.) accumulated the largest quantity of C (481 g m^{-2}) while that by castorbean (*Ricinus communis*) was 102 g m^{-2} (Wang et al. 2010b). While studying the long-term effect (10 years) of intercropping, Makumba et al. (2007) reported C sequestration of 0.8 to 4.8 Mg C ha^{-1} in gliricidia–maize system when compared to 0.4–1.0 Mg C ha^{-1} in sole maize system. A net decrease in soil C (6.0–7.0 Mg C ha^{-1}) in the upper 20 cm soil layer over initial C content was observed in the sole maize system. A total of 123–149 Mg C ha^{-1} was sequestered up to a depth of 200 cm through root biomass and pruning application in the gliricidia–maize system (Makumba et al. 2007). Makumba et al. (2006) reported that 11 years of intensive pruning of gliricidia trees may add 4.0–5.0 Mg DM ha^{-1} soil. They reported that SOC content in the upper 20 cm soil layer after 11 years of gliricidia pruning application was 3.0 g kg^{-1} higher in the gliricidia–maize system than in the sole maize system. Russell et al. (2005) reported that cropping systems that contained alfalfa in a rotation (corn–oat–alfalfa) had significantly ($p \leq 0.05$) higher SOC stock than systems without alfalfa (continuous corn and corn–soybean). Furthermore, the highest soil CO_2 emission was recorded from continuous corn cropping system than corn–soybean cropping system (Wilson and Kaisi 2008). The introduction of perennial legumes in semiarid climatic conditions has been known to reduce energy requirements by adding significant amounts of N to the soil (Entz et al. 2002). In the dry subhumid environmental conditions, alfalfa hay crop cultivation has been reported to contribute 84–137 kg N ha^{-1} (Kelner et al. 1997). Attempts must be made to plant more trees or crops with a higher root to shoot ratio, as greater will be then sequestered C into deep soils which will not be allowed to escape back from the soil to the atmosphere (Negash and Kanninen 2015).

Cropping systems significantly impacts soil C sequestration (Singh and Benbi 2020a) and the net ecosystem C budget of different wheat-based cropping systems (Singh and Benbi 2020b). Rice–wheat system had ~23% and 45% higher net primary production (NPP) compared with maize–wheat and cotton–wheat cropping system, respectively. The NPP through the aboveground residue biomass yield comprised the largest proportion (~48–50%) of the total NPP in three cropping systems, the highest being for rice and the lowest for cotton-based ecosystem (Fig. 2). Similarly, the NPP through different components was significantly higher for rice–wheat, followed by maize–wheat and the lowest for cotton–wheat cropping system. The net ecosystem C budget (NECB) was significantly lower for cotton, followed by maize and the highest for rice ecosystem (Fig. 3). It was negative for rice (scenario 1; continuously flooded), maize, and cotton, but positive for wheat. It was 2856, 2895, and 1801 kg C ha^{-1}, respectively, for wheat sown in rotation with rice, maize, and cotton. The NECB for a rice–wheat cropping system was 2427, 2448, and 2459 kg C ha^{-1}, respectively, under scenario 1 (continuously flooded), scenario

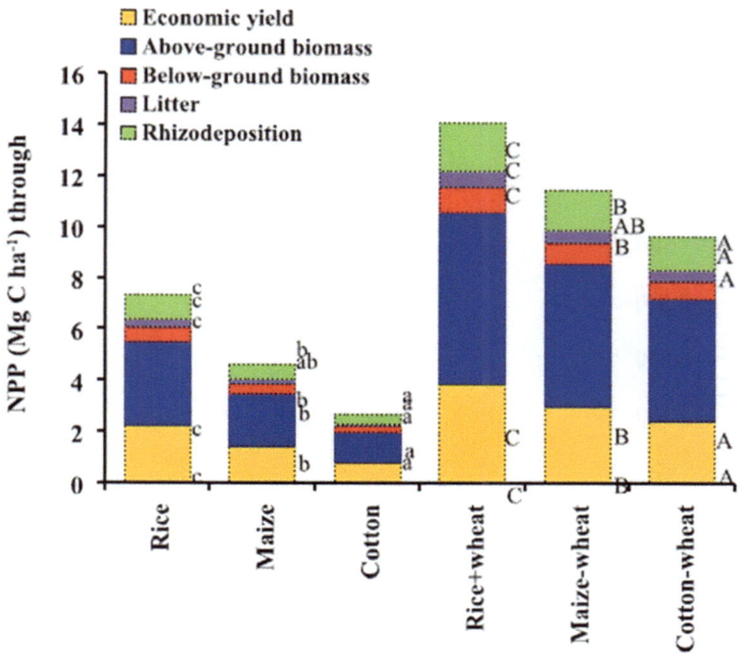

Fig. 2 Different components of net primary production (NPP) through different components in various wheat-based cropping system (Source: Singh and Benbi 2020b)

2 (intermittently flooded with single aeration), and scenario 3 (intermittently flooded with multiple aerations). The comparison of cropping systems revealed a significantly higher NECB for rice–wheat, compared with maize–wheat and cotton–wheat cropping system. Sharma et al. (2020a) reported increased SOC concentration due to regular cultivation of rice–wheat cropping sequence and that might be due to addition of higher fractions plant roots per year, flooded conditions which restrict the oxidizing microorganisms to oxidize the SOC into CO_2 and use of heavy chemical fertilizers.

3.4 Nutrient Management and C Sequestration

Nutrient management plays a significant role in soil C sequestration due to its impact on above-and below-ground biomass C input and cycling in ecosystems (Benbi et al. 2016; Singh and Benbi 2018a, 2021) (Table 1). The application of fertilizer nutrients increases the plant biomass and the quantity of crop residue returned into the soil (Sharma et al. 2020a; Singh et al. 2021a). Six et al. (2004) reported that improved N management is of utmost importance to realize complete benefits from C sequestered in soils to ensure mitigation of GHGs. The application of fertilizer-N had little effect

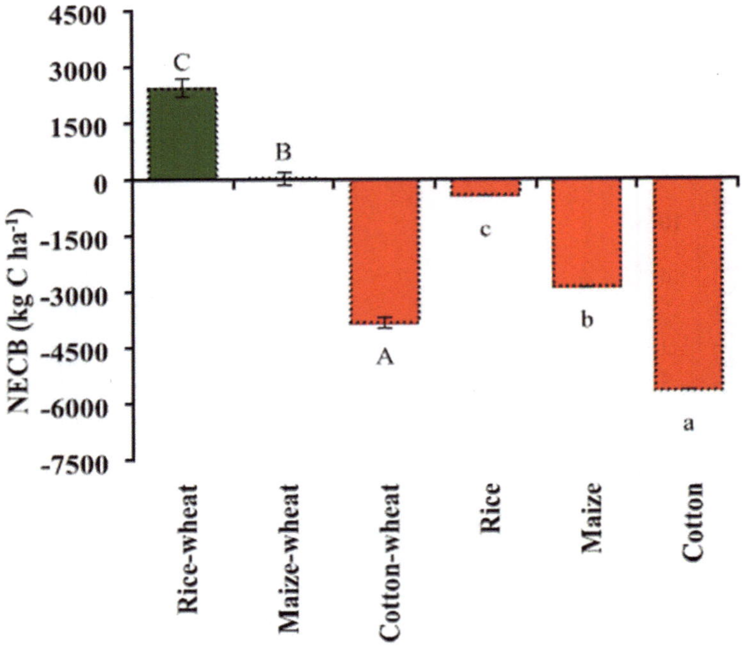

Fig. 3 Net ecosystem carbon budget (NECB) for different wheat-based cropping systems (Source Singh and Benbi 2020b)

on soil C sequestration although crop residue production was increased with N fertilization in spring wheat–winter wheat–sunflower and spring wheat–fallow cropping systems (Halvorson et al. 2002). Sainju et al. (2008) reported increased C input in soil due to cropping and N fertilization as a result of increased biomass in cotton–cotton–corn and rye/cotton–rye/cotton–corn cropping sequences. Nitrogen application in continuous corn and corn–soybean cropping system has been reported to check CO_2 emissions from the soil, and the effect was pronounced at higher N applied plots (Wilson and Kaisi 2008). The SOC in control (no-N) plots increased from 37.2 to 38.8 Mg ha^{-1} in plots receiving 270 kg N ha^{-1} (Wilson and Kaisi 2008). Halvorson and Reule (1999) reported that fertilizer-N application for long term under NT practiced in dryland annual cropping in Colorado results in C sequestration through enhanced C sequestration efficiency. Conjoint application of NPK + FYM significantly increased SOC stock from 6.33 to 7.33 Mg C ha^{-1} compared to NPK alone (Singh and Benbi 2020a). NPK alone also maintained SOC stocks of 6.16 t C ha^{-1} which was higher than the controlled plots. Furthermore, soil organic C stocks were reported to be significantly lower by 14–18% and 12–14% in surface and subsurface, respectively, in imbalanced plots, compared to balanced (NPK) application of fertilizer nutrients.

The average rate of change in soil organic C sequestration varied between 15 and 117 kg C ha^{-1} year^{-1} during long-term (10–25 years) fertilizer application (Table 3). Farmyard manure application combined with balanced fertilizer

Table 1 Effect of long-term manuring on soil organic C sequestration at experimental locations in India (Source: Swarup 1998; Rasool et al. 2007; Brar et al. 2013; Singh and Benbi 2018a)

Location (state)	Study period (years)	Soil type	Soil organic C concentration (g kg^{-1})				Rate of change over control (kg C ha^{-1} $year^{-1}$)
			Initial	Control	NPK	NPK + FYM	
Bangalore (Karnataka)	10	*Haplustalf*	4.5	4.8	5.9	8.4	101
Barrackpore	24	*Eutrochrept*	7.0	4.1	5.0	5.4	15
Bhubaneshwar (Odisha)	21	*Haplaquept*	2.6	3.7	5.7	8.1	59
Coimbatore (Tamil Nadu)	23	*Vertic Ustochrept*	3.0	4.3	4.9	6.2	23
Delhi (country capital)	25	*Ustochrept*	4.3	4.4	5.5	6.7	25
Hydrabad (Andhra Pradesh)	23	*Tropaquept*	5.0	4.6	5.3	8.0	41
Jabalpur (Madhya Pradesh)	25	*Chrmustert*	5.8	5.3	6.0	9.8	48
Ludhiana (Punjab)	25	*Typic Ustochrept*	2.0	2.5	3.3	3.8	15
Palampur (Himachal Pradesh)	22	*Hapludalf*	7.8	7.3	10	12	60
Pantnagar (Uttar Pradesh)	24	*Hapludoll*	13	5.0	8.3	15	117
Ludhiana (Punjab)	9	*Typic Ustochrept*	2.1	3.3	4.11	4.52	–
Gurdaspur (Punjab)	5	*Eutric Cambisols*	4.2	4.5	5.11	6.08	–
Ludhiana (Punjab)	32	*Typic Ustochrept*	2.9	3.0	5.30	–	–

application (NPK + FYM) resulted in an 8–80% increase in soil organic C concentration, compared to NPK alone. At the surface (0–15 cm) layer, NPK + FYM contained the significantly highest soil organic C concentration (7.7 g kg^{-1}) followed by NPK + crop residue (CR) (7.5 g kg^{-1}) and NPK + green manure (GM) (7.4 g kg^{-1}). There was a significant reduction in soil organic C concentration with the sole application of inorganic fertilizers (NPK) compared to those in the mixed organic and inorganic treatments. The lowest SOC concentration (3.6 g kg^{-1}) in 0–15 cm layer was observed in the treatment of continuous cropping of rice–wheat over 25 years without any amendments. Mean soil organic C concentration in the profile increased from 2.4 g kg^{-1} in control to 4.1 g kg^{-1} in NPK + FYM. All the treatments showed a higher accumulation of soil organic C in the surface layer. Significant variations in SOC content were also observed in the subsoil layers; mean soil organic C content decreased from 6.4 at surface 0–15 cm to 1.8 g kg^{-1} at 45–60 cm soil layer.

3.4.1 Fertilizer-N Management and C Sequestration

The C footprints are described as the quantity of GHGs expressed in terms of CO_{2e} emissions released in the atmosphere by an individual, organization, process, product, or event from within a specified boundary (Pandey et al. 2010). Jiang et al. (2019) reported that C footprints were positively correlated with fertilizer-N application rates, suggesting that GHGs emissions were strongly dependent on the rates of fertilizer-N application in rice cultivation. The C footprints in rice cultivation increased as the fertilizer-N application rates increased. The C footprints from N75 to N375 were increased by 15.1%, 31.2%, 39.7%, 41.8%, and 56.3%, respectively, compared to the control. The methane (CH_4) emissions from the rice fields were the largest contributor to total C footprints, which increased initially followed by a gradual decrease and increase with increasing fertilizer-N input rates. The ecosystem service values of C sequestration decreased from positive to negative with increasing N fertilizer rates, suggesting that rice soils were transformed from a net C sink (0–300 kg N ha^{-1}) to a net C source (375 kg N ha^{-1}). Zhang et al. (2017) reported that grain production has higher C footprints of 4052 kg CO_{2e} ha^{-1} or 0.48 kg CO_{2e} kg for maize, 5455 kg CO_{2e} ha^{-1} or 0.75 kg CO_{2e} kg^{-1} for wheat, and 11,881 kg CO_{2e} ha^{-1} or 1.60 kg CO_{2e} kg^{-1} for rice cultivation in China. Of the total CE emissions, fertilizer-N contributes 8–49%, straw burning 0–70%, energy consumption by agri-machinery 6–40%, energy consumption for irrigation 0–44%, and the CH_4 emissions from rice soils ~15–73%. On the other hand, the major C input was the returning of crop straw contributing ~41–90%, fertilizer-N application ~10–59%, and no-till farming practices contributing 0–10%. The C footprints concept has several implications. Therefore, the recent research focusing on the inclusion of soil organic C changes in addition to C footprint assessments for highlighting influences of crop production and soil management practices on total ecosystems' C budget (Pandey and Agrawal 2014; Singh and Benbi 2020a).

3.5 *Agricultural Waste Management and C Sequestration*

Agri-waste, namely, crop residue (CR), farm yard manure (FYM), poultry litter (PL), and green manures (GM) application such as surface mulch can play an important role in the maintenance of soil organic C levels and productivity through increasing recycling of mineral nutrients, increasing fertilizer use efficiency, and improving soil physical and chemical properties and decreasing soil erosion (Hargopal-Singh et al. 2013). Furthermore, the C sequestration potential of added crop residue varies largely according to the total C content of the residue and the rate of total C input in the soil system (Kaisi and Grote 2007). After 19 years of continuous rice–wheat cropping, a highly significant positive linear relationship ($R^2 = 0.980$) between stable C and cumulative C input from added organic sources showed that soil still has the potential to sequester more C with increasing C input from added organics (Ghosh et al. 2010). Fortuna et al. (2003) reported that compost application consecutively for 6 years has resulted in a 30% increase in resistant C pool and 10% in slow pool of C. Wang et al. (2010b) reported that mean net C remained in the crop residue after 127 days of decomposition period were 187 g m^{-2} and 91 g m^{-2}, respectively, which represents approximately 73% and 52% of total biomass C for winter- and summer-grown cover crops. Sainju et al. (2008) reported that the soil organic C in the upper 20 cm soil layer after 10 years was greater with PL than with NH_4NO_3 applied on an equivalent N basis, resulting in a C sequestration rate of 510 kg C ha^{-1} year^{-1} with PL as compared with -120 to +147 kg C ha^{-1} year^{-1} with NH_4NO_3. An increase in soil organic C due to PL application compared with inorganic N fertilization in NT and CT system suggests the supplementation of 1.7 t C ha^{-1} year^{-1} from PL that was applied to supply 100 kg N ha^{-1} year^{-1} (Sainju et al. 2008). Averaged across tillage and cropping system, PL sequestered C at an estimated rate of 461 kg C ha^{-1} year^{-1} when compared to 141 kg C ha^{-1} year^{-1} (Sainju et al. 2008). The conjoint application of inorganic and organic fertilizers in NPK + CR $_{\text{(Rice straw)}}$ and NPK + GM$_{(\textit{Sesbania sesban})}$-treated plots to rice–wheat cropping system practiced consecutively for 19 years in IGPs of West Bengal (India) has resulted in a significant increase in labile C fraction by 28% and 25%, respectively, over control (no-NPK/organics) (Ghosh et al. 2010). The highest value of labile C fraction in NPK + CR dressed plots has been ascribed to the effect of higher polysaccharide (cellulose and hemicellulose) content of crop residue that leads to higher production of labile C fractions compared to GM (Ghosh et al. 2010). NPK use for 45 years has resulted in ~3% increase in SOC content compared to ~115% increase in soil organic C in soils receiving NPK conjointly with FYM with an improvement in SOC from 4.95 to 7.30 t C ha^{-1} after 18 years that showed a C sequestration rate of 13 g C m^{-2} year^{-1} (Raji and Ogunwole 2006). The application of fertilizers and/or FYM increased the mean weight diameter of soil aggregates and thereby providing physical protection to soil organic C from decomposition (Banger et al. 2010). After 16 years of NPK application, there was a significant increase (~19.4%) in soil organic C under NPK-applied plots (0.430%) compared to control (0.360%). The integrated use of inorganic and organic fertilizers (NPK + FYM) and purely organic fertilizer application (FYM alone) has enhanced soil organic C by 33.4% and 36.3% over control (no-NPK/FYM). The concentration of water-soluble C (WSC), microbial biomass C (MBC), particulate organic matter (POM), and light fractions of C (LFC) were higher in organics that followed integrated system when compared to chemical (NPK) fertilizer dressed plots (Banger et al. 2010). A negative soil C sequestration range (−410 to −193 g C m^{-2}) was observed in different crops including grass, cereals, and pulses, and however, recapitulates that C loss from soils could not be compensated through C inputs through plant photosynthates (Mu et al. 2006).

4 Land-Use Management and Soil C Sequestration

Soil organic carbon is influenced by the different soil and land management systems (Collins et al. 2000). Approximately ~10–30% (~7.0 Pg C $year^{-1}$) of total global CO_2 emission results from land-use change that is associated with deforestation, crop biomass burning, and conversion from natural to agricultural ecosystem (Prentice et al. 2001). Dowuona and Adjetey (2010) compared C stored by a *Ferric Acrisol* in the savanna zone of Ghana under different land-use systems and fertilized plot fallow and Leucaena woodlot. They reported that soils under the Leucaena woodlot stored the largest amount of C. An increase of ~200% in MBC has been reported under the switchgrass (*Panicum virgatum* L.) cropping system than in the corn–soybean cropping system (Kaisi and Grote 2007). The studies of Wang et al. (2009) revealed decreased soil organic C stocks by ~9.83 Mg C ha^{-1} in the upper 30-cm soil layer 28 years after shifting from grassland (meadow steppe) and by ~21.9 Mg C ha^{-1} 42 years after shifting from grassland, which represents approximately ~10% and ~ 25% reductions, respectively.

Koul and Panwar (2008) found that in *Tarai* region of West Bengal (India), fallow land and agricultural fields sequester 5.86% and 4.73% C, respectively, compared to the natural forest having *Shorea robusta* canopy. In contrast, agroforestry, namely, tea garden and agrihorticulture contributed 24.2% and 9.1% C, respectively, compared to natural forests having *Shorea robusta* canopy (Koul and Panwar 2008). The rate of C sequestration in the fallow period was ~400% higher than the rate under continuous cropping (Raji and Ogunwole 2006). According to Lal (1999) of the 136 Pg C emitted due to land-use change, ~57.4% (~78 Pg C) was estimated to be the contribution of soil organic C pool. In a 45-year-old *Gmelina* forest, the soil C stock was 8987 g C m^{-2} compared to parts of the forest that were cleared and continuously cropped using conservation tillage practices for 15 years and had 75% lower C stock (1978 g C m^{-2}) (Anikwe 2010). Gerzabek et al. (2006) reported that soil organic C of grassland soils was 1.8 times greater than that of cropland plots in *Eutric Cambisols* of Sweden. Singh and Benbi (2018a) reported that the C preservation capacity of water-stable aggregates was significantly higher in soils under grasslands in contrast to eroded slopes, which had the lowest C preservation capacity.

The hilltop and cropland soils, however, did not differ significantly with regard to C preservation capacity of water-stable aggregates. The formation of water-stable aggregates has been related to soil organic C concentration (Benbi et al. 2016). Benbi et al. (2012) reported that the soils under rice–wheat system had significantly lower soil organic C concentrations compared to those under maize–wheat and agri-forestry land-use systems. As compared with the rice–wheat cropping system, the soils under agro-forestry and maize–wheat land-use systems had ~88% and 65% higher in soil organic C concentration, respectively. In another study, Benbi et al. (2015) reported that the soils under different land-use systems did not differ in C:N ratios except that under agro-forestry compared to maize–wheat and the rice–wheat cropping systems. The total organic C pool was higher ($p < 0.05$) in the uncultivated soils than the cultivated soils, but differences among cropping systems were non-significant. For restoring the degraded and waste lands, different improved practices, namely, zero tillage, legume intercropping, retaining crop residues on the soil surface, cover crops, integrated nutrient management, and agro-forestry, must be adopted (FAO 2007). Human activities accelerate the pace of different physico-chemical reactions in the soil (Scharlemann et al. 2014; Smith et al. 2008). West and Marland (2003) estimated the net C flux for the United States with a change in the tillage intensity (Table 2).

Globally, net emissions of 1.1 ± 0.7 Pg C year^{-1} were recorded from 2000 to 2010 which were mainly due to extensive forest tree cuttings for agricultural production (Don et al. 2011; Houghton 2003). The Indian tropical conditions are responsible for the higher C turnover and sequestration (Malhi et al. 2004). The warmer climatic conditions are considered responsible for the higher C emissions than the colder regions (Zech et al. 1997). Furthermore, C stocks decreased to ~25–30% on shifting from the forest to the cultivation (Don et al. 2011); therefore, land use must be sustainable if soil C stocks are to be maintained (NETL 2010). A shift to agricultural land use from the forest lands in year zero resulted in higher soil organic C stocks with improved management practices of reduced tillage, mulching, and so on (Sanderman and Baldock 2010) (Fig. 4). Instead, its management of the mined lands, forest lands, rangeland, grassland, and agricultural lands is also very important for sequestering the C in the terrestrial ecosystems.

5 Legumes and C Sequestration

Growing of the leguminous crops, namely, lentil, guar, moong in between the conventional rice-based cropping sequences, is considered as one of the most important interventions for improved C sequestration and reduced C footprints of the system as a whole including the intervening periods (Poeplau and Don 2015). Legumes are a better fit for the diversification option due to their shorter stay in the field, ability to tolerate different stresses, and finally ability to sequester C (Ghosh et al. 2020). Poeplau and Don (2015) delineated an annual organic C sequestration of 0.32 t ha^{-1} with the inclusion of leguminous crops. Lal (2010) reported enrichment

Table 2 Average net carbon flux for the United States with change in tillage from conventional tillage (CT) to zero tillage (ZT) (Source: West and Marland 2003)

	Conventional tillage (CT, kg C ha^{-1} $year^{-1}$)	Zero tillage (ZT, kg C ha^{-1} $year^{-1}$)
Sequestered C in soil	0	−337
Emission of C from farm machinery	69	23
C emissions from agril. Input	99	114
Net C flux	168	−200
Relative net C flux	0	−368

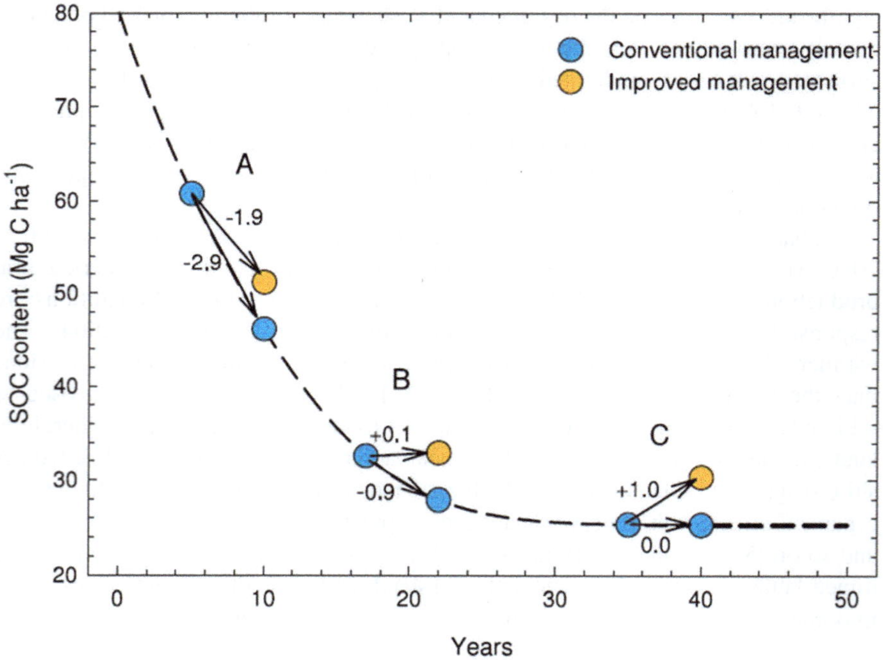

Fig. 4 Relationship between SOC stocks and aging by hypothetical field trial (Source: Sanderman and Baldock 2010)

of soil organic C under proper crop rotation with legumes due to higher exchange efficiency of residue-derived C to soil C pool. Legumes on the long run assured N availability which produces higher biomass C. These crops promote C impounding potential of succeeding crop gown in the rotation, improving the microbial functions, and biomass fabrication by successive crop and finally improving soil quality (Yadav et al. 2017). The inclusion of leguminous crops in cropping systems leads to increased nutrient use efficiencies and enhanced root biomass which eventually enhances C inputs and labile C pools in the soil (Lal 2010).

Legume crops such as green manure, cover crops, and forage promote C and N in the soil system, thereby improving soil physical, chemical, and biological properties that enhance the crop grain yields (Bedoussac et al. 2015) while reducing the C footprints. The increased mechanical operations and C footprints due to the inclusion of cover crops (Saini and Bhatt 2020) are offset by the positive prevalence of cover crops on the mitigation of GHGs with increased C sequestration. Once C is lost into the atmosphere as GHG, it contributes toward GWP (Singh and Benbi 2020b). Crop diversifications with legumes not only reduce the C footprints but also restore soil organic C stocks due to increased photosynthetic activities throughout the year (Dhakal et al. 2015). Nonetheless, the C:N ratio and concentration of nutrients in crop residues of the legumes are relatively higher than in other cereals crops (Srinivasarao et al. 2011).

6 Crop Cultivars and Soil C Sequestration

Carbon input into the soil depends inter alia on crop cultivars and resultant biomass. The crop cultivars with deep roots and large leaf foliage had higher C sequestration and low C footprints. The terrestrial and atmospheric C pools strongly interact with one another. The development of new plant cultivars using biotechnological interventions with similar yield potential and lesser stay in the field lowers fertilizer-N requirements and leads to lesser emission of CO_2 and ammonia (NH_4) gas which finally reduces both C footprints (Lal et al. 1998). The selection of crop cultivars had a significant effect on soil C status, water, and energy footprints. The crop cultivars that had a short duration in the field had low irrigation water requirements and high-water productivity (Singh et al. 2020c). The short- and medium-duration crop cultivars had higher water productivity due to reduced soil water evaporation (Singh and Saini 2011). In northwestern India, a long duration rice variety (Pusa-44) that takes 160 days to mature consumes a significantly higher amount of irrigation water than the short duration varieties, namely, PR-126 and PR-127 (PAU 2020) which take only around 123 and 137 days and thus have potential to save water by 15–20% (Balwinder-Singh et al. 2015; PAU 2020). The results of 550 farmer's field demonstrations conducted in southwestern Punjab (India) revealed that grain yield of rice variety PR-122 was significantly higher by ~11.1% and 14.5%, compared to PR-126 and PR-124, respectively (Singh et al. 2020c). They have reported that the production efficiency of 54.5 kg ha^{-1} day^{-1} was higher for PR-126, compared to PR-124 (50.0 kg ha^{-1} day^{-1}) and PR-122 (50.6 kg ha^{-1} day^{-1}). The economic efficiency of PR-124 was lower by ~Rs. 111.9 ha^{-1} day^{-1} and Rs. 43.6 ha^{-1} day^{-1} than the PR-126 and PR-122, respectively. The water use efficiency was higher for PR-126, compared with the other two genotypes. Therefore, the adoption of short-duration rice cultivars, demanding lower volumes of irrigation water when compared to the longer duration one (Jalota and Arora 2002; Arora et al. 2008; Jalota et al. 2009) had lower C footprints (Singh and Benbi 2020b). The comparison of three different irrigation management regimes, namely, Scenario 1 (continuously flooded), Scenario 2 (intermittently flooded with single aeration), and Scenario 3 (intermittently flooded with multiple aerations) revealed that the global warming potential (GWP) for rice cultivation in northwestern India was 2.94, 2.41, and 2.13 Mg CO_{2e} ha^{-1} $year^{-1}$, respectively (Singh and Benbi 2020b). Nonetheless, the inclusion of suitable legumes or perennial forages such as alfalfa (*Medicago sativa* L.) reduces the C footprints in comparison to the annual sequences due to higher belowground C input (Sainju and Lenssen 2011). C sequestration in soils had a key role in reducing the C footprint of crop cultivation and in improving soil health (Benbi et al. 2016; Singh and Benbi 2018a). Attempts must be made to sequester a greater fraction of the atmospheric C back in plant biomass for a longer period of time for better C stabilization; otherwise, the sequestered C will be released back into soil within a shorter period and responsible for causing different adverse effects, namely, global warming and so on.

7 Crop Diversification and C Sequestration

Crop residues in terms of both quantity and quality affect the soil organic matter (Srinivasarao et al. 2013). Among different options purposed for improving the sustainability cropping systems, crop diversification seems to be quite effective in decreasing the C footprints and enhanced C sequestration. For example, if in a rice-based cropping system, rice is being replaced with maize or other oilseed or pulse crops; then, energy footprints (as no intensive tillage and fertilization) (Singh et al. 2019a, b), water footprints (as no puddling and reduced conditions), and ultimately the C footprints could be sustainably reduced (Singh et al. 2020a; Singh and Benbi 2020b). Therefore, for reduced C emissions and C footprints, rice crops must be diversified with the vegetables, pulses, maize, and oilseed crops. The inclusion of legumes or pulse crops in cereal-based cropping systems enriches the soil with different nutrient and root exudates that promoted the soil structure and aggregation. Therefore, any crop diversification option which adds a good quantity of crop residues in the soil back will certainly improve the SOC stocks within different soil (Singh et al. 2018b; Singh and Singh 2017; Dhillon et al. 2020), which is important for the good soil health and land productivity (Majumder et al. 2008). Green manuring, keeping less fallowing and winter cover, and adding legumes that fix atmospheric N are required for reducing the C footprints (Flynn et al. 2009). The buildup of SOC concentration with crop diversification must concentrate on increased crop biomass and organic matter humification. The atmospheric C is sequestered in the soil through the roots as organic acid, phenolic acid, and amino acid. According to Mandal et al. (2007), up to ~82% of the applied or added C is lost to the atmosphere, which needs to be cut down through different options including crop diversification as the highest C fixation reported in the rice–wheat–jute cropping system (535 kg $ha^{-1}year^{-1}$) followed by rice–mustard–sesame (414 kg $ha^{-1}year^{-1}$) than rice–fallow–rice (402 kg $ha^{-1}year^{-1}$) at CRRI, Cuttack. The addition of pulse crops in cereal-based cropping systems could serve the

Table 3 Effect of inclusion of pulses in rice-based cropping systems on soil organic carbon fractions of different oxidizability in *Inceptisols* of Indo-Gangetic Plain (IGP) zone (Kanpur), India (Source: Ghosh et al. 2012)

Cropping system	Very labile C (%)		Labile C (%)		Less labile C (%)		Non-labile C (%)	
	0–20 cm	20–40 cm	0–20 cm	20–40 cm	0–20 cm	20–40 cm	0–20 cm	20–40 cm
Rice–wheat–fallow	0.11	0.12	0.09	0.17	0.23	0.15	0.14	0.08
Rice–chickpea–fallow	0.17	0.14	0.10	0.14	0.21	0.13	0.15	0.07
Rice–wheat–mungbean	0.18	0.18	0.11	0.09	0.27	0.07	0.15	0.08
Rice–chickpea–fallow–rice–wheat–fallow (2 years)	0.18	0.15	0.13	0.12	0.17	0.12	0.14	0.08
LSD ($p < 0.05$)	0.021	0.019	0.02	NS	0.04	0.01	NS	NS

purpose and have a significant role in improving soil health (Porpavai et al. 2011) by fixing a higher proportion of atmospheric C back into the soil in different C pools of variable oxidizability (Table 3). The less labile C was the highest SOC fraction in soils under different cropping systems (Ghosh et al. 2012). They have reported that higher crop biomass maintained greater SOC under rice–wheat–mungbean and rice–wheat–rice–chickpea cropping systems in the IGPs. Under the crop diversification, the incorporated legumes fix biological N_2 and produce root exudates besides leaf litter, and the deep root system helps in improving the C economy of soils when compared to conventional systems. The maize–wheat–mungbean and pigeonpea–wheat systems are reported to significantly increase the SOC concentration by ~11% and 10% and the MBC by ~10% and 15%, respectively, (Venkatesh et al. 2013). Therefore, crop diversification sustainable options with legumes will certainly serve the purpose to improve root and crop biomass additions in soils which further promoted higher C stocks and hence soil health with reduced C footprints in the soils. But, still, there is a need to give more attention to this option for all kinds of lands with divergent soil textural classes in different agroclimatic regions of the country. Identifying sustainable diversification systems for higher C sequestration and increased productivity will certainly help to mitigate the climate change effects on one side while improving the livelihoods on the other (Bhatt et al. 2021).

8 Soil Conservation Practices and C Sequestration

In the hilly tracts, the landscapes are generally prone to soil erosion by both wind and water (Arora et al. 2008; Singh and Benbi 2018b). These landscapes generally had coarse-textured soils low in organic matter and poor in fertility. The high-intensity rains on steeper slopes on soils with lesser amounts of organic matter results in the detaching of the soil particles from their parent material and transported with the water or wind to remote distances (Singh and Benbi 2018b). The ecosystems require an urgent need to control soil erosion to improve C sequestration, soil health, and soil productivity. In northwestern India, accelerated erosion has been taking place during the last 50 years, and these landscape positions have been losing 0.12 Mg C ha^{-1} $year^{-1}$ through soil erosion (Singh and Benbi 2018b). In the light of scientific knowledge, the following indigenous techniques for efficient soil and water management have been suggested to improve land and water productivity (Bhatt and Kukal 2016). Soil organic C potential and C sequestration rates of degraded lands revealed the highest C sequestration potential (Table 4).

9 Crop Residue Management and C Sequestration

In India, rice-based cropping systems are intensively cultivated on ~90% of the area in Indo-Gangetic Plains (IGPs) (Janaiah and Hossain 2003; Bhatt et al. 2021). The impact of residue removal on soil organic C concentration and stocks across different soils in the U.S. Corn Belt region has been shown in Table 5. In the intensive cultivation in these fertile plains, a large proportion of total crop residue generated (500–550 Mt. $year^{-1}$) is burnt in situ in open fields (MOA 2012; Singh et al. 2020a). The open field burning of crop residue is adversely affecting the sustainability of rice–wheat system due to increased C footprints due to decreased C sustainability which has been a great challenge in front of agricultural scientists (Bisen and Rahangdale 2017). The data illustrate that cereal-based cropping systems had the largest (~58%) share toward total crop residue produced, of which ~25% is produced under the rice–wheat cropping sequences (Sarkar et al. 1999).

The rice straw burning contributes ~0.05% of the total GHGs emissions in India, which has been the major concern for human and environmental health (Gadde et al. 2009). Apart from the loss of C, up to 80% loss of N and S, 25% of P, and 21% of K occur during burning of crop residues (Ponnamperuma 1984; Yadvinder-Singh et al. 2005). It is estimated that one tone of rice straw contained 5–8 kg of N (Dobermann and Fairhurst 2002); therefore, an annual rice–wheat cropping system with an

Table 4 Soil organic C sequestration potential and C sequestration rates through restoration of degraded soils (Source: Srinivasarao et al. 2013)

Degradation process	Area (Mha)	Soil organic C sequestration rate (kg ha^{-1} $year^{-1}$)	Total soil organic C sequestration potential (Tg C $year^{-1}$)
Water erosion	32.8	80–120	2.62–3.94
Wind erosion	10.8	40–60	0.43–0.65
Soil fertility decline	29.4	120–150	3.53–4.41
Water logging	3.1	40–60	0.12–0.19
Salinization	4.1	120–150	0.49–0.62
Lowering of water table	0.2	40–60	0.01–0.012
Total			7.20–9.82

Table 5 Impacts of residue removal on soil organic C concentration and stocks across different soils in the U.S. Corn Belt region

Soil series (Reference)	Soil depth (cm)	Tillage system	Management duration (years)	Rate of residue addition	Soil organic C concentration (g kg^{-1})
Marshall silty clay loam (Larson et al. 1972)	0–15	Plow	11	0 2 t ha^{-1} 4 t ha^{-1} 8 t ha^{-1} 16 t ha^{-1}	15.8 16.3 17.7 18.6 23.2
Crosby silt loam (Duiker and Lal 1999)	0–10	Plow tillage	7	0 2% 4% 8% 16%	8.3 10.3 9.9 11.2 11.5
Crosby silt loam (Duiker and Lal 1999)	0–10	Ridge tillage	7	0 2% 4% 8% 16%	9.4 11.1 10.7 14.1 14.9
Crosby silt loam (Duiker and Lal 1999)	0–10	No-tillage	7	0 2% 4% 8% 16%	9.1 11.5 10.3 13.3 15.4
Raub silt loam	0–15	Plow	11	0 100%	17.7 16.0
Rayne silt loam (Blanco-Canqui and Lal 2007)	0–2	No-tillage	2.5	0 25% 50% 75% 100%	6.42 6.04 5.57 5.31 4.47
Celina silt loam (Blanco-Canqui and Lal 2007)	0–2	No-tillage	2.5	0 25% 50% 75% 100%	7.37 6.10 5.74 5.40 5.47
Hoytville clay loam (Blanco-Canqui and Lal 2007)	0–2	No-tillage	2.5	0 25% 50% 75% 100%	6.07 5.38 5.35 5.22 4.83
Crosby silt loam	0–50	No-tillage	10	0 8% 16%	82.5 94.1 104.9

average production of 7 t ha^{-1} of rice +4 t ha^{-1} of wheat grains removed more than ~300 kg N, ~30 kg P and ~ 300 kg K ha^{-1} (Singh and Singh 2001). Therefore, burning the crop residues has not been a viable option and therefore requires immediate attention.

10 Biochar Application and C Sequestration

Biochar prepared through pyrolysis, gasification, and hydrothermal carbonization has ~70% of C, which might be emitted into the atmosphere. Biochar application in soils enhance the SOC and thereby favors the soil's physicochemical and biological properties and soil health (Sohi et al. 2010; Day et al. 2005; Srinivasarao et al. 2013). The small-scale biochar production proves to be economical and sustainable (Pratt and Moran 2010). Being a fine-grained, soft, C-rich source with highly porous structure and high surface area, biochars are considered important regarding C sequestration and reducing C footprints in crop production (https://www.pau.edu/content/pf/pp_kharif.pdf). The biochars are also considered as efficient liming materials for the reclamation of acidic soils. The biochars had the potential to mitigate ~12–50% of anthropogenic C emissions depending upon the material used, pyrolysis conditions, and energetic performance of the biochar production system (Cayuela et al. 2010). Quite often, biochars prepared with thermal decomposition of crop or plant biomass waste in three ways, namely, pyrolysis, gasification, and hydrothermal carbonization, where the material is heated up in the absence of the O_2 which produced volatile gas leaving behind C-rich biochar (Sohi et al. 2009) which on application helps in improving the water and nutrient holding capacity of soils.

Prali Char is prepared in a pyramid or dome type kiln made up of bricks and clay which is dome-shaped (height = 14 ft, diameter = 10 ft) and can accommodate 12 t of rice straw. The whole process of making *prali char* usually takes ~10–12 h. On average, *prali char* contains 30–36% C, 0.5–0.6% N, 0.16–0.22% P, and 1.6–2.2% K. Field application of this in rice and wheat at 5 t ha^{-1} saves 40 kg N ha^{-1} besides increasing crop productivity and improving soil health. It uses perlite as an insulator between the two fire brick walls to check the heat loss. The drums are filled with agricultural residues with the provision to escape syngas. Temperature is hiked to 300–400 °C for proper heating of residues. This method requires only ~2 hours for the preparation of good quality biochar.

11 Nitrogen Transformation Inhibitor and C Sequestration

The misuse or overuse of chemical fertilizers has been the major cause of soil, water, and air pollution (Fowler et al. 2013; Neubauer and Megonigal 2015; IPCC 2014). Nitrogen fertilizer is one of the main source of nitrous oxide (N_2O) production in agricultural land (Bouwman et al. 2002) and is further affected by the form (Dobbie and Smith 2003), amount (Ma et al. 2010), and fertilization method (Lin et al. 2010; Liu et al. 2011). Globally, N use efficiency of cereals is ~33% on an average (Raun and Johnson 1999). The N recovery efficiency of fertilizers in lowland rice is reported to be ~40% (Fageria 2014; Fageria et al. 2007) which highlighted a significant loss to the biosphere through leaching, volatilization, and denitrification causing environmental pollution. Around 70% of all plant nutrients at a global level are received from fertilizers (Ayoub 1999; Khalil et al. 2011). Therefore, improving N use efficiency has significance in achieving sustainability which can be ensured following different management strategies. For example, the use of N transformation inhibitor and slow release fertilizers, e.g., neem-coated urea or polycoated urea (Shaviv and Mikkelsen 1993) has been advocated due to improved N use efficiency, land productivity (Shoji et al. 2001), mitigation of N_2O emissions, and reduce the C footprints of the system (Jiang et al. 2010).

12 Agroforestry and C Sequestration

Agroforestry enhances C sequestration rates and thus the soil health (Benbi et al. 2012; Dhillon et al. 2020; Sharma et al. 2021a). Dhillon et al. (2020) reported that total organic C stocks in soils were related to the age of the agroforestry system. They have reported that total organic C was higher in soils under 20-year agroforestry compared to the soils under the relatively younger agroforestry system. The organic C stocks in surface soil under the 20-year agroforestry system were significantly higher, although TOC stocks in soils under 10 and 15 year of agroforestry did not differ significantly. In the subsurface soils, TOC stocks were significantly higher under agroforestry older than 10 years, while TOC stocks did not differ significantly under 10-, 15-, and 20-year agroforestry systems. The C storage capacity of different agroforestry models is detailed in Table 6.

13 Conclusions

The present review focused on effects of conventional land management practices, namely, intensive tillage, residue burning in open, puddling operations, use of higher N fertilizers, and so on on the soil C stocks and GHGs emissions in comparison to

Table 6 Carbon storage capacity of the soils under different agroforestry models

Agro forestry model	C storage capacity (t C ha^{-1})	Source
Agri-silviculture	13.4	Verma et al. (2008)
Silviculture	31.2	Verma et al. (2008)
Agri-horticulture	12.3	Verma et al. (2008)
Silvo-pastoralism	6.6	Kumar et al. (1998)
Block plantation	24.1–31.1	Swami et al. (2003)
Popular based agroforestry	2.3	Chauhan et al. (2011)

the recommended technologies. Carbon stocks are the core of the terrestrial ecosystem which must be enhanced by sequestering the higher fraction of C back in the soil for a longer period of time and by reducing GHGs emissions. C sequestration is enhanced by converting the degraded or waste lands under forest/agroforestry. In rice-based cropping sequences, anaerobic conditions resulted in the production of GHGs, which complicate the systems' sustainability. Scientists across the regions suggested many technological interventions for enhancing C sequestration with reduced C footprints in the agricultural sector. Among others, conservation tillage, zero tillage in standing rice stubbles, N and agri-waste management, land use management, legume inclusion, crop cultivars, crop diversification, biochar application, and water and nutrient use efficiency within a cropping system are important. The government and private agencies and other farmer's welfare organizations need to come forward for farmers' welfare by providing subsidies on agri-machinery for in situ crop residue management to enhance C sustainability due to reduced C footprints.

Acknowledgment The authors fully acknowledge the support received from different scientists who shared their research for compiling this important review paper.

References

Agnelli A, Ascher J, Corti G, Ceccherini MT, Nannipieri P, Pietramellara G (2004) Distribution of microbial communities in a forest soil profile investigated by microbial biomass, soil respiration and DGGE of total and extracellular DNA. Soil Biol Biochem 36(5):859–868

Anikwe MAN (2010) Carbon storage in soils of southeastern Nigeria under different management practices. Carbon Balance Manag 5:5. (http://cbmjournal.com/content/5/1/5)

Arora S, Hadda MS, Bhatt R (2008) Tillage and mulching in relation to soil moisture storage and maize yield in Foothill Region. J Soil Water Conserv 7:51–56

Ayoub AT (1999) Fertilizers and the environment. Nutr Cycl Agroecosyst 55:17–121

Balwinder-Singh, Eberbach PL, Humphreys E (2014) Simulation of the evaporation of soil water beneath a wheat crop canopy. Agric Water Manag 135:19–26

Balwinder-Singh, Eberbach PL, Humphreys E, Kukal SS (2011) The effect of rice straw mulch on evapotranspiration, transpiration and soil evaporation of irrigated wheat in Punjab, India. Agric Water Manag 98:1847–1855

Balwinder-Singh, Humphreys E, Yadav S, Gaydon DS (2015) Options for increasing the productivity of the rice-wheat system of north-west India while reducing groundwater depletion. Part 1. Rice variety duration, sowing date and inclusion of mungbean. Field Crops Res 173:68–80

Banger K, Toor GS, Biswas A, Sidhu SS, Sudhir K (2010) Soil organic carbon fractions after 16-years of application of fertilizers and organic manure in a Typic Rhodalfs in semi-arid tropics. Nutr Cycl Agroecosyst 86:391–399

Bangroo SA, Kirmani NA, Ali T, Wani MA, Bhat MA, Bhat MI (2011) Adapting agriculture for enhancing eco-efficiency through soil carbon sequestration in agro-ecosystem. Res J Agric Sci 2:164–169

Bationo A, Kihara J, Vanlauwe B, Waswa B, Kimetu J (2007) Soil organic carbon dynamics, functions and management in West African agro-ecosystems. Agric Syst 94(1):13–25

Beare MH, Cabrera ML, Hendrix PF, Coleman DC (1994) Aggregate-protected and unprotected pools of organic matter in conventional and no-tillage soils. Soil Sci Soc Am J 58:787–795

Bedoussac L, Journet E-P, Hauggaard-Nielsen H, Naudin C, Corre H, Jensen ES, Prieur L, Justes E (2015) Ecological principles underlying the increase of productivity achieved by cereal-grain legume intercrops in organic farming. A review. Agron Sustain Dev 35:911–935

Benbi DK, Brar K, Toor AS, Singh P (2015) Total and labile pools of organic carbon in cultivated and uncultivated soils in northern India. Geoderma 237-238:149–158

Benbi DK, Brar K, Toor AS, Singh P, Singh H (2012) Soil carbon pools under poplar based agroforestry, rice-wheat and maize-wheat cropping in semiarid India. Nutr Cycl Agroecosyst 92(1):107–118

Benbi DK, Singh P, Toor AS, Verma G (2016) Manure and fertilizer application effects on aggregate and mineral-associated organic carbon in a loamy soil under rice-wheat system. Commun Soil Sci Plant Anal 47:1828–1844

Bhandari AL, Ladha JK, Pathak H, Padre AT, Dawe D, Gupta RK (2002) Yield and soil nutrient changes in a long-term rice–wheat rotation in India. Soil Sci Soc Am J 66:162–170

Bhatt R, Hussain A, Singh P (2019) Scientific interventions to improve land and water productivity for climate-smart agriculture in South-Asia. Chapter-24. In: Mirza H (ed) Agronomic crops Volume-2: management practices. Springer, pp 449–458. ISBN= 978-981-32-9782-1, ISBN= 978-981-32-9782-8 (eBook)

Bhatt R, Kukal SS (2016) Tillage and establishment method impacts on land and irrigation water productivity of wheat-rice system in North-West India. Exp Agric 53(2):178–201

Bhatt R, Singh P, Hussain A, Timsina J (2021) Rice-wheat system in the north-west Indo-Gangetic Plains of South Asia: issues and technological interventions for increasing productivity and sustainability. Paddy Water Environ. (in press). https://doi.org/10.1007/s10333-021-00846-7

Bijay-Singh, Shah YH, Beebout J, Singh Y, Buresh RJ (2008) Crop residue management for low land rice based cropping systems in Asia. Adv Agron 98:117–199

Bisen N, Rahangdale CP (2017) Crop residues management option for sustainable soil health in rice-wheat system: a review. Int J Chem Stud 5(4):1038–1042

Blanco-Canqui H, Lal R (2007) Soil and crop response to harvesting corn residues for biofuel production. Geoderma 141:355–362

Bouwman AF, Boumans LJM, Batjes NH (2002) Emissions of N_2O and NO from fertilized fields: summary of available measurement data. Global Biogeochem Cycles 16(6):1–13

Bowen GD, Rovira AD (1999) The rhizosphere and its management to improve plant growth. Adv Agron 66:1–102

Brar BS, Singh K, Dheri GS, Kumar B (2013) Carbon sequestration and soil carbon pools in a rice–wheat cropping system: effect of long-term use of inorganic fertilizers and organic manure. Soil Till Res 128:30–36

Bronson KF, Zobeck TM, Chua TT, Acosta-Martinez V, Scott van Pelt R, Booker JD (2004) Carbon and nitrogen pools in southern hill plain cropland and grassland soils. Soil Sci Soc Am J 68:1695–1704

Buyanovsky GA, Wagner GH, Gantzer CJ (1987) Soil respiration in winter wheat ecosystem. Soil Sci Soc Am J 50:338–344

Calegari A, Hargrove WL, Rheinheimer DDS, Ralisch R, Tessier D, de Tourdonnet S, de Fatima GM (2008) Impact of long-term no-tillage and cropping system management on soil organic carbon in an Oxisol: a model for sustainability. Agron J 100:1013–1019

Campbell CA, McConkey BG, Zenter RP, Dyck FB, Selles F, Curtin D (1995) Carbon sequestration in a brown chernozem as affected by tillage and rotation. Can J Soil Sci 75:449–458

Campbell CA, VandenBygaart AJ, Zetner RP, McConkey BG, Smith W, Lemke R, Grant B, Jefferson PG (2007) Quantifying carbon sequestration in a minimum tillage crop rotation study in semiarid southwestern Saskatchewan. Can J Soil Sci 87:235–250

CAST (2004) Council for agricultural science and technology; climate change and greenhouse mitigation: challenges and opportunities for agriculture. CAST, Ames, IA

Cayuela ML, Van Zwieten L, Singh BP, Jeffery S, Roig A, Sánchez-Monedero MA (2010) Biochar's role in mitigating soil nitrous oxide emissions: a review and meta-analysis. Agric Ecosyst Environ 191:5–16

Chauhan SK, Gupta N, Walia R, Yadav S, Chauhan R, Mangat PS (2011) Biomass and carbon sequestration potential of poplar-wheat intercropping system in irrigated agroecosystem in India. J Agric Sci Technol 1:575–586

Chevallier TE, Blanchart E, Albrecht A, Feller C (2004) The physical protection of soil organic carbon in aggregates: a mechanism of carbon storage in a vertisol under pasture and marketgardening (Martinique, West Indies). Agric Ecol Environ 103:375–387

Cole CV, Duxbury J, Freney O, Heinemeyer K, Minami A, Moisre K, Paustian N, Rosenberg N, Sampson D, Sauerbeck H, Zhao Q (1997) Global estimates of potential mitigation of greenhouse gas emission by agriculture. Nutr Cycl Agroecosyst 49:221–228

Collins HP, Elliott ET, Paustian K, Bundy LG, Dick WA, Huggins DR, Smucker AJM, Paul EA (2000) Soil carbon and fluxes in long-term corn belt agroecosystem. Soil Biol Biochem 32:157–168

Davidson EA, Ackerman IL (1993) Changes in soil carbon inventories following cultivation of previously untilled conditions. Biogeochem 20:161–193

Day D, Evans RJ, Lee JW, Reicosky D (2005) Economical CO_2, SO_2, and NO_2 capture from fossil-fuel utilization with combined renewable hydrogen production and large-scale carbon sequestration. Energy 30:2558–2579

Dhakal Y, Meena RS, De N, Verma SK, Singh A (2015) Growth, yield and nutrient content of mungbean (*Vigna radiata* L.) in response to INM in eastern Uttar Pradesh, India. Bangladesh. J Bot 44(3):479–482

Dhillon NK, Singh P, Singh H (2020) Soil organic carbon, phosphorus and potassium in soils under poplar based agro-forestry in Punjab. Int J Curr Microbiol Appl Sci. (in press). https://doi.org/10.20546/ijcmas.2020.907.xxx

Dobbie KE, Smith KA (2003) Impact of different forms of N fertilizer on N_2O emissions from intensive grassland. Nutr Cycl Agroecosyst 67:37–46

Dobermann A, Fairhurst TH (2002) Rice straw management. Better Crops Int 16:7–9

Don A, Schumacher J, Freibauer A (2011) Impact of tropical land-use change on soil organic carbon stocks—a meta-analysis. Glob Chang Biol 17:1658–1670

Dong W, Hu C, Chen S, Zhang Y (2008) Tillage and residue management effects on soil carbon and CO_2 emission in a wheat-corn double-cropping system. Nutr Cycl Agroecosyst 83(1):27–37. https://doi.org/10.1007/s10705-008-9195-x

Dowuona GNN, Adjetey ET (2010) Assessment of carbon storage in some savanna soils under different land use systems in Ghana. ICID+18, August 16–20, 2010, Fortaleza—Ceará, Brazil. Pp. 1–16

Duiker SW, Lal L (1999) Crop residue and tillage effects on C sequestration in a Luvisol in Central Ohio. Soil Till Res 52:73–81

Entz MH, Baron VS, Carr PM, Meyer DW, Smith SR, McCaughey WP (2002) Potential of forages to diversify cropping systems in the Nothern Great Plains. Agron J 94:240–250

Eswaran H, Reich PR, Kimble JM, Beinroth FH, Padammabhan E, Moncharoen P (2000) Global carbon stocks. In: Lal R, Kimble JM, Eswaran H, Steward BA (eds) Global climate change and pedogenic carbonates. Lewis Publishers, Boca Raton, pp 15–25

Fabrizzi KP, Moron A, Garcia FO (2003) Soil carbon and nitrogen organic factions in degraded vs. non-degraded Mollisols in Argentina. Soil Sci Soc Am J 67:1831–1841

Fageria NK (2014) Mineral nutrition of rice. CRC Press, Boca Raton, FL

Fageria NK, Santos AB, Cutrim VA (2007) Yield and nitrogen use efficiency of lowland rice genotypes as influenced by nitrogen fertilization. Pesquisa Agropecuaria Brasileira 42:1029–1034

FAO (2007) The state of food and agriculture. Paying farmers for environmental services

Fierer N, Schimel JP, Holden PA (2003) Variations in microbial community composition through two soil depth profiles. Soil Biol Biochem 35:167–176

Flynn DFB, Gogol-Prokurat M, Nogeire T, Molinari N, Richers BT, Lin BB, Simpson N, Mayfield MM, DeClerck F (2009) Loss of functional diversity under land use intensification across multiple taxa. Ecol Lett 12:22–33

Fortuna A, Harwood R, Kizikaya K, Paul EA (2003) Optimizing nutrient availability and potential carbon sequestration in an agroeconsystem. Soil Biol Biochem 35:1005–1013
Fowler D, Coyle M, Skiba U, Sutton MA, Cape JN, Reis S, Sheppard LJ (2013) The global nitrogen cycle in the twenty-first century. Philo Trans Royal Soc B 368:1621. https://doi.org/10.1098/rstb.2013.0164
Gadde B, Bonnet S, Menke C, Garivait S (2009) Air pollutant emissions from rice straw open field burning in India, Thailand and the Philippines. Environ Pollut 157:1554–1558
Gajri PR, Ghuman BS, Singh S, Mishra RD, Yadav DS, Singh H (2002) Tillage and residue management practices in rice–wheat system in Indo- Gangetic Plains-A Diagnostic Survey. Technical Report, National Agricultural Technology Project, Indian Council of Agricultural Research/Punjab Agricultural University, New Delhi/Ludhiana, India, p. 12
Gama-Rodrigues EF, Nair PKR, Nair VD, Gama-Rodrigues AC, Baligar VC, Machado RCR (2010) Carbon storage in soil size fractions under two cacao agroforestry systems in Bahia, Brazil. Environ Manag 45:274–283
Gan Y, Liang C, Campbell CA, Zentner RP, Lemke RL, Wang H, Yang C (2012) Carbon footprint of spring wheat in response to fallow frequency and soil carbon changes over 25 years on the semiarid Canadian prairie. Eur J Agron 43:175–184. https://doi.org/10.1016/j.eja.2012.07.004
Garcia-Franco N, Albaladejo J, Almagro M, Martínez-Mena M (2015) Beneficial effects of reduced tillage and green manure on soil aggregation and stabilization of organic carbon in a Mediterranean agro-ecosystem. Soil Till Res 153:66–75
Gerzabek MH, Antil RS, Kogel-Knabner I, Knicker H, Kirchmann H, Haberhauer G (2006) How are soil use and management reflected by soil organic matter characteristics: a spectroscopic approach. Eur J Soil Sci 57:485–494
Ghosh PK, Kumar S, Mandal MD, Mandal B, Srinivasan R (2020) Carbon management in tropical and sub-tropical terrestrial systems. https://doi.org/10.1007/978-981-13-9628-1_1
Ghosh PK, Venkatesh MS, Hazra KK, Kumar N (2012) Long-term effect of pulses and nutrient management on soil organic carbon dynamics and sustainability on an Inceptisol of indo-Gangetic plains of India. Exp Agric 48(4):473–487. https://doi.org/10.1017/S0014479712000130
Ghosh S, Wilson B, Ghoshal SK, Senapati N, Mandal B (2010) Management of soil quality and carbon sequestration with long-term application of organic amendments. 19th World Congress of Soil Science on 'Soil Solutions for a changing World' from 01–06th August, 2010, Brisbane, Australia
Guan D, Zhang Y, Kaisi MMA, Wang Q, Zhang M, Li Z (2015) Tillage practices effect on root distribution and water use efficiency of winter wheat under rain-fed condition in the North China plain. Soil Till Res 146:286–295
Gupta PK, Sahai S, Singh N, Dixit CK, Singh DP, Sharma C, Tiwari MK, Gupta RK, Garg SC (2004) Residue burning in rice–wheat cropping system: causes and implications. Curr Sci 87: 1713–1717
Halvorson AD, Reule CA (1999) Long-term nitrogen fertilization benefits soils carbon sequestration. Better Crops 83:16–20
Halvorson AD, Wienhold BJ, Black AL (2002) Tillage, nitrogen, and cropping system effect on soil carbon sequestration. Soil Sci Soc Am J 66:906–912
Hargopal-Singh, Singh P, Singh D (2013) Direct, residual and cumulative effects of mixed sludge generated by Coca-cola soft-drink industry on crop yield, soil fertility, and heavy-metal uptake in rice-wheat cropping sequence. Commun Soil Sci Plant Anal 44:3483–3505
Hira GS, Jalota SK, Arora VK (2004) Efficient management of water resources for sustainable cropping in Punjab. Research Bulletin. Department of Soils, Punjab Agricultural University, Ludhiana, pp. 20
Houghton RA (2003) Revised estimates of the annual net flux of carbon to the atmosphere from changes in land use and land management 1850–2000. Tellus B: Chem Phys Meteorol 55:378–390

Hu B, Wang K, Wu L, Yu SH, Antonietti M, Titirici MM (2010) Engineering carbon materials from the hydrothermal carbonization process of biomass. Adv Mater 22:813–828

Humphreys E, Kukal SS, Christen EW, Hira GS, Singh B, Yadav S, Sharma RK (2010) Halting the groundwater decline in north-West India-which crop technologies will be winners? Adv Agron 109:156–199

Hurisso TT, Norton JB, Norton U (2013) Soil profile carbon and nitrogen in prairie, perennial grass-legume mixture and wheat-fallow production in the central High Plains, USA. Agric Ecosyst Environ 181:179–187

Ingram JSI, Fernades ECM (2001) Managing carbon sequestration in soils: concepts and terminology. Agric Ecosyst Environ 87:111–117

IPCC (2007) IPCC, mitigation of climate change: contribution of Working Group III to the fourth assessment report of the intergovernmental panel on climate change

IPCC (2014) Climate change: synthesis report. In: Pachauri RK, Meyer LA (eds) Contribution of working groups I, II and III to the fifth assessment report of the intergovernmental panel on climate change Core writing team. IPCC, Geneva, p 151

Jalota SK, Arora VK (2002) Model-based assessment of water balance components under different cropping systems in north-west India. Agric Water Manag 57:75–87. https://doi.org/10.1016/S0378-3774(02)00049-5

Jalota SK, Singh KB, Chahal GBS, Gupta RK, Chakraborty S, Sood A, Ray SS, Panigrahy S (2009) Integrated effect of transplanting date, cultivar and irrigation on yield, water saving and water productivity of rice (Oryza sativa L.) in Indian Punjab: field and simulation study. Agric Water Manag 96:1096–1104

Janaiah A, Hossain M (2003) Farm-level sustainability of intensive rice–wheat system: socio-economic and policy perspectives. Addressing Resource Conservation Issues in Rice–Wheat Systems of South Asia, A Resource Book, Rice Wheat Consortium for Indo Gangetic Plains (CIMMYT)

Jiang JY, Hu ZH, Sun WJ, Huang Y (2010) Nitrous oxide emissions from Chinese cropland fertilized with a range of slow-release nitrogen compounds. Agric Ecosyst Environ 135:216–225

Jiang Z, Zhong Y, Yang J, Wu Y, Li H, Zheng L (2019) Effect of nitrogen fertilizer rates on carbon footprint and ecosystem service of carbon sequestration in rice production. Sci Total Environ 670:210–217

Jitareanu G, Ailincai C, Ailincai D, Raus L (2009) Impact of different tillage systems and organo-mineral fertilization on soil physical and chemical characteristics in the Moldavian plain. Cercetari Agron Moldovo 47:41–54

Jung WK, Kitchen NR, Sudduth KA, Kremer RJ (2008) Contrasting grain crop and grassland management effects on soil quality properties for a north-Central Missouri clay pan soil landscape. Soil Sci Plant Nutr 54:960–971

Kaisi MM, Grote JB (2007) Cropping system effects on improving soil carbon stocks of exposed subsoil. Soil Sci Soc Am J 71:1381–1388

Kaur T, Brar BS, Dhillon NS (2008) Soil organic matter dynamics as affected by long-term use of organic and inorganic fertilizers under maize-wheat cropping system. Nutr Cycl Agroecosyst 81:59–69

Kelner DJ, Vessey JK, Entz MH (1997) The nitrogen dynamics of 1-, 2-, and 3-year stands of alfalfa in a cropping system. Agric Ecosyst Environ 64:1–10

Khalil MI, Schmidhalter U, Gutser R, Heuwinkel H (2011) Comparative efficiency of urea fertilization via supergranules versus prills on nitrogen distribution, yield response and nitrogen use efficiency of spring wheat. J Plant Nutr 34:779–797

Koul DN, Panwar P (2008) Prioritization land-management options for carbon sequestration potential. Curr Sci 95:658–663

Kroodsma DA, Field CB (2006) Carbon sequestration in California agriculture, 1980-2000. Ecol Appl 16:1975–1985

Kumar BM, George SJ, Jamaludheen V, Suresh TK (1998) Comparison of biomass production, tree allometry and nutrient use efficiency of multipurpose trees grown in wood lot and silvopastoral experiments in Kerala, India. For Ecol Manag 112:145–163
Kumar R, Panday S, Panday A (2006) Plant roots and carbon sequestration. Curr Sci 91:885–890
Lal R (1999) Soil management and restoration for C sequestration to mitigate the accelerated greenhouse effect. Prog Environ Sci 1:307–326
Lal R (2003) Achieving soil carbon sequestration in the United States: a challenge to policy makers. Soil Sci 168:827–845
Lal R (2004a) Soil carbon sequestration to mitigate climate change. Geoderma 123:1–22
Lal R (2004b) Soil carbon sequestration impacts on global climate change and food security. Science 304(5677):1623–1627
Lal R (2010) Soil quality and ethics: the human dimension. In: Lal R, Stewart BA (eds) Food security and soil quality. Adv Soil Sci. Taylor & Francis (CRC Press), Boca Raton, pp 301–308
Lal R, Kimble J, Follet R (1997) Soil quality management for carbon sequestration. P. 1-8. In: Lal R et al (eds) Soil properties and their management for carbon sequestration. USDA, NRCS, Nat. Soil Survey Centre, Lincoln NE, Lal R (1997). Residue management, conservation tillage and soil restoration for mitigating greenhouse effect by CO_2 enrichment. Soil Till Res 43:81-107
Lal R, Kimble J, Follett RF (1998) Land use and terrestrial C pools in terrestrial ecosystem. In: Lal R et al (eds) Management of carbon sequestration in soil. CRC Press, Boca Raton, FL, pp 1–10
Larson WE, Clapp CE, Pierre WH, Morachan YB (1972) Effects of increasing amounts of organic residues on continuous corn: II. Organic carbon, nitrogen, phosphorus, and sulfur. Agron J 64: 204–208
Lefèvre R, Barré P, Moyano FE, Christensen G, Bardoux T, Eglin C, Girardin S, Houot T, Kätterer F, van Oort CC (2014) Higher temperature sensitivity for stable than for labile soil organic carbon—evidence from incubations of long-term bare fallow soils. Glob Chang Biol 20: 633–640. https://doi.org/10.1111/gcb.12402
Lin S, Iqbal J, Hu RG, Feng ML (2010) N_2O emissions from different land uses in mid-subtropical China. Agric Ecosysyst Environ 136:40–48
Liu M, David A, Ussiri N, Lal R (2016) Soil organic carbon and nitrogen fractions under different land uses and tillage practices. Commun Soil Sci Plant Anal 47(12):1528–1541
Liu YT, Li YE, Wan YF, Chen DL, Gao QZ, Li Y, Qin XB (2011) Nitrous oxide emissions from irrigated and fertilized spring maize in semi-arid northern China. Agric Ecosyst Environ 141: 287–295
Lopez-Garrido R, Diaz-Espejo A, Madejon E, Murillo JM, Moreno F (2009) Carbon losses by tillage under semi-arid Mediterranean rainfed agriculture (SW Spain). Spanish J Agric Res 7: 706–716
Ma BL, Wu TY, Tremblay N, Deen W, Morrison MJ, Mclauglin NB, Gregorich EG, Stewart G (2010) Nitrous oxide fluxes from corn fields. Glob Chang Biol 16:156–170
Majumder B, Mandal B, Gangopadhyay A, Mani PK, Kundu AL, Mazumdar D (2008) Organic amendments influence soil organic carbon pools and rice-wheat production. Soil Sci Soc Am J 72:775–785
Makumba W, Akinnifesi FK, Janssen B, Onema O (2007) Long-term impact of a gliricidia-maize intercropping system on carbon sequestration in southern Malawi. Agric Ecosyst Environ 118: 237–243
Makumba W, Janssen B, Oenema O, Akinnifesi FK, Mweta D, Kwesiga F (2006) The long-term effects of a gliricidia-maize intercropping system in southern Malavi, on gliricidia and maize yield, and soil properties. Agric Ecosyst Environ 116:85–92
Malhi Y, Baker TR, Phillips OL, Almeida S, Alvarez E, Arroyo L, Chave J, Zimczik CI, Di Fiore A, Higuchi N, Killeen TJ, Laurance SG, Laurance WF, Lewis SL, Montoya LMM, Monteagudo A, Neill DA, Vargas PN, Patino S (2004) The above-ground coarse wood productivity of 104 neo-tropical forest plots. Glob Chang Biol 10:563–591

Mandal B, Majumder B, Bandyopadhyay PK (2007) The potential of cropping systems and soil amendments for carbon sequestration in soils under long-term experiments in subtropical India. Glob Chang Biol 13:357–369

Minamikawa K, Sakai N (2007) Soil carbon budget in a single-cropping paddy field with rice straw application and water management based on soil redox potential. Soil Sci Plant Nutr 53:657–667

Minoshima H, Jackson LE, Cavagnaro TR, Ferris H (2007) Short-term fates of carbon-13-depleted cowpea shoots in no-till and standard tillage soils. Soil Sci Soc Am J 71:1859–1866

MOA (2012) (Ministry of Agriculture) Govt. of India, New Delhi. www.eands.dacnet.nic.in

Mu Z, Kimura D, Hatano R (2006) Estimation of global potential from upland cropping systems in Central Hokkaido, Japan. Soil Sci Plant Nutr 52:371–377

Naitam R, Bhattacharyya T (2004) Quasi-equilibrium of organic carbon in swell-shrink soils of the sub-humid tropics in India under forest, horticulture and agriculture systems. Australian J Soil Res 42:181–188

Negash M, Kanninen M (2015) Modeling biomass and soil carbon sequestration of indigenous agroforestry systems using CO_2FIX approach. Agric Ecosyst Environ:147–155. https://doi.org/10.1016/j.agee.2015.02.004

NETL (2010) Best practices for: Terrestrial Sequestration of Carbon Dioxide. pp:1–82

Neubauer SC, Megonigal JP (2015) Moving beyond global warming potentials to quantify the climatic role of ecosystems. Ecosystems 18:1000–1013. https://doi.org/10.1007/s10021-015-9879-4

Oades JM (1984) Soil organic matter and structural stability: mechanisms and implications for management. Plant Soil 49:37–62

Okonkwo CL, Mbagwu JSC, Egwu SO (2009) Changes in soil properties under allay cropping system of three leguminous crops. J Trop Agric Food Environ Ext 8:60–65

Paccard CG, Chiquinquir H, Ignacio MS, Pérez J, León P, González P, Espejo R (2015) Soil– water relationships in the upper soil layer in a Mediterranean Palexerult as affected by no-tillage under excess water conditions – Influence on crop yield. Soil Till Res 146:303–312

Pandey D, Agrawal M (2014) Greenhouse gas fluxes from sugarcane and pigeon pea cultivated soils. Agric Res 4(3):245–253

Pandey D, Agrawal M, Pandey JS (2010) Carbon footprint: current methods of estimation. Environ Monitor Assess 178(1–4):135–160

Patra PK, Saha N, Mukherjee R, Chakraborty A, Sarkar S, Mukherjee D (2010) Influenec of tillage techniques and organic matter on carbon and nitrogen transformation in the rice rhizosphere in an alluvial soils of West Bengal. Appl Ecol Environ Res 8:313–327

PAU (2020) The package of practices for the crops of Punjab kharif 2020. Half yearly Package published by Punjab Agricultural University, Ludhiana, Punjab, India

Paustian K, Andren O, Janzen HH, Lal R, Smith P, Tian G, Tiessen H, Van Noordwijk M, Woomer PL (1997a) Agricultural soils as a sink to mitigate CO_2 emission. Soil Use Manag 13:230–244

Paustian K, Collins HP, Paul EA (1997b) Management controls on soil carbon. In: Paul EA, Paustian K, Elliot ET, Cole CV (eds) Soil organic matter in temperate agro-ecosystems: long-term experiments in North America. CRC Press, Boca Raton, Florida, pp 15–49

Paustian K, Six J, Elliott ET, Hunt HW (2000) Management option for reducing CO_2 emissions from agricultural soils. Biochemist 48:147–163

Poeplau C, Don A (2015) C sequestration in agricultural soils via cultivation of cover crops—a meta-analysis. Agric Ecosyst Environ 200:33–41. https://doi.org/10.1016/j.agee.2014.10.024

Ponnamperuma FN (1984) Straw as a source of nutrients for wetland rice. In: Organic matter and Rice. International Rice Research Institute, Los Baños, pp 117–136

Porpavai S, Devasenapathy P, Siddeswaran K, Jayaraj T (2011) Impact of various rice based cropping systems on soil fertility. J Cereals Oilseeds 2(3):43–46

Pratt K, Moran D (2010) Evaluating the cost-effectiveness of global biochar mitigation potential. Biomass Bioenergy 34:1149–1158

Prentice IC, Farquhar GD, Fasham MJR (2001) The carbon cycle and atmospheric carbon dioxide. In: Houghton JT, Ding Y, Griggs DJ et al (eds) Climate change 2001: the scientific basis. Cambridge University Press, New York, pp 183–237

Raich JW, Potter CS (1995) Global patterns of carbon dioxide emission form soils. Global Biochem Cycl 9:23–36

Raji BA, Ogunwole JO (2006) Potential of soil carbon sequestration under various landuse in the sub-humid and semi-arid Savana of Nigeria: lessons from long-term experiments. Int J Soil Sci 1:33–43

Rasool R, Kukal SS, Hira GS (2007) Soil physical fertility and crop performance as affected by long term application of FYM and inorganic fertilizers in rice–wheat system. Soil Till Res 96:64–72

Raun WR, Johnson GV (1999) Improving nitrogen use efficiency for cereal production. Agron J 91: 357–363. https://doi.org/10.2134/agronj1999.00021962009100030001x

Russell AE, Laird DA, Parkin TB, Mallarino AP (2005) Impact of nitrogen fertilization and cropping system on carbon sequestration in Midwestern mollisols. Soil Sci Soc Am J 69:413–422

Saini J, Bhatt R (2020) Global warming -causes, impacts and mitigation strategies in agriculture. Curr J Appl Sci Technol 39(7):93–107. https://doi.org/10.9734/CJAST/2020/v39i730580

Sainju UM, Jabro JD, Caesar-TonThat T (2010) Tillage, cropping sequence, and nitrogen fertilization effects on dryland soil carbon dioxide emission and carbon content. J Environ Qual 39: 935–945. https://doi.org/10.2134/jeq2009.0223

Sainju UM, Lenssen AW (2011) Dryland soil carbon dynamics under alfalfa and durum-forage cropping sequences. Soil Till Res 113:30–37. https://doi.org/10.1016/j.still.2011.02.002

Sainju UM, Senwo ZN, Nyakatawa EZ, Tazisong IA, Reddy KC (2008) Tillage, cropping system, and nitrogen fertilizer source effect on soil carbon sequestration and fractions. J Environ Qual 37:880–888

Sanderman J, Baldock JA (2010) Accounting for soil carbon sequestration in national inventories: a soil scientist's perspective. Environ Res Lett 5(3):1–7

Sarkar A, Yadav RL, Gangwar B, Bhatia PC (1999) Crop residues in India. Technical Bulletin. Directorate of Cropping System Research, Modipuram, India

Scharlemann JPW, Tanner EVJ, Hiederer R, Kapos V (2014) Global soil carbon: understanding and managing the largest terrestrial carbon pool. Carbon Manage 5:81–91

Schlesinger WH (1997) Biogeochemistry: an analysis of global change, 2nd edn. Academic Press Inc., San Diego, CA

Sharma S, Singh P, Angmo P, Dhaliwal SS (2021a) Micro-nutrient pools and their mobility in relation to land-use in a cold high altitude Himalayan mountainous region. Agroforestry Syst (in press)

Sharma S, Singh P, Choudhary OP, Neemisha (2021b) Nitrogen and rice straw incorporation impact nitrogen use efficiency, soil nitrogen pools and enzyme activity in rice-wheat system in North-Western India. Field Crops Res 266:108131

Sharma S, Singh P, Kumar S (2020a) Responses of soil carbon pools, enzymatic activity and crop yields to nitrogen and straw incorporation in a rice-wheat cropping system in North-Western India. Front Sustain Food Syst, Sec Climate-Smart Food Syst. https://doi.org/10.3389/fsufs.2020.532704

Sharma S, Singh P, Sodhi GPS (2020b) Soil organic carbon and biological indicators of uncultivated vis-à-vis intensively cultivated soils under rice–wheat and cotton–wheat cropping systems in South-western Punjab. Carbon Manage. https://doi.org/10.1080/17583004.2020.1840891

Shaviv A, Mikkelsen RL (1993) Slow release fertilizers for a safer environment maintaining high agronomic efficiency. Fertilizer Res 35:1–12

Shoji S, Delgado JA, Mosier A, Miura Y (2001) Use of controlled release fertilizers and nitrification inhibitors to increase nitrogen use efficiency and to conserve air and water quality. Commun Soil Sci Plant Anal 32(7):1051–1070. https://doi.org/10.1081/CSS-100104103

Sidhu HS, Singh M, Blackwell J, Humphreys E, Bector V, Singh Y, Singh M, Singh S (2008) Development of the Happy Seeder for direct drilling into combine harvested rice. In: Humphreys E, Roth CH (eds) Permanent Beds and Rice-residue Management for Rice–Wheat Systems in the Indo-Gangetic Plain. ACIAR Proc 159–70, http://www.aciar.gov.au/publication/term/18

Sidhu HS, Singh M, Humphreys E, Singh Y, Singh B, Dhillon SS, Blackwell J, Bector V, Singh M, Singh S (2007) The happy Seeder enables direct drilling of wheat into rice stubble. Aust J Exp Agric 47:844–854

Singh G, Singh P, Sodhi GPS, Tiwari D (2020b) Adoption status of rice residue management technologies in South-Western Punjab. Indian J Ext Edu 56(3):76–82

Singh H, Singh P (2017) Influence of land use on organic carbon and available nutrient status. Indian J Fertil 13:70–75

Singh P, Benbi DK (2016) Effect of inorganic fertilizers and farm yard manure on physical properties of soil under rice-wheat cropping. Agric Res J (PAU) 53(3):328–333

Singh P, Benbi DK (2018a) Nutrient management effects on organic carbon pools in a sandy loam soil under rice-wheat cropping. Arch Agron Soil Sci 64(13):1879–1891

Singh P, Benbi DK (2018b) Soil organic carbon pool changes in relation to slope position and land-use in Indian lower Himalayas. Catena 166:171–180

Singh P, Benbi DK (2020a) Modeling soil organic carbon with DNDC and RothC models in different wheat-based cropping systems in North-Western India. Commun Soil Sci Plant Anal 51(9):1184–1203

Singh P, Benbi DK (2020b) Nutrient management impacts on net ecosystem carbon budget and energy flow nexus in intensively cultivated cropland ecosystems of North-Western India. Paddy Water Environ. (in press). https://doi.org/10.1007/s10333-020-00812-9

Singh P, Benbi DK (2021) Nutrient management effects on carbon input through root and shoot biomass in a rice-wheat system. Agric Res J (PAU). (in press)

Singh P, Benbi DK, Verma G (2021a) Nutrient management impacts on nutrient use efficiency and energy, carbon, and net ecosystem economic budget of rice-wheat cropping system in North-Western India. J Soil Sci Plant Nutr. https://doi.org/10.1007/s42729-020-00383-y

Singh P, Saini SP (2011) Effect of rice straw mulching and irrigation intervals on sugarcane (*Saccharum officinarum*) yield and water productivity in sub-tropics of Punjab. Crop Res (An International J) 41:88–93

Singh P, Singh G, Sodhi GPS (2019a) Energy auditing and optimization approach for improving energy efficiency of rice cultivation in South-Western Punjab. Energy 174:169–179

Singh P, Singh G, Sodhi GPS (2019b) Applying DEA optimization approach for energy auditing in wheat cultivation under rice-wheat and cotton-wheat cropping systems in North-Western India. Energy 181:18–28

Singh P, Singh G, Sodhi GPS (2020a) Energy and carbon footprints of wheat establishment following different rice residue management strategies vis-à-vis conventional tillage coupled with rice residue burning in North-Western India. Energy 200:117554

Singh P, Singh G, Sodhi GPS (2020c) On-farm participatory assessment of short and medium duration rice genotypes in South-Western Punjab. Indian J Ext Edu 56(3):88–94

Singh P, Singh G, Sodhi GPS, Sharma S (2021b) Energy optimization in wheat establishment following rice residue management with Happy Seeder technology for reduced carbn footprints in North-Western India. Energy:120680

Singh Y, Singh B (2001) Efficient management of primary nutrients in the rice-wheat system. In: Katoke PK (ed) Rice-wheat cropping system of South Asia: efficient production management. Food Products Press, Binghamton

Six J, Feller C, Denef K, Ogle SM, de Moraes JC, Albrecht A (2002) Soil organic matter, biota and aggregation in temperate and tropical soils-effect of no-tillage. Agronomie 22:755–775

Six J, Ogle SM, Breidt EJ, Conant RT, Mosier AR, Paustian K (2004) The potential to mitigate global warming with no-tillage management is only realized when practice in the long term. Glob Chang Biol 10:155–160

Smith P, Martino D, Cai Z, Gwary D, Janzen H, Kumar P, McCarl B, Ogle S, O'Mara F, Rice C, Scholes B, Sirotenko O, Howden M, McAllister T, Pan G, Romanenkov V, Schneider U, Towprayoon S, Wattenbach M, Smith J (2008) Greenhouse gas mitigation in agriculture. Philos Trans R Soc London, Series B 363:789–813

Sohi S, Eliza L, Evelyn K, Roland B (2009) Bio-char, climate change and soil: a review to guide future research. CSIRO Land Water Sci Rep 05(09):1834–6618

Sohi SP, Krull E, Lopez-Capel E, Bol R (2010) A review of biochar and its use and function in soil. Adv Agron:47–82. https://doi.org/10.1016/s0065-2113(10)05002-9

Srinivasarao C, Deshpande AN, Venkateswarlu B, Lal R, Singh AK, Kundu S, Vittal KPR, Mishra PK, Prasad JVNS, Mandal UK, Sharma KL (2011) Grain yield and carbon sequestration potential of post monsoon sorghum cultivation in Vertisols in the semi arid tropics of Central India. Geoderma 175–176:90–97

Srinivasarao C, Vankateswarlu B, Lal R, Singh AK, Sumanta K (2013) Sustainable management of soils of dryland ecosystems for enhancing agronomic productivity and sequestering carbon. Adv Agron 121:253–325

Sundermeier A, Reeder R, Lal R (2008) Soil carbon sequestration fundamentals. Rep. No. OSU Factsheet AEX-510–05. OSUE, Columbus, OH. http://ohioline.osu.edu/aex-fact/0510.html

Swami SL, Puri S, Singh AK (2003) Growth, biomass, carbon storage and nutrient distribution in Gmelina arboreaRoxb. Stands on red lateritic soils in Central India. Bioresour Technol 90:109–126

Swarup A (1998) Emerging soil fertility management issues for sustainable crop productivity in irrigated systems. In: Swarup A, Reddy DD, Prasad RN (eds) Long-term soil fertility management through integrated plant nutrient supply. Indian Inst Soil Sci, Bhopal, pp 54–68

Swarup A, Manna MC, Singh GB (2000) Impact of land use and management practices on organic carbon dynamics in soils of India. In: Lal R, Kimble JM, Stewart BA (eds) Global climate change and tropical ecosystems. CRC/Lewis Publishers, Boca Raton, FL, pp 261–281

Taylor JP, Wilson B, Mills MS, Burns RG (2002) Comparison of microbial numbers and enzymatic activities in surface soils and sub-soils using various techniques. Soil Biol Biochem 34:387–401

Tripathi RS, Raju R, Thimmappa K (2013) Impact of zero tillage on economics of wheat production in Haryana. Agric Econ Res Rev 26(1):101–108

Velayutham M, Pal DK, Bhattacharyya T (2000) Organic carbon stock in soils of India. In: Lal R, Kimble JM, Stewart BA (eds) Global climate change and tropical ecosystems. CRC/Lewis Publishers, Boca Raton, FL, pp 71–97

Venkatesh MS, Hazra KK, Ghosh PK, Praharaj CS, Kumar N (2013) Long-term effect of pulses and nutrient management on soil carbon sequestration in Indo-Gangetic plains of India. Canadian J Soil Sci 93(1):127–136

Verma KS, Kumar S, Bhardwaj DR (2008) Soil organic carbon stocks and carbon sequestration potential of agroforestry systems in H.P. Himalaya region of India. J Tree Sci 27(1):14–27

Wang Q, Li Y, Alva A (2010a) Cropping system to improve carbon sequestration for mitigation of climate change. J Environ Prot 1:207–215

Wang Q, Li Y, Alva A (2010b) Growing cover crops to improve biomass accumulation and carbon sequestration: a phytotron study. J Environ Prot 1:73–84

Wang Q, Zhang L, Li L, Bai Y, Cao J, Han X (2009) Changes in carbon and nitrogen of Chernozem soil along a cultivation chronosequence in a semi-arid grassland. Eur J Soil Sci 60:916–923. https://doi.org/10.1111/j.1365-2389.2009.01174.x

West T, Marland G (2003) Net carbon flux from agriculture: carbon emissions, carbon sequestration, crop yield, and land-use change. Biogeochem 63(1):73–83

West TO, Post WM (2002) Soil organic carbon sequestration rates by tillage and crop rotation; a global data analysis. Soil Sci Soc Am J 66:1930–1946

Wilson HM, Kaisi MM (2008) Crop rotation and nitrogen fertilization effect on soil CO_2 emission in central lowa. Appl Soil Ecol 39:264–270

Wright AL, Dou F, Hons FM (2007) Crop species and tillage effect on soil carbon sequestration in subsurface soils. Soil Sci 172:124–131

Wright AL, Hons FM (2005) Tillage impacts on soil aggregates and carbon and nitrogen sequestration under wheat cropping sequence. Soil Till Res 84:67–75

Yadav GS, Lal R, Meena RS, Babu S, Das A, Bhomik SN, Datta M, Layak J, Saha P (2017) Conservation tillage and nutrient management effects on productivity and soil carbon sequestration under double cropping of rice in North Eastern region of India. Ecol Indic 105:303–315

Yadvinder-Singh B-S, Ladha JK, Khind CS, Gupta RK, Meelu OP, Pasuquin (2004) Long-term effects of organic inputs on yield and soil fertility in the rice-wheat rotation. Soil Sci Soc Am J 68:845–853

Yadvinder-Singh, Bijay-Singh, Timsina J (2005) Crop residue management for nutrient cycling and improving soil productivity in rice-based cropping systems in the tropics. Adv Agron 85:269–407

Yadvinder-Singh, Sidhu HS, Singh M, Humphreys E, Kukal SS, Brar NK (2008) Straw mulch, irrigation water and fertilizer N management effects on yield, water use and N use efficiency of wheat sown after rice. In: Humphreys E, Roth CH (eds) Permanent beds and rice-residue management for rice–wheat systems in the Indo-Gangetic Plain. Proceedings of a workshop held at PAU, Ludhiana, India from 7–9th September 2006. ACIAR Proceedings No. 127, pp. 171– 81. http://www.aciar.gov.au/publication/term/18

Zech W, Senesi N, Guggenberger G, Kaiser K, Lehmann J, Miano TM, Miltner A, Schroth G (1997) Factors controlling humification and mineralization of soil organic matter in the tropics. Geoderma 79:117–161. https://doi.org/10.1016/S0016-7061(97)00040-2

Zentner RP, Basnyat P, Brandt SA, Thomas AG, Ulrich D, Campbell CA, Nagy CN, Frick B, Lemke R, Malhi SS, Olfert O, Fernandez MR (2011) Effects of input management and crop diversity on economic returns and riskiness of cropping systems in the semi-arid Canadian prairie. Renewable Agric Food Syst 26:208–223

Zhang D, Shen J, Zhang F, Yu'e L, Zhang W (2017) Carbon footprint of grain production in China. Sci Rep 7(1):4126. https://doi.org/10.1038/s41598-017-04182-x

Zhangliu D, Ren T, Huc C, Zhang Q (2015) Transition from intensive tillage to no-till enhances carbon sequestration in microaggregates of surface soil in the North China plain. Soil Till Res 146:26–31

Saline Toxicity and Antioxidant Response in *Oryza sativa*: An Updated Review

Mahmuda Binte Monsur, Juel Datta, Md. Motiar Rohman, Mirza Hasanuzzaman, Akbar Hossain, Mohammad Sohidul Islam, Maham Asif Bukhari, Tasmiya Jabeen, Muhammad Mubeen, Wajid Nasim, Abdul Rehman, Disna Ratnasekera, Analía Llanes, Habib-ur-Rehman Athar, Muhammad Aamir Iqbal, Sharif Ahmed, Rafaqat Ali Gill, Adel Hadifa, Akihiro Ueda, Liu Liyun, Milan Skalicky, Marian Brestic, Murat Erman, Yüksel Kaya, and Ayman E. L. Sabagh

M. B. Monsur
Department of Genetics and Plant Breeding, Bangabandhu Sheikh Mujibur Rahman Agricultural University, Gazipur, Bangladesh

J. Datta
State Key Laboratory of Rice Biology, China National Rice Research Institute, Hangzhou, China

M. M. Rohman
Molecular Breeding Lab, Plant Breeding Division, Bangladesh Agricultural Research Institute, Gazipur, Bangladesh

M. Hasanuzzaman
Department of Agronomy, Sher-e-Bangla Agricultural University, Dhaka, Bangladesh

A. Hossain
Bangladesh Wheat and Maize Research Institute, Dinajpur, Bangladesh

M. S. Islam · T. Jabeen
Department of Agronomy, Hajee Mohammad Danesh Science and Technology University, Dinajpur, Bangladesh

M. A. Bukhari · M. Mubeen
Department of Environmental Sciences, COMSATS University Islamabad, Vehari, Pakistan

W. Nasim
Department of Agronomy, University College of Agriculture and Environmental Sciences, The Islamia University of Bahawalpur (IUB), Bahawalpur, Pakistan

A. Rehman
Department of Crop Sciences and Biotechnology, Dankook University, Cheonan-si, South Korea

D. Ratnasekera
Department of Agric. Biology, Faculty of Agriculture, University of Ruhuna, Matara, Sri Lanka

A. Llanes
Plant Physiology Laboratory, Department of Natural Sciences, FCEFQyN, Universidad Nacional de Río Cuarto, Río Cuarto, Argentina

M. Hasanuzzaman et al. (eds.), *Managing Plant Production Under Changing Environment*, https://doi.org/10.1007/978-981-16-5059-8_4

Abstract Abiotic stresses such as drought, waterlogging, extreme temperatures, salinity, and mineral toxicity negatively impact the growth and development, yield, and seed quality of crop plants. Presently, abiotic stresses are severely affecting crop yields, resulting in higher economic losses for the farmers. One of the severe consequences of abiotic stresses is the overproduction of reactive oxygen species (ROS), which results in oxidative stress. However, plants possess antioxidative defense machinery to protect against oxidative stress. The underlying mechanisms of antioxidant defense in rice plants have been published in many papers in recent decades. In this review, we aim at summarizing the updated information on physiological interventions in making rice plants more tolerant to salt-induced oxidative stress. We also focused on the understanding of the physiological mechanisms in rice under salinity stress that could facilitate the development of salt-tolerant cultivars. This review aims to know the activity of antioxidant enzymes and other proteins response in saline-stress conditions, and which proteomics and molecular markers were used to assess oxidative stress and antioxidants to discriminate cultivars, genotypes, or species for salt tolerance were also included.

Keywords Antioxidants · Lipid peroxidation · Proline · Rice · Salt tolerance

H.-R. Athar
Institute of Pure and Applied Biology, Bahauddin Zakariya University, Multan, Pakistan

M. A. Iqbal
Department of Agronomy, Faculty of Agriculture, University of Poonch Rawalakot, Rawalakot, Pakistan

S. Ahmed
International Rice Research Institute, Bangladesh office, Dhaka, Bangladesh

R. A. Gill
Oil Crops Research Institute, Chinese Academy of Agriculture Sciences, Wuhan, China

A. Hadifa
Rice Research and Training Center (RRTC), Field Crops Research Institute, Agricultural Research Center, Sakha, Kafr Elsheikh, Egypt

A. Ueda · L. Liyun
Graduate School of Integrated Sciences for Life, Hiroshima University, Higashi-Hiroshima, Japan

M. Skalicky · M. Brestic
Department of Botany and Plant Physiology, Faculty of Agrobiology, Food and Natural resources, Czech University of Life Sciences Prague, Prague, Czechia

M. Erman · Y. Kaya
Department of Field Crops, Faculty of Agriculture, Siirt University, Siirt, Turkey

A. E. L. Sabagh (✉)
Department of Field Crops, Faculty of Agriculture, Siirt University, Siirt, Turkey

Department of Agronomy, Faculty of Agriculture, Kafrelsheikh University, Kafr Elsheikh, Egypt
e-mail: ayman.elsabagh@agr.kfs.edu.eg

1 Introduction

Among abiotic stresses, salinity has emerged as the most challenging constraint to the sustainability of modern farming systems worldwide (Yassin et al. 2019; EL Sabagh et al. 2019a, b). It adversely affects numerous physiological processes of plants such as ionic imbalance leading to toxicity, nutrition-related disorders, and metabolic processes disruption leading to oxidative stress (Hossain et al. 2020; EL Sabagh et al. 2020). As a result, plant growth is negatively affected mainly owing to the disorganization of membranes and a significant reduction in the cell division as well as the expansion of the cells (Egrão et al. 2017). However, at the cellular level, plants have evolved a coherent antioxidative defense mechanism, which minimizes the toxicity caused by salinity stress (Monsur et al. 2020; Liu et al. 2020; Hakim et al. 2021). Improving plant antioxidant defense systems through exogenous application of different hormones, trace elements, osmoprotectants, signaling molecules, and polyamines have the potential to bolster the plant's tolerance against abiotic stresses (Rohman et al. 2016). Globally, rice (*Oryza sativa* L.) constitutes the leading cereal crop with a variety of agricultural uses and diversified economic significance (Iqbal et al. 2015; Ahmed et al. 2021; Gaballah et al. 2021), as it continues to remain the staple food of more than 50% human populace (Ali et al. 2015; Iqbal 2018). However, rice has been found to be highly susceptible to even moderate levels of salinity, which drastically hampers the growth and productivity of 5.5 M ha of rice-cultivated area in saline, alkaline, or saline-sodic. Around 48 M ha of the agricultural area is uncultivated in South Asia and Southeast Asia due to salinity (Hairmansis et al. 2017), especially lowland rice (Kibria et al. 2017). For instance, it is estimated that around one-fifth of the total area under rice cultivation is salt affected (Qadir et al. 2014).

It has been reported that changes in various physiological and metabolic processes owing to osmotic stress reduced the crop yield under salinity conditions (Rahnama et al. 2010; James et al. 2011). Indeed, salinity affects several physiological pathways, including photosynthesis, respiration, nitrogen fixation, and carbohydrate metabolism (Kordrostami et al. 2017; Polash et al. 2019). It leads to the reactive oxygen species (ROS), which is the main source of damaging cells during the stresses, such as superoxide radical ($O_2^{\bullet-}$), hydrogen peroxide (H_2O_2), and hydroxyl radicals ($OH^{\bullet}$). Highly cytotoxic characteristics of these ROS enable them to react with numerous biomolecules, including proteins, nucleic acid, lipids, and so on and result in peroxidation of lipids, denaturation of proteins along with severe damage to DNA (Dhanyalakshmi et al. 2013). In the most severe cases, ROS can produce DNA damages, causing severe breaks in double-strand DNA as well as interstrand cross-links (Manova and Gruszka 2015). It deserves mentioning that ROS might be produced by plants due to electron leakage between the electron transport chains of the photosynthetic and respiratory electron with oxygen (Hossain and Dietz 2016). Another important source of ROS has been found to be photorespiration, which resulted from *Rubisco oxygenase* activity. Thus, the peroxisomes have been established to be the important sites of producing various ROS through oxidation pathways of fatty acids and photorespiration (Corpas 2015). The photorespiration rate is mainly governed by prevailing temperature and CO_2–O_2 ratio. Several experiments have been conducted regarding toxicity management in cellular systems. However, abiotic stress tolerance mechanisms are not clear which necessitate increasing our understanding regarding numerous physiological and biochemical characteristics along with molecular characteristics of plants leading to bolstering the strives for enhancing salinity tolerance (Gill and Tuteja 2010; Karuppanapandian et al. 2011; Foyer and Noctor 2011) along with non-enzymatic components such as ascorbate (ASC) and glutathione (GSH) (Gill and Tuteja 2010). Superoxide dismutase (SOD) and ascorbate peroxidase (APX) also play an important role in ROS detoxification in cells. APX reduces H_2O_2 as a specific electron donor. Nonetheless, several rice cultivars show increasing activity of scavenging enzymes such as catalase (CAT), which enhances the levels of different antioxidants, including ASC and GSH, under saline conditions. Salinity showed different effects on the components of antioxidant defense systems in plants (El-Shabrawi et al. 2010; Hasanuzzaman et al. 2014; Kibria et al. 2017). Besides, Vighi et al. (2017) reported that rice plants exposed to NaCl (150 mM) for 3 days showed a significant increment in the activity of SOD and CAT. Similarly, rice plants under NaCl (up to 100 mM) treatments increased the activity of SOD, APX, CAT, and glutathione reductase (GR) in both salt-tolerance and salt-sensitive cultivars of rice (Chawla et al. 2013). Kim et al. (2014) reported that the enzyme maintains the ascorbic acid contents and ROS levels. However, most of the studies describe the total activities of a few enzymatic antioxidants, while assessment of isoenzyme patterns alterations in contrasting genotypes remains unexplored. It is of the utmost importance for adjusting plant cells to change and adapt to cellular redox environments. Besides, a knowledge gap exists on the ROS role in signaling, which gets mediated by NADPH oxidases (NOX), especially under salt stress. Besides the toxic impact of ROS, it has been widely recognized that ROS might serve as signals provided at low concentrations. These sorts of signals are generated by ROS-producing systems (NOX), while the site of their location is the plasma membrane (Mittler et al. 2011; Uzilday et al. 2017). These are also considered as the prime sources of endogenous $O_2^{\bullet-}$.

Exogenous application of protectants in rice plants under saline conditions remains effective to ameliorate the damages caused by salinity. The underlying mechanisms of protectants include regulation of enzyme activities, which has recently got more attention from the researcher to conduct further exploring studies for rice. The defense mechanism and pathways of signal transduction remain unclear, necessitating executing further studies. Furthermore, application rate, time, and techniques of exogenous protectants in plants under salt stress need to be studied in depth for the exploration of the underlying protection mechanisms. Molecular markers have been established and developed as reliable tools to evaluate the genetic variations and to elucidate the interspecies and intraspecies relationships (Tehrim et al. 2012). There are numerous molecular techniques for analyzing the genetic variations among plant species. It has been declared as an effective method for determining genetic variations (Rabbani et al. 2008; Pervaiz et al. 2009).

In this chapter, we identify the knowledge and research gaps of physiological mechanisms adopted by rice under saline conditions. The objective of the chapter is (1) to study the oxidative damage between tolerance and susceptible rice genotypes under saline conditions, (2) to understand the possible protective role of antioxidative and glyoxalase systems in rice under salinity, and (3) to identify stress-inducible proteins.

2 Salt-Stress Tolerance Mechanism

Many advances have been made through studies on various rice cultivars under salinity stress conditions. Lee et al. (2013) stated the ROS-scavenging mechanisms in roots and leaves of two rice cultivars (a salt-sensitive IR 29 and a salt-resistant Pokkali) under salinity stress. Pokkali cultivar, which was not treated with salt, presented higher H_2O_2 scavenging enzyme actions in the seedlings such as APX and CAT activities. Nevertheless, such enzymatic activities were significantly observed in IR29 due to salt stress. While the H_2O_2 level was lesser in Pokkali when compared to IR29, the reverse was observed under the salt treatment. However, the diminished amount of H_2O_2 in IR29 under salt stress did not reveal a further scavenging action of total cell extracts for H_2O_2, O_2^-, and OH^- species.

Physiological responses and the proteomics variations due to salt stress in the leaves of rice cultivars were examined by Lee et al. (2010). Leaf water content and shoot growth were more decreased in the cultivar Dalseongaengmi-44 (a salt-sensitive) when compared to Dongjin (a salt-tolerant) under salt-stress conditions. The cultivar Dalseongaengmi-44 showed a more increment in the accumulation of Na^+ in leaves than Dongjincultivar (showing a better salt-tolerance mechanism). Likewise, Lee et al. (2010) conducted a comparative examination of proteins of leaf utilizing two-dimensional gel electrophoresis (2-DGE), and the result of this investigation uncovered that 23 proteins were upregulated in rice when plants were subjected to salinity stress. Ten of the recognized proteins were already known as salt-responsive proteins. Under saline conditions, the fragmentation process was enhanced in proteins having Rubisco large chain (Lee et al. 2010). The outcomes of their proteomics and physiological analyses gave valuable information that can improve the molecular basis of the response of rice under salt stress. On the other hand, Rahman et al. (2016) mentioned the antioxidant defense regulation, ion homeostasis, and systems of glyoxalase in seedlings of rice variety BRRI dhan47. The 12-day-old hydroponically developed rice seedlings were treated with 150 mM NaCl solution alone and in combination with 0.5 mM solution of $MnSO_4$. Salt stress caused disturbance of homeostasis of ions due to K^+ and Na^+ influx. They noticed lower water uptake and higher accumulation of Na^+ due to salt toxicity resulted in chlorosis, osmotic stress, and, finally, inhibition of growth. Salt-induced osmotic stress and toxicity of ions subsequently were a cause of oxidative stress disturbing the glyoxalase and defense systems of antioxidants by excessive production of methylglyoxal (MG) and ROS. Salt stress–induced damage was enhanced due to an increasing time period of stress. Ascorbic acid (ASA) flavonoid and by enhancing the activities of dehydroascorbate reductase (DHAR), monodehydroascorbate reductase (MDHAR), and CAT, in the seedlings having salt treatment. Manganese (Mn) supplements also strengthened MG detoxification by enhancing the glyoxalase in salt-induced seedlings (Rahman et al. 2016). So, the application of Mn provided salinity resistance by synchronized ion homeostasis, glyoxalase, and antioxidant defense systems of antioxidants in salt-stressed plants.

Generation of ROS and then the scavenging system in rice varieties under various levels of salinity stress were studied by Kaur et al. (2016). Salinity-resistant cultivars from Indian origin can bear salinity upto 6–8 dS m^{-1} and demonstrated reduced ROS showing lower levels of O_2^- and H_2O_2 than in salt-sensitive cultivars. Here, several enzymatic- and nonenzymatic antioxidants activities were recorded higher in resistant plants. Further analysis showed that the antioxidant enzyme's transcript level was according to the activities of the abovementioned enzymes. Proline (Pro) contents were more in tolerant cultivars, while the chlorophyll (Chl) and malondialdehyde (MDA) contents were increased in susceptible varieties.

Similarly, Kabir et al. (2015) illustrated the salt-stress–damaged antioxidant defense systems revealed by biochemical and physiological characteristics in tolerant and sensitive rice genotypes under saline conditions. Chl, Pro, antioxidant enzymes (peroxidase and ascorbate peroxidase, and catalase) activities, and K^+/Na^+ ratio were remarkably increased in salt-tolerant genotypes, and these may be the essential ingredients of salt-tolerance mechanisms. In another study, results showed that the overproduction of Pro protects rice plants against salinity in salt-tolerant varieties (Bhusan et al. 2016). Higher salinity level considerably reduced ASA and Chl contents, the ratio of K^+ /Na^+ and actions of antioxidant enzyme guaiacol peroxidase (POX) in both varieties, and exogenous application of Pro significantly increased the values of those traits.

Dhanyalakshmi et al. (2013) concluded that the morpho-physiological traits like the shoot length, dry weight, and photosynthetic rate (Pn) of rice as well as its biochemical traits like the activity of APX and SOD, concentration of MDA, and ratio of Na^+/K^+ were significantly reduced with increasing the salt concentration in growing media. They also noticed that a low ratio of Na^+/K^+ in combination with enhanced antioxidant enzyme activity, and reduced peroxidation of lipids may influence salt tolerance ability in rice varieties.

Expression of gene and specific antioxidative enzymes activity like tocopherol, ASA, DHA, and lipid peroxidation were accounted for in salt-sensitive (PB1) and salt-tolerant (CSR10) rice genotype under 200 mM NaCl stress (Turan and Tripathy 2012). The result revealed that the total amount of ASC and the contents of tocopherol were reduced in the seedlings grown under salt stress, and salt-resistant variety (CSR10) showed more tolerance because of its early readiness for combating oxidative stress through regulation of gene expression, antioxidative enzyme activity, and increased redox status.

The antioxidant system was also analyzed in two different rice varieties that differed in salt tolerance (El-Shabrawi et al. 2010). Under salinity stress, cultivar like Pokkali (salt-tolerant) demonstrated a significantly lower level of H_2O_2 in comparison with IR64 (salt-sensitive), and the authors illustrated such physiological behavior to more enzymatic activities in Pokkali, which had a direct or indirect effect on the detoxification of H_2O_2. They also observed higher antioxidant enzyme activity in several forms in the tissues of Pokkali when compared to IR64 cultivar, which suggested that Pokkalli acquired an effective antioxidant defense system that could withstand oxidative stress. In addition, Pokkali showed a higher ratio of GSH/GSSG. Further, the activity of the MG detoxification system (glyoxalase I and II) was also established significantly higher in Pokkali than in IR64. As a decrease in the glutathione may be involved in ASA-GSH and the pathway of MG detoxification, it may become an interaction point between two pathways (El-Shabrawi et al. 2010). They also recommended that glutathione and ascorbate homeostasis collectively, which are also prepared by enzymes glyoxalase, can be regarded as biomarkers in Pokkali for salt resistance. All these studies conclude that the status of defense systems of antioxidants (including non-enzymes and enzymes antioxidants) and ROS could be suggested as a rapid and resistant biomarker for screening under saline conditions in rice.

3 Mechanisms of ROS Generation and Oxidative Damage

It is reported that in plant cells, organelles with high oxidize metabolic activity like chloroplast, peroxisomes, and mitochondria are key sources of ROS. Salt persuades oxidative stress through a series of actions. It triggers stomatal closure (Zou et al. 2015), leading to decreased CO_2 influx and their fixation in the Calvin cycle while enhanced the electron transport from PSII to PSI (Ahmad et al. 2010a, 2011). In both normal and stressed conditions, ROS generation can be induced in the apoplast of the plant cell. This ROS generation is performed by NADPH oxidases, cell wall peroxidases, and amino oxidases (Qi et al. 2017).

Moreover, the formation of $^1O_2$ is also found by photoexcitation of Chl and as a result of its reaction with O_2. The generation of these ROS influences the expression of several genes and control several important cycles under abiotic stresses (Gill and Tuteja 2010).

4 Antioxidant Defense Mechanisms

The acclimatization of plants exposed to ROS stresses has been attributed to protection, mainly through the antioxidant defense mechanism of the plants. The cellular-level damages, which cause structural and functional alterations, are reported due to the production of ROS when plants exposure to unfavorable environmental conditions (Ferreira et al. 2007).

The damages caused by oxidative stresses could be successfully controlled by the antioxidant defense system, protecting plants under various stresses (Singh et al.

Table 1 Major ROS scavenging antioxidant enzymes

Enzymatic antioxidants	Enzyme code	Reactions catalyzed
Superoxide dismutase (SOD)	EC 1.15.1.1	$O_2^{\bullet -} + O_2^{\bullet -} + 2H^+ \rightarrow 2H_2O_2 + O_2$
Catalase (CAT)	EC 1.11.1.6	$H_2O_2 \rightarrow H_2O + 1/2\ O_2$
Ascorbate peroxidase (APX)	EC 1.11.1.11	H_2O_2 + AA → $2H_2O$ + DHA
Glutathione peroxidase (GPX)	EC 1.11.1.7	H_2O_2 + GSH → H_2O + GSSG
Monodehydroascorbate reductase (MDHAR)	EC 1.6.5.4	MDHA+NAD(P)H → AA+ NAD $(P)^+$
Dehydroascorbate reductase (DHAR)	EC 1.8.5.1	DHA + 2GSH → AA +GSSG
Glutathione reductase (GR)	EC 1.6.4.2	GSSG +NAD(P)H→ 2GSH + NAD $(P)^+$

2008). The protection mechanism of the ROS-induced damages is regulated primarily by CAT, SOD, APX, GPX, MDHAR, DHAR, GR, and so on. Major ROS scavenging enzymes are presented in Table 1.

5 Superoxide Dismutase (SOD)

Scavenging of $O_2^{\bullet -}$ is catalyzed by SOD, converting $O_2^{\bullet -}$ to H_2O_2 and O_2, decreasing the damage because of $OH^{\bullet}$ formation via the metal-catalyzed Habere-Weiss type reaction with a very speedy reaction rate. There are three known types of SODs classified based on metal cofactor: the copper/zinc (Cu/Zn-SOD), the manganese (Mn-SOD), and the iron (Fe-SOD) localized in different cellular compartments (Mittler 2002). The subcellular localization of each SOD, as reported by Alscher et al. (2002), is Cu/Zn-SODs in peroxisomes, chloroplasts, cytosol, and extracellular space, Mn-SOD in peroxisomes and mitochondria, and Fe-SOD in chloroplasts. The SOD-mediated preventive mechanisms of oxidative stress caused by biotic and abiotic stresses showed a critical role in the survival of plants under a stressed environment. The increased SOD activity has been observed under salt stress in rice (Guan et al. 2017). Meanwhile, Rossatto et al. (2017) reported that the highly responsive genes under salinity induced ROS production and reduced the lipid peroxidation in rice. The overexpressed plants showed increased SOD activity resulting in the enhanced ROS detoxifying capacity, thereby limiting the salinity-induced oxidative damage (Guan et al. 2017). The transgenic rice line overexpressing *OsCu/Zn-SOD* gene exhibited higher SOD activity with better morphological traits compared to nontransgenic lines demonstrating that the higher activity of Cu/Zn-SOD isoform is vital for cellular molecules protection (Xu et al. 2013), as it is localized in mitochondria, cytosol, and chloroplast. High SOD and its different isoform activity in salt-tolerant rice genotypes exhibit the inherent ability of the plant to overexpress SOD at physiological and molecular levels irrespective of ROS production (Islam et al. 2019).

6 Catalases (CAT)

Catalases are tetramers containing porphyrin heme groups with high potential to directly catalyze the decomposition of H_2O_2 into H_2O and O_2 with three different isoforms (CAT_1, CAT_2, CAT_3). These CAT isoforms are critical for H_2O_2 scavenging and prevent oxidative stress in plants under saline environments (Garg and Manchanda 2009; Zou et al. 2015). CAT is an enzyme with the highest conversion rates. Thus, a molecule of this enzyme can catalyze around 6 million H_2O_2 molecules into H_2O and O_2 per minute. CAT is essential in oxidative stress, which occurs by the action of oxidases involved in peroxisome processes like oxidation of fatty acids, degradation of purines, and several photorespiration reactions (Sharma and Ahmad 2014; Sofo et al. 2015).

There are reports of increases in the activity of CAT in plants when the concentration of NaCl increases at the medium level. Nounjan et al. (2012) informed that the CAT antioxidant enzyme activity is modified in rice plants during salt stress. They showed that when salt stress increased, low CAT activity in plants growing in salt stress conditions enhanced the oxidative damage by H_2O_2. Higher activity of CAT under salt stress conditions helps in effective detoxification of H_2O_2, a major contributor to oxidative damage induced by salinity (Islam et al. 2019). According to some reports, increased CAT activity was observed in different rice cultivars under salinity (Abdallah et al. 2016; Kibria et al. 2017). For instance, rice cultivar (*Oryza sativa*) BRS AG exposed to salt stress showed enhanced CAT activity due to enhanced transcription level of *OsCATA*, *OsCATB*, and *OsCATC* which helped in detoxification of ROS (e.g., H_2O_2) particularly and reduced the oxidative damage (Rossatto et al. 2017). Similarly, Kim et al. (2007) found increased expression of *OsCATB* in cultivar Nipponbare exposed to saline treatments.

7 Ascorbate Peroxidase (APX)

APX is an enzyme that utilizes ASH as a donor of electrons and is involved in the removal of H_2O_2 that occurs in the water–water and ASH–GSH cycles. Until now, the APX family has been reported to possess at least five different isoforms such as the thylakoid (tAPX), glyoxysome membrane (gmAPX), stromal chloroplast (sAPX), and cytosolic (cAPX) isoforms. Eight APX genes encoding APX enzymes have been identified and located in different cell compartments of rice plants. For instance, *OsAPx1* and *OsAPx2* genes encode for two cAPX enzyme isoforms; some gmAPX enzymes are encoded by *OsAPx3* and *OsAPx4* genes, while four genes (*OsAPx5*, *OsAPx6*, *OsAPx7*, *OsAPx8*) encode chloroplast isoforms of APX. The mitochondrial APX isoform is coded by *OsAPX6* (Teixeira et al. 2004). The cytosolic isoforms of APX (*OsAPX1* and *OsAPX2*) are present in abundance, showing higher activity, and the expression of cytosolic APX isomers shows a strong response to salt stress. Silencing (single or double) of cytosolic APX

(APX1 and 2) could hamper the normal functioning of the antioxidant system and increase plant vulnerability to stress (Islam et al. 2019). APX is characterized by a higher affinity for H_2O_2 (μM range) than other enzymes like peroxidases and catalases (mM range), which could indicate essential roles in the ROS homeostasis of plants during stressful conditions. In this regard, an overexpression of APX was reported to enhance plant growth tolerance to a saline environment. The function of the *OsAPX2* gene was demonstrated using a T-DNA knockout rice mutant under drought, salinity, and cold treatments. APX2 enzyme has essential functions in the protection of rice plants exposed to stressful conditions through scavenging ROS (Zhang et al. 2013). In addition, an increased APX activity was observed in shoot tissues from salt-sensitive rice cultivars (IR29) grown under salt stress conditions (Lee et al. 2013). Likewise, Guan et al. (2010) informed an enhanced *OsAPX4* expression in rice cultivar cv. Nipponbare on exposure to NaCl (100 mM) treatments. Furthermore, an increase of *OsAPX1* and *OsAPX2* gene expressions were reported in IR29 rice cultivar in response to salinity, demonstrating that different APX isoforms have important roles in H_2O_2 scavenging (Lee et al. 2015). Moreover, cytosolic APX isoform *OsAPx1* was highly responsive to ROS production induced by salinity, and it helped membrane stability through reduced lipid peroxidation (Rossatto et al. 2017).

8 Glutathione Reductase (GR)

GR is an important enzyme of the plant antioxidant defense system. It is located in most of the cell compartments like cytosol, chloroplast, and mitochondria. GR triggers the glutathione disulfide (GSSG) reduction using NADPH to reduce glutathione (GSH), which helps in H_2O_2 scavenging via ASA–GSH pathway (Hasanuzzaman et al. 2017). Maintenance of cellular redox balance is important to regulate antioxidant machinery and to prevent oxidative damage. However, cellular redox balance is monitored by the ratio of reduced and oxidized glutathione (GSH/GSSG ratio). Thus, GR plays an important role in monitoring redox balance and regulation of antioxidants to prevent oxidative damage. Overexpression of genes regulating GR activity enhances salt tolerance in different plant species, while knockdown or silencing of GR genes plants increase the susceptibility and sensitivity to oxidative stress (Wu et al. 2013). The isoforms of GR enzyme (GR1 and GR3) express in chloroplast and mitochondria and are vital for salt tolerance. For instance, the enhanced expression of the *OsGR3* transcript level responsible for GR3 activity induces salt tolerance in rice (Wu et al. 2013). In another study, Wu et al. (2015) demonstrated that salinity stress induces oxidative stress and limits the PSII quantum yield through narrowing the GSH–GSSG ratio. Nevertheless, the GR3 complementation rice lines showed salt tolerance through enhanced PSII efficiency and lower oxidative stress, indicating the role of GR3 in the maintenance of cellular redox balance and ROS scavenging. Salt stress triggers the expression of *OsGR1* and *OsGR2*, two isoforms of GR under salt stress, and the expression of these two GR isoforms is higher in salt-tolerant genotypes (Kordrostami et al. 2017). For instance, Turan and Tripathy (2013) found increased GR activity in two rice cultivars, namely, PB1 and CSR10 on exposure to high salt stress (200 mM). The salt-tolerant cultivar (cv. CSR10) exhibited early readiness against oxidative stress owing to a higher expression of APX encoding genes *OsGR2*. High biomass and grain production are linked with enhanced GSH levels in rice (Park et al. 2017), as high GSH/GSSG enhances intrinsic stress tolerance through ROS detoxification, which induces salt tolerance and a positive effect on plant growth and development (Talaat 2014).

9 Monodehydroascorbate Reductase (MDHAR)

The MDHAR is a vital enzyme of the plant antioxidant defense system and plays a crucial role against oxidative stresses by intracellular maintenance of AA in reduced form. The MDA reductase is pervasive and present in cytosol, mitochondria, peroxisomes, and chloroplast (Hossain et al. 1984; Dalton et al. 1993; Jimenez et al. 1997). It is a FAD enzyme and is highly specific for electron acceptor from monodehydroascorbate (MDHA) and prefers NADH over NADPH for electron donation. Since MDHAR regenerates AA, it is localized with APX in the cell organelles producing high ROS (i.e., mitochondria and peroxisomes), where APX oxidizes AA and scavenge H_2O_2 (Mittler et al. 2012). In plants, MDA reductase is abundant in roots, seeds, and shoots (Das and Roychoudhury 2014).

$$\mathrm{MDHA} + \mathrm{NADPH} \rightarrow \mathrm{AA} + \mathrm{NADP}^{+}$$

The reduction of MDHAR initiates with the formation of charge-transfer complex, followed by successive electron donation to the MDHA, leading to the production of two AA molecules through semiquinone form [E-FAD-NADP $(P)^{+}$]. In thylakoid, disproportionation by redFd (photo-induced ferrodoxin) is well established as it is more effective in MDHA reduction than $NADP^{+}$ as MDHAR does not involve MDHA reduction in the thylakoid scavenging system.

The high activity of MDHAR is associated with stress tolerance in plants. The activity of MDHAR has been found to induce resistance against salt stress in plants (Kavitha et al. 2010). Das and Roychoudhury (2014) observed the role of MDHAR against diverse abiotic stress (salinity, drought, and UV radiation) in finger millet and elucidated the correlation between *mdar* gene and enzyme activity under the various magnitude of oxidative stress. The enzymes (GR, MDHAR, and DHAR), which regenerated ASH, are found to exhibit higher activities in salt-stressed rice plants. For instance, Sultana et al. (2012) reported that transgenic rice lines harboring AeMDHAR, which have been introduced from mangrove plants, showed better attributes such as tolerance to salt at germination and early seedling growth, a higher tiller number, a lower rate of sterility, longer panicle, and bold grains compared to untransformed nontransgenic plants when subjected to salinity conditions. In another study, Rehman et al. (2016) demonstrated that higher activities of MDHAR confer tolerance to rice plants exposed to salt stress. In addition, the expression of the MDHAR controlling gene (*OsMDHAR*) from rice to other species such as *Saccharomyces cerevisiae* improved tolerance to ROS and fermentative capacity (Kim et al. 2016).

10 Dehydroascorbatereductase (DHAR)

The ASH is the key to stress tolerance, which is produced from DHAR under the oxidized state. Overexpression of DHAR also enhances plant tolerance against various abiotic stresses.

DHAR regenerates another agent to the cellular AA pool, apart from MDHAR. For maintaining the redox state in the plant cell, it is critical for regulating the AA pool size in both symplast and apoplast. DHAR is located abundantly in seeds, roots, and both green and etiolated shoots (Das and Roychoudhury 2014; Chang et al. 2017).

$$\mathrm{DHA} + 2\mathrm{GSH} \rightarrow \mathrm{AA} + \mathrm{GSSG}$$

Some studies suggest that the overexpression of DHAR induces salt tolerance in rice. Thus, Kim et al. (2014) developed transgenic rice with overexpression of *OsDHAR1*. They found that *OsDHAR1* overexpression enhanced the DHAR activity and AsA–DHA ratio, leading to higher AsA level and lower DHA content. The ascorbate-glutathione system enzymes (DHAR, APX, and GR) also increased irrespective of salt stress in homozygous transgenic rice. The overexpression of *OsDHAR1* reduced the oxidative damage as evident from low H_2O_2 and MDA levels, which helped rice to adapt to salt stress conditions through maintaining AsA level, ionic, and redox homeostasis.

11 Glutathione *S*-Transferases (GST)

Plant GST gene families are large and highly diverse, which found 25 members in soybean, 42 members in maize, and 54 members in Arabidopsis (Dixon et al. 2002; Sappl et al. 2004; Diaz-Vivancos et al. 2015). These are generally cytoplasmic proteins but are found as microsomal, plastidic, nuclear, and apoplastic isoforms. Plant GSTs act as herbicide detoxification, hormone homeostasis, vacuolar sequestration of anthocyanin, tyrosine metabolism, hydroxyperoxide detoxification, and apoptosis regulation (Dixon et al. 2010). GSTs also remove cytotoxic or genotoxic compounds, which can damage the DNA, RNA, and proteins. With the help of GSH, GSTs can reduce peroxides and produce scavengers of cytotoxic and genotoxic compounds. Overexpression of GST genes into low-temperature sensitive rice plants enhanced their tolerance to low temperature (Toshikazu et al. 2002). Coexpression of GST and CAT1 genes in rice provided increased tolerance to oxidative stress (Zhao and Zhang 2006).

The overexpression of OsGSTU4 in Arabidopsis plants, which is transformed from rice, resulted in better growth and higher GST activity under various abiotic stress conditions (Sharma et al. 2014). Recently, Li et al. (2018) performed an RNA-Seq analysis and found that GST proteins were involved in copper tolerance in rice plants. The enhanced tolerance in the B1139 rice cultivar was due to an increase in the detoxification of ROS by GST proteins.

12 Glutathione Peroxidase (GPX)

The GPX is one of the important ROS scavengers inside the cell. A GPX gene (*PgGPX*) was found in *Pennisetum glauccum* cDNA library under abiotic stress. Enzyme kinetics data revealed that the *PgGPX* belongs to the functional peroxiredoxin group and has a preference toward thioredoxin rather than glutathione as an electron donor. Moreover, its activity depends on the divalent cations, especially Cd^{2+}, and the homology model shows the presence of Cd^{2+} at the binding site in the protein. Site-directed mutagenesis study of *PgGPX* protein revealed that Cysteine residues take part in a vital role for enzymatic activity and structural folding. Expression analysis suggested that *PgGPX* transcript is highly upregulated in response to salinity and drought stresses. When expressed ectopically, the *PgGPX* enhanced the tolerance capacity in prokaryotic *E. coli* and rice plants against multiple abiotic stresses. Transgenic rice plants accumulated a lesser amount of MDA and H_2O_2 and a higher amount of Pro when compared to wild-type plants under salinity and drought stresses, which indicates the suppression of lipid peroxidation and ROS generation in transgenic lines. Likewise, transgenic plants maintained better photosynthesis efficiency and a higher level of antioxidant enzyme activity when compared to wild-type plants under stress conditions (Islam et al. 2015).

13 Nonenzymatic Antioxidants

13.1 Ascorbic Acid (Vitamin C)

Among these antioxidants, ascorbic acid (AsA) is the furthermost water-soluble and performs to limit or minimize the stress-induced excessive accumulation of ROS in plants (Khan et al. 2008). It is also known as ascorbate that manufactures in the cytosol of plants by the transformation of d-glucose. Besides assisting in numerous physiological developments, including growth, differentiation, and metabolism in plants, AsA significantly eliminates free radicals, thus lessening the impairment caused by oxidative stress. AsA further assists in protecting the cell membrane, and also performs as a co-factor of violaxanthin-de-epoxidase, thereby supporting the debauchery of additional excitation energy. Several studies found an elevation level of AsA in plant leaves under stress conditions (Mohamed et al. 2010).

Similarly, Munir and Aftab (2011) detected that improved levels of AsA coupled with an increased level of CAT, POD, and SOD stimulated the growth and development of plants under salt stress-induced osmotic stress. Several earlier findings (Shalata and Neumann 2001; Khan and Ashraf 2008; Wang et al. 2014) revealed that foliar AsA stimulates numerous enzymatic actions and reduces the damage caused by stress-induced excessive ROS. For example, tomato seedling treated with exogenous AsA helps to reduce lipid peroxidation, thus recovers plants from salt stress (Polash et al. 2019). Similarly, Hamada and Al-Hakimi (2009) stated that foliar AsA limits the adverse effects of salt stress on membrane integrity, net photosynthetic rate, and also bio-synthesis pigments in sunflower plants. Khafagy et al. (2009) found that salt-prompted oxidative stress reduced in leaf Chl a and Chl b concentrations in chili, however, enhanced with AsA pretreatments. Azzedine et al. (2011) and Polash et al. (2019) stated that the foliar use of AsA alleviated the hostile effect of salinity stress through improving leaf area, Chl, and carotene contents.

13.2 Glutathione (GSH)

Glutathione (GSH) is measured as the furthermost significant intracellular protection contrary to ROS-induced oxidative damage. Generally, it is present in plant tissues, including cytosol, vacuole, chloroplasts, mitochondria, endoplasmic reticulum, peroxisomes, and in the apoplast (Mittler and Zilinskas 1992; Jiménez et al. 1998). The GSH has a crucial role in numerous physio-biochemical processes (Xiang et al. 2001) and also works as a messenger to stimulate stress-responsive genes (Mullineaux and Rausch 2005). The GSH is another robust antioxidant documented in plants that counteract the damage of principal cellular components as a result of the stress-induced excessive production of ROS (Pompella et al. 2003). GSH also defends proteins from denaturation and function as a substrate for GPX and GST, which controls the elimination of ROS (Noctor et al. 2002). It also contributes to the renaissance of AsA via AsA–GSH cycle. A study stated by Aly-Salama and Al-Mutawa (2009) reported that foliar application of GSH assisted in preserving the penetrability of the plasma membrane and feasibility of cells under salt-induced osmotic stress in onion. Other researchers observed that treating with a combination of GSH and AsA assisted in increasing the morphological parameters, antioxidant activity, and mineral ion content while exposed to salinize environment (Rawia et al. 2011; Polash et al. 2019).

13.3 Proline (Pro)

Plants follow several physiological mechanisms to survive against abiotic stresses. Plants that can survive against salinity-induced oxidative stress could accrue osmolytes against (Lehmann et al. 2010), such as Pro, which plays a significant role in enhancing cellular-osmolarity during salinity stress (Szabados and Savoure 2010; Muzammil et al. 2018). Pro performs as a signal transduction pathway that controls stress-responsive genes for the direction of several biochemical processes to adopt against salt stress (Heuer 2003; Trovato et al. 2019). Pro efficiently reduced the salt-induced excess production of ROS levels, ultimately limiting cell death (Chen and Dickman 2005; Ashraf and Foolad 2007; Trovato et al. 2008).

Deivanai et al. (2011) demonstrated that pretreatment with onemM Pro enhanced growth during salt stress in rice seedlings by increasing the actions of SOD, CAT, and POD. Ahmad et al. (2010b) reported that Pro supplement in olive trees appeared to expand the adaptability of plants against salt stress by regulating enzymatic activities of antioxidants and enhancing the photosynthetic activity. Exogenous supplementation of Pro repressed the Na^+ uptake and its translocation in rice (Nounjan et al. 2012), whereas the K^+ content was fairly increased, leading to a high K^+/Na^+ ratio (Polash et al. 2019). In the recent era, scientists have been showing much interest in demonstrating the positive role of Ca^{2+} in lessening the antagonistic effect of NaCl-induced salt stress (Roy et al. 2019). Application of Ca^{2+} activates calcium-dependent abiotic stress signaling pathways. This leads to an increase in photosynthetic capacity and reduces oxidative damage by changing the antioxidant-defense mechanism and Pro biosynthesis (Parvin et al. 2015; Roy et al. 2019).

13.4 Carotenoid (Car)

Carotenoids (Cars) are biosynthesized from carotenes and xanthophylls and are one of the significant pigments of the photosynthetic antenna. In plants, Cars are synthesized in all photosynthetic green tissues. Cars contribute to harvest light energy during photosynthesis (Holt et al. 2005; Zakar et al. 2016). Additionally, Cars are also linked in the resistance mechanism against several stress-induced oxidative stress in plants, including salt stress also (Bouvier et al. 2005; Santabarbara et al. 2013; Nagy et al. 2015; Campos et al. 2016). However, several researchers (Davenport et al. 2005) observed that during salt stress, imbalance accumulation of Na^+ in plant cells affects the photosynthetic components such as enzymes, Chl, and Cars. At the same time, Abdallah et al. (2016) observed that expression of some related Cars was encouraged by NaCl, which ultimately improved the accumulation of β-carotene, lutein, and quercetin 3-β-d-glucoside in *Solanum nigrum* L. under salt stress. Finally, they revealed that prospective antioxidant attributes of Cars and flavonoids and their connected key genes are proficiently complicated in the constraint of salt-induced oxidative injuries (Abdallah et al. 2016).

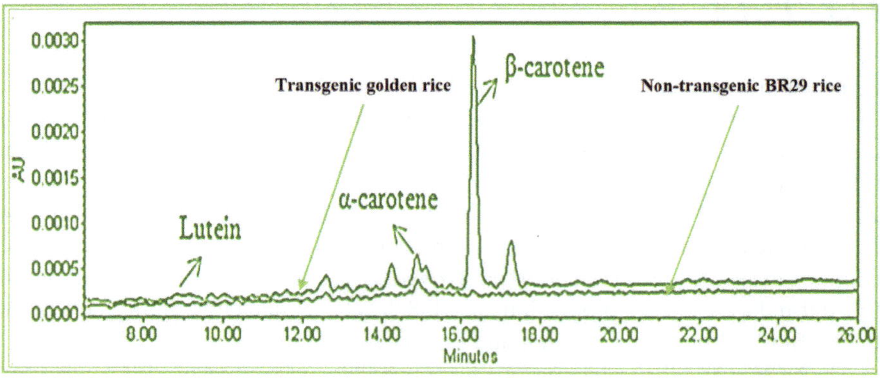

Fig. 1 A representative HPLCof the Crt extract of transgenic golden rice and non-transgenic BR29 showing β-carotene, α-carotene, and lutein peak (showed by arrows). Modified from: Gayen et al. (2016)

The carotenoid synthesis was improved in transgenic golden rice after genetic manipulation of endosperm-specific pathways (Al-Babili et al. 1999; Rayan and Abbott 2015). Recently, transgenic golden rice was established by Agrobacterium-mediated transformation of β-carotene biosynthetic (provitamin A) pathways into rice endosperm (Gayen et al. 2016). The carotenoid content of rice seeds was measured by HPLC (High-performance liquid chromatography) analysis: the outer layers of raw (non-parboiled) brown rice contains carotene, lutein, and/or lycopene (Gayen et al. 2016; Fig. 1).

14 Conclusion

Numerous physiological and biochemical responses are involved in rice plants under saline conditions. These changes include a reduction in photosynthetic capacity, plant water status, changes in osmoregulation, accumulation of toxic ions, and ROS generation in roots and leaves. Production of ROS, such as $O_2^{\bullet -}$, H_2O_2, and secondary metabolites of ROS like MDA increased in rice under salinity. ROS and oxidative damage due to salt stress was much higher in salt-sensitive rice cultivars than in salt-tolerant cultivars. Moreover, salt-tolerant cultivars have a greater antioxidant capacity, which is regulated by the ratio of GSH/GSSG. The most prominent antioxidants in rice are SOD, POD, CAT, GR, and GST. Moreover, activities of antioxidants such as SOD, DHAR, and MDHAR are much higher in roots than in leaves of salt-tolerant rice genotypes. It is important to mention here that the accumulation of Pro is associated with an increase in the antioxidant capacity of rice plants under salt stress. It is not known whether the accumulation of Pro improves the plant antioxidant capacity, or it is the consequence of plant salt tolerance. However, advanced molecular analyses will help us in understanding the mechanism of salt-induced oxidative damage. It is recommended to study how activation of antioxidants in different cellular parts such as chloroplast, mitochondria, peroxisomes occurs to scavenge ROS, and what kind of cellular Redox signaling takes place to modulate other biochemical processes to induce salt tolerance in rice.

Conflict of Interest No conflict of interest is hereby declared.

References

Abdallah MS, Abdelgawad ZA, El-Bassiouny HMS (2016) Alleviation of the adverse effects of salinity stress using trehalose in two rice varieties. South Afr J Bot 103:275–282

Ahmad P, Jaleel CA, Salem MA, Nabi G, Sharma S (2010a) Roles of enzymatic and non-enzymatic antioxidants in plants during abiotic stress. Crit Rev Biotechnol 30:161–175. https://doi.org/10.3109/07388550903524243

Ahmad P, Jaleel CA, Sharma S (2010b) Antioxidativedefence system, lipid peroxidation, proline metabolizing enzymes and biochemical activity in two genotypes of Morus alba L. subjected to NaCl stress. Russ J Plant Physiol 57:509–517

Ahmad P, Nabi G, Ashraf M (2011) Cadmium-induced oxidative damage in mustard [*Brassica juncea* (L.) Czern. &Coss.] plants can be alleviated by salicylic acid. South Afr J Bot 77:36–44. https://doi.org/10.1016/j.sajb.2010.05.003

Ahmed S, Alam MJ, Hossain A, Islam AKMM, Awan TH, Soufan W, Qahtan AA, Okla MK, EL Sabagh A (2021) Interactive effect of weeding regimes, rice cultivars, and seeding rates influence the rice-weed competition under dry direct-seeded condition. Sustainability 13(1): 317. https://doi.org/10.3390/su13010317

Al-Babili S, Hartung W, Kleinig H, Beyer P (1999) CPTA modulates levels of carotenogenic proteins and their mRNAs and affects carotenoid and ABA content as well as chromoplast structure in *Narcissus pseudonarcissus* flowers. Plant Biol 1:607–612

Al-Hakimi HAM (2001) Counteraction of salinity stress on wheat plants by grain soaking in ascorbic acid, thiamine or sodium salicylate. Biol Planta 44:253–261

Ali MA, Kim PJ, Inubushi K (2015) Mitigating yield-scaled greenhouse gas emissions through combined application of soil amendments: a comparative study between temperate and subtropical rice paddy soils. Sci Total Environ 529:140–148

Alscher RG, Erturk N, Heath LS (2002) Role of superoxide dismutases (SODs) in controlling oxidative stress in plants. J Exp Bot 53:1331–1341

Aly-Salama KH, Al-Mutawa MM (2009) Glutathione-triggered mitigation in salt-induced alterations in plasmalemma of onion epidermal cells. Int J Agric Biol 11(5):639–642

Ashraf M, Foolad M (2007) Roles of glycine betaine and proline in improving plant abiotic stress resistance. Environ Exp Bot 59:206–216. https://doi.org/10.1016/j.envexpbot.2005.12.006

Azzedine F, Gherroucha H, Baka M (2011) Improvement of salt tolerance in durum wheat by ascorbic acid application. J Stress Physiol Biochem 7:27–37

Bhusan D, Das DK, Hossain M, Murata Y, Hoque MA (2016) Improvement of salt tolerance in rice (*Oryza sativa* L.) by increasing antioxidant defense systems using exogenous application of proline. Aust J Crop Sci 10(1):50–56

Bouvier F, Isner JC, Dogbo O, Camara B (2005) Oxidative tailoring of carotenoids: a prospect towards novel functions in plants. Trends Plant Sci 10:187–194

Campos LM, Rios EA, Guapyassu L, Midlej V, Atella GC, Herculano-Houzel S, Benchimol M, Mermelstein C, Costa ML (2016) Alterations in zebrafish development induced by simvastatin:

comprehensive morphological and physiological study, focusing on muscle. Exp Biol Med 241(17):1950–1960

Chang YF, Broyles CN, Brook FA, Davies MJ, Turtle CW, Takeharu Nagai T, Daniels MJ (2017) Non-invasive phenotyping and drug testing in single cardiomyocytes or beta-cells by calcium imaging and optogenetics. PLoS One 12(4):e0174181. https://doi.org/10.1371/journal.pone.0174181

Chawla S, Jain S, Jain V (2013) Salinity induced oxidative stress and antioxidant system in salt-tolerant and salt-sensitive cultivars of rice (*Oryza sativa* L.). J Plant Biochem Biotechnol 22:27–34. https://doi.org/10.1007/s00709-011-0365-3

Chen C, Dickman MB (2005) Proline suppresses apoptosis in the fungal pathogen *Colletotrichum trifolii*. Proc Natl Acad Sci U S A 102:3459–3464

Corpas FJ (2015) What is the role of hydrogen peroxide in plant peroxisomes? Plant Biol (Stuttg) 17:1099–1103

Dalton DA, Baird LM, Langeberg L, Taugher CY, Anyan WR, Vance CP, Sarath G (1993) Subcellular localization of oxygen defense enzymes in soybean (*Glycine max* [L.] Merr.) root nodules. Plant Physiol 102:481–489

Das K, Roychoudhury A (2014) Reactive oxygen species (ROS) and response of antioxidants as ROS-scavengers during environmental stress in plants. Front Environ Sci 2:53. https://doi.org/10.3389/fenvs.2014.00053

Davenport R, James RA, Zakrisson-Plogander A, Tester M, Munns R (2005) Control of sodium transport in durum wheat. Plant Physiol 137:807–818

Deivanai S, Xavier R, Vinod V, Timalata K, Lim OF (2011) Role of exogenous proline in ameliorating salt stress at early stage in two rice cultivars. J Stress Physiol Biochem 7:157–174

Dhanyalakshmi KH, Vijayalakshmi C, Boominathan P (2013) Evaluation of physiological and biochemical responses of rice (*Oryza sativa*L.) varieties to salt stress. Indian J Agric Res 47(2): 91–99

Diaz-Vivancos P, de Simone A, Kiddle G, Foyer CH (2015) Glutathione-linking cell proliferation to oxidative stress. Free Radical Biol Med 89:1154–1164. https://doi.org/10.1016/j.freeradbiomed.2015.09.023

Dixon D, Davis B, Edwards R (2002) Functional divergence in the glutathione transferase super-family in plants: identification of two classes with putative functions in redox homeostasis in Arabidopsis thaliana. J Biol Chem 277(34):30859–30869. https://doi.org/10.1074/jbc.M202919200

Dixon DP, Skipsey M, Edwards R (2010) Roles for glutathione transferases in plant secondary metabolism. Phytochemistry 71:338–350

Egrão S, Schmöckel SM, Tester M (2017) Evaluating physiological responses of plants to salinity stress. Ann Bot 119(1):1–11

EL Sabagh A, Hossain A, Barutçular C, Islam MS, Ratnasekera D, Kumar N, Meena RS, Gharib HS, Saneoka H, Teixeira da Silva JA (2019a) Drought and salinity stress management for higher and sustainable canola (‘*Brassica napus*’ L.) production: a critical review. Aust J Crop Sci 13(1):88–96

EL Sabagh A, Hossain A, Islam MS, Barutçular C, Ratnasekera D, Kumar N, Meena RS, Gharib HS, Saneoka H, Teixeira da Silva JA (2019b) Sustainable soybean production and abiotic stress management in saline environments: a critical review. Aust J Crop Sci 13(2):228–236

EL Sabagh A et al (2020) Drought and heat stress in cotton (*Gossypium hirsutum* L.): consequences and their possible mitigation strategies. In: Hasanuzzaman M (ed) Agronomic crops. Springer, Singapore. https://doi.org/10.1007/978-981-15-0025-1_30

El-Shabrawi H, Kumar B, Kaul T, Reddy MK, Singla-Pareek SL, Sopory SK (2010) Redox homeostasis, antioxidant defense, and methylglyoxal detoxification as markers for salt tolerance in Pokkali rice. Protoplasma 245(1-4):85–96

Ferreira RB, Monteiro S, Freitas R, Santos CN, Chen Z, Batista LM, Duarte J, Borges A, Teixeira AR (2007) The role of plant defence proteins in fungal pathogenesis. Mol Plant Pathol 8 (5):677–700. https://doi.org/10.1111/j.1364-3703.2007.00419.x

Foyer CH, Noctor G (2011) Ascorbate and glutathione: the heart of the redox hub. Plant Physiol 155(1):2–18. https://doi.org/10.1104/pp.110.167569

Gaballah MM, Metwally AM, Skalicky M, Hassan MM, Brestic M, EL Sabagh A, Fayed AM (2021) Genetic diversity of selected rice genotypes under water stress conditions. Plan Theory 10(1):27. https://doi.org/10.3390/plants10010027

Garg N, Manchanda G (2009) ROS generation in plants: boon or bane. Plant Biosyst 143(1):81–96

Gayen D, Ghosh S, Paul S, Sarkar SN, Datta SK, Datta K (2016) Metabolic regulation of carotenoid-enriched golden rice line. Front Plant Sci 7:1622. https://doi.org/10.3389/fpls.2016.01622

Gill SS, Tuteja N (2010) Reactive oxygen species and antioxidant machinery in abiotic stress tolerance in crop plants. Plant Physiol Biochem 48(12):909–930

Guan G, Xia D, Liu S (2010) OsAPX4 gene response to several environmental stresses in rice (*Oryza sativa* L.). Afr J Biotechnol 9:5908–5913

Guan Q, Liao X, He M, Li X, Wang Z, Ma H, Yu S, Liu S (2017) Tolerance analysis of chloroplast OsCu/Zn-SOD overexpressing rice under NaCl and $NaHCO_3$ stress. PLoS One 12(10): e0186052. https://doi.org/10.1371/journal.pone.0186052

Hairmansis A, Nafisah, Jamil A (2017) Towards developing salinity tolerant rice adaptable for coastal regions in Indonesia. In: 2nd International Conference on Sustainable Agriculture and Food Security: A Comprehensive Approach, KnE Life Sciences, pp. 72–79. doi:https://doi.org/10.18502/kls.v2i6.1021

Hakim MA, Juraimi AS, Karim SMR, Khan MSI, Islam MS, Choudhury MK, Soufan W, Alharby H, Bamagoos A, Iqbal MA, Hnilicka F, Kubes J, Habib Ur Rahman M, Saud S, Hassan MM, EL Sabagh A (2021) Effectiveness of herbicide to control rice weeds in diverse saline environments. Sustainability 13(4):2053. https://doi.org/10.3390/su13042053

Hamada AM, Al-Hakimi AM (2009) Exogenous ascorbic acid or thiamine increases the resistance of sun flower and maize plants to salt stress. Acta Agronomica Hungarica 57:335–347

Hasanuzzaman M, Alam M, Rahman A, Hasanuzzaman M, Nahar K, Fujita M (2014) Exogenous proline and glycine betaine mediated upregulation of antioxidant defense and glyoxalase systems provides better protection against salt-induced oxidative stress in two rice (*Oryza sativa* L.) varieties. Biomed Res Int 2014:1–17

Hasanuzzaman M, Nahar N, Hossain MS, Mahmud JA, Rahman A, Inafuku M, Oku H, Fujita M (2017) Coordinated actions of glyoxalase and antioxidant defense systems in conferring abiotic stress tolerance in plants. Int J Mol Sci 18:200. https://doi.org/10.3390/ijms18010200

Heuer B (2003) Influence of exogenous application of proline and glycinebetaine on growth of salt-stressed tomato plants. Plant Sci 165:693–699. https://doi.org/10.1016/S0168-9452(03)00222-X

Holt NE, Zigmantas D, Valkunas L, Li XP, Niyogi KK (2005) Carotenoid cation formation and the regulation of photosynthetic light harvesting. Sci 307(5708):433–436. https://doi.org/10.1126/science.1105833

Hossain A, EL Sabagh A, Erman M, Fahad S, Islam T, Bhatt R, Hasanuzzaman M (2020) Nutrient management for improving abiotic stress tolerance in legumes of the family Fabaceae. In: Hasanuzzaman M, Araújo S, Gill S (eds) The plant family Fabaceae. Springer, Singapore, pp 393–415

Hossain MA, Nakanoand Y, Asada K (1984) Monodehydroascorbate reductase in spinach chloroplast and its participation in regeneration of ascorbate for scavenging hydrogen peroxide. Plant Cell Physiol 25:388–395

Hossain MS, Dietz KJ (2016) Tuning of redox regulatory mechanisms, reactive oxygen species and redox homeostasis under salinity stress. Front Plant Sci 7:548. https://doi.org/10.3389/fpls.2016.00548

Iqbal MA, Iqbal A, Afzal S, Akbar N, Abbas RN, Khan HZ (2015) In Pakistan, agricultural mechanization status, and future prospects. American-Eurasian J Agric Environ Sci 15:122–128

Iqbal MZ (2018) The effects of 2016-2017 rice price increase on household welfare and poverty in rural Bangladesh (October 12, 2018). Available at SSRN: https://ssrn.com/abstract=3265448 or https://doi.org/10.2139/ssrn.3265448

Islam F, Wang J, Farooq MA, Yang C, Jan M, Mwamba TM, Hannan F, Xu L, Zhou W (2019) Rice responses and tolerance to salt stress: deciphering the physiological and molecular mechanisms of salinity adaptation. In: Advances in Rice research for abiotic stress tolerance. Woodhead Publishing, UK, pp 791–819

Islam T, Manna M, Kaul T, Pandey S, Reddy CS, Reddy M (2015) Genome-wide dissection of Arabidopsis and rice for the identification and expression analysis of glutathione peroxidases reveals their stress-specific and overlapping response patterns. Plant Mol Biol Rep 33:1413–1427

James RA, Blake C, Byrt CS, Munns R (2011) Major genes for Na^+ exclusion, Nax1 and Nax2 (wheat HKT1; 4 and HKT1; 5), decrease Na^+ accumulation in bread wheat leaves under saline and waterlogged conditions. J Exp Bot 62(8):2939–2947

Jimenez A, Hernandez JA, Del Rio LA, Sevilla F (1997) Evidence for the presence of the ascorbate-glutathione cycle in mitochondria and peroxisomes of pea leaves. Plant Physiol 114:275–228

Jiménez A, Hernández JA, Pastori GM, del Río LA, Sevilla F (1998) Role of the ascorbate-glutathione cycle of mitochondria and peroxisomes in the senescence of pea leaves. Plant Physiol 118:1327–1335

Kabir AH, Rahman MM, Haider SA, Paul NK (2015) Mechanisms associated with differential tolerance to Fe deficiency in okra (*Abelmoschus esculentus* Moench). Environ Exp Bot 112:16–26. https://doi.org/10.1016/j.envexpbot.2014.11.011

Karuppanapandian T, Moon JC, Kim C, Manoharan K, Kim W (2011) Reactive oxygen species in plants: their generation, signal transduction, and scavenging mechanisms. Aust J Crop Sci 5(6): 709–725

Kaur N, Dhawan M, Sharma I, Pati PK (2016) Interdependency of reactive oxygen species generating and scavenging system in salt sensitive and salt tolerant cultivars of rice. BMC Plant Biol 16(1):131. https://doi.org/10.1186/s12870-016-0824-2

Kavitha K, George S, Venkataraman G, Parida AA (2010) Salt-inducible chloroplastic monodehydroascorbate reductase from halophyte Avicennia marina confers salt stress tolerance on transgenic plants. Biochimie 92:1321–1329

Khafagy MA, Arafa AA, El-Banna MF (2009) Glycinebetaine and ascorbic acid can alleviate the harmful effects of NaCl salinity in sweet pepper. Aust J Crop Sci 3:257–267

Khan A, Ashraf M (2008) Exogenously applied ascorbic acid alleviates salt-induced oxidative stress in wheat. Environ Exp Bot 63(1-3):224–231

Khan MJ, Drochner W, Steingass H, Islam KMS (2008) Nutritive evaluation of some tree leaves from Bangladesh for feeding ruminant animals. Indian J Anim Sci 78(11):1273–1277

Kibria MG, Hossain M, Murata Y, Hoque MA (2017) Antioxidant defense mechanisms of salinity tolerance in rice genotypes. Rice Sci 24(3):155–162

Kim D, Shibato J, Agrawal JK, Fujihara S, Iwahashi H, Kim DH, Shim LS, Rakwal R (2007) Gene transcription in the leaves of rice undergoing salt-induced morphological changes (*Oryza sativa* L.). Mol Cell 24:45–59

Kim IS, Kim YS, Kim YH, Park AK, Kim HW, Lee JH, Yoon HS (2016) Potential application of the *Oryza sativa* Monodehydroascorbate reductase gene (OsMDHAR) to improve the stress tolerance and fermentative capacity of *Saccharomyces cerevisiae*. PLoS One 11(7):e0158841. https://doi.org/10.1371/journal.pone.0158841

Kim YS, Kim IS, Shin SY, Park TH, Park HM, Kim YH, Lee GS, Kang HG, Lee SH, Yoon HS (2014) Overexpression of dehydroascorbate reductase confers enhanced tolerance to salt stress in rice plants (*Oryza sativa* L. japonica). J Agron Crop Sci 200(6):444–456. https://doi.org/10.1111/jac.12078

Kordrostami M, Rabiei B, Kumleh HH (2017) Different physio-biochemical and transcriptomic reactions of rice (*Oryza sativa* L.) cultivars differing in terms of salt sensitivity under salinity stress. Environ Sci Pollut Res 24(8):7184–7196. https://doi.org/10.1007/s11356-017-8411-0

Lee J, Kim HR, Lee C (2010) Trial-to-trial variability of spike response of V1 and saccadic response time. J Neurophysiol 104(5):2556–2572. https://doi.org/10.1152/jn.01040.2009

Lee MH, Cho EJ, Wi SG, Bae H, Kim JE, Cho JY, Lee S, Kim JH, Chung BY (2013) Divergences in morphological changes and antioxidant responses in salt-tolerant and salt-sensitive rice seedlings after salt stress. Plant Physiol Biochem 70:325–335. https://doi.org/10.1016/j.plaphy.2013.05.047

Lee S, Chung MS, Kim JE, Lee GW, Jeong YS, Lee MH, Hong SH, Lee SS, Kim JH, Chung BY (2015) Liquid chromatography-tandem mass spectrometry-assisted identification of two salinity-inducible ascorbate peroxidases in a salt-sensitive rice cultivar (*Oryza sativa* L. cv. 'IR29'). Plant Growth Regul 75:143–153

Lehmann S, Funck D, Szabados L, Rentsch D (2010) Proline metabolism and transport in plant development. Amino Acids 39:949–962. https://doi.org/10.1007/s00726-010-0525-3

Li J, Zeng L, Cheng Y, Lu G, Fu G, Ma H, Liu Q, Zhang X, Zou X, Li C (2018) Exogenous melatonin alleviates damage from drought stress in *Brassica napus* L. (rapeseed) seedlings. Acta Physiol Planta 40:1–11

Liu L, Nakamura Y, Taliman NA, EL Sabagh A, Moghaieb RE, Saneoka H (2020) Differences in the growth and physiological responses of the leaves of Peucedanum japonicum and Hordeum vulgare exposed to salinity. Agriculture 10:317

Manova V, Gruszka D (2015) DNA damage and repair in plants-from models to crops. Front Plant Sci 6:885. https://doi.org/10.3389/fpls.2015.00885

Mittler R (2002) Oxidative stress, antioxidants and stress tolerance. Trends Plant Sci 7(9):405–410

Mittler R, Andrija Finka A, Goloubinoff P (2012) How do plants feel the heat? Trends Biochem Sci 37(3):118–125. https://doi.org/10.1016/j.tibs.2011.11.007

Mittler R, Vanderauwera S, Suzuki N, Miller G, Tognetti VB, Vandepoele K et al (2011) ROS signaling: the new wave. Trends Plant Sci 16(6):300–309

Mittler R, Zilinskas BA (1992) Molecular cloning and characterization of a gene encoding pea cytosolic ascorbate peroxidase. J Biol Chem 267(30):21802–21807

Mohamed MA, Matter MA, Saker MM (2010) Effect of salt stress on some defense mechanisms of transgenic and wild potato clones (*Solanum tuberosum* L.) grown in vitro. Nature 12:181–193

Monsur MB et al (2020) Oxidative stress tolerance mechanism in rice under salinity. Phyton 89(3): 497–517

Mullineaux PM, Rausch T (2005) Glutathione, photosynthesis and the redox regulation of stress-responsive gene expression. Photosynth Res 86:459–474. https://doi.org/10.1007/s11120-005-8811-8

Munir N, Aftab F (2011) Enhancement of salt tolerance in sugarcane by ascorbic acid pre-treatment. Afr J Biotechnol 10:18362–18370

Muzammil S, Shrestha A, Dadshani S, Pillen K, Siddique S, Leon J, Naz AA (2018) An ancestral allele of Pyrroline-5-carboxylate synthase1 promotes proline accumulation and drought adaptation in cultivated barley. Plant Physiol 178:771–782. https://doi.org/10.1104/pp.18.00169

Nagy V, Cole T, Van Campenhout C, Khoung TM, Leung C, Vermeiren S, Novatchkova M, Wenzel D, Cike D, Polyansky AA, Kozieradzki I, Meixner A, Bellefroid EJ, Neely GG, Penninger JM (2015) The evolutionarily conserved transcription factor PRDM12 controls sensory neuron development and pain perception. Cell Cycle 14(12):1799–1808

Noctor G, Gomez L, Vanacker H, Foyer CH (2002) Interactions between biosynthesis, compartmentation and transport in the control of glutathione homeostasis and signalling. J Exp Bot 53: 283–304

Nounjan N, Nghia PT, Theerakulpisut P (2012) Exogenous proline and trehalose promote recovery of rice seedlings from salt-stress and differentially modulate antioxidant enzymes and expression of related genes. J Plant Physiol 169:596–604

Park SI, Kim YS, Kim JJ, Mok JE, Kim YH, Park HM, Kim IS, Yoon HS (2017) Improved stress tolerance and productivity in transgenic rice plants constitutively expressing the *Oryza sativa* glutathione synthetase OsGS under paddy field conditions. J Plant Physiol 215:3947

Parvin K, Ahamed KU, Islam MM, Haque MN (2015) Response of tomato plant under salt stress: role of exogenous calcium. J Plant Sci 10(6):222–233

Pervaiz Z, Rabbani M, Pearce S, Malik S (2009) Determination of genetic variability of Asian rice (*Oryza sativa* L.) varieties using microsatellite markers. Afr J Biotechnol 8(21):5641–5651

Polash MAS, Sakil MA, Hossain MA (2019) Plants responses and their physiological and biochemical defense mechanisms against salinity: a review. Trop Plant Res 6(2):250–274

Pompella A, Visvikis A, Paolicchi A, De Tata V, Casini AF (2003) The changing faces of glutathione, a cellular protagonist. Biochem Pharmacol 66:1499–1503

Qadir M, Quillérou E, Nangia V, Murtaza G, Singh M, Thomas RJ, Drechsel P, Noble AD (2014) Economics of salt-induced land degradation and restoration. Nat Resour For 38:282–295. https://doi.org/10.1111/1477-8947.12054

Qi J, Wang J, Gong Z, Zhou JM (2017) Apoplastic ROS signaling in plant immunity. Curr Opin Plant Biol 38:92–100. https://doi.org/10.1016/j.pbi.2017.04.022

Rabbani MA, Pervaiz ZH, Masood MS (2008) Genetic diversity analysis of traditional and improved cultivars of Pakistani rice (*Oryza sativa* L.) using RAPD markers. Electron J Biotechnol 11(3):1–10

Rahman MA, Thomson MJ, Shah-E-Alam M, de Ocampo M, Egdane J, Ismail AM (2016) Exploring novel genetic sources of salinity tolerance in rice through molecular and physiological characterization. Ann Bot 117:1083–1097

Rahnama A, James RA, Poustini K, Munns R (2010) Stomatal conductance as a screen for osmotic stress tolerance in durum wheat growing in saline soil. Func Plant Biol 37(3):255–263

Rawia AE, Lobna ST, Soad MMI (2011) Alleviation of adverse effects of salinity on growth, and chemical constituents of marigold plants by using glutathione and ascorbate. J Appl Sci Res 7(5):714–721

Rayan AM, Abbott LC (2015) Compositional analysis of genetically modified corn events (NK603, MON88017×MON810 and MON89034×MON88017) compared to conventional corn. Food Chem 176:99–105. https://doi.org/10.1016/j.foodchem.2014.12.044

Rehman S, Guo S, Hou Y (2016) Rational design of Si/SiO_2@Hierarchical porous carbon spheres as efficient polysulfide reservoirs for high-performance Li-S battery. Adv Mater 28:16. https://doi.org/10.1002/adma.201506111

Rohman MM, Talukder MZA, Hossain MG, Uddin MS, Amiruzzaman M, Biswas A et al (2016) Saline sensitivity leads to oxidative stress and increases the antioxidants in presence of proline and betaine in maize (*Zea mays* L.) inbred. Plant Om J 9(1):35–47

Rossatto T, do Amaral MN, Benitez LC, Vighi IL, Braga EJB, de Magalhaes Júnior AM, Maia MAC, da Silva Pinto L (2017) Gene expression and activity of antioxidant enzymes in rice plants, cv. BRS AG, under saline stress. Physiol Mol Biol Plants 23(4):865–875

Roy PR, Tahjib-Ul-Arif M, Polash MAS, Hossen MZ, Hossain MA (2019) Physiological mechanisms of exogenous calcium on alleviating salinity-induced stress in rice (*Oryza sativa* L.). Physiol Mol Biol Plants 25(3):611–624

Santabarbara S, Casazza AP, Ali K, Economou CK, Wannathong T, Zito Z, Redding KE, Rappaport F, Purton S (2013) The requirement for carotenoids in the assembly and function of the photosynthetic complexes in *Chlamydomonas reinhardtii*. Plant Physiol 161:535–546. https://doi.org/10.1104/pp.112.205260

Sappl PG, Oñate-Sánchez L, Singh KB, Millar AH (2004) Proteomic analysis of glutathione S-transferases of *Arabidopsis thaliana* reveals differential salicylic acid-induced expression of the plant-specific phi and tau classes. Plant Mol Biol 54(2):205–219

Shalata A, Neumann PM (2001) Exogenous ascorbic acid (vitamin C) increases resistance to salt stress and reduces lipid peroxidation. J Exp Bot 52:2207–2211

Sharma I, Ahmad P (2014) Catalase: a versatile antioxidant in plants (chapter 4). In: Oxidative damage to plants antioxidant: networks and signaling. Academic Press, pp 131–148. https://doi.org/10.1016/B978-0-12-799963-0.00004-6

Sharma R, Sahoo A, Devendran R, Jain M (2014) Over-expression of a rice tau class glutathione S-transferase gene improves tolerance to salinity and oxidative stresses in *Arabidopsis*. PLoS One 9(3):e92900. https://doi.org/10.1371/journal.pone.0092900

Singh S, Anjum NA, Khan NA, Nazar R (2008) Metal-binding peptides and antioxidant defence system in plants: significance in cadmium tolerance. Abiotic stress and plant responses. IK International, New Delhi, pp 159–189

Sofo A, Scopa A, Nuzzaci M, Vitti A (2015) Ascorbate peroxidase and catalase activities and their genetic regulation in plants subjected to drought and salinity stresses. Int J Mol Sci 16:13561–13578. https://doi.org/10.3390/ijms160613561

Sultana S, Khew CY, Morshed MM, Namasivayam P, Napis S, Ho CL (2012) Overexpression of monodehydroascorbate reductase from a mangrove plant (AeMDHAR) confers salt tolerance on rice. J Plant Physiol 169(3):311–318. https://doi.org/10.1016/j.jplph.2011.09.004

Szabados L, Savoure A (2010) Proline: a multifunctional amino acid. Trends Plant Sci 15:89–97. https://doi.org/10.1016/j.tplants.2009.11.009

Talaat NB (2014) Effective microorganisms enhance the scavenging capacity of the ascorbateglutathione cycle in common bean (*Phaseolus vulgaris* L.) plants grown in salty soils. Plant Physiol Biochem 80:136–143

Tehrim S, Pervaiz ZH, Rabbani MA (2012) Molecular characterization of traditional and improved rice cultivars based on random amplified polymorphic DNAs (RAPDs) markers. Afr J Biotechnol 11(45):10297–10304

Teixeira FK, Menezes-Benavente L, Margis R, Margis-Pinheiro M (2004) Analysis of the molecular evolutionary history of the ascorbate peroxidase gene family: inferences from the rice genome. J Mol Evol 59:761–770

Toshikazu T, Yamamoto T, Miyauchi M (2002) Interpretation of actinide transmutation in thermal and fast reactors. Progress Nucl Energy 40(3-4):449–456. https://doi.org/10.1016/S0149-1970(02)00037-9

Trovato M, Forlani G, Signorelli S, Funck D (2019) Proline metabolism and its functions in development and stress tolerance. In: Hossain M, Kumar V, Burritt D, Fujita M, Mäkelä P (eds) Osmoprotectant-mediated abiotic stress tolerance in plants. Springer, New York. https://doi.org/10.1007/978-3-030-27423-8_2

Trovato M, Mattioli R, Costantino P (2008) Multiple roles of proline in plant stress tolerance and development. Rend Lincei 19:325–346. https://doi.org/10.1007/s12210-008-0022-8

Turan S, Tripathy BC (2012) Salt and genotype impact on antioxidative enzymes and lipid peroxidation in two rice cultivars during de-etiolation. Protoplasma 250:209–222. https://doi.org/10.1007/s00709-012-0395-5

Turan S, Tripathy BC (2013) Salt and genotype impact on antioxidative enzymes and lipid peroxidation in two rice cultivars during de-etiolation. Protoplasma 250(1):209–222

Uzilday RÖ, Uzilday B, Yalçinkaya T, Türkan İ (2017) Mg deficiency changes the isoenzyme pattern of reactive oxygen species-relatedenzymes and regulates NADPH-oxidase-mediated ROS signaling in cotton. Turk J Biol 41(6):868–880

Vighi IL, Benitez LC, Amaral MN, Moraes GP, Auler PA, Rodrigues GS, Deuner S, Maia LC, Braga EJB (2017) Functional characterization of the antioxidant enzymes in rice plants exposed to salinity stress. Biol Planta 61:540–550. https://doi.org/10.1007/s10535-017-0727-6

Wang P, Du Y, Hou Y, Zhao Y, Hsu C, Yuan F, Zhu X, Tao WA, Song C, Zhu J (2014) Nitric oxide negatively regulates abscisic acid signaling in guard cells by S-nitrosylation of OST1. Proc Natl Acad Sci U S A 112:613–618

Wu TM, Lin WR, Kao CH, Hong CY (2015) Gene knockout of glutathione reductase 3 results in increased sensitivity to salt stress in rice. Plant Mol Biol 87(6):555–564. https://doi.org/10.1007/s11103-015-0290-5

Wu TM, Lin WR, Kao YT, Hsu YT, Yeh CH, Hong CY, Kao CH (2013) Identification and characterization of a novel chloroplast/mitochondria co-localized glutathione reductase 3 involved in salt stress response in rice. Plant Mol Biol 83(4–5):379–390. https://doi.org/10.1007/s11103-013-0095-3

Xiang C, Werner BL, Christensen ELM, Oliver DJ (2001) The biological functions of glutathione revisited in Arabidopsis transgenic plants with altered glutathione levels. 126:564–574. https://doi.org/10.1104/pp.126.2.564

Xu S, Zhu S, Jiang Y, Wang N, Wang R, Shen W, Yang J (2013) Hydrogen-rich water alleviates salt stress in rice during seed germination. Plant Soil 370:47–57

Yassin M, Mekawy AM, EL Sabagh A, Islam MS, Hossain A, Barutcular C, Alharby H, Bamagoos A, Liu L, Ueda A, Saneoka H (2019) Physiological and biochemical responses of two bread wheat (*Triticum aestivum* L.) genotypes grown under salinity stress. Appl Ecol Environ Res 17(2):5029–5041

Zakar T, Laczko-Dobos H, Toth TN, Gombos Z (2016) Carotenoids assist in cyanobacterial photosystem II assembly and function. Front Plant Sci 7:295. https://doi.org/10.3389/fpls.2016.00295

Zhang Z, Zhang Q, Wu J, Zheng X, Zheng S, Sun X, Qiu Q, Lu T (2013) Gene knockout study reveals that cytosolic ascorbate peroxidase 2 (OsAPX2) plays a critical role in growth and reproduction in rice under drought, salt and cold stresses. PLoS One 8(2):e57472. https://doi.org/10.1371/journal.pone.0057472

Zhao F, Zhang H (2006) Salt and paraquat stress tolerance results from co-expression of the Suaeda salsa glutathione S-transferase and catalase in transgenic rice. Plant Cell Tissue Organ Cult 86(3):349–358

Zou JJ, Li XD, Ratnasekera D, Wang C, Liu WX, Song LF, Zhang WZ, Wu WH (2015) Arabidopsis calcium-dependent protein kinase 8 and catalase 3 function in abscisic acid-mediated signaling and H_2O_2 homeostasis in stomatal guard cells under drought stress. Plant Cell 27:1445–1460. https://doi.org/10.1105/tpc.15.00144. 1-16

Soybean Plants Under Waterlogging Stress: Responses and Adaptation Mechanisms

Khadeja Sultana Sathi, Abdul Awal Chowdhury Masud, Taufika Islam Anee, Khussboo Rahman, Naznin Ahmed, and Mirza Hasanuzzaman

Abstract Waterlogging stress retards plant growth and development by inducing a number of physiochemical processes. Plants subjected to waterlogging suffer from substantial yield losses. In soybean, waterlogging stress creates partial or full deprivation of oxygen, which leads to severe morphophysiological decays in the plant. Excess accumulation of reactive oxygen species and poor antioxidant defense system become phenomenal under such conditions. As a consequence, water and nutrient uptake, stomatal conductance, photosynthesis rate, enzymatic activities, and hormonal balances are greatly disrupted. However, soybean develops few anatomical features among which the formation of the adventitious root is of great importance to counteract the detrimental effect of waterlogging stress. This chapter focuses on soybean plant responses to waterlogging conditions and different approaches how waterlogging tolerance or adaptation can be imparted in soybean through morphological and anatomical modifications as well as hormonal regulation and antioxidant balance.

Keywords Abiotic stress · Adventitious root · Aerenchyma · Flooding · Leguminous crop · Anaerobiosis

Abbreviations

ABA	abscisic acid
ADH	alcohol dehydrogenase
APX	ascorbate peroxidase
CAT	catalase
Chl	chlorophyll
CySNO	*S*-nitroso L-cysteine
DAS	days after sowing

K. S. Sathi · A. A. C. Masud · T. I. Anee · K. Rahman · N. Ahmed · M. Hasanuzzaman (✉)
Department of Agronomy, Faculty of Agriculture, Sher-e-Bangla Agricultural University, Dhaka, Bangladesh

M. Hasanuzzaman et al. (eds.), *Managing Plant Production Under Changing Environment*, https://doi.org/10.1007/978-981-16-5059-8_5

DAT	days after treatment
EBR	24-epibrassinolide
Fe	iron
GA	gibberellic acid
GR	glutathione reductase
GB	glycinebetaine
H_2O_2	hydrogen peroxide
JA	jasmonic acid
LDH	lactate dehydrogenase
MDA	malondialdehyde
MSI	membrane stability index
NADH	nicotinamide adenine dinucleotide hydride
PDC	pyruvate decarboxylase
POD	peroxidase
Pro	proline
QTL	quantitative trait locus
ROS	reactive oxygen species
RWC	relative water content
SA	salicylic acid
SNO	*S*-nitrosothiol
SNP	sodium nitroprusside
SOD	superoxide dismutase
WSL	waterlogged-susceptible line
WTL	waterlogged-tolerant line

1 Introduction

The detrimental effect of nonliving factors in a given situation on the living organism is collectively referred to as abiotic stress. Plants being immobile in nature are generally exposed to a number of environmental stresses, for example, drought, waterlogging, salinity, extreme temperatures, metal/metalloid toxicity, high/low light, UV radiation, atmospheric pollutants, and so on (Hasanuzzaman et al. 2012, 2017a, 2018). Acquaah (2007) reported that abiotic stress lowered agricultural production by more than 50%. A wide range of morphological, physiological, biochemical, and molecular alterations are triggered by abiotic stress that adversely perturbs crop growth and production. It may lead to an unendurable metabolic pressure on cells that decreases growth and leads to plant death in extreme cases, since the stress becomes excessive and lasts for an expanded period (Hasanuzzaman et al. 2012). Waterlogging or flooding is the condition of the soil where the water table outreaches the root zone of the soil or when soil becomes fully saturated and becomes unsuitable for crop production. Flooding is of two kinds: (i) waterlogging (only the root systems are under anaerobic condition) and (ii) submergence;

submergence may be of two kinds: (a) partial submergence (all roots are immersed in water and just a portion of shoots are under water) and (b) complete submergence (the whole plant is under the water). Waterlogging occurs as a result of excess precipitation, tides or storms, floods, and sometimes due to improper drainage practices (Kim et al. 2015). The oxygen availability in plants gets declined under waterlogged conditions which primarily causes hypoxia (O_2 deficiency) that disturbs aerobic respiration and subsequently causes anoxia (O_2 absence) and inhibits respiration (Wegner 2010). Basic metabolic, developmental, and physiological processes in plants such as cell division, growth, respiration, water and nutrient uptake, and transportation require $O_{2.}$ So, this oxygen deprivation results in inhibited growth, N-fixation, yield along with chlorosis, necrosis, and defoliation that may sometimes lead to plant death. Accumulation of ethanol in waterlogged plants disrupts plant developmental processes to a large extent (Hasanuzzaman et al. 2017b). Waterlogging-induced changes in plant morphological, physiological, or biochemical attributes along with the response mechanisms in plants have been studied by a number of researchers worldwide. Akhtar and Nazir (2013) observed a reduction in growth and development of plants such as lack of oxygen and deficit of micronutrients under waterlogging stress. Flooding stress posed detrimental effects on seed germination, growth, yield, and root anatomy in plants (Al-Amri 2019; Rajendran et al. 2019; Ploschuk et al. 2020). Oxygen deficiency in the soil triggers the accumulation of ethylene, phytotoxic minerals (Mn^{2+}, Fe^{2+}), reactive oxygen species (ROS), reactive nitrogen species (RNS), etc. in plants (Voesenek and Bailey-Serres 2015). These ROS create oxidative stress leading to cellular damages like lipid peroxidation, nucleic acid damage, enzyme inactivation, and programmed cell death (Anjum et al. 2015). Plants exposed to waterlogging try to diffuse oxygen in aerial parts of plants for adapting to the stress condition. For adaptation under stress conditions, plants modify morphological structures that include leaves hyponasty, shoot elongation, the formation of aerenchyma in root cells, formation of lenticels in the stem, and commencing adventitious roots in the waterlogged stem. These workers also reported that waterlogging stress triggered ethylene production, which drives the plant to epinasty, chlorosis, and senescence of leaves (Voesenek and Bailey-Serres 2015).

Soybean (*Glycine max* L.) is a popular legume crop that is especially known for its versatile uses as oil, food, and forage. Also, it has a high nutritional value containing significant amounts of protein as well as oil for what it is being extensively cultivated worldwide (Kim et al. 2015). As a cultivated crop for obtaining a good yield both in quantity and quality, soybean growers have to focus on the optimization of environmental conditions, irrigation, drainage, fertilizer, plant protection, and other cultivation factors. Although it is possible to control most of the cultivation factors by humans, controlling the environmental factors is almost impossible. Hence, crops face different disastrous situations that are uninvited. In the present era of global warming and subsequent unexpected climate events, flooding or waterlogging has turned into an important issue for environmentalists and researchers worldwide. Crop species, growth stages, stress duration, water levels, and so on have been considered in different aspects in a wide range of experiments. A number of such studies and relevant findings have been discussed in Sect. 2 of this chapter. However, there are many original articles, review articles, and books/book chapters available that contain elaborative studies of waterlogging/ flooding causes and effects, sometimes adaptive mechanisms. But very few crop-specific studies have been done so far mostly limited to submerged crops or subaquatic plants (rice plants). Other dryland crops like soybean also need attention in this regard as the fall victim to such condition which may result in the deterioration of protein and oil consumption sources worldwide. Recent findings demonstrate the ability of dryland species like soybean to aerenchyma formation besides adventitious root and hypertrophied lenticels (Yamauchi et al. 2013; Kim et al. 2015). Therefore, this chapter has been designed to focus on the possible responses and tolerance of soybean to waterlogging stress. The objective of this chapter is to review the recent research works on effects or responses of soybean to waterlogging stress and include the updated information on the understanding of possible adaptive or tolerance mechanisms of soybean to waterlogging stress.

2 Plant Responses to Waterlogging

2.1 Growth

One of the primary impacts of waterlogging is the suppression of plant growth due to a lack of oxygen in the root zone which may cause eventual plant death. In addition, this deficit oxygen condition substantially hampers plant growth, development, and survival as the root respiration substantially hampered due to O_2 shortage which accelerates an energy-deficit state (Garcia et al. 2020). Plant responses to waterlogging vary depending on plant species, growth stages, stress duration, and so on. In *Gossypium hirsutum*, the highest reduction of plant height (30%) and leaf area (40%) was found upon exposure to waterlogging stress during the flowering stage for 10 d compared to control (Wang et al. 2017). González et al. (2009) demonstrated that waterlogging stress reduced leaf area and specific leaf area (SLA) by 36% and 26% than the control quinoa (*Chenopodium quinoa* Wild.) plants. While working with 15 genotypes of *Zea mays* under waterlogging stress, Lone and Warsi (2009) found that nearly all the tested genotypes showed depletion in the plant height and ear height. In another experiment, Ren et al. (2014) observed that the overall growth and development of *Z. mays* were significantly affected due to waterlogging. In *Solanum lycopersicum*, increased adventitious root formation was observed under waterlogging (Ezin et al. 2010). Palta et al. (2010) reported that leaf area (56%), the number of branches (50%), and root growth were decreased in *Cicer arietinum* cultivars at flooding stress in comparison with the control plants. In *Vigna radiata*, plant height (33%), leaf area (31%), the number of leaves (33%), the number of branches (34%), and dry matter (30%) were considerably reduced by flooding stress (4 d) compared to control (Prasanna and Rao 2014). In some research, the exaggerated effect of waterlogging has been recorded on some water-loving plants. While working on two rice genotypes (Puzhuthiikar and IR72593), Anandan et al. (2015) found that the genotype Puzhuthiikar exhibited a noteworthy increment of leaf blade length, sheath length, and area but a reduction in leaf blade area than the genotype IR72593 under prolonged flooding stress. Duhan et al. (2018) carried out an experiment with four genotypes of *Cajanus cajan* and found that total plant biomass was declined by 22–28% under 8 d of waterlogging. Among the tested genotypes, ICPH 2431 exhibited a minimum and UPAS 120 exhibited a maximum decline in plant biomass at flooding stress. Waterlogging-tolerant genotypes exhibited less decline in plant height, tillers, shoot, and root biomasses than the waterlogging-sensitive genotypes of *Hordeum vulgare* (Luan et al. 2018). Li et al. (2018) performed a study with 18 different *Z. mays* cultivars and observed plant height, dry weight, root length, root hairs, root surface area, and root volume were decreased under flooding stress. According to Barickman et al. (2019) under the same duration plant height, leaf number, leaf area, fresh mass, and dry mass of *Cucumis sativus* decreased by 12, 11, 38, 35 and 28% respectively, over non-waterlogged plants. Liu et al. (2020) found that waterlogging (7 d) reduced root length (11.8 and 16.0%), root dry weight (10 and 11%), and shoot dry weight (13 and 25%) under flooding stress in lowland (YueFu) and upland (IRAT109) *Oryza sativa*, respectively.

2.2 *Physiology*

Waterlogging stress has a significant influence on plant physiology and metabolism. Lone and Warsi (2009) reported that waterlogging remarkably reduced transpiration rate, stomatal conductance, and SPAD value in *Z. mays* genotypes. The net photosynthetic rate of wheat showed a decreased manner when the plants were exposed to hypoxia (Zheng et al. 2009). Flooding impedes the photosynthetic rate in plants. Akhtar and Nazir (2013) also stated that in C_3 plants, waterlogging stress impelled stomatal closure. Kumar et al. (2013) performed research with four genotypes of *V. radiata*, including two tolerant genotypes (T-44 and MH-96-1) and two susceptible to waterlogging (MH-1 K-24 and Pusa Baisakhi). Plants were exposed to 3, 6, and 9 d of waterlogging at the vegetative stage. Relative water content (RWC), membrane stability index (MSI), photosynthetic rate, chlorophyll (Chl), and carotenoid contents were reduced at varied periods of waterlogging stress. Nevertheless, the effects were more noticeable in sensitive genotypes than the tolerant genotypes. Leaf water potential, net photosynthesis, and Chl concentration showed a decreasing manner due to recurrent flooding in *Cichorium intybus* (Vandoorne et al. 2014). In *Saccharum officinarum*, RWC and proline (Pro) content increased significantly while Chl *a*, Chl *b*, and carotenoids contents declined under waterlogged treatment (Bajpai and Chandra 2015).

During waterlogging conditions of nine wild solanaceous plants, there observed a rapid stomatal closure and reduced photosynthesis and stomatal conductance. Among them, *Solanum torvum* species appeared photosynthetically better under waterlogging and had greater stomatal conductance as well (Kumar et al. 2018). According to Wang et al. (2017), due to the reduction of O_2 level and increased respiration rate in root tissue, the highest reduction of SPAD value and the photosynthetic rate was observed at the flowering stage at 6, 8, and 10 d of waterlogging stress. Duhan et al. (2018) performed an experiment with four genotypes (ICPH-2431, PARAS, UPAS-120, H09-33) of *C. cajan* plants and found that Pro and membrane injury (MI) extended under waterlogging treatment. Alizadeh-Vaskasi et al. (2018) noticed that waterlogging treatments declined Chl *a* and *b* and carotenoids and enhanced Pro contents in three *Triticum aestivum* genotypes (N-93-19, N-93-9, and N-92-9) in the tillering and stem elongation stages. While working with six *Z. mays* genotypes, Akter et al. (2018) found the total Chl content was decreased by 12%, 9% and 8% in CML54 × CML487; 8%, 5%, and 3% in P18; 30%, 13%, and 43% in CML 54; 18%, 20%, and 14% in CML 486 × CML 487; 19%, 14%, and 27% in CML 486 followed by 12%, 10%, and 13% in CML 487, respectively, on the 2, 4, and 6 d of waterlogging stress. Waterlogging treatment significantly reduced MSI, Chl content, and fluorescence in four *Vigna mungo* genotypes (Uttara, T-44, IC530491, IC519330) under prolonged waterlogging (10 d) at the vegetative stage. In comparison with Uttara, MSI and Chl content was greater in IC530491, IC519330, and T44 (Ruchi et al. 2019). Anee et al. (2019) conducted an experiment with sesame plants that were exposed to waterlogging for 2, 4, 6, and 8 d and showed a reduction of RWC (75%), Pro content (20%), Chl *a*, Chl (*a* + *b*) and carotenoid content under waterlogging compared to their respective controls for up to 8 d.

2.3 Yield Loss

Waterlogging stress significantly reduced yield at both vegetative and reproductive stages except for some aquatic plants (Table 1). Lone and Warsi (2009) conducted experiments with 15 genotypes of *Z. mays*. Among them, 5 were parents and 10 were their single crosses in both winter and summer seasons. They observed that excess soil moisture exhibited a drastic reduction in grain yield of all the genotypes. Yield reduction was higher in the winter trial than in the summer trial. In winter trials, the reduction of yield ranged between 19% in YHPP45 (tolerant) and 53% in Pop 3121 × YHPP45. While the reduction of yield observed highest in Tarun83 (susceptible genotype) which is 66% and lowest in YHPP45 (tolerant) which is 2% in the summer trial. Forty-d-old *S. lycopersicum* plants (two cultivars and two wild related species) were exposed to continuous waterlogging for 2, 4, 6, and 8 d duration by Ezin et al. (2010). Waterlogging for 8 d showed a drastic reduction in the yield of all the genotypes. Yield reduction was observed by 23%, 69%, 90%, and 100% in CLN2498E, CA4, LA1421, and LA1579, genotypes, respectively upon exposure to 8 d of waterlogging. Among the four genotypes, LA1579 was waterlogging sensitive, CLN2498E and CA4 showed high tolerance, and LA1421 showed tolerance in some extent. Waterlogging for 12 d reduced seed yield of kabuli cultivar (Almaz)

Table 1 Yield reduction in plants under waterlogging conditions

Crop species	Stress duration	Yield reduction	References
Z. mays	7 d of submergence at knee height stage	19% (tolerant genotype) and 66% (susceptible genotype)	Lone and Warsi (2009)
S. lycopersicum cv. CLN2498E, CA4, LA1421, LA1579	8 d of waterlogging	23%, 69%, 90% and 100%	Ezin et al. (2010)
C. arietinum cv. Kabuli (Almaz) and desi (Rupali)	12 d of waterlogging	54% and 44%	Palta et al. (2010)
T. aestivum	30 d of waterlogging	45%	Rasaei et al. (2012)
V. radiata	3, 6, and 9 d of waterlogging	20%, 34% and 52%	Kumar et al. (2013)
T. aestivum	28 d of waterlogging	56%	Amri et al. (2014)
Sesamum indicum cv. BARI til 2 and BARI til 3	12, 24 and 36 h of waterlogging	24%, 38% and 39% in BARI til 2; 29%, 46% and 53% in BARI til 3, respectively	Sarkar et al. (2016)
S. indicum cv. BD 6992	3 d of waterlogging	7%	Saha et al. (2016)
T. aestivum cv. N-93-9, N-92-9 and N-93-19 genotypes	21 d of waterlogging	60%, 56% and 37%	Alizadeh-Vaskasi et al. (2018)
Brassica napus	At early stem elongation and floral bud initiation stage for 2 weeks	25% and 15%	Wollmer et al. (2018)
T. aestivum	14 d waterlogging at 65 and 85 DAS	14% and 29%	Ploschuk et al. (2020)
C. cajan cv. UPAS 120 and ICPH 2431	8 d of waterlogging	62% and 27%	Duhan et al. (2018)

and desi cultivar (Rupali) of *C. arietinum* by 54% and 44%, respectively (Palta et al. 2010).

Yaduvanshi et al. (2010) worked with eight *T. aestivum* genotypes exposing to 15 d of waterlogging. They detected grain yield reduction of the genotypes due to waterlogging stress. Yield reduced in the genotypes by 2% (KRL), 3% (KRL 200), 9% (KRL 146), 10% (NW 1076), 12% (KRL 3-4), 100% (HD 2009), 162% (PBW 343), 190% (Brookton) were recorded. In another experiment, Rasaei et al. (2012) detected a yield reduction of *T. aestivum* for 10, 20, and 30 d of waterlogging. They reported that the highest yield reduction (45%) was observed for 30 d of waterlogging. In *V. radiata*, sensitive genotypes showed 70% (Pusa Baisakhi) and 85% (MH–1 K–24) yield reduction when exposed to 9 d of waterlogging (Kumar et al. 2013).

Ren et al. (2014) worked with 2 summer *Z. mays* genotypes (cv. Denghai 605 (DH605) and Zhengdan 958 (ZD958)) in field condition exposing 3 and 6 d of waterlogging at three-leaves stage (V3), six-leaves stage (V6), and 10th day after tasseling stage (10VT). Waterlogging (6 d) reduced yield by 23%, 32%, 20%, 24%, 8%, and 18% in DH605 genotype and 21%, 35%, 15%, 33%, 7%, and 12% in ZD958 genotype at V3-3, V3-6, V6-3, V6-6, 10VT-3, and 10VT-stages when compared to their control plants. In *S. indicum,* Sarkar et al. (2016) reported that upon exposure to 36 h of waterlogging yield declined by 39% in BARI til 2 and 53% in BARI til 3 compared to their control plants. Saha et al. (2016) conducted an experiment with four *S. indicum* genotypes imposing 3 d of waterlogging. They noticed yield reduction was minimum in BD 7012 (24%) and maximum in BD 6980 (44%) genotypes. Ball number plant^{-1}, weight of single boll, and seed cotton yield of *G. hirsutum* were greatly reduced upon exposure to long-term (10 d) waterlogging stress at flowering stage and the maximum yield reduction was 38% at this stage (Wang et al. 2017). Grain yield declined by 60%, 56%, and 37% in N-93-9, N-92-9, and N-93-19 genotypes, respectively upon exposure to waterlogging for 21 d in comparison with their respective control plants of *T. aestivum* (Alizadeh-Vaskasi et al. 2018). Duhan et al. (2018) stated that between two genotypes of *C. cajan*, UPAS 120 showed 62% yield reduction while in ICPH 2431 the reduction was 27% due to waterlogging stress (8 d).

3 Soybean Plant Responses to Waterlogging

Soybean crops are usually not tolerant to waterlogging stress (Tougou et al. 2012). In different parts of the world, soybean growth and production have decreased significantly (Van Nguyen et al. 2017). In different flooding cycles, soybean plants reacted differently to flooding stress (Wu et al. 2017). In soybean, waterlogging stress switched aerobic respiration toward anaerobic. Anaerobic respiration triggers alcoholic (C_2H_5OH) fermentation by generating waterlogged-inducible proteins that help in the generation of NAD^+ and conveyed sharp increment activity of alcohol dehydrogenase (ADH) in soybean plants (Komatsu et al. 2011). Furthermore, a reduction of gas exchange and photosynthetic activity was observed in the leaves of soybean (Mutava et al. 2015; Garcia et al. 2020). Soybean plants faced nutrient deficiency, limited water uptake, reduced N_2 fixation, and restricted root permeability resulting in chlorosis, necrosis, and defoliation in plants and ultimately causing yield loss (Fig. 1). According to Oosterhuis et al. (1990), waterlogging can diminish the yield of soybeans by 17–43% at the vegetative growth stage and by 50–56% at the reproductive stage.

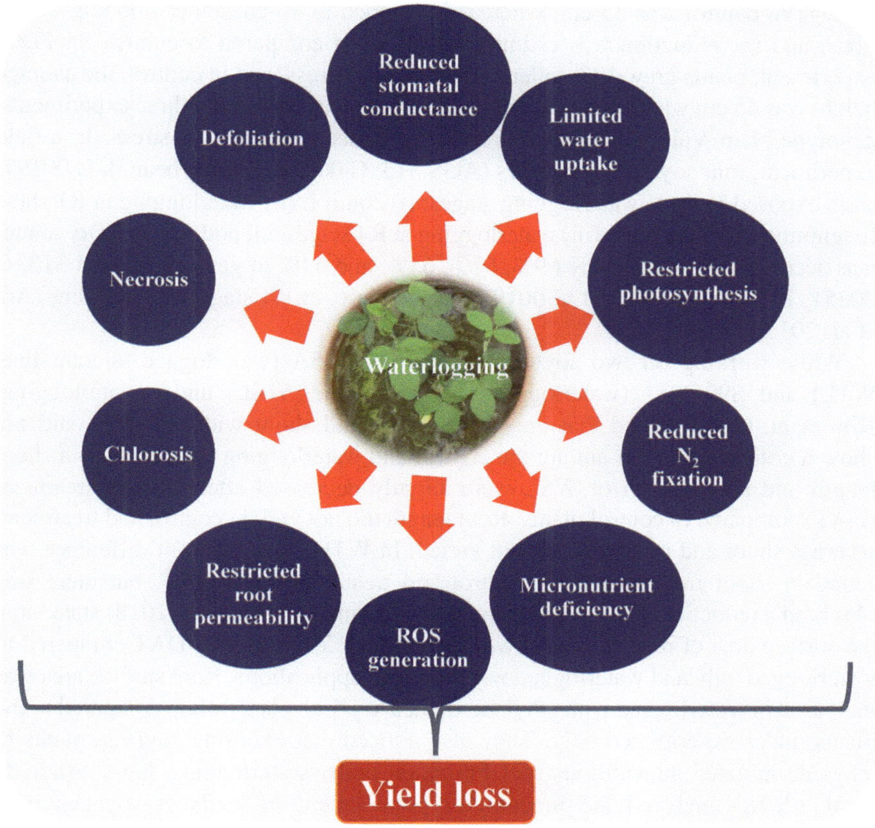

Fig. 1 Effects of waterlogging stress on soybean plants

3.1 Growth

Waterlogging for a certain period of time has an adverse effect on the growth and development of soybean plants. Youn et al. (2008) performed a study with supernodulating mutants (SS2-2 and Sakukei 4) and their wild type (Sinpaldalkong 2 and Enrei) of soybean. Waterlogging stress for about 15 d was imposed at the beginning of the flowering (R1) stage. After the removal of water, they observed that root dry mass was reduced in supernodulating mutants and wild types by 62–67% and 41–45%, respectively. Thirty days after removal of water presented 64–75% and 51–64% reduction of root dry mass in supernodulating mutants and wild types, respectively. VanToai et al. (2010) conducted both screen house (SP05 and SU05) and field tests (FD06) with 21 soybean genotypes to analyze waterlogging tolerance. Plant height was affected in both tests. In SP05 experiment, the height of the plant ranged from 22 to 69 cm with an average of 47 cm in control plants of 21 genotypes when waterlogging decreased the average to 30 cm. In SU05 experiment, the average in control was 55 cm, whereas it dropped to 40 cm under flooding conditions, and the reduction was estimated 27% when compared to control. In FD06 experiment, plants grew 13% taller under flooding stress, and in control, the average height was 45 cm whereas 50 cm in waterlogging treatment. In all three experiments, genotype Nam Vang and VND2 grew taller under waterlogging stress. In a field experiment, four soybean genotypes (AGS 313, G 00351, BD Soybean-4, G 00197) were exposed to three waterlogging stages: (i) control, (ii) waterlogging at R1 stage (beginning to flower), and (iii) waterlogging at R4 stage (full pod) for 7 d. Dry matter was accumulated in shoot by 69%, 67%, 65%, and 54% in genotypes AGS 313, G 00351, BD Soybean-4, and G 00197, respectively, at R4 stage waterlogging (Ara et al. 2015).

While working on two soybean lines, PI408105A (waterlogged-tolerant line, WTL) and S99-2281 (waterlogged-susceptible line, WSL) under waterlogging, Kim et al. (2015) found that the shoot length and shoot width of WTL did not show a notable variation among the control and waterlogging treatments, but shoot length and shoot width of WSL were slightly decreased after 10 d of treatment (DAT) compared to control plants. Root length did not vary in control and treatment, whereas shoot and root fresh weight varied. In WTL, no significant difference was found in shoot fresh weight in control and treatment for 5 DAT, but there was observed a reduction at 10 DAT. In another experiment, Kim et al. (2018) stated that the surface area of root in control was increased at 5, 10, and 15 DAT compared to waterlogged only and waterlogged with ethylene applications. Root surface area was increased in waterlogged with ethylene-treated soybean plants when compared to the plants under waterlogged only. They also noticed that exposing soybean plants to ethylene initiated adventitious root formation under waterlogging stress. Andrade et al. (2018) analyzed the function of pretreatment in seeds of soybean with hydrogen peroxide (70 mM H_2O_2 solution) for stimulating the resistance of soybean under flooding stress. Accumulation of biomass in shoots and roots and stem diameter enhanced in soybean plants as a result of seed pretreatment with H_2O_2.

We also observed that 15-d-old soybean plant under waterlogging stress for about 12 and at 25 DAS showed a decline in plant height, number of leaves, branches, and leaf area in waterlogged plant in comparison with the control (Fig. 2; unpublished data).

3.2 Nutrient Uptake

Waterlogging stress creates hypoxic or anoxic condition in the plant root zone which causes wilting although the plant is surrounded by surplus water. As a result, essential nutrients like N, K, Ca, and Mg availability are greatly hampered under such anaerobic conditions (Akhtar and Nazir 2013). Furthermore, downregulation of different plant physiological processes, namely, photosynthesis, respiration, carbohydrate partitioning, and hormonal imbalance stunted crop growth and yield significantly under flooding stress (Kim et al. 2015; Valliyodan et al. 2017). In addition,

Fig. 2 Growth reduction in soybean plants (genotype BINA soybean-2, at 25 DAS) under waterlogging stress

proteolytic activity is enhanced under the submerged condition which leads to a lower accumulation of carbohydrates involved in protein decomposition. Thus, under prolonged waterlogged conditions N content increases in plant tissues, whereas the amount of P and K contents decline at multiple crop growth stages as observed in *O. sativa* (Yamuangmorn et al. 2020). However, in soybean, N and K contents were significantly increased under waterlogging stress when soybean plants were supplemented with ethephon (ETP) as a phytohormone. The total macro element content was decreased in both waterlogged and ETP-treated plants compared to control at 10 and 15 DAT. Content of C was higher in overall soybean plants compared to other elements, while P and Fe contents were increased under stressed conditions (Kim et al. 2018). Voesenek and Bailey-Serres (2015) stated that waterlogging stress causes O_2 deficiency in the soil, which triggered the accumulation of phytotoxic mineral nutrients (Mn^{2+}, Fe^{2+}).

Smethurst et al. (2005) observed that it is not Na and Fe toxicity responsible for the detrimental effect of waterlogging rather reduced nutrient constituents (K, P, Cu, Ca, Mg, Zn, and B) of leaves and roots that were notably decreased in *Medicago sativa* due to flooding stress. Plants uptaking both the macronutrients and the micronutrients got disturbed due to waterlogging stress, which is also documented by Akhtar and Nazir (2013). Similarly, Rhine et al. (2010) found that the contents of N, P, and K in soybean leaves were lower when waterlogging was imposed for 8 d. Due to hypoxic conditions in the soil, plant root activity was greatly hampered. As a result, a remarkable reduction in N content and N accumulation in the plant was observed (Sigua et al. 2012). In soybean, N uptake is reduced due to inadequate root permeability resulting from reduced hydraulic conductivity (Jitsuyama 2017). Fatimah and Nurhidayati (2020) showed that in Grobogan variety of soybean, total leaf N concentration in control reached 2.45% whereas in waterlogging treatment G1 (100%), G2 (150%), and G3 (200%) reduced around 1.31%; 1.16%, and 1.34%, respectively. Conversely, excess absorption of Fe might be exponentially increased under waterlogging stress which is a limiting factor for soybean yield in the reproductive stage (Lapaz et al. 2020).

3.3 *Physiology*

Waterlogging stress has a significant influence on plant physiology and metabolism. Photosynthesis is the most common and vital physiological process in plants responsible for the proper growth and nourishment of the crop species. Hypoxic or anoxic condition due to waterlogging severely impaired leaf gas exchange parameter along with stunted root growth compared to shoot (Fig. 3).

Leaf gas exchange was decreased under 7 d of waterlogging in five soybean genotypes (Garcia et al. 2020). Yordanova and Popova (2007) reported that photosynthesis and Chl content dropped remarkably under prolonged waterlogging stress. Reduction of photosynthetic activity in soybean plants under hypoxic conditions was also observed by Mutava et al. (2015). While working on 40 soybean genotypes, Wu et al. (2017) observed that 3 d of waterlogging notably reduced leaf Chl content resulting in color variation in leaves. However, Andrade et al. (2018) analyzed the role of pretreatment of soybean seeds with hydrogen peroxide (70 mM H_2O_2 solution) for stimulating the tolerance of soybean seedlings under flooding stress. The 12-day-old soybean seedlings were exposed to waterlogging (0, 16, and 32 d).

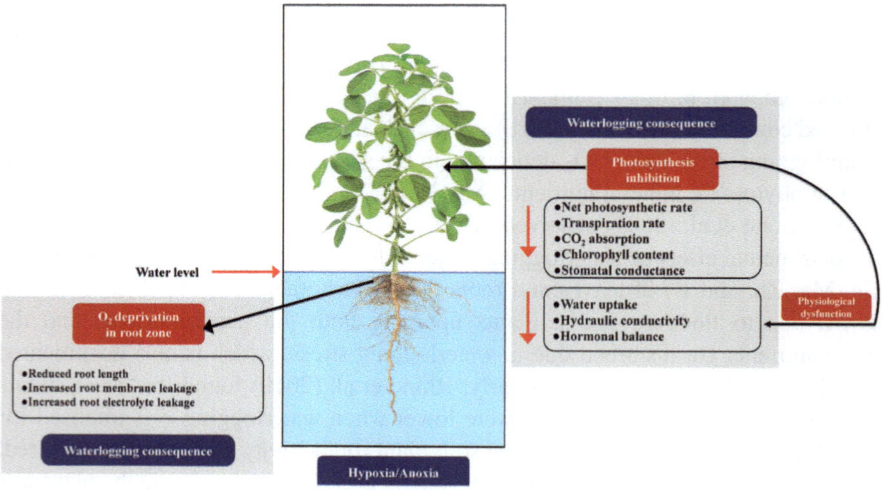

Fig. 3 Plant physiological changes under waterlogging stress

Since, 16 d of waterlogging, gas exchange and Chl content were higher in the pretreated plants than in the control. At 32 d, they tended to decrease while pretreated plants showed the least electrolyte leakage in the root cells. Furthermore, in soybean Kim et al. (2018) applied ethephon (ETP) on waterlogging stressed plants and found that application of ETP lessened the effect of flooding stress and enhanced photosynthetic activity as well as bioactive GA_4 content in comparison to untreated plants. Due to waterlogging stress, several proteins governing glucose degradation, sucrose accumulation, signal transduction, cell wall relaxing, and alcohol fermentation were changed in soybean plants (Komatsu et al. 2015). While working on two soybean lines, PI408105A (WTL) and S99-2281 (WSL) in waterlogging stress (5 cm from the soil surface), Kim et al. (2015) found that Pro contents were not significantly different at 5 DAT but showed a significant reduction at 10 DAT in both WTL and WSL. Shin et al. (2017) worked with 6 Korean *Z. mays* lines (KS85, KS124, KS140, KS141, KS163, KS1644) on which they imposed waterlogging for about 30 d at V3 stage. After 30 d of waterlogging treatment, they observed that SPAD values were reduced significantly and among the six lines, KS140 performed better. Alizadeh-Vaskasi et al. (2018) noticed that waterlogging treatments declined Chl *a* and *b* and carotenoids and enhanced Pro contents in three *T. aestivum* genotypes (N-93-19, N-93-9, and N-92-9) in the tillering and stem elongation stages. da-Silva and do Amarante (2020a) reported that Chl *a* and *b* declined after 120 h of waterlogging in soybean and the reduction was observed 2.8- and 1.4-fold lower than the control at 240 h of waterlogging. Meanwhile, Chl *a* and *b* were partly recovered after 120 h redistribution of oxygen. Likewise, 48 h of waterlogging reduced carotenoid levels, but a considerable increment of carotenoids levels was noticed after 120 h redistribution of oxygen.

3.4 Metabolic Responses

Under normal conditions, plants run aerobic respiration where glucose breaks down into pyruvic acid. These reactions occur in the cytosol of plants cell. After that, this pyruvic acid forms CO_2, water, and energy by a series of reactions with the presence of oxygen which took place in the mitochondria of the cell. Again, in absence of oxygen, plants cannot run aerobic respiration rather it switches to anaerobic which is also known as fermentation. Due to the absence of oxygen, pyruvic acid formed in the cytosol could not enter the TCA cycle. However, anaerobic respiration can undergo two fermentative pathways; one is lactate and another is ethanolic fermentation pathways and the ultimate by-products are lactate/ethanol, water, CO_2, and less amount of energy. The lactate fermentation pathway undergoes one-step reaction and produces lactate from pyruvate molecule which is catalyzed by lactate dehydrogenase (LDH) enzymes and meanwhile, NADH is oxidized into NAD^+. The ethanolic fermentation pathway undergoes two-step reactions, where pyruvate is decarboxylated to form acetaldehyde catalyzed by pyruvate decarboxylase (PDC) enzyme and acetaldehyde further converted to ethanol by alcohol dehydrogenase

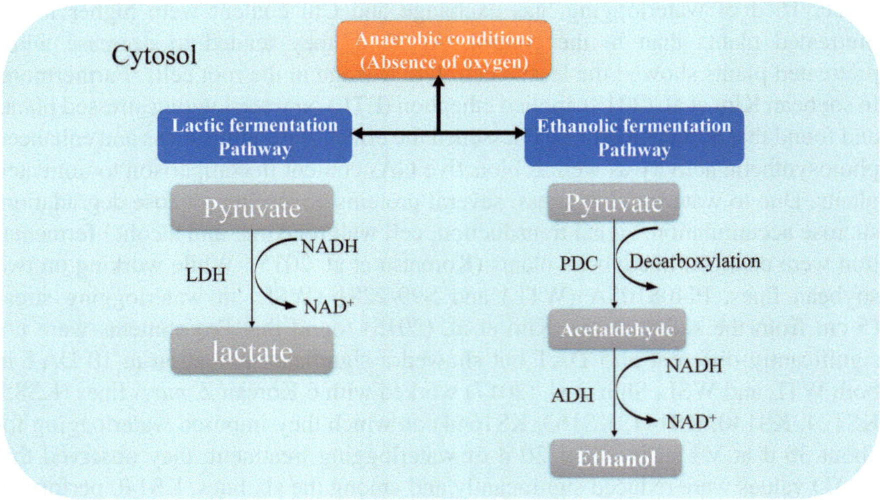

Fig. 4 Metabolic pathways induced under limited oxygen conditions in plants. *ADH* alcohol dehydrogenase, *LDH* lactate dehydrogenase, *NADH* nicotinamide adenine dinucleotide hydride, *NAD*$^+$ oxidized NADH, *PDC* pyruvate decarboxylase

(ADH) enzyme activity and regenerating NAD$^+$ (Fig. 4). Borella et al. (2014) worked with two soybean cultivars (Fundacep 53 RR and BRS Macota) imposing hypoxia stress for 24 and 72 h at R2 growth stages. They observed increased activity of fermentative enzymes both in roots and in nodules under hypoxia condition. The activity of ADH, PDC, LDH, and the content of lactate and ethanol increased both in roots and nodules under hypoxia stress and notably decreased when the plants returned to control condition. Pyruvate content, level of total soluble sugar, sucrose, and starch increased in roots but decreased in nodules of both soybean cultivars. Overall, the Fundacep 53 RR cultivar performed better than the BRS Macota cultivar.

3.5 Phenology

Plant phenology drastically altered at waterlogging conditions and these changes become much more severe at a longer duration of waterlogging. Kuswantoro (2015) stated that due to lack of nutrient availability and decay of root, day to flowering and day to day maturity delayed in waterlogged soybean plants compared to control plants. Miura et al. (2012) also observed a delay in flowering at 21 d of waterlogging stress. Beutler et al. (2014) conducted an experiment where soybean plants were exposed to 2, 4, 8, 16, and 32 d of waterlogging at flowering stage (R_2) and pod filling stage (R_5) and found a reduction in branch number with an extension of waterlogging duration. Sakazono et al. (2014) recorded a reduction in branch and

Fig. 5 Delay in maturation of soybean plants (genotype GC-840, 75-d-old) exposed to 12 d of waterlogging condition at 15 DAS

flower number in soybean at waterlogged conditions. Waterlogging substantially delayed the formation of pods in soybean plants which ultimately caused a delay in maturation (Cho and Yamakawa 2006). VanToai et al. (2001) reported a delay in maturation when soybean plants were exposed to waterlogging at vegetative stress. However, waterlogging at the reproductive stage accelerates senescence in soybean (GC-840) plants. We exposed 15-d-old soybean plant in waterlogging stress for 12 d, and at 75 DAS, we observed delay in pod maturation in waterlogging treated plants whereas, in control one, plants were already reached toward maturity (Fig. 5; unpublished data).

3.6 *Reproductive Development*

Waterlogging greatly hampers the reproductive development in soybean. The number of pods plant^{-1} and seeds pod^{-1} of soybean reduced at waterlogged condition (Ara et al. 2015). Similar results were recorded by Miura et al. (2012) in soybean plants where at 21 d of waterlogging condition in flowering stage reduced pod number by 54% corresponding to control. While working with 16 genotypes of soybean, Kuswantoro (2015) recorded a reduction in the number of branches, reproductive nodes, filled grains plant^{-1} at 5 cm standing water condition. VanToai et al. (2010) also observed significant relation of these parameters with the yield of soybean. At 6 d of waterlogging, the number of pods decreased by 23% corresponding to control (Sullivan et al. 2001). Similar results were also observed by Cho and Yamakawa (2006) in Saebyeolkong, Sobaeg-namulkong, and Pungsan-namulkong cultivars of soybean after 9 d of waterlogging.

3.7 Yield Attributes and Yield

Waterlogging negatively affects the yield of soybean. At the vegetative stage 17–43% and at reproductive stage 50–56% yield decline was observed in soybean at waterlogging stress. (Mustafa and Komatsu 2014). Cho and Yamakawa (2006) recorded 38%, 44%, and 66% seed yield reduction in Saebyeolkong, Sobaeg-namulkong, and Pungsan-namulkong cultivars of soybean, respectively, at 9 d of waterlogged condition. Rhine et al. (2010) worked with five soybean cultivars that were tolerant to excess soil water. They reported that short-term flooding (48–96 h) did not pose significant crop injury, and most of the cultivars were able to withstand the stress. Cultivars suffered less at V5 growth stage whether suffered most at R5 growth stages upon exposure to flooding. In addition, flooding at R5 stage for 8 d reduced yield by 20% to 39% in the soybean cultivars. The severity of yield reduction increased with the increase in waterlogging period. At 14 d of waterlogging condition, soybean yield of 21 soybean genotypes reduced by 53-62% and 74% corresponding to control both in the screen house and field tests (VanToai et al. 2010). Kuswantoro (2015) and Wu et al. (2017) also recorded similar results in 16 and 40 consecutive soybean genotypes at waterlogging stress.

Soybean plants showed reduced yield for about 17–40% and 40–57% at vegetative stage and reproductive stage, respectively, under waterlogging stressed condition than the non-stressed condition (Nguyen et al. 2012). Miao et al. (2012) found that in soybean waterlogging is more severe in flowering and pod filling stages (Hefeng 50 and Kenfeng 16 variety) and causes a significant reduction in the pod number which ultimately caused yield loss. The average yield reduction was 15–50% in both the cultivars. Similar results were also noticed by Beutler et al. (2014) where soybean plants were exposed to 4, 8, 16, and 32 d of waterlogging at flowering stage (R_2) and grain filling stage (R_5), and an increasing trend of yield reduction was recorded with the extent of waterlogging duration. The highest reduction of yield at flowering and pod filling stages (41% and 36%) was recorded at 32 d of waterlogging which indicated the pod filling stage as the vulnerable stage of soybean to waterlogged condition. Waterlogging also reduced 100-seed weight and seed per pod in soybean, and this reduction is much more severe in the R4 (full pod) stage compare to the R1 (blooming) stage (Ara et al. 2015).

3.8 *Quality*

Soybean is considered a highly valued food due to the presence of higher protein, oil, and other nutrient contents. Oleic acid content is notified as a seed quality index parameter of soybean seed which is negatively correlated with linoleic and linolenic acid content (Sudarić et al. 2019). A higher amount of oleic acid helps to improve oil quality and also has good nutritional attributes, whereas linoleic and linolenic acids are responsible for instability and flavor problems in soybean oil (Oliva et al. 2006). However, antioxidant and antifungal activities of soybean oil are controlled by isoflavones, for example, daidzein, genistein, and glycitein (Kim et al. 2012). VanToai et al. (2012) analyzed the change in seed composition of 5 waterlogging-tolerant soybean plant introductions (PIs) (PI086449-0, PI398395, PI416753, PI423838, and PI567251) and a cultivar (Williams) which was sensitive to waterlogging stress. They recorded no change in protein and oil content of soybean seed at the waterlogged condition in all genotypes. Under waterlogging stress, they found that resistant cultivars had higher levels of oleic and stearic acids and lower levels of linoleic and linolenic acids, daidzein, genistein, and glycitein. However, a composite indicator, seed quality index (SQI), was increased by 4% in the PIs, but SQI was decreased by 5% in the check cultivar. As very little research work has been done regarding the effect of waterlogging stress on the quality of soybean seed, further work should be done to establish and find out the relation of seed quality parameters of soybean and waterlogging stress.

4 Plant Adaptation to Waterlogging Stress

Plants exposed to waterlogging tries to diffuse oxygen in aerial parts of plants to adapt to the stress condition. Plants modify their morphological structures including leaves hyponasty, shoot elongation, formation of aerenchyma in root cells, formation of lenticels in the stem, commencing adventitious roots in the waterlogged stem, and so on. Several research works supported that waterlogging stress-triggered ethylene production, which drives the plant to epinasty, chlorosis, and senescence of leaf (Voesenek and Bailey-Serres 2015). Luan et al. (2018) observed morphological and anatomical adaptations in 7 barley genotypes under waterlogging stress where they observed that tolerant genotypes (TX9425, Yerong, TF58) displayed a higher number of adventitious roots under waterlogging stress conditions than the sensitive genotypes (Franklin, NasoNijo, TF57). Increased ethylene content, lower abscisic acid (ABA) level, and less accumulation of $O_2^{\bullet-}$ resulted in more intercellular spaces and better-integrated chloroplast membrane structures in the leaves of waterlogging-tolerant cultivars. In addition, to maintain lower redox potential nonsymbiotic-hemoglobins and nitric oxide are responsible for enhancing the hypoxic stress tolerance and signaling in plants (Sairam et al. 2008). The adverse effect of waterlogging stress can be reduced by increasing the rooting area that accelerates aerobic respiration. Fatimah and Nurhidayati (2020) found no adventitious roots formation at 100% (G1) waterlogging stress. However, in 150% (G2) and 200% (G3), waterlogging stress adventitious roots were increased by the values 18.00 and 7.14, respectively. The aerenchyma cells in adventitious roots enable O_2 movement thus enhancing aerobic respiration. Besides this, phytohormones like auxin and ethylene play a tremendous role in adventitious root formation (Verstraeten et al. 2014; Herzog et al. 2016).

4.1 Morphological and Anatomical Modification

Plants show some morphological and anatomical modifications to thrive against the adverse situations created by waterlogging stress. The formation of adventitious root, aerenchymatous cell, is one of the adaptive mechanisms showed by soybean plants under waterlogging stress. Aerenchyma formation was observed in Manokin soybean cultivar when exposed to 8 d of waterlogging (Rhine et al. 2010). In submerged stems, roots, and nodules of soybean, Thomas et al. (2005) reported adventitious roots and aerenchyma development. Waterlogging treatment for 7 d initiated schizogenous aerenchyma in the cortex of adventitious roots produced in the stem. As stress duration increased (day 14), there formed lysigenous aerenchyma in the cortex. After 21 d of treatment, there observed cell division in the pericycle which caused the displacement of endodermis. Adventitious root length varied to 2-5 mm long on 3 d of waterlogging treatment which increased to 1 cm by day 4 and 5. After 21 d of treatment, prominent adventitious roots were observed. In nodules, there was observed prominent white friable aerenchymatous cells.

While working on two soybean lines, PI408105A (WTL) and S99-2281 (WSL) in waterlogging stress, Kim et al. (2015) observed adventitious roots formation were better in WTL than the WSL. They also observed several endogenous hormones showed pronounced differences under waterlogging in both 5 and 10 d after treatment (DAT). The amount of ethylene produced was enhanced by 8.6- and 4-fold in WTL and 1.6- and 2.5-fold in WSL at 5 DAT and 10 DAT, respectively. Ethylene is a key hormone that is involved in plant stress response, thus inducing adventitious roots formation. These authors found that methionine content declined notably which is the precursor of ethylene. In WTL, the decline in methionine content was greater than in WSL. On this, they explained WTL efficiently used methionine in several physiological processes; thus, the line shows lower methionine content than the WSL. When exposed to waterlogging stress, WTL contained less ABA than WSL. This downregulation of ABA suppressed suberization of root cells which helps in aerenchyma cell development. They also observed higher salicylic acid (SA) content in WTL than WSL. The content of SA urged to programmed cell death (PCD) which resulted in increment of lipid peroxidation led to aerenchyma cell formation improving oxygen supply under stress conditions. Furthermore, the endogenous hormones acted significantly in inducing aerenchyma and adventitious roots formation enhancing tolerance under waterlogging stress. Even after a shorter duration of waterlogging (72 h), prominent aerenchyma formation was observed in soybean roots (Shimamura et al. 2014). As a major adaptive feature of plants against waterlogging stress, aerenchyma formation is regarded as it facilitates plants with transportation of atmospheric oxygen to the root tips from the shoot (Evans 2003).

4.2 *Stomatal Regulation*

Waterlogging hampered stomatal opening or closing and the absorption of CO_2. The restricted CO_2 led to a decrease in photosynthesis which is crucial for the growth and development of plants. The activities of photosynthesis-related enzymes were blocked, and the ability of leaves to synthesize Chl diminished, resulting in leaf senescence and then a fall in photosynthetic rate, eventually leading to plant mortality (Zhao et al. 2018). Ensuring higher water potentiality under waterlogging, soybean cultivar (BMX Potência) resulted in similar stomatal conductance as controlled plants compared to other cultivars and showed higher biomass accumulation (Garcia et al. 2020). According to Pereira et al. (2020) waterlogging resulted in the reduction of intercellular CO_2 concentration, stomatal density, stomatal functionality, and stomatal index that can be reverted with the pretreatment of 24-epibrassinosteroid (10 nM) which improved waterlogging tolerance in plants.

4.3 *Hormonal Regulation*

Phytohormone plays an integral part in imparting plant responses to abiotic stresses. Recent studies revealed that there exists an intricate relationship between plant hormone and regulation of ROS production during abiotic stress (Choudhury et al. 2017; Fig. 6). For example, ROS produced under stress conditions may alter the auxin gradient in plants that can meanwhile decrease the auxin-mediated signaling processes in plants (Xia et al. 2015). A hypoxic condition under submerged conditions often triggers multiple physiological and cellular processes. Different genes modulate the activity of phytohormones like ethylene, auxin, brassinosteroids (BRs), ABA, and gibberellic acid (GA) which play a fundamental role in the survivability of plants in the face of flooding stress. Ethylene level enhances under the prolonged waterlogged condition which provides signaling to regulate the early responses to flooding (Song et al. 2018). Furthermore, *O. sativa* plants slowly releasing ethylene in root resulted in a positive response to ethylene while decreasing ABA level. This reduced ABA level modulates tissue sensitivity to GA, thus enhancing intermodal length under waterlogged conditions (Voesenek and Sasidharan 2013; Phukan et al. 2016). The formation of adventitious roots and higher ethylene production can improve waterlogging stress tolerance in soybean (Fatimah and Nurhidayati 2020). The experimental result revealed that in severe waterlogging treatment, ethylene production reached 14,878 ppm which is 5 times higher than that of control.

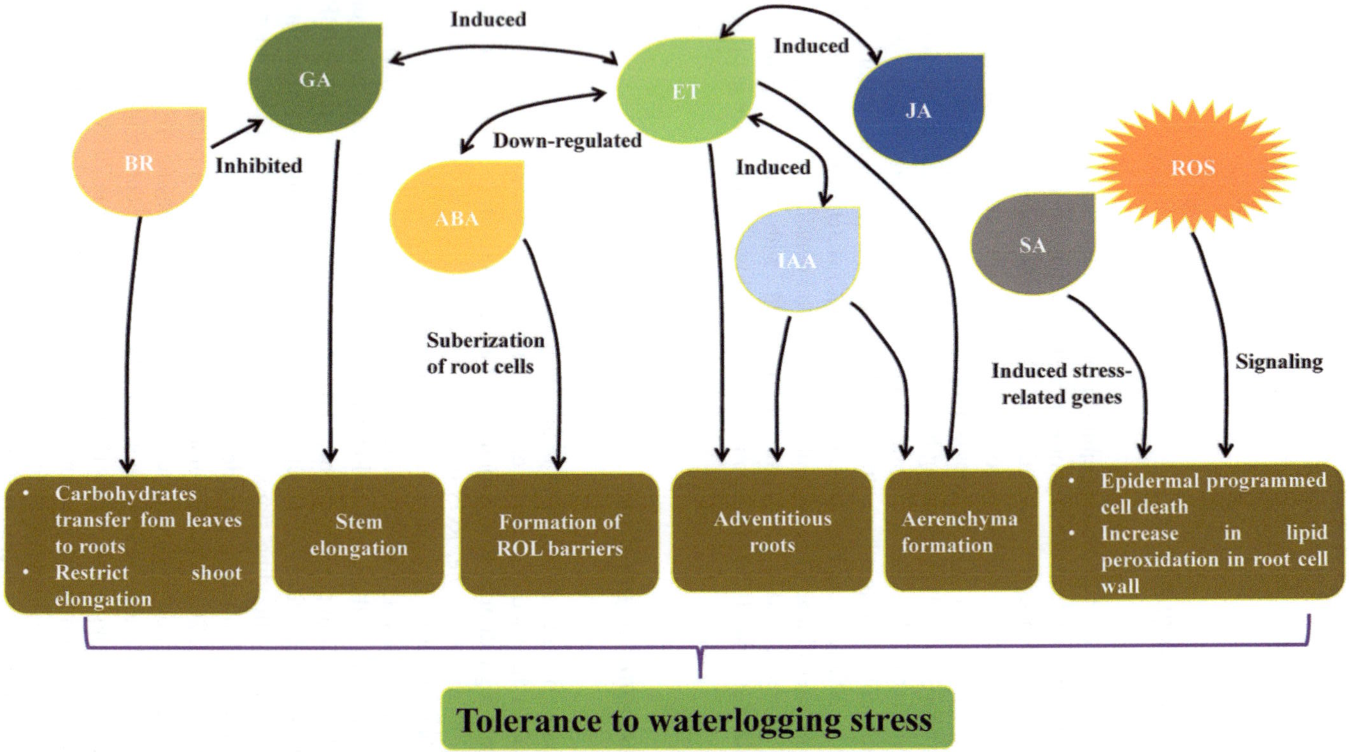

Fig. 6 Endogenous phytohormones and ROS signaling-induced modification in soybean morphology under waterlogging. Here, *ABA* abscisic acid, *BR* brassinosteroids, *ET* ethylene, *GA* gibberellic acid, *IAA* auxin, *JA* jasmonic acid, *ROS* reactive oxygen species, *SA* salicylic acid

Flooding condition in soybean enhances endogenous ABA content although in the control plant it tends to decline. In addition, when ABA was applied exogenously, it resulted in better cell transcription which determines the active role of ABA in enhancing submergence tolerance (Komatsu et al. 2013). Higher formation of O_2 and H_2O_2 under flooding stress happens due to ABA which causes rapid stomatal closure over expression of stress-concerned genes (Jannat et al. 2011; Rodríguez-Gamir et al. 2011). Plants exposed to long time flooding conditions are visible with extensive leaf senescence. Waterlogging affects the synthesis and/or signaling pathways of the hormones which in turn accelerate the expression of stress-responsive genes, and the ultimate consequence is leaf senescence. Plant hormones such as ABA, jasmonic acid (JA), ethylene, and SA are associated with the mitigation of this flood-induced adverse effect on plants. Among them, JA and SA are two vital phytohormones of leaf senescence (Lim et al. 2007). Kim et al. (2015) noticed the regulation of the endogenous phytohormones in two varying soybean genotypes upon exposure to waterlogging and observed that ABA content reduced by 38 and 49% in WTL and 16 and 26% in WSL at 5 and 10 DAT, respectively. Therefore, it is evident that ABA reduced much in WTL than WSL due to all waterlogging treatments. Under flooding condition, soybean roots formed aerenchyma cells that help plants to survive in submerged condition. Aerenchyma cells developed well when the biosynthesis of suberin is checked by the reduced production of ABA (Shimamura et al. 2014; Zhou et al. 2020). Kim et al. (2015) found well-developed aerenchyma cells in WTL roots as well as a significant drop in whole plant ABA levels when compared to WSL and control. 24-epibrassinolide (EBR) is a polyhydroxylated steroid, which is a biodegradable molecule that may play a role in metabolic and physiological processes. Pereira et al. (2020) reported that EBR enhances the antioxidant enzymes thus resulted in less destruction of chloroplast membranes. In addition, EBR enhances the efficacy of photosystem II while improving the gas exchange rate in soybean. Besides this, a higher density of stomata and increased Chl content and electron transport rate was also observed under 8 d prolonged waterlogging condition. Kim et al. (2018) performed two experiments (EP I and EP II). In EP I, they applied several plant growth regulators (PGRs), namely, ABA, ethylene, gibberellins, indole-acetic acid, kinetin, JA, and SA and found ethylene was the most promising one in resisting waterlogging stress in soybean plants than the other PGRs. In EP II, they applied different concentrations of ethylene (50, 100, and 200 μM) and found that 200 μM provides better support under waterlogging stress.

Under waterlogging stress, ethylene accumulation increased which induced ROS signaling that lead to epidermal programmed cell death thus forming aerenchyma and adventitious roots. Abscisic acid plays a key role in regulating water potential by regulating stomatal conductance and forming radial oxygen loss barrier by suberization of root cells. Abscisic acid inhibited the formation of adventitious roots. However, aerenchyma cells developed well when the biosynthesis of suberin is checked by the reduced production of ABA. The accumulation of ethylene downregulated ABA biosynthesis and induced GA content which stimulates stem elongation. Auxin and ethylene promote each other's accumulation and forming adventitious roots. Salicylic acid-induced stress-related genes urged to epidermal programmed cell death and increase in lipid peroxidation thus forming aerenchyma cells in roots. Jasmonic acid interacts with ethylene and increased ethylene accumulation and induced aerenchyma development. Brassinosteroid inhibited GA and restrict shoot elongation (Fig. 6). According to Khan et al. (2018), gibberellins can help soybean plants recover from the negative impacts of short-term flooding (3 and 6 h) stress. Two soybean cultivars Daewon (normal) and Pungsannamul (waterlogging sensitive) were subjected to flooding, and GAs (GA_{4+7}) were applied exogenously to mitigate the flooding stress. They analyzed different endogenous hormonal regulations under flooding and flooding with GAs-treated plants. Flooding for 6 h induced significant production of endogenous bioactive GA1 and GA4 contents, whereas nonsignificant production was noticed at 3 h of flooding. The exogenous application of GAs induced higher JA synthesis while inhibited ABA upregulation, production of superoxide anions, and decreased the content of reduced glutathione (GSH). In the sensitive Pungsannamul, *S*-nitrosothiol (SNO) related gene, namely, *S*-nitrosoglutathione reductase (GSNOR1), NO overproducer1 (NOX1), and nitrate reductase (NR) expressions were less exhibited, which indicated the loss of potential functions under flooding stress in sensitive cultivars. ABA-related transcript ABA-receptor (ABAR2) and Timing of CAB expression1 (TOC1) reduced their transcriptional accumulation when added GAs. In another experiment, Khan et al. (2019) exogenously applied sodium nitroprusside (SNP) and *S*-nitroso L-cysteine (CySNO) as an NO donor and observed regulatory activities of NO in short-term flooding in soybean. Same cultivars and flooding durations were maintained, and 250 μM SNP and 250 μM CySNO were applied exogenously to mitigate the flooding stress. NO is a signaling molecule that interacts with plant hormones and endogenous molecules of plants. SNP- and CySNO-treated Daewon plants showed a lower accumulation of superoxide anions than the flooded condition. The reduction was observed at 13% and 20% after 3 and 6 h of flooding, respectively. SNP- and CySNO-treated Pungsannamul plants decreased superoxide anions by 15% and 44% after 3 h and 11% and 46% after 6 h of flooding stress. During flooding stress, glutathione synthesis was increased, but the application of SNP and CySNO reduced glutathione content which suggested that plants might adopt some defensive mechanisms to alleviate oxidative stress. They observed ABA synthesis was highly activated in Daewon than the sensitive Pungsannamul cultivar and so, ABA content was declined by 10–26% at 3 h, 19–21% at 6 h of flooding treated with SNP and CySNO. They elucidated that the protein functions are regulated in plants by *S*-nitrosylation where NO binds with cysteine thiol forming *S*-nitrosothiol (SNO). SNO-related genes GSNOR1, NOX1, and NR were significantly increased and ABA-related transcript ABAR2 and TOC1 reduced their transcriptional accumulation when treated with SNP and CySNO.

4.4 Antioxidant Defense

Due to the overproduction of ROS under flooding/waterlogging stress, plants have to suffer from oxidative damages (Park and Lee 2019). This ROS production is detrimental to plants which can affect various plant physiological systems (Hasanuzzaman et al. 2020; Sachdev et al. 2021). Various types of harmful ions and free radicals are generated through metabolic action which causes oxidative damage to plants (Hasanuzzaman et al. 2020). However, plants have inbuilt natural antioxidant defense systems to counteract the negative effects of ROS. Antioxidants, both enzymatic and nonenzymatic, make up this defense system. (Hasanuzzaman et al. 2012). All these antioxidants act coordinately in scavenging ROS, which gives protection to tissues from oxidative stress. In two *Sorghum bicolor* genotypes JN01 (waterlogging-tolerant) and JN31 (waterlogging-sensitive), Zhang et al. (2019) reported 2.5-fold higher malondialdehyde (MDA) formation in waterlogging-sensitive genotype JZ31 where only 1.8-fold increase of MDA was recorded in waterlogging-tolerant genotype JN01. An increase of 54% and 208% MDA and H_2O_2 contents, respectively, was reported by Rasheed et al. (2018) in *S. lycopersicum* cv. Roma under waterlogging condition for 15 d. A similar increase of MDA, H_2O_2 contents, and $O_2^{\bullet-}$ generation was observed in *Deschampsia antarctica* and *H. vulgare* (Luan et al. 2018; Park and Lee 2019). Waterlogging increased fermentation, oxidative stress, and lowered ATP levels in soybean roots (da-Silva and do Amarante 2020b). Zhang et al. (2007) stated that waterlogging for about 18 d increased lipid peroxidation in the membrane of two *H. vulgare* genotypes: Xiumai 3 (tolerant) and Gerdner (sensitive). Upon exposure to waterlogging, the activity of superoxide dismutase (SOD) increased which was greater in the sensitive genotype than in the tolerant one. Both peroxidase (POD) and catalase (CAT) activity was increased in the tolerant than the sensitive genotype where glutathione reductase (GR) activity was increased in both the genotypes. While working on two genotypes (ICPL 84023 and ICP 7035) of *C. cajan,* Sairam et al. (2009) observed that H_2O_2 and $OH^{\bullet}$ contents increased under 6 d of flooding stress. SOD, ascorbate peroxidase (APX), GR, CAT activity increased upon exposure to waterlogging. When applied flooding stress for about 10 d, H_2O_2 content increased 290% than the control *Allium fistulosum* plants. The activity of SOD, POD, CAT, and GR also enhanced when exposed to flooding stress (Yiu et al. 2009). Research work was conducted by Zheng et al. (2009) with two wheat genotypes (Huaimai 17 and Yangmai 12) under hypoxia stress. They noticed an enhancement of lipid peroxidation and a reduction of ATP synthesis in the chloroplasts after 5 d of waterlogging stress. Sairam et al. (2011) performed an experiment with *V. luteola*, a highly tolerant wild species, and two mung bean (*V. radiata*) varieties: T 44 (tolerant) and Pusa Baisakhi (susceptible). *Vigna luteola* and T 44 showed an increased SOD and APX gene expression, while Pusa Baisakhi showed a little expression when exposed to waterlogging for 8 d than their control plants. Luan et al. (2018) carried out an experiment with seven different *H. vulgare* genotypes under waterlogging stress. Waterlogging for 21 d increased superoxide radical ($O_2^{\bullet-}$) in leaves by 9%, 8%, 29%, 28%, 27%, and 20% in genotypes TX9425, Yerong, YYXT, Franklin, NasoNijo, and TF57, respectively, while no notable variation was found in TF58. In cape gooseberry plants, MDA content increased for 8 d of waterlogging stressed plants, which was inoculated with *Fusarium oxysporum* f. sp. Physali (Foph) (Chávez-Arias et al. 2019). Anee et al. (2019) found that MDA (39%), H_2O_2 content and AsA increased, GR (23%) activity decreased in sesame (cv. BARI Til-4) plants when exposed to 8 d of flooding. They also reported an increase in APX (61%), MDHAR (55%), DHAR (59%), GPX (47%), CAT (33%), and GSSG activity upon exposure to 8 d of waterlogging in sesame plants. Luan et al. (2018) experimented with two genotypes (TF57, TF58) of *H. vulgare* and reported that SOD, POD, and CAT activities were increased in both waterlogging sensitive and tolerant genotype under 21 d of prolonged waterlogging stress. After 8 d of waterlogging in *S. indicum* cv. BARI Til-4, non-enzymatic antioxidants GSH and GSSG increased by 45% and 150%, respectively, but AsA concentration reduced by 38% (Anee et al. 2019). In *V. mungo*, MDA content and SOD activity also increased under 10 d of waterlogging treatment (Bansal et al. 2019). Kumar et al. (2018) found MDA content to be decreased under 7 d of waterlogging stress in *S. torvum* while working with 9 wild solanaceous species. Also, SOD, CAT, and POD activities increased under flooding stress. Liu et al. (2020) worked on two *O. sativa* genotypes (YueFu, YF) and IRAT109, IR), where they found that IAA, ethylene, and H_2O_2 content were increased under 7 d of flooding. According to da-Silva and do Amarante (2020a), SOD activity in soybean roots dropped after 12 h of waterlogging stress, but it rose continually in the leaves after 3 h. After 24 and 120 h of submergence, CAT activity increased marginally in the root and leaves, respectively, while the APX activity increased both in root and leaves after 12 h of waterlogging. Furthermore, the MDA content also elevated both in leaves and roots compared to control after 24 h of waterlogging stress. In addition, da-Silva and do Amarante (2020b) reported that short-time nitrate supplementation in waterlogged soybean plants lowers the activity of fermentative and antioxidant enzymes as well as the fermentation substrates and ROS. They also observed that, compared to control, enzymatic antioxidants activity greatly increased in plants under 10 d of waterlogging stress. However, SOD activity did not alter with nitrate supply while CAT and APX greatly reduced with nitrate fertilization. Although the increased levels of H_2O_2, $O_2^{\bullet -}$, and MDA were observed due to waterlogging stress, the presence of nitrate in waterlogged plants resulted in decreased levels of ROS and MDA. Pereira et al. (2020) reported that soybean plants exposed to 8 d of waterlogging resulted in a higher amount of stress indicators. When compared to waterlogging +0 nM EBR, the application of 10 nM EBR attenuated this effect and lowered O_2, H_2O_2, MDA, and electrolyte leakage (EL) to 52%, 39%, 27% and 5%, respectively. Furthermore, exogenous EBR boosts the activity of SOD, APX, and peroxidases (POX) by 25%, 30%, and 59%, respectively, in waterlogged plants.

5 Genotypic Differences Under Waterlogging

Flooding tolerance is related to several genes in soybean plants. Breeding for improving the genetic resources of soybean and developing waterlogged tolerant genotypes is of great importance. Many authors performed studies with several soybean genotypes to show the varietal differences among them. On this basis, further breeding programs could be performed to identify quantitative trait loci (QTLs) for soybean genotypes. Van Nguyen et al. (2017) aimed to find QTLs for soybean root development under hypoxic conditions. They found out 11 QTLs associated with root development and 7 QTLs associated with hypoxia tolerance of these root traits. For the validation of the QTLs, these authors developed a line derived from the Iyodaizu genotype which was tolerant to both hypoxia and waterlogging stress conditions. Plant height was affected in both tests. In all three experiments, VND2, NamVang, and ATF15-1 genotypes responded as tolerant against flooding stress. Rhine et al. (2010) carried out a 3-year cultivar screening trial to assess the tolerance of soybean genotypes to flooded conditions. Each year, there was performed a screening of about 360 soybean cultivars. In this study, the authors found five cultivars that showed a range of tolerances under flooding stress. The cultivars are Manokin, P94B73, Mersch- Denver, Desloy 4710, and DK4868. While working on 40 soybean genotypes under different levels of waterlogging stress (3, 6, 9, and 12 d), Wu et al. (2017) observed that the R1 stage was more sensitive than the V5 stage in plants. They used plant survival rate and foliar damage score as indicators to waterlogging tolerance. These workers also stated some optimum flooding duration, which may be used for screening out of tolerant genotypes. Flooding for about 9 and 6 d in V5 and R1 stages, respectively, were found to be distinguishable for the screening of soybean genotypes. Suematsu et al. (2017) worked with 162 soybean accessions in experiment-1 under normoxia and hypoxia conditions where they observed root length (RL), root surface area (RSA), root volume (RV), changes in average root diameter, shoot dry weight (SDW), and root dry weight (RDW). After that, in experiment-2 they selected 11 accessions (Tachinagaha, Miyagishirome, Komame, Nattou kotsubu, U 1155-4, Okjo, M 652, Iyodaizu, Maetsue zairai 90B, Kokubu 7, M 42) by performing principal components analysis. These selected 11 soybean accessions were tolerant to hypoxic situations. Mean RL was 843 and 327 cm, RSA was 112 and 59 cm^2, RV was 1.2 and 0.87 cm^3, SDW 303 and 271 mg, and RDW was 83 and 57 mg in control and waterlogged plants, respectively. Under waterlogging treatment for 7 d, all the parameters showed a decrease except for root diameter.

Rajendran et al. (2019) carried out an experiment to assess the germination of 128 soybean genotypes under different waterlogging levels (3, 5, 7, 9, and 11 d). A 2-d delay was observed in coleoptiles emergence of genotypes for 3 d flooding. The increase rose to 11 d when waterlogged for 9.7 d where only two genotypes showed 70% or more germination. These workers found two genotypes, WT3 and WT8, which had more than 70% germination and less than 5 d delayed coleoptiles emergence when compared with control. At 9 d of waterlogging treatment,

78 genotypes showed 50% germination and 5.7 d delayed coleoptiles emergence, whereas 73 genotypes showed no germination. At 11 d of waterlogging treatment, almost inhibited germination of all the 128 genotypes except 10 that showed 50–59% germination and delayed in final emergence.

Garcia et al. (2020) worked on three soybean genotypes (PELBR10-6000, PELBR11-6028, and PELBR11-6042) and two cultivars (TEC IRGA 6070 and BMX Potência) under flooding stress. They observed that all the genotypes vanquished waterlogging stress following discrete mechanisms. Flooded PELBR10-6000 exceeded CO_2 assimilation rate than the control plant levels by triggering fermentative enzymes and alanine aminotransferase. Cultivars BMX Potência showed similar mechanisms and restored metabolic activities to control level till the end of the recovery period. TEC IRGA 6070 did not delay blooming and PELBR11-6028 and PELBR11-6042 stimulated antioxidant defenses.

6 Conclusion

Flooding is a common natural disaster intrinsically connected with global climate change which causes massive crop loss every year. Crop production is greatly hampered due to frequent submerged conditions. Soybean, a semi-tolerant crop to waterlogging, has evolved with few morphological adaptations under prolonged submerged conditions. Although few tolerance mechanisms with adaptive features and hormonal balance work against the stress condition, this hypoxic condition causes severe oxidative stress at times for which soybean plants experience irreversible stressed conditions leading to plant death. However, the application of different exogenous protectants and hormones elevate the negative effect of submerged stress through upregulating antioxidant enzymes and several physiological processes. Indeed, the actual reason underlying the waterlogging stress tolerance mechanism is yet to reveal. Therefore, we summarized the concomitant research findings with the approaches to investigate the physiological and biochemical response of soybean under waterlogging stress. Furthermore, the molecular basis of this stress with genomic interventions in improving tolerant crop varieties can be explored focusing on the transgenic approaches and signaling mechanisms.

References

Acquaah G (2007) Principles of plant genetics and breeding. Blackwell, Oxford, p 385

Akhtar I, Nazir N (2013) Effect of waterlogging and drought stress in plants. Int J Water Res Environ Sci 2:34–40

Akter T, Ali MR, Rohman MM, Uddin MS (2018) Comparative analysis of biochemical and physiological responses of maize genotypes under waterlogging stress. 13th Asian Maize Conference and Expert Consultation on Maize for Food, Feed, Nutrition and Environmental Security. CIMMYT, Mexico, D.F, Oct. 8-10, Ludhiana, India

Al-Amri SM (2019) Differential response of faba bean (*vicia faba* L.) plants to water deficit and waterlogging stresses. Appl Ecol Environ Res 17(3):6287–6298
Alizadeh-Vaskasi F, Pirdashti H, Cherati Araei A, Saadatmand S (2018) Waterlogging effects on some antioxidant enzymes activities and yield of three wheat promising lines. Acta Agric Slov 111(3):621–631
Amri M, El Ouni MHM, Salem B (2014) Waterlogging affect the development, yield and components, chlorophyll content and chlorophyll fluorescence of six bread wheat genotypes (*Triticum aestivum* L.). Bulg J Agric Sci 20:647–657
Anandan A, Pradhan SK, Das SK, Behera L, Sangeetha G (2015) Differential responses of rice genotypes and physiological mechanism under prolonged Deepwater flooding. Field Crop Res 172:153–163
Andrade CA, de Souza KRD, de Oliveira SM, da Silva DM, Alves JD (2018) Hydrogen peroxide promotes the tolerance of soybeans to waterlogging. Sci Hortic 232:40–45
Anee TI, Nahar K, Rahman A, Mahmud JA, Bhuiyan TF, Alam MU, Fujita M, Hasanuzzaman M (2019) Oxidative damage and antioxidant defense in *Sesamum indicum* after different waterlogging durations. Plants 8(7):196. https://doi.org/10.3390/plants8070196
Anjum NA, Sofo A, Scopa A, Roychoudhury A, Gill SS, Iqbal M, Lukatkin AS, Pereira E, Duarte AC, Ahmad I (2015) Lipids and proteins-major targets of oxidative modifications in abiotic stressed plants. Environ Sci Pollut Res 22:4099–4121
Ara R, Mannan MA, Khaliq QA, Uddin Miah MM (2015) Waterlogging tolerance of soybean. Bangladesh Agron J 18(2):105–109
Bajpai S, Chandra R (2015) Effect of waterlogging stress on growth characteristics and sod gene expression in sugarcane. Int J Sci Res 5(1):1–8
Bansal R, Sharma S, Tripathi K, Kumar A (2019) Waterlogging tolerance in black gram [*Vigna mungo* (L.) Hepper] is associated with chlorophyll content and membrane integrity. Indian J Biochem Biophys 56(1):81–85
Barickman TC, Simpson CR, Sams CE (2019) Waterlogging causes early modification in the physiological performance, carotenoids, chlorophylls, proline, and soluble sugars of cucumber plants. Plants 8(6):160. https://doi.org/10.3390/plants8060160
Beutler AN, Giacomeli R, Albertom CM, Silva VN, da Silva Neto GF, Machado GA, Santos ATL (2014) Soil hydric excess and soybean yield and development in Brazil. Aust J Crop Sci 8: 1461–1466
Borella J, De Oliveira DDC, De Oliveira ACB, Braga EJB (2014) Waterlogging-induced changes in fermentative metabolism in roots and nodules of soybean genotypes. Sci Agric 71:499–508
Chávez-Arias CC, Gómez-Caro S, Restrepo-Díaz H (2019) Physiological, biochemical and chlorophyll fluorescence parameters of *Physalis peruviana* L. seedlings exposed to different short-term waterlogging periods and Fusarium wilt infection. Agronomy 9(5):213. https://doi.org/10.3390/agronomy9050213
Cho JW, Yamakawa T (2006) Effects on growth and seed yield of small seed soybean cultivars of flooding conditions in paddy field. J Fac Agr Kyushu Univ 51(2):189–193
Choudhury FK, Rivero RM, Blumwald E, Mittler R (2017) Reactive oxygen species, abiotic stress and stress combination. Plant J 90(5):856–867
da-Silva CJ, do Amarante L (2020a) Time-course biochemical analyses of soybean plants during waterlogging and reoxygenation. Environ Exp Bot 180:104242. https://doi.org/10.1016/j.envexpbot.2020.104242
da-Silva CJ, do Amarante L (2020b) Short-term nitrate supply decreases fermentation and oxidative stress caused by waterlogging in soybean plants. Environ Exp Bot 176:104078. https://doi.org/10.1016/j.envexpbot.2020.104078
Duhan S, Kumari A, Bala S, Sharma N, Sheokand S (2018) Effects of waterlogging, salinity and their combination on stress indices and yield attributes in pigeonpea (*Cajanus cajan* L. Millsp.) genotypes. Ind J Plant Physiol 23(1):65–76
Evans DE (2003) Aerenchyma formation. New Phytol 161:35–49

Ezin V, Pena RDL, Ahanchede A (2010) Flooding tolerance of tomato genotypes during vegetative and reproductive stages. Braz J Plant Physiol 22:131–142

Fatimah VS, Nurhidayati T (2020) Morphophysiological characteristic responses of soybean (*Glycine max* L.) grobogan variety in waterlogging stress. Ecol Environ Conserv 26:S132–S138

Garcia N, da-Silva CJ, Cocco KLT, Pomagualli D, de Oliveira FK, da Silva JVL, de Oliveira ACB, do Amarante L (2020) Waterlogging tolerance of five soybean genotypes through different physiological and biochemical mechanisms. Environ Exp Bot 172:103975. https://doi.org/10.1016/j.envexpbot.2020.103975

González JA, Gallardo M, Hilal M, Rosa M, Prado FE (2009) Physiological responses of quinoa (*Chenopodium quinoa* Willd.) to drought and waterlogging stresses: dry matter partitioning. Bot Stud 50:35–42

Hasanuzzaman M, Bhuyan MHM, Zulfiqar F, Raza A, Mohsin SM, Mahmud JA, Fujita M, Fotopoulos V (2020) Reactive oxygen species and antioxidant defense in plants under abiotic stress: revisiting the crucial role of a universal defense regulator. Antioxidants 9(8):681. https://doi.org/10.3390/antiox9080681

Hasanuzzaman M, Hossain MA, Teixeira da Silva JA, Fujita M (2012) Plant responses and tolerance to abiotic oxidative stress: antioxidant defense is a key factor. In: Bandi V, Shanker AK, Shanker C, Mandapaka M (eds) Crop stress and its management: perspectives and strategies. Springer, Berlin, pp 261–316

Hasanuzzaman M, Islam MT, Nahar K, Anee TI (2018) Drought stress tolerance in wheat: omics approaches in enhancing antioxidant defense. In: Zargar SM (ed) Abiotic stress-mediated sensing and signaling in plants: an omics perspective. Springer, New York, pp 267–307

Hasanuzzaman M, Nahar K, Hossain MS, Anee TI, Parvin K, Fujita M (2017a) Nitric oxide pretreatment enhances antioxidant defense and glyoxalase system to confer PEG-induced oxidative stress in rapeseed. J Plant Interact 12:323–331

Hasanuzzaman M, Mahmud JA, Nahar K, Inafuku M, Oku H, Fujita M (2017b) Plant responses, adaptation and ROS metabolism in plants exposed to waterlogging stress. In: Khan MIR, Khan NA, Ismail AM (eds) Reactive oxygen species and antioxidant systems: role and regulation under abiotic stress. Springer, Singapore, pp 257–281

Herzog M, Striker GG, Colmer TD, Pedersen O (2016) Mechanisms of waterlogging tolerance in wheat–a review of root and shoot physiology. Plant Cell Environ 39(5):1068–1086

Jannat R, Uraji M, Morofuji M, Islam MM, Bloom RE, Nakamura Y, McClung CR, Schroeder JI, Mori IC, Murata Y (2011) Roles of intracellular hydrogen peroxide accumulation in abscisic acid signaling in Arabidopsis guard cells. J Plant Physiol 168(16):1919–1926

Jitsuyama Y (2017) Hypoxia-responsive root hydraulic conductivity influences soybean cultivar-specific waterlogging tolerance. Am J Plant Sci 8(4):770–790

Khan MA, Hamayun M, Iqbal A, Khan SA, Hussain A, Asaf S, Khan AL, Yun BW, Lee IJ (2018) Gibberellin application ameliorates the adverse impact of short-term flooding on *Glycine max* L. Biochem J 475(18):2893–2905

Khan MA, Khan AL, Imran QM, Asaf S, Lee S, Yun B, Hamayun M, Kim T, Lee I (2019) Exogenous application of nitric oxide donors regulates short-term flooding stress in soybean. PeerJ 7:e7741. https://doi.org/10.7717/peerj.7741

Kim EH, Ro HM, Kim SL, Kim HS, Chung IM (2012) Analysis of isoflavone, phenolic, soyasapogenol, and tocopherol compounds in soybean (*Glycine max* (L.) Merrill) germplasms of different seed weights and origins. J Agric Food Chem 60:6045–6055

Kim Y, Seo CW, Khan AL, Mun BG, Shahzad R, Ko JW, Yun BW, Lee IJ (2018) Ethylene mitigates waterlogging stress by regulating glutathione biosynthesis-related transcripts in soybeans. Bio Rxiv. https://doi.org/10.1101/252312

Kim YH, Hwang SJ, Waqas M, Khan AL, Lee JH, Lee JD, Nguyen HT, Lee IJ (2015) Comparative analysis of endogenous hormones level in two soybean (*Glycine max* L.) lines differing in waterlogging tolerance. Front Plant Sci 6:714. https://doi.org/10.3389/fpls.2015.00714

Komatsu S, Deschamps T, Hiraga S, Kato M, Chiba M, Hashiguchi A, Tougou M, Shimamura S, Yasue H (2011) Characterization of a novel flooding stress-responsive alcohol dehydrogenase expressed in soybean roots. Plant Mol Biol 77:309–322

Komatsu S, Han C, Nanjo Y, Altaf-Un-Nahar M, Wang K, He D, Yang P (2013) Label-free quantitative proteomic analysis of abscisic acid effect in early-stage soybean under flooding. J Proteome Res 12(11):4769–4784

Komatsu S, Sakata K, Nanjo Y (2015) 'Omics' techniques and their use to identify how soybean responds to flooding. J Anal Sci Technol 6:9. https://doi.org/10.1186/s40543-015-0052-7

Kumar KM, Sujatha KB, Rajashree V, Kalarani MK (2018) Study on gas exchange and antioxidant system of solanaceous species under water logged conditions. J Agric Ecol 6:54–63

Kumar P, Pal M, Joshi R, Sairam RK (2013) Yield, growth and physiological responses of mung bean [*Vigna radiata* (L.) Wilczek] genotypes to waterlogging at vegetative stage. Physiol Mol Biol Plants 19:209–220

Kuswantoro H (2015) Agronomical characters of some soybean germplasm under waterlogging condition. J Agron 14(2):93–97

Lapaz AM, de Camargos LS, Yoshida CHP, Firmino AC, de Figueiredo PAM, Aguilar JV, Nicolai AB, de Paiva WDS, Cruz VH, Tomaz RS (2020) Response of soybean to soil waterlogging associated with iron excess in the reproductive stage. Physiol Mol Biol Plants 26:1635–1648

Li W, Mo W, Ashraf U, Li G, Wen T, Abrar M, Gao L, Liu J, Hu J (2018) Evaluation of physiological indices of waterlogging tolerance of different maize varieties in South China. Appl Ecol Environ Res 16:2059–2072

Lim PO, Kim HJ, Gil Nam H (2007) Leaf senescence. Annu Rev Plant Biol 58:115–136

Liu J, Hasanuzzaman M, Suna H, Zhanga J, Penga T, Suna H, Xina Z, Zhaoa Q (2020) Comparative morphological and transcriptomic responses of lowland and upland rice to root-zone hypoxia. Environ Exp Bot 169:103916. https://doi.org/10.1016/j.envexpbot.2019.103916

Lone AA, Warsi MZK (2009) Response of maize (*Zea mays* L.) to excess soil moisture (ESM) tolerance at different stages of life cycle. Bot Res Int 2:211–217

Luan H, Guo B, Pan Y, Lv C, Shen H, Xu R (2018) Morpho-anatomical and physiological responses to waterlogging stress in different barley (*Hordeum vulgare* L.) genotypes. Plant Growth Regul 85:399–409

Miao S, Shi H, Jian J, Judong L, Xiaobing L, Guanghua W (2012) Effects of short-term drought and flooding on soybean nodulation and yield at key nodulation stage under pot culture. J Food Agric Environ 10:819–824

Miura K, Ogawa A, Matsushima K, Morita H (2012) Root and shoot growth under flooded soil in wild groundnut (*Glycine soja*) as a genetic resource of waterlogging tolerance for soybean (*Glycine max*). Pak J Weed Sci Res 18:427–433

Mustafa G, Komatsu S (2014) Quantitative proteomics reveals the effects of protein glycosylation in soybean root under flooding stress. Front Plant Sci 18:627. https://doi.org/10.3389/fpls.2014.00627

Mutava RN, Prince SJK, Syed NH, Song L, Valliyodan B, Chen W, Nguyen HT (2015) Understanding abiotic stress tolerance mechanisms in soybean: a comparitive evaluation of soybean response to drought and flooding stress. Plant Physiol Biochem 86:109–120

Nguyen VT, Vuong TD, VanToai T, Lee JD, Wu X, Rouf Mian MA, Dorrance AE, Shannon JG, Nguyen HT (2012) Mapping of quantitative trait loci associated with resistance to *Phytophthora sojae* and flooding tolerance in soybean. Crop Sci 52:2481–2493

Oliva ML, Shannon JG, Sleper DA, Ellersieck MR, Cardinal AJ, Paris RL, Lee JD (2006) Stability of fatty acid profile in soybean genotypes with modified seed oil composition. Crop Sci 46:2069–2075

Oosterhuis DM, Scott HD, Hampton RE, Wullschleter SD (1990) Physiological response of two soybean [*Glycine max* L. Merr] cultivars to short-term flooding. Environ Exp Bot 30(1):85–92

Palta JA, Ganjealic A, Turnerb NC, Siddique KHM (2010) Effects of transient subsurface waterlogging on root growth, plant biomass and yield of chickpea. Agric Water Manag 97:1469–1476

Park JS, Lee EJ (2019) Waterlogging induced oxidative stress and the mortality of the Antarctic plant, *Deschampsia antarctica*. J Ecol Environ 43(1):1–8

Pereira YC, da Silva FR, da Silva BRS, Cruz FJR, Marques DJ, Lobato AKDS (2020) 24-epibrassinolide induces protection against waterlogging and alleviates impacts on the root structures, photosynthetic machinery and biomass in soybean. Plant Signal Behav 15:11. https://doi.org/10.1080/15592324.2020.1805885

Phukan UJ, Mishra S, Shukla RK (2016) Waterlogging and submergence stress: affects and acclimation. Crit Rev Biotechnol 36(5):956–966

Ploschuk RA, Miralles DJ, Colmer TD, Striker GG (2020) Waterlogging differentially affects yield and its components in wheat, barley, rapeseed and field pea depending on the timing of occurrence. J Agron Crop Sci 206(3):363–375

Prasanna YL, Rao GR (2014) Effect of waterlogging on growth and seed yield in greengram genotypes. Int J Food Agric Vet Sci 4:124–128

Rajendran A, Lal SK, Jain SK, Raju D (2019) Screening of soybean genotypes for pre-germination anaerobic stress tolerance to waterlogging. J Pharmacogn Phytochem 2:01–03

Rasaei A, Ghobadi ME, Jalali-Honarmand S, Ghobadi M, Saeidi M (2012) Impacts of waterlogging on shoot apex development and recovery effects of nitrogen on grain yield of wheat. Eur J Exp Biol 2:1000–1007

Rasheed R, Iqbal M, Ashraf MA, Hussain I, Shafiq F, Yousaf A, Zaheer A (2018) Glycine betaine counteracts the inhibitory effects of waterlogging on growth, photosynthetic pigments, oxidative defence system, nutrient composition, and fruit quality in tomato. J Hortic Sci Biotechnol 93(4):385–391

Ren B, Zhang J, Li X, Fan X, Dong S, Liu P, Zhao B (2014) Effects of waterlogging on the yield and growth of summer maize under field conditions. Can J Plant Sci 94:23–31

Rhine M, Stevens G, Shannon G, Wrather A, Sleper D (2010) Yield and nutritional responses to waterlogging of soybean cultivars. Irrig Sci 28:135–142

Rodríguez-Gamir J, Ancillo G, González-Mas MC, Primo-Millo E, Iglesias DJ, Forner-Giner MA (2011) Root signalling and modulation of stomatal closure in flooded citrus seedlings. Plant Physiol Biochem 49(6):636–645

Ruchi B, Shivani S, Kuldeep T, Ashok K (2019) Waterlogging tolerance in black gram [*Vigna mungo* (L.) Hepper] is associated with chlorophyll content and membrane integrity. Indian J Biochem Biophys 56:81–85

Sachdev S, Ansari SA, Ansari MI, Fujita M, Hasanuzzaman M (2021) Abiotic stress and reactive oxygen species: generation, signaling, and defense mechanisms. Antioxidants 10(2):277. https://doi.org/10.3390/antiox10020277

Saha RR, Ahmed F, Mokarroma N, Rohman MM, Golder PC (2016) Physiological and biochemical changes in waterlog tolerant sesame genotypes. SAARC J Agric 14(2):31–45

Sairam RK, Dharmar K, Lekshmy S, Chinnusamy V (2011) Expression of antioxidant defense genes in mung bean (*Vigna radiata* L.) roots under water-logging is associated with hypoxia tolerance. Acta Physiol Plant 33(3):735–744

Sairam RK, Kumutha D, Ezhilmathi K, Chinnusamy V, Meena RC (2009) Waterlogging induced oxidative stress and antioxidant enzymes activity in pigeon pea. Biol Plant 53:493–504

Sairam RK, Kumutha D, Ezhilmathi K, Deshmukh PS, Srivastava GC (2008) Physiology and biochemistry of waterlogging tolerance in plants. Biologia Plant 52(3):401–412

Sakazono S, Nagata T, Matsuo R, Kajihara S, Watanabe M, Ishimoto M, Shimamura S, Harada K, Takahashia R, Mochizuki T (2014) Variation in root development response to flooding among 92 soybean lines during early growth stages. Plant Prod Sci 17(3):228–236

Sarkar PK, Khatun A, Singha A (2016) Effect of duration of water-logging on crop stand and yield of sesame. Int J Innov App Stud 14(1):1–6

Shimamura S, Yoshioka T, Yamamoto R, Hiraga S, Nakamura T, Shimada S, Komatsu S (2014) Role of abscisic acid in flood induced secondary aerenchyma formation in soybean (*Glycine max*) hypocotyls. Plant Prod Sci 17(2):131–137. https://doi.org/10.1626/pps.17.131

Shin S, Jung GH, Kim SG, Son BY, Kim SG, Lee JS, Kim JT, Bae HH, Kwon Y, Shim KB, Lee JE, Baek SB, Jeon WT (2017) Effect of prolonged waterlogging on growth and yield of characteristics of maize (*Zea mays* L.) at early vegetative stage. J Korean Soc Grassl Forage Sci 37(4): 271–276

Sigua G, Williams M, Chase C Jr, Albano J, Kongchum M (2012) Yield and uptake of bahiagrass under flooded environment as affected by nitrogen fertilization. Agric Sci 3:491–500

Smethurst CF, Garnet T, Shabala S (2005) Nutrition and chlorophyll fluorescence responses of lucerne (*Medicago sativa*) to waterlogging subsequent recovery. Plant Soil 270:31–45

Song L, Valliyodan B, Prince S, Wan J, Nguyen HT (2018) Characterization of the XTH gene family: new insight to the roles in soybean flooding tolerance. Int J Mol Sci 19(9):2705. https://doi.org/10.3390/ijms19092705

Sudarić A, Kočar MM, Duvnjak T, Zdunić Z, Kulundžić AM (2019) Improving seed quality of soybean suitable for growing in europe. In: Sudarić A (ed) Soybean for human consumption and animal feed. IntechOpen, London. https://doi.org/10.5772/intechopen.89922

Suematsu K, Abiko T, Nguyen VL, Mochizuki T (2017) Phenotypic variation in root development of 162 soybean accessions under hypoxia condition at the seedling stage. Plant Prod Sci 20(3): 323–335

Sullivan M, Van Toai TT, Fausey N, Beuerlein J, Parkinson R, Soboyejo A (2001) Evaluating on-farm flooding impacts on soybean. Crop Sci 41:93–100

Thomas AL, Guerreiro SMC, Sodek L (2005) Aerenchyma formation and recovery from hypoxia of the flooded root system of nodulated soybean. Ann Bot 96(7):1191–1198

Tougou M, Hashiguchi A, Yukawa K, Nanjo Y, Hiraga S, Nakamura T, Nishizawa K, Komatsu S (2012) Responses to flooding stress in soybean seedlings with the alcohol dehydrogenase transgene. Plant Biotechnol 29:301–305

Valliyodan B, Ye H, Song L, Murphy M, Shannon JG, Nguyen HT (2017) Genetic diversity and genomic strategies for improving drought and waterlogging tolerance in soybeans. J Exp Bot 68(8):1835–1849

Van Nguyen L, Takahashi R, Githiri SM, Rodriguez TO, Tsutsumi N, Kajihara S, Mochizuki T (2017) Mapping quantitative trait loci for root development under hypoxia conditions in soybean (*Glycine max* L. Merr.). Theor Appl Genet 130:743–755

Vandoorne B, Descamps C, Mathieu AS, Van den Ende W, Vergauwen R, Javaux M, Lutts S (2014) Long term intermittent flooding stress affects plant growth and inulin synthesis of *Cichorium intybus* (var. *sativum*). Plant Soil 376:291–305

VanToai TT, Hoa TTC, Hue NTN, Nguyen HT, Shannon GJ, Rahman MA (2010) Flooding tolerance of soybean [*Glycine max* (L.) merr.] germplasm from Southeast Asia under field and screen-house environments. Open Agric J 4:38–46

VanToai TT, Lee JD, Goulart PFP, Shannon GJ, Alves JD, Nguyen HT, Yu O, Rahman M, Islam R (2012) Soybean (*Glycine max* L. Merr.) seed composition response to soil flooding stress. J Food Agric Environ 10(1):795–804

VanToai TT, St. Martin SK, Chase K, Boru G, Schnipke V, Schmitthenner AF, Lark KG (2001) Identification of a QTL associated with tolerance of soybean to soil waterlogging. Crop Sci 41: 1247–1252

Verstraeten I, Schotte S, Geelen D (2014) Hypocotyl adventitious root organogenesis differs from lateral root development. Front Plant Sci 5:495. https://doi.org/10.3389/fpls.2014.00495

Voesenek LACJ, Bailey-Serres J (2015) Flood adaptive traits and processes: an overview. New Phytol 206(1):57–73

Voesenek LACJ, Sasidharan R (2013) Ethylene–and oxygen signalling–drive plant survival during flooding. Plant Biol 15(3):426–435

Wang X, Deng Z, Zhang W, Meng Z, Chang X, Lv M (2017) Effect of waterlogging duration at different growth stages on the growth, yield and quality of cotton. PLoS One 12(1):e0169029. https://doi.org/10.1371/journal.pone.0169029

Wegner LH (2010) Oxygen transport in waterlogged plants. In: Mancuso S, Shabala S (eds) Waterlogging signalling and tolerance in plants. Springer, Berlin, pp 3–22

Wollmer AC, Pitann B, Mühling KH (2018) Waterlogging events during stem elongation or flowering affect yield of oilseed rape (*Brassica napus* L.) but not seed quality. J Agron Crop Sci 204(2):165–174

Wu C, Zeng A, Chen P, Florez Palacios L, Hummer W, Mokua J, Klepadlo M, Yan L, Ma Q, Cheng Y (2017) An effective field screening method for flood tolerance in soybean. Plant Breed 136: 710–719

Xia XJ, Zhou YH, Shi K, Zhou J, Foyer CH, Yu JQ (2015) Interplay between reactive oxygen species and hormones in the control of plant development and stress tolerance. J Exp Bot 66(10):2839–2856

Yaduvanshi NPS, Setter TL, Sharma SK, Singh KN, Kulshreshtha N (2010) Waterlogging effects on wheat yield, redox potential, manganese and iron in different soils of India. Paper presented at the 19th world congress of soil Science, 1-6 August, Brisbane, Australia, pp 45–48

Yamauchi T, Shimamura S, Nakazono M, Mochizuki T (2013) Aerenchyma formation in crop species: a review. Field Crop Res 152:8–16. https://doi.org/10.1016/j.fcr.2012.12.008

Yamuangmorn S, Rinsinjoy R, Lordkaew S, Dell B (2020) Responses of grain yield and nutrient content to combined zinc and nitrogen fertilizer in upland and wetland rice varieties grown in waterlogged and well-drained condition. J Soil Sci Plant Nutr 20(4):2112–2122

Yiu JC, Liu CW, Fang DYT, Lai YS (2009) Waterlogging tolerance of welsh onion (*Allium fistulosum* L.) enhanced by exogenous spermidine and spermine. Plant Physiol Biochem 47(8):710–716

Yordanova RY, Popova LP (2007) Flooding-induced changes in photosynthesis and oxidative status in maize plants. Acta Physiol Plant 29(6):535–541

Youn JT, Van K, Lee JE, Kim WH, Yun HT, Kwon YU, Ryu YH, Lee SH (2008) Waterlogging effects on nitrogen accumulation and N_2 fixation of supernodulating soybean mutants. J Crop Sci Biotechnol 11:111–118

Zhang G, Tanakamaru K, Abe J, Morita S (2007) Influence of waterlogging on some anti-oxidative enzymatic activities of two barley genotypes differing in anoxia tolerance. Acta Physiol Plant 29:171–176

Zhang R, Zhou Y, Yue Z, Chen X, Cao X, Xu X, Xing Y, Jiang B, Ai X, Huang R (2019) Changes in photosynthesis, chloroplast ultrastructure, and antioxidant metabolism in leaves of sorghum under waterlogging stress. Photosynthetica 57(4):1076–1083

Zhao T, Aleem M, Sharmin RA (2018) Adaptation to water stress in soybean: morphology to genetics. In: Andjelkovic V (ed) Plant, abiotic stress and responses to climate change. Intech Open, London, pp 33–68

Zheng C, Jiang D, Liu F, Dai T, Jing Q, Cao W (2009) Effects of salt and waterlogging stresses and their combination on leaf photosynthesis, chloroplast ATP synthesis, and antioxidant capacity in wheat. Plant Sci 176:575–582

Zhou W, Chen F, Meng Y, Chandrasekaran U, Luo X, Yang W, Shu K (2020) Plant waterlogging/ flooding stress responses: from seed germination to maturation. Plant Physiol Biochem 148: 228–236

Niger (*Guizotia abyssinica* (L. f.) Cass.) an Oilseed Crop under Biotic Stress

Vinod Kumar Goyal, Nisha Agrawal, Meeta Jain, Rajani Bisen, and Rachana Tripathi

Abstract Niger is one of the wild-growing, eco-friendly and less labour needed crops. It is one of the best sources for animal and human consumption. This crop is diversified from Ethiopia but is now gaining attention to increasing production in other countries. Nonetheless, Niger is not produced globally till now due to lack of knowledge about its beneficial characters other than the oilseeds crop. Niger is a newly emerging crop with many advantageous characters that need more study and understanding. Crop improvement with breeding programmes for biotic and abiotic resistance to better adaptation needs more emphasis on its study. Productivity and yield compromising due to stresses are concerned areas of study with more focus on biotic conditions. Insect pests, pathogens and weeds harshly decrease the productivity and growth of the Niger crop. Niger is a good crop that needs less consumption with more giving products used in various fields. But it is not that familiar when compared to other oilseed crops. Here we concern a brief introduction to biotic conditions and their preventive measures for great productivity. We also describe methods of controlling biotic stresses through chemical, plant breeding, biotechnology, mechanical and cropping methods.

Keywords Biotic stress · Germplasm · Insect · Pathogen · Production · Oilseed

1 Introduction

Environmental harsh conditions and adverse circumstances affect the proper growth of plants. Agricultural threats due to various biotic and abiotic stresses hinder the crop, affect its productivity and yield. Stresses are the ecological conditions that are not suitable for plant development and responsible for damaging plants. Plants meet

V. K. Goyal · R. Bisen
Jawahar Lal Nehru Krishi Vishwavidyalaya, Jabalpur, MP, India

N. Agrawal · M. Jain (✉) · R. Tripathi
School of Biochemistry, Devi Ahilya University, Takshashila Campus, Indore, MP, India

M. Hasanuzzaman et al. (eds.), *Managing Plant Production Under Changing Environment*, https://doi.org/10.1007/978-981-16-5059-8_6

different stresses and respond accordingly for adaptation in the stress condition. Stress conditions are of two types based on the non-living (abiotic) circumstances or the living (biotic) factors responsible for them. Biotic stresses are due to any pathogen, insects, pest and compound synthesized by the living system to damage the crops. On the other hand, the abiotic stresses resulted because of environmental conditions, such as high or low-temperature, water deficit or over water condition, harsh light cues and chemical excessive or deficit surroundings. These stresses have a significant impact on the productivity of crops and yield including Niger (*Guizotia abyssinica* (L.f.) Cass.) crop. To survive under unfavourable environmental conditions, plants themselves develop sophisticated strategies; products of some stress-inducible genes directly counteract these stressful situations in plants. Transfer of these genes into plants confirmed their adaptation in stressful conditions and decreases the impact of biotic stresses on crop productivity. Various stresses induced genes and transcription factors that activate molecules and confer regulation of stress conditions. Stresses altered the expression pattern and send signals to plant tissue to respond accordingly by altering the plant growth. The new emerging techniques of plant tissue culture, biotechnology and sustainable crop production make it possible to overcome stresses.

Niger (*Guizotia abyssinica*) is a minor oilseed crop rich in protein and various nutrients consumed by humans as an edible crop. Basically, it originated in Ethiopia in 1774 (Baagoe 1974) but now is cultivated in tropical and subtropical countries and African and Asian countries also. In India, Niger crop ranks first in production area and export to the other countries (Bisen et al. 2015). Niger is classified in the Plantae kingdom with the order Asterales, family Asteraceae and genus Guizotia. Niger seed contains a rich source of oil percentage of 40% (Getinet and Teklewold 1995) with high (75–80%) linoleic fatty acid. It contributes 50% Ethiopian oil production and around 3% Indian oilseed production and best example of soil preservation and sustains land. Niger is a wild crop that grows easily in low-fertility areas with high tolerance to stress. Its oil is used for various manufacturing purposes like in paint, soap formulation, illuminant and as a food source due to its slow drying property (Patil and Joshi 1978; Patil and Patil 1981). Sometimes, it is used as a substitute for ghee and olive oil (Ranganatha et al. 2014). Niger has 30% protein content and around 23% content of crude fibre (Seegeler 1983). It has medicinal properties used in birth control, syphilis treatment and cough treatment with other ingredients (Belayneh 1991). It has also been noted that niger seed is beneficial for human consumption because it has a small number of tocopherols, steroids and phospholipids that are good for cardiovascular diseases and cancer protection (Ramadan and Morsel 2002; Kumar and Bisen 2019).

In this chapter, we deal with the various aspects of biotic stresses that affect the Niger crop productivity. We also emphasize the understanding regarding insects, pests and pathogens that harm the crop. It has better tolerance to diseases, insects, pests, pathogens and weeds when compared to other oilseed crops (Ranganatha et al. 2014). But preventive measures and different harms to the crop are to be understood for making it a globally important crop.

2 Biotic Stresses on Niger Crop

Different biotic stresses in the form of attacks of insects, weeds, pests, diseases and pathogens threats the Niger crop. In Table 1, different biotic stresses are mentioned (Getinet et al. 1996). The crop is significantly harmed by weeds that grow near the crop and decrease the fertility of the soil through the absorption of nutrients. Some common weeds that attack the Niger crop are *Cuscuta* weed, (*Commelina benghalensis, Cyperus iria, Cynodon dactylon, Ocimum sanctum, Parthenium hysterophorus, Spilanthes calva, Celosia argentea* and *Eleusine indica*). On the other hand, many insect pests harm the leaves, branches, stems and roots that reduce the photosynthetic rate and affect starch production needed to nourish the plant that ultimately decreases productivity. Despite this, many diseases even dry up the whole plant and contaminate the nearby plants resulted in complete crop damage. Insects-pests include the caterpillar (*Condica conducta, Diacrisia oblique*), semi looper (*Plusia orichalcea*), larvae (*Agrotis ipsilon*) and butterflies (*Dioxyna sororcula*), while pathogens include many disease-causing fungi, bacteria, parasites, viruses and nematodes (Fig. 1). Kumar and Bisen (2016) described the correlation of yield with branches, flowering and seed weight that directly relate to yield. These traits are disturbed by the biotic stresses and affect the crop yield significantly. In Ethiopia, various diseases harm the crop that might not be seen in the Indian agricultural field. Similarly, various Indian diseases may not harm the agricultural field of other countries due to genetic diversity (Abebe 1992). Common diseases caused in Ethiopia and India due to pathogens, pests and weeds are shown in Table 1.

3 Effects of Biotic Stresses on Niger Crop

During biotic stress condition, plant shows interaction with microbes that lead to micro-associated molecular changes. Stress-related alternation in plants metabolism occurs based on genotype present in that situation (Lenk et al. 2018). Further, proper signalling molecules and low-molecular-weight metabolites for plant development in the stress condition provide resistance in various crops (Dawid and Hille 2018). As a result, the resistant cultivar with metabolic changes is suitable for biotic stress. The chemicals used to control the diseases and pests through suppressing the stress condition without comprising the yield may harm the soil nutrient quality. So, cultivars with resistant genes or adapted crops are the best choice for the development of environmental considerable cultivars through using breeding techniques or biotechnological tools.

Table 1 Different biotic stresses trigger by Niger crop

Biotic stress	Name	Tissue/Part of plant damage	References
Insect–Pest	*Achaea janata* (Grasshopper)	–	Sharma (1990)
Insect–Pest	*Agrotis ipsilon* (Cut worm)	Dried twigs and leaves	Sharma (1990)
Insect–Pest	*Condica conducta* (Caterpillar) *Spilosoma oblique*	Leaves	ICAR (1992)
Insect–Pest	*Chrotogonus* sp. (grasshopper) *Plusia orichalcea* *Prospalta capensis* *Luxus brachyrrhinus* *Sphaeroderma guizotae*	Leaves and defoliates it	Jakhmola (1981); Chavan (1961); Bayeh and Medhin (1992); Jakhmola (1981); Haile (1993)
Insect–Pest	*Chrysodeixis circumflexa* (Plusia worm)	–	Bayeh and Medhin (1992)
Insect–Pest	*Decaria abolominalis* (Chrysomelid beetle)	–	Sharma (1990); Getinet et al. (1996)
Insect–Pest	*Dioxyna sororcula* (Niger fly)	Flower heads, interfering in seed-set and pollination	Getinet et al. (1996); Schmutterer (1971); Jakhmola (1981)
Insect–Pest	*Diacrisia oblique* (Hairy caterpillar)	Leaves	ICAR (1992); Getinet et al. (1996)
Insect–Pest	*Haplothrips articulosus* (Niger flower thrips)	–	Schmutterer (1971); Getinet et al. (1996)
Insect–Pest	*Gryllus bimaculatus* (Cricket)	–	Bayeh and Medhin (1992)
Insect–Pest	*Liroleucon carhami* (aphid)	Bud	ICAR (1992)
Insect–Pest	*Meligethes species* (Black pollen beetle)	Flower heads	Getinet et al. (1996)
Insect–Pest	*Medicogryllus spp.* (Crickets)	–	Bayeh and Medhin (1992)
Insect–Pest	*Piezotrachelus milkoi* (Apionid weevil)	–	Getinet et al. (1996); Bayeh and Medhin (1992)
Insect–Pest	*Perigaea capensis* (Caterpillar)	Leaves	Jakhmola (1981); Chavan (1961)

(continued)

Table 1 (continued)

Biotic stress	Name	Tissue/Part of plant damage	References
Insect–Pest	*Synaptothrips sp.* (Thrips)	–	Bayeh and Medhin (1992)
Insect–Pest	*Taylorilygus pallidulus* (Mirid bug)	–	Bayeh and Medhin (1992)
Insect–Pest	*Trichoplusia orichalcea* (Golden plusia)	–	Bayeh and Medhin (1992)
Pathogen	*Alternaria dauci*	On seeds and leaf	Stewart and Yirgu (1964)
Pathogen	*Anguina amsinckia*	Leaf gall	Stewart and Yirgu (1964)
Pathogen	*Alternaria porri species* (dauci)	Leaf spot	Yirgu (1964)
Pathogen	*Alternaria species*	Stem and leaf blight	Yitbarek and Truwork (1992)
Pathogen	*Aspergillus species*	Stem and leaf blight	Kolte (1985)
Pathogen	*Bremia lactucae*	Downy mildew	Stewart and Yirgu (1964)
Pathogen	*Cercospora guizoticola*	Leaves/leaf spot	Yirgu (1964)
Pathogen	*Coniothyrium species*	–	Kolte (1985)
Pathogen	*Cladosporium species*	Leaves/leaf spot	Yirgu (1964)
Pathogen	*Emericella species*	Leaves/leaf spot	Kolte (1985)
Pathogen	*Epicoccum nigrum*	–	Yirgu (1964)
Pathogen	*Erysiphe cichoraceurum*	–	Yirgu (1964)
Pathogen	*Ozonium taxanum var. Parasiticum*	Ozonium wilt	Kolte (1985)
Pathogen	*Macrophomina phaseolina*	Ozonium wilt	Chavan (1961); Yirgu (1964)
Pathogen	*Phoma species*	Stem lesion, wilting	Yitbarek and Truwork (1992)
Pathogen	*Phyllosticta* spp.	Tar spot	Yirgu (1964)
Pathogen	*Plasmopara halstedii*	Downy mildew	Yitbarek and Truwork (1992)
Pathogen	*Penicillium* spp.	–	Yirgu (1964)
Pathogen	*Puccinia guizotiae*	Rust	Yirgu (1964)
Pathogen	*Rhizoctonia solani*	Root rot	Yirgu (1964)
Pathogen	*Rhizoctonia bataticola*	Seed rot	Yitbarek and Truwork (1992)
Pathogen	*Sclerotium rolfsii*	Seed rot	Kolte (1985)
Pathogen	*Septoria species*	–	Stewart and Yirgu (1964)
Pathogen	*Sphaerotheca species*	Powdery mildew	Yirgu (1964)

(continued)

Table 1 (continued)

Biotic stress	Name	Tissue/Part of plant damage	References
Pathogen	*Xanthomonas campestris pv. Guizotiae*	Leaf spot	Yirgu (1964)
Pathogen	*Xanthomonas campestris pv. guizota var. Indicus*	–	Kolte (1985)
Weed	*Cuscuta chinensis/ C. Hyalina*	Seed	Misar et al. (1981)
Weed	*Cuscuta campestris*	Branches, flowers	Fessehaie (1992); Sharma and Sengar (1989)

Fig. 1 Biotic stresses that affect plant

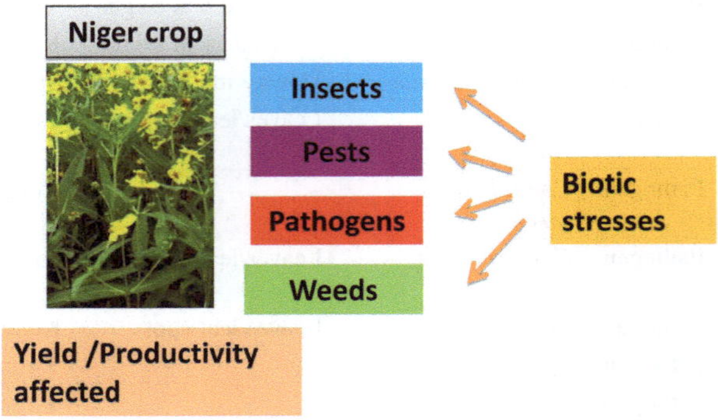

3.1 *Major Insects–Pests Affect Niger Crop*

Twenty-four insects have been recorded that attack the Niger crop mostly in India and Ethiopia. Different species of insect pests including caterpillar, semi loopar, aphids, grasshopper and butterflies that adhere to the Niger crop are mentioned subsequently with proper measurement to control it. Description of some major insect pests attacking Niger crop is shown in Fig. 2.

Niger capsule (*Dioxyma sarorcula* and *Eutretosoma* spp.) is one of the common flies and is also known as Niger fly. They feed on seed and pulp inside the capitula. It lays eggs on the florets disc (Getinet et al. 1996). This fly mainly feeds on the flowers and lays eggs on them. The Niger fly is a serious insect pest that feeds inquisitive in seeds and affects pollination that adversely affects the overall yield reported in Ethiopia and India (Schmutterer 1971; Jakhmola 1981; Getinet et al. 1996). Studies show that the spray of Quinalphos or Acephate overcome the problem associated with this fly in Niger crop.

Niger caterpillar (*Condica conducta*) destroys the leaf part by feeding and defoliates it (ICAR 1992). Recommended cure to overcome this stress is by using

Fig. 2 Major insect-pests that affects Niger crop

Phorate, 5% NSKE, insecticide-based Nimbecidin and the spray of Chloropyriphos/Quinalphos/Triazophosare. It is a very effective measure against this caterpillar.

Cutworms (*Agrotis ipsilon*) Sharma (1990) described that it is the moth that affects dried twigs by hiding itself in it or lays eggs on the leaves. Despite this, larvae attack the plant at the ground level. Phorate, 5% NSKE, insecticide-based Nimbecidin and the spray of Chloropyriphos/Quinalphos/Triazophos are very effective measures against this.

Bihar Hairy caterpillar (*Spilosoma obliqua*) The caterpillars remain sociable in early stages and affect leaves and cause serious loss in yield (ICAR 1992; Getinet et al. 1996). Spray with NSKE 5%, Nimbecidin, Chloropyriphos, Triazophos, Quinalphos, Acephate and/or Indoxacarb are some common preventive measures for this caterpillar.

Semilooper (*Plusia orichalcea*) destroys the crop by feeding on the leaves and defoliates (ICAR 1992). By applying the spray with NSKE 5%, Nimbecidin, Chloropyriphos, Triazophos, Quinalphos, Acephate and/or Indoxacarb are effective measures to be ensured.

Aphids (*Uroleucon carthami*) affect bud formation in the initial stage (ICAR 1992). Spray with NSKE 5%, Nimbecidin, dimethoate/Quinalphos/Dichlorvos/Triazophos/Imidacloprid control the pest without affecting the plant growth.

Surface grasshopper (*Chrotogonus* sp.) affects the leaves and defoliates them. Usually, they destroy the crop and are active in the early stages (Jakhmola 1981). Control the pest in the early stages by applying 4% phosalone or 5% malathion.

3.2 Major Pathogens Affecting Niger Crop

As in the modern cultivars, various pathogens cause severe problems to the Niger crop such as stem and leaf blight. There are various diseases caused by pathogens which considerably affect the overall crop productivity and yield. Most fungi are pathogens that adversely affect the crop than other pathogens. *Aspergillus sps., Alternaria, Cercospora* and *Fusarium* are the common fungi that infect the crop. Some common pathogen infects the niger crop and their preventive measures are listed in Fig. 3.

Cercospora leaf spot (*Cercospora guizoticola*) starts appearing with a small spot with brown colour and then grey on the centre of the leaves and appear as defoliation (Yirgu 1964). It is a seed-borne disease that spreads through contact of one plant to another with the help of wind, water and so on. Chemical Treatments involve the Thiram + Carbendazim on pre-seed treatment, or the Carbendazim+ Mancozeb can manage the disease (Gupta 2017).

Alternaria leaf spot (*Alternaria* sp.) brown to black spots with rings mainly in the form of stem and leaf blight (Yitbarek and Truwork 1992). It is a fungi pathogen that greatly reduced the leaf area of photosynthesis and reduces the CO_2 assimilation. It affects the early maturing plants by reducing the photosynthetic rate (Yitbarek and Truwork 1992). Among the fungi, it is also a seed-borne disease

Fig. 3 Major pathogen affects Niger crop

and causes leaf spot disease in plants. Niger blight of *Alternaria* sp. causes leaf spots that are serious threats to the Niger crop. Thiram + Carbendazim pre-treatment on the seeds or the Carbendazim+ Mancozeb control the disease (Saharan et al. 2005).

***Aspergillus sp.* (*Aspergillus niger, A. flavus, Penicillium* sp., *Alternaria alternata, Rhizoctonia solani* and *R. bataticola*)** spots with rings mainly stem and leaf blight (Kolte 1985). Control the pathogen by applying Thiram + Carbendazim. Fungi *Aspergillus sp.* mostly occurred due to humidity present in seed and not properly dried (Hussain and Usman 2019).

Powdery mildew (*Sphaerotheca* sp.) spots with small powdery mildew show on the leaf surface, then it gradually reaches various parts of the plant like in the lamina and stems ensuing defoliation (Yirgu 1964). Sulphur or Carbendazimor Karathane (0.1%) is the best preventive spray against powdery mildew (Sharma 1982, 1989).

Stem/root rot (*Macrophomina phaseolina*) roots are infected and appear blackish with black sclerotia and become brittle. It extends the blackening from the ground level to the upward part of the plant on the stem (Chavan 1961; Yirgu 1964). Preventive measures for this disease through pre-treatment of the seed with Thiram (0.2%) + Carbendazim(0.1%). It can also be controlled by applying *Trichoderma viride* with FYM (Farmyard manure) in the field before sowing (Gupta et al. 2018).

Leaf spot (*Fusarium* sp.) infects the leaf and causes leaf spot disease on the niger crop. It is soil-borne and present in the soil for several years. It causes the yellowing of leaves and ultimately the leaves die. On some level, fungi affect the root and stem part also. A spray with benomyl and carbendazim significantly reduces the effects of this fungus on plants.

3.3 Major Weeds Affect Niger Crop

Weeds are the major host for the survival, reproduction and shelter house for pests, insects and nematodes. Removal of weeds ultimately causes the secondary host to remove and yield may be increased. Some major weeds that attack the Niger crop were described below with the picture mentioned on (Figs. 4 and 5) source of images PC Unit (Sesame and Niger) Annual report.

***Cuscuta* weed (*Cuscuta chinensis/C. hyalina/C. campestris*)** threatens the niger plants to become stunted, pale yellow with the small size of flowers. Cuscuta weed (parasitic weed) is severe for the niger and its infested the crop at the initial stage of Niger growth. It affects the plant by reducing the branches, height of plants, flower size and seeds significantly. *Cuscuta* seeds are very small in size and mixed with crop seed or soil of the field. *Cuscuta* weeds were removed by sieving before the sowing. Consequently before sowing by steeping of Niger seed in brine solution can be done or removal of seedlings infected with *Cuscuta.* Pre-sowing applies Fluchloralin on the soil or pre-emergence of Niger apply Pendimethalin. Yield loss of Niger crop could be controlled by the herbicide Chlorpropham as reported by Tosh and Patro (1975) and Propyzamide (Tosh et al. 1977, 1978).

Fig. 4 Some major weeds affecting Niger crop

Commelina benghalensis weed likely affects the productivity of the vast crop including the niger crop. In Ethiopia, it is found as a weed (Caton et al. 2004). It is broad leaves weed that affects the yield of crop by reducing the flower size with a decrease in number. Bentazon, actanilde and dinitroaniline are the treatments for this weed.

Cyperus iria is the major weed flora found on the soil. Narrow leaves grass-type weeds affect the growth of roots and shoot harms the plant height and proper vegetative growth. Butachlor/propanil or Almix is the best measure ensured to control the weed.

Cynodon dactylon is also known as crab grass. It grows above 15° C and is found in all soil type, host for many insects, virus and nematodes. Threats to the agriculture commodity by decreasing the vegetative growth. It is prevented through the use of herbicides Chlorpropham as reported by Tosh and Patro (1975) and Propyzamide (Tosh et al. 1977, 1978).

Fig. 5 Showing weeds that affect the Niger crop

Ocimum sanctum is commonly called holy basil or Tulsi. Narrow leaves weeds spread easily in any soil type and similarly affects the agriculture field by decreasing the vegetative growth of the niger crop.

Parthenium hysterophorus carrot grass spread very easily in any soil type. Weed produces an allelopathic compound that suppresses the growth of the plant and also absorbs the nutrients from the soil. It inhibits fruit set on the plant. Prevent the crop using the herbicides Chlorpropham as reported by Tosh and Patro (1975) and Propyzamide (Tosh et al. 1977, 1978). Paraquat + diquat or picloram are also useful herbicides to control the unwanted growth of the carrot grass.

Spilanthes calva is the weed with broad leaves and control through the spray of Oxyflorfen, pendimethalin (Gnanvel and Anbhazhagan 2006).

Celosia argentea a narrow leaf weed reduces crop yields by promoting the growth of microorganisms and also inhibits the germinability of surrounding crop plant seeds by producing allelochemicals into the soil. It is controlled by Oxyflorfen, Pendimethalin (Shiva Shankar and Subramanyam 2011).

Eleusine indica is the major weed flora found in the soil. It is grass and affects the niger crop. Narrow leaves weeds and control through Oxyflorfen, Pendimethalin, Trifluralin and Metalachlor (Bhan and Kolhe 2008).

4 Preventive Measures of Niger Crop from Biotic Stress

Niger crop is a wild-type crop and easily grown in harsh conditions. However, different biotic conditions affect its productivity. To overcome such biotic stress, various mechanisms need to adapt the crop in stress. Various mechanisms are adapted to develop better crop through the utilization of plant breeding methods, new biotechnological techniques, chemical methods and cropping sense. First, the plant breeding method is used through which resistant varieties to particular disease-resistant or the insect/pests tolerance needs to be visualized. Then, through the crossing method, many varieties should be developed with resistant genes and be tagged using marker-assisted breeding and QTLs analysis (Agrawal et al. 2020). Niger is a self-incompatible crop and cross-pollination is the best suit for the development of resistant plants (Doggett 1987). Niger crop is attacked by many pathogens and insects, but genetically controlled cultivars were disseminated and give a new way of resistance breeding. Many potential lines were identified and developed (Rajpurohit et al. 2005; Anonymous 2014). The resistance lines are used for the development of new resistant lines in the disease susceptible lines. The wild species of Niger is the most excellent sources of tolerance genes and the introgressed of these genes into cultivated species is the best way for the development of resistant cultivars (Getinet et al. 1996). The second technique is through biotechnology various approaches like transgenic crop, GM (Genetically Modified Crop) the diploid haploid technology and so on (Gupta et al. 2018) used to develop resistant lines. The anther and haploid cultures may also be compatible methods to develop homozygous inbred lines within a short duration of time. The third mechanism of tolerance to biotic stresses is through chemical ways like the spray of various insecticides, pesticides and chemicals to destroy the harmful organism (Hussain and Usman 2019). During seed treatment, various fungicides and insecticides have been used before sowing (Venkata Ramana 1995; Misar et al. 1981). However, based on the diseases caused and disease stage, different sprays should be performed from time to time and in regular intervals to destroy the stress. Another way of preventive measure includes crop rotation, removal of the weeds in proper time and disease-causing plants should be removed on regular basis (Hussain and Usman 2019). It also ensures that the seeds should be properly dried before sowing through sunlight or artificial means. Fungus-containing seeds should be avoided to sow. Excessive irrigation before seed germination and after emergence extra water in the field was also avoided. Good cultivation practices are one of the best options to avoid biotic stress. One more method useful for protection against insect pests includes the mechanical method. It is very effective in trapping lepidopterous insects and cut-worms. It destroys the eggs laid and larvae of caterpillars that affect the niger crop.

5 Conclusion and Future Prospects

Niger is one of the cheapest growing crops without fertilizer and herbicide requirement. It is a good source of oil and needs less management. It also preserves soil and land conservation. Niger oil extraction is free from any harmful substances and easy to extract. Indeed, oil recovery of Niger crop is lesser than the other oilseed crops like soybean and so on. The understanding related to plant breeding is comparatively less concern on this crop. Overwhelming population feeding is one of the biggest challenges for researchers. The Niger (wild crop) is the best alternative source of oil content that provides great attention towards the cultivation of the crop. However, new technology and breeding accelerate its productivity but, not that much. To increase the yield of the crop we must have to understand the Niger crop deeply. Different stresses that affect the crop yield needs to be better understood. But Niger crop acreage productivity is lesser due to less study on the crop. We need to understand the different mechanisms, biotic and abiotic stresses affecting crop yield, etc. Wild species of Niger is the best sources of resistant gene and the best way of introgressed into cultivated species for better adaptation. To increase focus on this crop, we need understanding and knowledge about the crop. Studies on disease aspects need more attention and promise to understand.

References

Abebe D (1992) Ethiopia's oilseed genetic resources in Oilseed Research and Development in Ethiopia. Proceedings of The First National Oilseed Workshop, Addis Abeba. 13–23

Agrawal N, Tripathi R, Jain M (2020) Molecular marker tools for breeding program in crops. https://doi.org/10.1007/978-981-15-2172-0_20

Anonymous (2014) 4th Advance estimates, agriculture statistics division, Directorate of Economics and Statistics, New Delhi

Baagoe J (1974) The genus *Guizotia* (Compositae). A taxonomic revision. Bot Tidsskrift 69:1–39

Bayeh M, Medhin TG (1992) Insect pests of Niger, linseed and Brassica. In Oilseed Research and Development in Ethiopia. Proceedings of the First National Oilseed Workshop, Addis Abeba. pp 174–177

Belayneh H (1991) Oilcrop germplasm: a vital resource for the plant breeder. In: Engels JMM, Hawkes JG, Worede M (eds) Plant genetic resources of Ethiopia. Cambridge University Press, Cambridge, pp 344–354

Bhan A, Kolhe SS (2008) Impact of integrated weed management on the performance of sunflower, weed dynamics and mycoflora. Indian J Weed Sci 40(1&2):112–112

Bisen R, Panday AK, Jain S, Sahu R (2015) Prog Res, 10 (III), 1536–1539

Caton BP, Mortimer M, Hill JE (2004) A practical field guide to weeds of rice in Asia. International Rice Research Institute, pp 22–23

Chavan VM (1961) Niger and Safflower. Indian Central Oilseeds Committee, Hyderabad

Dawid C, Hille K (2018) Functional metabolomics—a useful tool to characterize stress-induced metabolome alterations opening new avenues towards tailoring food crop quality. Agronomy 8: 138

Doggett H (1987) Niger/Noug research methodology. Oil Crops: Niger and Rapeseed/Mustard. In: Proceedings of the Third Oil Crops Network Workshop held at Addis Abeba, Ethiopia. 210–219

Fessehaie R (1992) Weed science research on Niger, linseed and rapeseed. *in* Oilseed research and development in Ethiopia. Proceedings of the First National Oilseed Workshop, Addis Abeba. pp. 136–151

Getinet A, Sharma SM, Heller J, Engels JMM (1996) Niger Guizotia abyssinica (L. f.) Cass. Promoting the conservation and use of underutilized and neglected crops. 5. Institute of Plant Genetics and Crop Plant Research, Gatersleben/International Plant Genetic Resources Institute, Rome. 59p

Getinet A, Teklewold A (1995) An agronomic and seed-quality evaluation of Niger (*Guizotia abyssinica* Cass.) germplasm grown in Ethiopia. Plant Breed 114:375–376

Gnanvel T, Anbhazhagan R (2006) Integrated weed management in sesame (*Sesamum indicum* L.). Agric Sci Digest 26(1):67–68

Gupta KN (2017) Management of Alternaria and cercospora leaf spot in Niger. Bioinfolet 14(1):111

Gupta KN, Bisen R, Tiwari A (2018) A review: current status of Niger diseases and their integrated management. Int J Chem Stud 6(6):2131–2135

Haile W (1993) Niger leaf miner (*Sphaeroderma guizotiae* Selman), important pest of Niger indigenous to northwestern Ethiopia. Oil Crops Newsl 10:66

Hussain F, Usman F (2019) Fungal Biotic stresses in plants and its control strategy, Abiotic and Biotic stress in plants, Alexandre Bosco de Oliverira, IntechOpen

Indian Council of Agricultural Research (ICAR) (1992) Niger: Package of practices for increasing production. Extension Bulletin No. VII, Directorate of Oilseeds Research, ICAR

Jakhmola SS (1981) Niger grain fly, *Diozina sororcula* (Wiedemann), a serious pest of Niger in Central India. J Bombay Nat His Soc 80:439–440

Kolte SJ (1985) Diseases of annual edible oilseed crops. Vol. III. Sunflower, safflower and Niger diseases. CRC Press Inc., Roca Baton, FL

Kumar V, Bisen R (2016) Genetic study for yield and yield attributing traits in Niger germplasm. Int J Agric Sci 8(56):3044–3046

Kumar V, Bisen R (2019) Principal component analysis of Niger germplasm. J Entomol Zool Stud 7(6):1204–1207

Lenk M, Wenig M, Mengel F, Haubler F, Vlot A (2018) Arabidopsis thaliana immunity-related compounds modulate disease susceptibility in barley. Agronomy 8:142

Misar A, Tosh GC, Mohanty DC, Patro GK (1981) Herbicidal and selective affect of pronamide for control of dodder in niger. Proc 8th Asian Pacific Weed Sci. Soc. Conf, Banglore, pp 255–257

Patil CB, Joshi BP (1978) Niger yields can be doubled. Indian Farming 27:9

Patil CB, Patil BB (1981) Niger cultivation in Maharashtra. Indian Farming:13–14

Rajpurohit TS, Nema S, Khare MN (2005) Current status of diseases of sesame and Niger and their management Paper presented in National seminar on Strategies for enhancing production and export of sesame and Niger. pp 44–45

Ramadan MF, Morsel JT (2002) Proximate neutral lipid composition of Niger (*Guizotia abyssinica* Cass.) seed. Czech J Food Sci 20:98–104

Ranganatha ARG, Panse RK, Panday AK, Deshmukh MR (2014) Strategies for Maximizing Sesame and Niger Production. In: Recent Advances in Weed Management. Directorate of Weed Science Research, Jabalpur

Saharan GS, Mehta N, Sangwan MS (2005) Diseases of oil-seeds. Indus Publishing Company, New Delhi, pp 475–479

Schmutterer H (1971) Contribution to the knowledge of the crop pest fauna in Ethiopia. Angewandte Entomologie 67:371–389

Seegeler CJP (1983) Oil plants in Ethiopia. Their taxonomy and agricultural significance. Centre for Agricultural Publication and Documentation, PUDOC, Wageningen

Sharma SM (1982) Improved technology for sesamum and Niger. Indian Farming 32:72–77

Sharma SM (1989) Niger seed in India. Three Meetings, Oilseeds held at Pantnagar and Hyderabad, India. IDRC/CRDI/CIID.159–165

Sharma SM (1990) New potential areas of Niger in India. In Proceedings of the three meetings held at Pantnagar and Hyderabad India, (A. Omran, ed.). IDRC-MR 252e. Pp. 169–170

Sharma SM, Sengar RBS (1989) Control of diseases of Niger. Indian Farming 29:13–14
Shiva Shankar K, Subramanyam D (2011) Weed flora and yield of sunflower (*Helianthus annus L.*) as influenced by pre and post-emergence application of herbicides. Indian J Weed Sci 43(1&2): 105–109
Stewart RB, Yirgu D (1964) Index of plant diseases in Ethiopia. Haile Sellassie I University, College of Agriculture, Experiment Bulletin No.30
Tosh GC, Patro GK (1975) Control of dodder (*Cuscuta chinensis* Damk) in Niger (*Guizotia abyssinica* L. F. Cass) with Chlorpropham. Weed Res 15:207–209
Tosh GC, Patro GK, Misra A (1977) Effect of Pronamide and Chlorpropham on cuscuta in Niger. Weed Abstracts 1978. Abstract No. 4024
Tosh GC, Patro GK, Misra A (1978) Control of cuscuta in Niger seed. Weed Abstract No. 1202
Venkata Ramana A (1995) Retrospect and prospect of niger research in Andhra Pradesh (In) Niger production Technology (Ed. Reddy PS, Sharma SM, Singh Mev 1995)
Yirgu D (1964) Some diseases of *Guizotia abyssinica* in Ethiopia. Plant Dis Rep 48:672
Yitbarek S, Truwork A (1992) Field evaluation of fungicides on Niger for the control of shot hole (*Septoria* sp.). Oil Crops Newsl 9:26–29

Role of Phytohormones in Antioxidant Metabolism in Plants under Salinity and Water Stress

Md. Motiar Rohman, Sadia Haque Tonny, Sadia Sabrina Alam, Sumaiya Haque Omy, Afsana Hoque Akhi, Hasan Uz Zaman Raihan, Tahmina Akter, Khursheda Parvin, and Mirza Hasanuzzaman

Abstract Abiotic stresses including drought, waterlogging, salinity, heat, cold, ultraviolet radiation, limiting nutrients, and environmental toxicities severely affect plant growth and productivity. Upon such stresses, the production of reactive oxygen species (ROS) in plant cells increases beyond its safe level. Excessive ROS accumulation results in gradual oxidative stress and finally causes cell death, whereas ROS act as versatile signal molecules when remaining below the toxicity threshold and a balance exists between ROS production and scavenging by antioxidants. Plant hormones also act as signaling biomolecules translocated from the synthesis site to their activation site for regulating plant responses to abiotic stresses. Both the classical phytohormones, such as auxins, cytokinins, ethylene, gibberellins, abscisic acid, and the recently identified brassinosteroids, salicylates, jasmonates, and strigolactones are important, which act at very low concentration to mitigate physiological, morphological, and metabolic impairments for increasing surveillance. Phytohormone signaling pathways and ROS mitigating systems are interrelated and they interact with each other to mount a stress response. In this chapter, we critically review and summarize the phytohormones-mediated antioxidant regulation in plants under two major environmental problems, salinity and water stress.

Keywords Abiotic stress · Antioxidant · ROS · Plant hormone · Salinity

M. M. Rohman (✉) · S. H. Tonny · S. S. Alam · S. H. Omy · A. H. Akhi · H. U. Z. Raihan · T. Akter
Molecular Breeding Lab, Plant Breeding Division, Bangladesh Agricultural Research Institute, Gazipur, Bangladesh

K. Parvin
Department of Horticulture, Sher-e-Bangla Agricultural University, Dhaka, Bangladesh

M. Hasanuzzaman (✉)
Department of Agronomy, Sher-e-Bangla Agricultural University, Dhaka, Bangladesh

M. Hasanuzzaman et al. (eds.), *Managing Plant Production Under Changing Environment*, https://doi.org/10.1007/978-981-16-5059-8_7

1 Introduction

Various abiotic stresses including salinity, drought, flooding, extreme temperature, nutrient deficiency, and metal toxicities limit crop growth, development, and production worldwide. Among them, salinity and water stress, where scarcity of water brings drought and excessive water causes flooding/waterlogging, are the most detrimental stresses that alter the physiological, morphological, biochemical, and metabolic changes in plant cells (Pereira 2016; Raza et al. 2019; Hasanuzzaman et al. 2020). Salinity and drought are responsible for the deterioration of major crop production in a range of 50–70% (Mittler 2006). Only excess saline or sodium contents have already polluted more than 800 million hectares of cultivable land (Munns 2011). Salinity affected crop growth in two ways: first, its excess accumulation causes cellular damage by ion toxicity, and second, increased ion concentration of soil water causes osmotic stress, henceforth, hindered plant water uptake. On the other hand, due to global climate change, crop production is being affected by drought than any other stresses and has been increasing over time. For instance, drought-stressed area has become doubled in the period of 1970–2000 (Rohman et al. 2019). On the other hand, flooding (also called waterlogging) is also another crucial stress disturbing crop growth and yield production, where approximately 13% of the total global land including about 10% of the agricultural land are affected by this stress (Cramer et al. 2011). Therefore, these three stresses are potentially hazardous for crop production and cause a higher magnitude of oxidative stress (Hernández et al. 2012; Tewari and Mishra 2018; Naveed et al. 2020). During stresses, disruption occurs in the equilibrium between ROS generation and antioxidant capacity resulting in a higher accumulation of excess ROS and thus causes oxidative stress (Gill and Tuteja 2010). ROS are constituted of both kinds of radical and non-radical components, whereas radical, superoxide anion ($O_2^{\bullet -}$), hydroperoxyl radical ($HO_2^{\bullet}$), alkoxy radical ($RO^{\bullet}$), and hydroxyl radical ($^{\bullet}OH$) are notable and hydrogen peroxide (H_2O_2) and singlet oxygen ($^1O_2$) are non-radical types (Gill and Tuteja 2010; Mehla et al. 2017). It is well known that drought stress-mediated overproduced ROS are highly reactive to DNA, RNA, proteins, lipid, pigments, and cell membrane to cause their functional damage to cell death. Plants have their own mechanism, consisting of both enzymatic and nonenzymatic antioxidants, to minimize the oxidative damage through ROS (Gill and Tuteja 2010).

It is established that plants have their internal mechanism for extreme ROS to keep them under toxic levels known as antioxidants defense system consisting of enzymatic and nonenzymatic types. Enzymatic antioxidants are superoxide dismutase (SOD), catalase (CAT), ascorbate peroxidase (APX), glutathione peroxidase (GPX), guaiacol peroxidase (GPOX), glutathione reductase (GR), monodehydroascorbate reductase (MDHAR), dehydroascorbate reductase (DHAR), and glutathione *S*-transferase (GST), while nonenzymatic antioxidants are ascorbic acid (AsA), reduced glutathione (GSH), carotenoids, α-tocopherol, phenolic acids, flavonoids, and so on (Gill and Tuteja 2010). In plants, cellular antioxidants work very systematically with cyclic order to diminish extra ROS

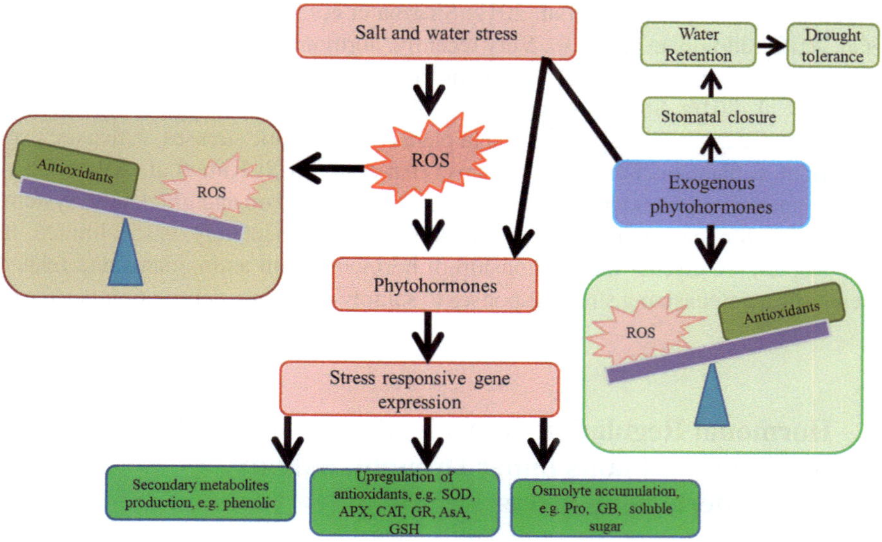

Fig. 1 Phytohormones-mediated plant tolerance to salinity and water stress

where SOD provides the first-line protection by dismutating $O_2^{\bullet-}$ into H_2O_2 (Hasanuzzaman et al. 2020). This H_2O_2 is further metabolized to water through the direct activity of CAT or through entering into the AsA–GSH cycle where the activity of APX, GPX, MDHAR, DHAR, and GR maintains AsA and GSH to scavenge the toxic H_2O_2 (Apel and Hirt 2004; Gill and Tuteja 2010). On the other hand, GSTs are also vital enzymes to detoxify H_2O_2 with possession of additional xenobiotic detoxification properties (Hasanuzzaman et al. 2020). Besides, plants accumulate several compatible solutes, which have crucial roles to guard the cellular protection system under environmental stresses (Gill and Tuteja 2010). Among the solutes, proline, glycine betaine (GB), sugars (mannitol, sorbitol, and trehalose), polyamines, nitric oxide, and several hormones are important (Sharma et al. 2019).

As a diverse group of signaling molecules, phytohormones can be translocated from a place of origin to other parts acting as one of the endogenous factors for interceding plant stress responses (verma et al. 2016). Accordingly, they have the ability to control versatile plant physiological processes to ever-changing environments by mediating growth, development, source/sink manipulation, and nutritional distribution (Fig. 1; Javid et al. 2011; Fahad et al. 2015).

In addition, in the last two decades, significant information has been generated on biosynthesis and the role of the hormone in the growth, development, and stress mitigation of different plants. Among them, the major produced by plants are auxin (IAA), cytokinins (CKs), abscisic acid (ABA), ethylene (ET), gibberellins (GAs), salicylic acid (SA), brassinosteroids (BRs), and jasmonates (JAs). Therefore, phytohormones have the potentiality to change the antioxidants under different abiotic stresses including saline and water stress (Rao et al. 2002; Anuradha and Rao 2001; Bray 2002; Chakrabarti and Mukherji 2003; Sedghi et al. 2012; Soares et al. 2016;

Zayed et al. 2017; Vwioko et al. 2017; Sharma et al. 2018; Kim et al. 2018; Salah et al. 2019; Siddiqui et al. 2020). Very recently, hormonal relations with antioxidants were related to different osmolytes and metabolites have been reviewed under stress (Wani et al. 2016; Sharma et al. 2019). Moreover, external stimuli change both hormonal and antioxidant phenomena in plants under abiotic stresses, which are very important in plant-tolerance mechanisms (Han et al. 2013; Kim et al. 2014). However, plant hormones modulated strengthening of antioxidant and glyoxalase system, particularly for scavenging ROS and methylglyoxal, respectively, is still limited. In this chapter, we discuss the relationship of hormones with antioxidants and related metabolites emphasizing stress responses to salinity and water deprivation or excess.

2 Hormonal Regulation on Antioxidants and Related Attributes in Plants under Drought, Salinity, and Waterlogging Stress

2.1 Abscisic Acid

Abscisic acid (ABA) is a largely occurring essential phytohormone, which was primarily identified as an abscission-promoting hormone. Far ahead, it was discovered that ethylene biosynthesis is partly responsible as the indirect cause of this kind of abscission (Cracker and Abeles 1969). ABA is a 15 carbon ring ($C_{15}H_{20}O_4$) isoprenoid plant hormone, found in cyanobacteria, algae, mosses, sponges, lichens to plants (Wasilewska et al. 2008; Cutler et al. 2010; Mehrotra et al. 2014). This hormone was able to provide protective functions in lower plants from salt and osmotic stress-induced photoinhibition and oxidative in *Chlamydomonas reinhardtii* cells (Saradhi et al. 2000).

It has been reported that ABA can regulate stomatal movement in mosses and sporophytes. Besides, 100 algae species respond variously on applied stress because of containing ABA (reviewed by Hartung 2010). Conversely, it was also revealed that the quantity of ABA in lower plants is also very low (Mehrotra et al. 2014).

Abscisic acid is synthesized in chloroplasts and other plastids through 2-C-methyl-d-erythritol-4-phosphate (MEP) pathway, which is the mevalonic-acid independent pathway. ABA is generated from the cleavage of C_{40} carotenoids, which originates from the MEP pathway (Nambara and Marion-Poll 2005). The ABA biosynthesis from the zeaxanthin (C_{40} carotenoid) starts in the plastids, which ends by producing ABA via oxidation of abscisic aldehyde in the cytosol (Seo and Koshiba 2002). It is noteworthy that, ABA biosynthesis and signaling pathways are directly influenced by antioxidants like GSH at the plant's translational level. Moreover, it was reported that ABA showed the root to shoot transportation capability under drought stress (Sauter et al. 2001). Interestingly, ABA and auxin were first identified than other hormones in GSH-treated plants followed by JA as an indication of earlier plant responses upon abiotic stress than biotic ones.

More than 8-h GSH treatment increases ABA synthesis in roots of *Arabidopsis thaliana*. At that 8-h GSH treatment level ABA-responsive genes including *OPEN STOMATA 1* (*OST1/SnRK2.6*), *ARABIDOPSIS THALIANA PROTEIN PHOSPHATASE 2CA* (*PP2CA*), the ABA co-receptor *ABSCISIC ACID RESPONSIVE ELEMENTS-BINDING FACTOR 2 (ABF2)* and *ABF3* were upregulated. The ABA content was not altered in both roots and shoots after 8-h GSH treatment, whereas 24-h treatment duration notably increased the shoot ABA levels. Thereafter, Cheng et al. (2015) suggested that GSH treatment for 8-h caused the ABA synthesis in roots, but 24-h durated treatment transported this produced ABA to shoots. ABA transportation occurs in both passive and mediated ways by ABA transporters. The dealings between anabolic and catabolic pathways maintain the endogenous ABA. ABA glucosyl ester and phaseic acid are deposited in the cell vacuole or apoplast pool as the catabolism products.

Under drought stress, glucosidase enzyme (such as AtBG1-) mediated-ABA accumulation in guard cells causes stomatal closure in *A. thaliana* (Lee et al. 2006). Although, specific transporters mediated ABA movement between cells is not clearly known yet, but ABA can be moved to outside of cells by the changing of pH (Seo and Koshiba 2011). However, the existence of carrier-mediated and indirect ABA transport mechanisms is suggested.

Several ABA transporter genes under the family of ATP-binding cassette (ABC) were found in vascular tissues of *A. thaliana* such as *AtABCG25* and *AtABCG40* (Kuromori and Shinozaki 2010). Both genes are associated with ABA signaling, where *AtABCG25* and *AtABCG40* regulate ABA movement from inside to the outside of the cell and outside to the inside of the cells, respectively.

Abscisic acid regulates stomatal opening (Hartung et al. 1998) and root hydraulic conductivity (Hose et al. 2000, 2001). It is already established that ABA improved plant stress tolerance like drought and salinity (Giraudat et al. 1994). In addition, ABA works as an important regulator for seed development with enhancing dormancy, seedling establishment, and root branching (Larson 1988; Alscher et al. 1997). ABA also has essential functions in plant-pathogen reactions in a pathosystem-dependent manner (Wasilewska et al. 2008; Cutler et al. 2010).

Thus, ABA controls many significant developmental processes in plants upon stresses to induce tolerance responses (Giraudat et al. 1994). Both ABA and ROS accumulation prevent seed germination and growth of drought-stressed rice (Liu et al. 2019). Previously, Frey et al. (2012) reported that ABA hormone and antioxidant enzyme activity were changed during seed germination processes. Conversely, it was also found that ABA prevents seed germination, while ABA-deficient mutant seeds germinate earlier than wild genotypes as well as transgenic lines with overexpressed ABA also showed late germination (Qin and Zeevaart 2002; Okamoto et al. 2006; Okamoto et al. 2010). Moreover, drought-mediated ABA accumulation causes the preventation of seed germination and consequent seedling establishment (Toh et al. 2008; Liu et al. 2015; Wen 2015; Wen et al. 2016). It was might be because of the ROS involvement in ABA-induced regulation of seed germination (Liu et al. 2010; Ye et al. 2012). Expression of key gene *OsNCED3* in ABA synthesis was induced in drought-stressed rice. On the other hand, drought caused the reduction of antioxidant enzyme activities including SOD, POD, CAT, and APX during seed germination, supported by the expression of *OsCu/ZnSOD*, *OsCATc*, and *OsAPX2* (Liu et al. 2019).

Stress-responsive genes are regulating by ABA. While Cis-elements like ABA-responsive element (ABRE) containing a core sequence PyACGTGGC having one or two coupling elements (CE) were found to upstream of stress-responsive genes. There are many members of bZIP transcription factor family known as ABA-responsive element-binding protein (AREB) (Shen et al. 2004). Drought stress activates many ABA-dependent and ABA-independent events and much effort has been devoted to identifying the components in the major gene networks in order to understand how the stress information may be processed and integrated ("cross-talk") (Shinozaki and Yamaguchi-Shinozaki 2000). Therefore, dehydration, high salinity, or ABA treatments caused the activation of bZIP transcription factors like AREB1/ABF2, AREB2/ABF4, and ABF3 in plant cell, which might be the causes of improved stress tolerance (Yoshida et al. 2010). Drought and high salt-stressed soybean showed an increment in ABA synthesis with higher expression of GmbZIP1—a member of AREB subfamily, which later caused the stomatal closure leading to enhancement of abiotic stress tolerance (Gao et al. 2011). Moreover, besides ABA-induced stress-responsive gene expression, ABA also shows an antagonistic effect on seed germination by suppressing the gibberellin pathway (Fernando and Schroeder 2016). Thus, ABA plays a role in promoting dormancy, which prevents preharvest sprouting. Upon stress, the activities of POX, CAT, and SOD were enhanced along with increasing the endogenous ABA content in *Physcomitrella patens* (Vujicic et al. 2017). Again, Foliar application of ABA (20 μm) showed the significant lowering of H_2O_2 and MDA contents because of increased AsA:DHA ratio and enzymatic antioxidant activities in *Triticum aestivum* under water stress (Kaur and Zhawar 2018).

Chandrasekar et al. (2000) reported that drought stress increased ABA accumulation in a tolerant variety of wheat than susceptible one. Consequently, exogenous ABA application increased the activity of SOD, CAT, APX, and GR enzymes with the higher contents of ascorbate, reduced glutathione, α-tocopherol, and carotenoids (Jiang and Zhang 2001). On the other hand, ABA pretreated maize seedling showed the reduction of the catalytic iron content with augmented antioxidant enzyme activity under water stress (Jiang and Zhang 2002).

2.2 *Cytokinin*

Cytokinins (CKs) are another group of important phytohormones whose chemical nature, metabolism, and signal transduction pathways were broadly studied due to their potential roles in plant growth and development. Previously many researchers had been narrated about plant organ and tissue-specific biological actions of CKs (Bielach et al. 2012; Bishopp et al. 2011; Chiang et al. 2012; Efroni et al. 2013; Zwack and Rashotte 2015). These are adenine derivatives consisting of isoprenoid or aromatic side chains where isoprenoid CKs are very common and renowned as isopentenyladenine (iP)-, trans-zeatin (tZ)-, cis-zeatin (cZ)-, or dihydrozeatin (DHZ)-type derivatives. Conversely, aromatic CKs as, for example, N6-(meta-hydroxybenzyl) adenine (BA) are rarely found in plants (Faiss et al. 1997). Variation of isoprenoid type of CKs not only occurred due to their differences in biochemical properties, biological functions, metabolic conversions, and transportability but also with plant tissue (Pavlů et al. 2018).

There are different types of enzymes involved in CK biosynthesis, degradation, and interconversion to maintain CK homeostasis (Thu et al. 2017). Moreover, a coordinating network of CK with other hormones like ABA, JAs, SA, and ET exists in conferring tolerance under environmental stress (Efroni et al. 2013; Thu et al. 2017; Artner and Benkova 2019; Verma et al. 2016).

Stomatal functions including stomatal conductance and stomatal density are also regulated by CKs (Farber et al. 2016) and thus they can control the regulation of osmotic status (Hai et al. 2020).

It is reported that CKs are able to inhibit ROS-producing enzyme-like xanthine oxidase or enhance antioxidant enzymes activities like SOD and CAT; thus suggesting the potential role of CKs in ROS metabolism (Hönig et al. 2018). Exogenous application of CKs (sprayed-2-(chloro-4-pyridyl)-N-phenyl urea (CPPU) as a source of CK) revealed the improvement in photosynthesis through improving chlorophyll (Chl) pigments, stomatal conductance, and RuBisCO activity in drought-stressed rice (Gujjar et al. 2020). Besides exogenous CKs-induced betterment in plant growth, with higher membrane stability, leaf relative water content (RWC), the content of Chl and soluble sugars, increased antioxidant activities were also reported (Chang et al. 2016; Kumari et al. 2018; Samea-Andabjadid et al. 2018). Transgenic tobacco with overexpressed *Arabidopsis AtCKX2* showed the enhanced activities of SOD, GR, and APX (Mýtinová et al. 2011) while *AtCKX1* overexpressed tobacco showed the promotion of CAT, APX, or SOD encoded gene (Lubovská et al. 2014). Barley with the ectopic expression of *CKX* enhanced the nonenzymatic antioxidants responses to attain higher drought stress tolerance (Nakabayashi et al. 2014). Transgenic creeping bent grass with *Agrobacterium IPT* gene had lower ROS along with higher SOD, CAT, APX, and DHAR activities (Xu and Huang 2017). A similar result was also reported in brinjal with an indication of enhancement in ROS-scavenging enzymes activities (Xiao et al. 2017a, b).

CKs also enhance saline tolerance through maintaining higher photosynthesis, soluble sugars, and proline concentration (Gashaw et al. 2014). Exogenous application of kinetin (one kind of CK) improved Chl and carotenoid contents, photosynthesis, enzymatic antioxidants (SOD, CAT, APX, MDHAR, DHAR, GR, and GST activities), and nonenzymatic antioxidants (ASA and GSH), along with reduced ROS generation and lipid peroxidation in salt-stressed tomato (Ahanger et al. 2018). Kinetin as foliar spray also counteracted the inhibitory effect of salinity and waterlogging stresses by improving Chl and growth parameters in both *Vigna sinensis* and *Zea mays* (Younis et al. 2003). A summary table of CK-mediated regulation on antioxidants and related attributes is summarized in Table 1.

2.3 Auxin

Although the biosynthesis of auxin is still not clear, three major auxins such as naphthalene acetic acid (NAA), indole butyric acid (IBA), and indole-3-acetic acid (IAA) are notable. Auxins have potential roles in plant growth including cellular differentiation and development under environmental stress conditions. Auxin transportation due to passive and active mechanisms called auxin maxima is responsible for the polar cell growth and morphogenesis of plants (Grunewald and Friml 2010). This is accomplished in cell membranes by the differential localization of PIN-FORMED (PIN) proteins, which are rate-limiting auxin efflux transporters. Contrariwise, AUXIN RESISTANT 1 (AUX1) has an important role in inducing auxin import (Benjamins and Scheres 2008). The function of PIN and AUX protein for getting auxin response is also associated with their polar localization in the plasma membranes (Sauer et al. 2006). Auxin mediated such regulation of plant growth and development like cell elongation is closely related to ROS, while auxin-mediated ROS generation and cellular redox state play the signaling role developmental processes (Schopfer 2001). However, little information is available about the status of auxin levels and its relations in plants under waterlogging stress. But it is reported that IAA is responsible to increase the root and shoot growth of salt-stressed plants (Sheng and Xia 2006; Egamberdieva 2009). Thereafter, Fahad and Bano (2012) revealed that IAA levels were significantly reduced in salt-stressed maize and suggested that hormonal balance is very critically required for stress response as well as regulating plant growth and development under stress condition. However, information on auxin-mediated antioxidants upregulation is still limited. Exogenous IAA-treated drought-stressed white clover showed the upregulated expression of auxin-responsive genes (*GH3.1, GH3.9, IAA8*) as well as improved RWC and Chl as stress-tolerance responses (Zhang et al. 2020). They also reported about higher ABA and JA content with upregulated drought stress-responsive genes (*bZIP11, DREB2, MYB14, MYB48, WRKY2, WRKY56, WRKY108715,* and *RD22*) and downregulated expressions of auxin-responsive genes (*GH3.3, GH3.6, IAA27*) and leaf senescence genes (*SAG101* and *SAG102*). On the other hand, the transgenic lines of *Arabidopsis* (iaaM-OX) showed higher endogenous IAA level where IAA pretreatment modulated the expression levels of *RAB18*, *RD22*, *RD29A*, *RD29B*, *DREB2A*, and *DREB2B* whose are renowned multiple abiotic stress-responsive genes (Shi et al. 2014). They also reported that, exogenously IAA treated plants of transgenic iaaM-OX line had lower H_2O_2 and $O_2^{\bullet-}$ ROS with elevated SOD, CAT, POD, and GR activities under drought stress than non-treated wild plants. Similarly, it was found that auxin is helpful in improving POD, SOD, APX, and CAT when applied as foliar on salt-stressed *Helianthus annuus* (Zayed et al. 2017). Consequently, this auxin-mediated higher antioxidant activity was associated with lowering salt toxicity showing increased biomass production, membrane stability, and plastid pigment content.

Drought-stressed transgenic potato (by overexpressing *AtYUC6*) showed elevated auxin levels and better tolerance phenotypes through regulating ROS homeostasis

Table 1 Regulations of antioxidants and related attributes by ABA, gibberellins, auxins, and ethylene in plants under salinity and water stress

Phytohormone	Stress	Crops	Responses	References
Abscisic acid	Drought, salinity	*Agrostis stolonifera* and *Poa pratensis*	Suppressed electrolyte leakage (EL) and MDA. Increased APX, POD, and SOD activities	Yang et al. (2012a, b, c)
Abscisic acid (ABA)	Drought	*Camellia sinensis*	Upregulated glycolysis and photosystem II Increased APX, GST activities with higher chlorophyll (Chl), and proline contents	Zhou et al. (2014)
ABA	Drought	*Triticum aestivum*	Improved SOD activity with higher RWC where POD was reduced	Bano et al. (2012)
ABA	Drought	*Oryza sativa*	Increased $O_2^{\bullet-}$, H_2O_2, and MDA generation during seed germination, were reduced the activity of SOD, CAT, and APX	Liu et al. (2019)
ABA	Drought	*Cynodon dactylon*	Increased RWC with reduction of EL. Improved SOD and CAT activities	Lu et al. (2009)
ABA	Drought	*Lolium perenne*	Reduced MDA, EL, and H_2O_2. Provoked the activity of CAT, SOD, POD, and APX with higher RWC	Mohammadi et al. (2017)
ABA	Drought and salinity	*Arabidopsis thaliana*	Accelerated gene expression and stomatal closure	Jakab et al. (2005)
ABA	Drought and salinity	*A. thaliana*	Increased stress tolerance with association of GSH indicated by 100% survival rate in stressed condition	Chen et al. (2012)
ABA	Drought	*T. aestivum*	Reduced H_2O_2 and thiobarbituric acid (TBA) content Increased SOD, POD, and CAT activities	Agarwal et al. (2005)
ABA	Drought	*A. thaliana*	Regulated stomatal closure	Brossa et al. (2011)
ABA	Water	*Capsicum annuum*	Increased peroxidase contents	Choi and Hwang (2011)
ABA	Drought	*Brassica juncea*	Improved activities of enzymatic antioxidants (SOD, CAT, APX, and GR) with higher nonenzymatic antioxidants content like ascorbic acid combined with NO	Sahay et al. (2019)

(continued)

Table 1 (continued)

Phytohormone	Stress	Crops	Responses	References
ABA	Salinity	*Pseudomonas syringae* pv *lachrymans*	Stronger H_2O_2 accumulation and Fe-SOD activation. Changed SA/ABA balance with reduction and increase of CAT and SOD, respectively	Chojak-Koźniewska et al. (2017)
ABA	Salinity	*Medicago sativa*	Increased ABA accumulation Alleviated the detrimental effect of salinity by improving plant growth and nitrogen fixation Improved the antioxidants activity like SOD, CAT, and GR with reduction of MDA generation	Palma et al. (2014)
Cytokinin (CK)	Salt	*Solanum lycopersicum*	Increased both active and inactive cytokinin levels Decreased oxidative stress level	Keshishian et al. (2018)
Auxin	Drought	*A. thaliana*	Regulated IBA and IAA homeostasis With improved survival rate significantly	Tognetti et al. (2010)
Auxin	Salinity	*A. thaliana*	Increased glutaredoxins activity	Sharma et al. (2015)
Auxin	Oxidative stress	*A. thaliana*	Increased the activity of APX, CAT, and POD	Kim et al. (2011)
Gibberellin (GA)	Salinity	*S. lycopersicum*	Decreased CAT, POD, and PPO activities. Lowered the generation of lipid peroxidation with higher glutathione content	Halo et al. (2015)
GA	Salinity	*Cucumis sativus*	Reduced antioxidant activity, MDA, and EL contents Lowered RWC, photosynthesis rate, and nitrogen assimilation	Khan et al. (2012)
GA	Salinity and osmotic	*C. sativus*	Negative correlated with growth-promoting rhizobacteria for POD, CAT, PPO	Kang et al. (2014)
Ethylene (ET)	Salinity	*C. sativus*	Decreased H_2O_2, MDA content Improved SOD and CAT activities	Shakar et al. (2016)
ET	Salinity	*A. thaliana*	Decreased H_2O_2 generation	Lin et al. (2013)

(Park et al. 2013). Overexpression of *TaSAUR75* improved the tolerance of *Arabidopsis* by lowering H_2O_2 accumulation under both drought and salt stress (Guo et al. 2018). A summary table about auxin-mediated regulation of antioxidants and related attributes is summarized in Table 1.

2.4 Ethylene

Ethylene (ET) is colorless and odorless gaseous phytohormone, which possesses key functions in plant growth and development (Riyazuddin et al. 2020). Due to its gaseous characteristic, ethylene is produced principally at or near to the site of action. The conversion of methionine via *S*-adenosyl-L-methionine (AdoMet) is catalyzed by ACC (1-aminocyclopropane-1-carboxylic acid) synthase and the cyclic nonprotein amino acid ACC is catalyzed by ACC oxidase, which ultimately produces ethylene. Signaling components of ethylene were identified in *A. thaliana* in 1990s (Fahad et al. 2015). Signal transduction of ethylene involves the combined action of various receptors and other components (viz. CTR1, SIMKK, MAP kinase 4/5, 6, EBF1/2, etc.). Different plant species possess different numbers of ethylene receptors. For example, five receptor genes namely *ETR1* (*Ethylene Response1*), *ETR2*, *ERS1* (*Ethylene Response Sensor1*), *ERS2,* and *EIN4* (*Ethylene Insensitive4*) are found in *Arabidopsis*. Seven ethylene receptors such as *LeETR1*, *LeETR2*, *NR*, *LeETR4*, *LeETR5*, *LeETR6,* and *SlETR7* have been identified from tomato. Moreover, rice and tobacco possess one ethylene receptor named OsERS1 and NTHK1, respectively (Fahad et al. 2015; Riyazuddin et al. 2020). These receptors have enabled ethylene as a key modulator of abiotic stress regulation, specifically, salinity stress. Ethylene has been found to have a positive impact on salinity stress tolerance in various plant species like *Arabidopsis*, maize, tomato, and grapevines (Freitas et al. 2018; Gharbi et al. 2017; Riyazuddin et al. 2020; Xu et al. 2019). Salt stress causes the increment in ethylene biosynthesis by the modulation of ACC and ACC oxidase in various plants. Therefore, this overproduced ethylene has been reported to enhance salt tolerance in maize, *Arabidopsis*, wild tomato (*Solanum chilense*), and drought stress in soybean (Arraes et al. 2015; Freitas et al. 2018; Gharbi et al. 2017; Yang et al. 2013).

Ethylene or ACC enhances ROS scavengers and thus mitigating salt toxicity by different mechanisms. Ethylene Insensitive 3 (EIN3) protein prompted the salt tolerance mechanism to decrease the ROS accumulation decrease under stress conditions. As we know that, oxidative stress is originated from the imbalance in the formation and scavenging of ROS. In this EIN3 protein-mediated pathway, EIN3 enhances the responses of various peroxidases that scavenge ROS and maintain the equilibrium of ROS generation and scavenging. Additionally, EIN3/EIL1 (Ethylene Insensitive 3-Like 1) transcription factors enhanced salt tolerance by regulating innumerable *SIED* genes whose overexpression prevents ROS accumulation under stress (Peng et al. 2014). Salicylic acid application suppressed ethylene level, but optimal level of ethylene is maintained which is responsible for inhibiting oxidative stress as well as regulates photosynthesis in salt-stressed mungbean (*Vigna radiata* L.) (Khan et al. 2014). This result supports that the regulation of ET precursor ACC is effective to maintain optimal ET level for reducing oxidative stress in salt-affected *Arabidopsis* (Wang et al. 2009). Therefore, there is a relationship between ethylene and antioxidants response against ROS, which can be correlated by the increased AsA content from EIN2–1, EIN3–1, and EIN4 mutant *Arabidopsis* (Gergoff et al. 2010). Interestingly, ethylene has been found to have both beneficial and detrimental effect on plants under salinity stress as they cause higher ROS production by decreasing the antioxidant enzymes activities (reduced POD and GR activities). ERFs are also involved in ROS formation and signaling. It has been reported that ERF1 reduces the expression of SOD and peroxidase in *Tamarix hispida* under both saline and drought stress conditions resulting in increased ROS accumulation with reduced scavenging capacity. Collectively, these findings suggest the role of ethylene as biphasic and complex under abiotic stresses (Li et al. 2014a, b; Wang et al. 2014a, b).

Waterlogging negatively affects various plants, which results in stunted growth and yield loss due to lower nutrient uptake, reduction of photosynthesis, and imbalance in phytohormones. In waterlogging condition, plants develop aerenchyma in their roots, which trigger signal regulators like ROS, NO, and most importantly, ethylene as the acclimation strategies. Therefore, soybean plants (*Glycine max* L.) had been treated by ethephon (ET, donor source of ethylene) to mitigate the overproduction of ROS under waterlogging stress. The underlying mechanism regarding ET to mitigate ROS is, ETP up-regulates GST3 and GST8 which results in increased GST (DHAR2) that scavenge ROS as well as mitigated cell damage (Kim et al. 2018). Also, it has been reported that free radical formation under flooding stress triggers the production of ethylene from ACC that plays role in mitigating oxidative stress by enhancing antioxidants (CAT, POD, GR, GSH) activities (Beltrano et al. 1997; Kim et al. 2018).

Comparatively a smaller number of studies have been found about the effect of ethylene in mitigating oxidative stress under drought stress (Table 1). Ethylene acts against drought stress by overexpressing ERF proteins (namely, JERF1) that enhance crop tolerance through the regulation of stress-responsive genes in rice (Zhang et al. 2010). Ethylene promotes the stomatal closure under drought stress in tomato, *Arabidopsis, Solanum lycopersicum,* and so on, which might be because of ROS production in stomatal guard cells (Desikan et al. 2006). The opposite result has also been reported that is ethylene increases the stomatal conductance in mustard (Desikan et al. 2006; Madhavan et al. 1983).

2.5 Gibberellins

Gibberellins (GAs) are one of the essential phytohormones playing role in seed germination, photomorphogenesis, leaf expansion, stem elongation, and flowering (Ryu and Cho 2015). The GA signaling has been found to either enhance or suppress plant growth depending on the type of abiotic stress and their level and thus become one of the prime targets under stress conditions for getting growth modification. This hormone belongs to the large group of tetracyclic diterpenoid carboxylic acids where GA_1 and GA_4 are the most predominant bioactive forms. Biosynthesis of GAs starts in plastids from trans-geranylgeranyl diphosphate and finishes in cytosol via methylerythritol phosphate pathway in the endoplasmic reticulum by cytochrome P450 monooxygenases and then by soluble 2-oxogluterate-dependent dioxygenases. Three types of small families named GA 20-oxidase (GA20ox), GA 3-oxidase (GA3ox), and GA 2-oxidase (GA2ox) are the members of these dioxygenases. Among them, GA2ox enables GA turnover (Kasahara et al. 2002; Yamaguchi 2008; Hedden and Thomas 2012).

The exogenous GAs application is known to be effective for the enhancement in plants' growth and development by mitigating various abiotic stresses toxicity (Rhaman et al. 2021). External application of GAs enhanced the growth of drought-stressed wheat (*T. aestivum*) by improving water content, Chl content as well as balancing antioxidant defense system (Al Mahmud et al. 2019). In pepper (*Capsicum annuum*) plants, GAs in combination with poultry manure have been found to enhance the growth and stress tolerance under salinity (AlTaey 2017). Similarly, salt-affected tomato plants showed a higher leaf RWC, stomatal density, and Chl content by foliar application of GA_3 salt (Saeidi-Sar et al. 2013; Jayasinghe et al. 2019). Supplementation of GAs also can be effective to mitigate abiotic stresses when applied as seed-priming agents (Rhaman et al. 2021). Wheat and maize seeds primed with GAs 150 ppm and 5 mg L^{-1}, respectively, showed enhancement in germination, seedling growth parameters, tissue water content under salinity stress (Ghodrat and Rousta 2012; Abido et al. 2019). As seed-priming agent, GAs are also effective to alleviate the water shortage toxicity in *A. thaliana, Z. mays, Sorghum bicolor, Cicer arietinum* as well as flood stress in *Oryza sativa* (Rhaman et al. 2021).

Exogenous application of GA_3 alone or along with AsA reduces NaCl-induced accumulation of toxic H_2O_2 and MDA. The GA_3 modulated antioxidant enzymes activity in alleviating salt toxicity has been discussed by Maggio et al. (2010). The most important underlying mechanism of GAs in abiotic stress tolerance is the involvement of DELLA proteins. It is noteworthy that, DELLA activity restrains the accumulation of ROS that delays plant cell death in *A. thaliana* under salinity (Achard et al. 2008). Another defense mechanism of GAs against salt stress might be the regulation of hormonal homeostasis. The combination of GA_3 and AsA has been found to be more protective than GA_3 and AsA alone in salt-stressed *Phaseolus vulgaris* (Saeidi-Sar et al. 2013). This AsA is attributed to an increase in photosynthetic activity and antioxidative defense. A summary table of GAs mediated regulation on antioxidants and related attributes is mentioned in Table 1.

However, GAs-induced defense system under abiotic stresses GAs-induced in response to ROS is hardly found, especially in case of drought and waterlogging stress. More or less, some experiments have been done with salinity stress. So, more studies are required focusing on the antioxidant defense system by applying GAs under various abiotic stresses.

2.6 Salicylic Acid (SA)

Salicylic acid (SA) is another kind of phytohormones with phenolic nature, which are ubiquitously distributed in plants and have multiple functions in various developmental events like photosynthesis, ion uptake, enzyme activity, and stress tolerance (Ahmad et al. 2011; Moravcová et al. 2018). The SA is synthesized through a distinct isochorismate (IC) and phenylalanine ammonia-lyase (PAL) pathway. Both pathways are initiated with chorismic acid, while the end product is derived from the shikimic acid pathway (Ahmad et al. 2018). In several plant species, isochorismate synthase (ICS) catalyzes the conversion of chorismic acid into IC (Garcion et al. 2008; Wan et al. 2012). Moreover, isochorismate pyruvate lyase catalyzes the conversion of IC to SA, although the actual action is still ambiguous (Mercado-Blanco et al. 2001). In the phenylpropanoid pathway, PAL causes the deamination of phenylalanine to generate trans-cinnamic followed by the conversion of orthocoumaric acid or the benzoic acid to synthesis SA (Yalpani et al. 1993; Catinot et al. 2008).

Several research groups reported its potential role in improving antioxidant activity and photosynthesis (Nazar et al. 2011; Khan et al. 2014; Moravcová et al. 2018). Recently it was reported that SA has a vital role in the regulation of phytohormone, ROS, and antioxidant defense in different plant species under various environmental stresses like salinity, drought, and waterlogging (Abdelaal et al. 2020a, b; Ahanger et al. 2020a, b; Bashar 2018; Sharma et al. 2020; Torun et al. 2020; Loutfy et al. 2020). The SA accumulation causes the increment in plants tolerance under drought conditions. Water shortage increases the SA accumulation in *Phillyrea angustifolia* (Bandurska 2013), while drought tolerance of tomato and bean is improved by the exogenous application of SA (Senaratna et al. 2000). Mutant *Arabidopsis* with *adr1* and *myb96-1d* showed higher SA accumulation as well as higher drought tolerance (Seo et al. 2009; Seo and Park 2010). Mutant *siz1*, *cpr5*, and *acd6* in *Arabidopsis* exhibited elevated SA accumulation with drought tolerance by closing the stomatal opening (Bowling et al. 1997; Rate et al. 1999; Lee et al. 2007). Higher SA accumulation is effective to inhibit the stomatal opening, which reduces the ROS production in *Arabidopsis* (Okuma et al. 2014). Nowadays, exogenous application of SA has been frequently used to confer drought and salinity tolerance where antioxidant responses (both enzymatic and nonenzymatic) were upregulated in many crop species (Table 1). Exogenous SA mediated higher CAT, APX, and GPX activities with lower ROS were observed in drought-stressed rice (Sohag et al. 2020), while SOD and APX activities were increased in soybean (Razmi et al. 2017). Salicylic acid also regulated the antioxidant responses in *Fritillaria przewalskii* (Ma et al. 2019) and in maize (Shemi et al. 2021) under drought stress. Though this SA-mediated regulation of antioxidants under drought conditions is largely depend on doses and application methods. Similarly, the effect of SA to mitigate salt stress is also dose and time dependent for getting the balance between ROS and antioxidant defense system (Yu et al. 2020). Numerous studies reported about positive roles of SA to improve salt tolerance in different plants through the regulation of antioxidants for balancing ROS and ionic status (Table 2). SA also enhanced the expression level of selected genes (*bZIP62*, *DREB2*, *ERF3*, and *OLPb*) to conferring salinity tolerance in rosemary (El-Esawi et al. 2017). In addition, SA also stimulated the enzymatic antioxidants like CAT, SOD, and APX as well as nonenzymatic antioxidants including free and total ascorbate.

2.7 Jasmonates

Jasmonates (JAs) are lipid-derived phytohormones capable to regulate plant growth under changing environments through diverse signaling approaches coordinating with other hormones (Per et al. 2018). There are three key members of this family naming methyl jasmonate (MeJA), jasmonic acid (JA), and jasmonate isoleucine conjugate (JA-Ile) (Ruan et al. 2019). Jasmonates maintain cellular integrity in plants upon abiotic stresses by upregulating the antioxidant activities including both enzymatic and nonenzymatic components, osmoprotectants synthesis, remobilize the photoassimilates remobilization (Farhangi-Abriz and Ghassemi-Golezani 2019; Siddiqi and Husen 2019). In addition, JA accumulation increased as a response of various kinds of abiotic stresses like drought, salinity, waterlogging (Lehmann et al. 1995; Xiao et al. 2017a, b; Ghassemi-Golezani and Farhangi-Abriz 2018; Wang et al. 2009). External use of JA showed an enhancement in plant stress tolerances (Table 3). Moreover, stress tolerance mechanism is improved through JA production and its signaling approach. Kim et al. (2009) studied this by adopting a transgenic approach where *jasmonic acid carboxyl methyl transferase* gene (*AtJMT*) expression was increased in rice under drought stress and found higher JA level. Similarly, overexpressing *CmLOX10* promotes JA-mediated stomatal closure in *Cucumis melo* as well as enhances drought tolerance (Xing et al. 2020). Involvement of the transcription factor JIN1/MYC2 activated by jasmonate signal regulates ABA-dependent salt stress reactions in plants (Ryu and Cho 2015).

Lipoxygenases (LOXs) catalyze the oxidation of oxygenation of polyunsaturated fatty acids like linoleic acid is also an essential enzyme for JA synthesis containing a cis, cis-1, 4-pentadiene (Cao et al. 2016). According to the specific oxidation sites, they are classified into 9-LOXs and 13-LOXs in plants (Feussner and Wasternack 2002). On the other hand, LOXs act importantly in plant physiology and senescence (Seltmann et al. 2010; Yang et al. 2012a, b, c), as well as in responses against abiotic stresses (Hou et al. 2015). It is already reported that *13-LOX* genes regulating JA synthesis to improve abiotic stress tolerance in plants, where *Arabidopsis* LOX2 is associated to JA synthesis in osmotic stress, while *Arabidopsis* lox3 mutant displayed salt sensitivity (Xing et al. 2020). Exogenous MeJA application had been found to improve plant tolerances under stresses including salinity, drought, and waterlogging by scavenging ROS with the up-stimulation of enzymatic and nonenzymatic antioxidants responses (Kamal and Komatsu 2016; Sadeghipour 2017; Yastreb et al. 2018; Tayyab et al. 2020). Besides of increased activity of POD, CAT, and APX, MeJA also regulated osmolytes synthesis, stomatal

Table 2 Regulation of antioxidants and related attributes by salicylic acid in plants under salinity and water stress

Phytohormone	Stress	Crop	Responses	References
Salicylic acid (SA)	Salinity	*Triticum aestivum*	Increased SOD, CAT, and APX activities Reduced H_2O_2 and MDA accumulation	Alsahli et al. (2019)
SA	Salinity	*T. aestivum*	Decreased H_2O_2 and MDA content Reduced GSSG content Improved the activity of CAT, POD, GST, APX, DHAR, and MDHAR. Increased RWC and the homeostasis of AsA and GSH	Fardus et al. (2017)
SA	Salinity	*Vigna angularis*	Increased SOD, CAT, APX, DHAR, and GR activities Improved the content of AsA and GSH	Ahanger et al. (2020a)
SA	Salinity	*Raphanus sativus*	Improved photosynthesis. Enhanced the activity of SOD, CAT, and POD. Increased proline content Lowered the ROS generation and membrane damage	Bukhat et al. (2020)
SA	Salinity	*Rosmarinus offificinallis*	Increased the content of phenolic, chlorophyll (Chl), proline, and AsA Enhanced CAT, SOD, and APX activities	El-Esawi et al. (2017)
Sodium salicylate	Salinity	*T. aestivum*	Increased proline accumulation and K^+/Na^+	Al-hakimi and Hamada (2001)
SA	Salinity	*Brassica carinata*	Increased proline and activity of SOD, CAT, POD Improved Chl fluorescence (Fv/Fm), stomatal conductance (gs), net photosynthetic rate (Pn), transpiration rate (E)	Husen et al. (2018)
SA	Salinity	*Helianthus annuus*	Enhanced proline, soluble sugar, and RWC Decreased MDA content	Ebrahimian and Bybordi (2012)
SA	Salinity	*Capsicum annuum*	Decreased MDA, EL, $O_2^{\bullet-}$, and H_2O_2 production Increased D and PPO activities with reduction of CAT activity	Abdelaal et al. (2020b)

(continued)

Table 2 (continued)

Phytohormone	Stress	Crop	Responses	References
SA	Salinity	*Arabidopsis thaliana*	Lowered H_2O_2 generation and enhanced POD activity	Lee et al. (2010)
SA	Salinity	*Zea mays*	Increased SOD activity and decreased CAT activity	Gautam and Singh (2011)
SA	Salinity	*Hordeum vulgare*	Improved SOD, POD, APX, and GR activities	Torun et al. (2020)
SA	Salinity	*T. aestivum*	Increased Chl, tillers number, and K^+/Na^+	Suhaib et al. (2018)
SA	Salinity	*Z. mays*	Enhanced K^+/Na^+ ratio, soluble sugar, with lowering proline	El-Katony et al. (2019)
SA	Salinity	*Solanum lycopersicum*	Increased proline and soluble sugars accumulation	Souri and Tohidloo (2019)
SA	Salinity	*A. thaliana*	Increased activities of SOD CAT, APX, GPX Reduced H_2O_2 concentration	Yu et al. (2020)
SA	Salinity	*Vicia faba*	Increased IAA and IBA with reduction of ABA Increased SOD, POD, CAT APX, and GR activities Reduced H_2O_2 accumulation	Ahmad et al. (2018)
SA	Salinity	*Nitraria tangutorum*	Upregulated activities of SOD, POD, CAT, and APX	Liu et al. (2016)
SA	Salinity	*Dianthus superbus* L.	Enhanced SOD, POD, and CAT activities. Decreased ROS and MDA concentration	Ma et al. (2017)
SA	Salinity	*Z. mays*	Regulated phytochromes and various organic and inorganic osmolytes. Improved IAA, and GA_3 contents, but decreased ABA	Elhakem (2020)
SA	Salinity	*S. tuberosum*	Improved SOD, CAT, and POD activities Reduced ROS. Regulated osmotic adjustment (proline, phenolic contents) Improved water relation and gaseous exchange	Faried et al. (2017)
SA	Salinity	*Torreya grandis* cv. Merrillii	Increased SOD, CAT, and POD activities Reduced MDA content	Li et al. (2014a, b)
SA	Drought	*Fritillaria przewalskii*	Increased RWC, soluble carbohydrate, proline, MDA. Reduced SOD, CAT, CAT, GR, and APX activities	Ma et al. (2019)

(continued)

Table 2 (continued)

Phytohormone	Stress	Crop	Responses	References
SA	Drought	*Cicer arietinum*	Reduced MDA and H_2O_2 generation Increased proline and chl *a* content Enhanced the activity of POD, CAT, and SOD	Hussain et al. (2020)
SA	Drought	*T. aestivum*	Enhanced SOD, CAT, APX, and POD activities Increased accumulation of phenol and flavonoids with reduction of H_2O_2	Hassanein et al. (2015)
SA	Drought	*T. aestivum*	Increased ascorbate oxidase (AO), POD, and phenylalanine ammonia lyase (PAL), and PPO reduced	Aldesuquy and Ghanem (2015)
SA	Drought	*B. campestris*	Reduced MDA content Increased CAT, APX, GR activities	Cha et al. (2020)
SA	Drought	*T. aestivum*	Improved photosynthetic performance, membrane permeability Enhanced activity of SOD and CAT	Khalvandi et al. (2021)
SA	Drought	*T. aestivum*	Improved RWC, Chl, carotenoids, and proline content Increased SOD, POD, APX, and CAT activities Reduced MDA content	Azmat et al. (2020)
SA	Drought	*Z. mays*	Reduced the accumulation of H_2O_2, $O_2^{\bullet-}$, and MDA Improved RWC, soluble sugar, and Chl content Enhanced SOD, POD, and CAT activities	Shemi et al. (2021)
SA	Drought	*Hordeum vulgare*	Increased RWC, MDA, and EL, H_2O_2, and $O_2^{\bullet-}$ levels Decreased CAT, POD, and PPO activities	Abdelaal et al. (2020a)
SA	Waterlogging	*S. lycopersicum*	Elevated Chl, protein, and sugar contents	Singh et al. (2017)
SA	Drought	*Glycine max*	Reduced $O_2^{\bullet-}$, H_2O_2, and MDA Increased CAT and SOD, APX activities Improved protein and proline	Razmi et al. (2017)
SA	Drought	*T. aestivum*	Enhanced POD, APX, CAT, and SOD activities Decreased H_2O_2 and MDA generation	Maghsoudi et al. (2019)

(continued)

Table 2 (continued)

Phytohormone	Stress	Crop	Responses	References
SA	Salinity, drought	*T. aestivum*	Decreased proline content and CAT activity Increased APX and GPX activities with higher GSH level	Loutfy et al. (2020)
SA	Salinity, drought	*H. vulgare*	Reduced MDA and Na^+/K^+ Enhanced proline, soluble carbohydrate, and total phenolic compounds	Fayez and Bazaid (2014)

movement, and chlorophyll contents in barley plants (Pazirandeh et al. 2013). Conversely, Anjum et al. (2016) revealed that MeJA can improve the drought tolerance in wheat via better growth and yield. A dose-dependent seed treatment of cowpea by MeJA increased salt tolerance where 50 μM MJ showed better plants growth including higher Chl content, stomatal conductance, photosynthesis, accumulation of proteins; proline; and sugar with better RWC under salinity (Sadeghipour 2017). There is an interaction between JA signaling and ascorbate–glutathione pathway to regulate reoxygenation responses in waterlogging–stressed *Arabidopsis* to improve their tolerance (Xiao et al. 2017a, b).

2.8 *Brassinosteroids (BRs)*

Polyhydroxy steroidal type new ubiquitous phytohormones, which have direct significance on growth-promoting factor, are known as brassinosteroids (BRs) (Vardhini 2012a, b; Bajguz and Piotrowska-Niczyporuk 2014). These were extracted from pollens of *Brassica napus* (Grove 1979). There are three classifications of BRs based on carbon number in the structure such as C27, C28, or C29 with alkyl functional group (Vardhini and Anjum 2015). Three bioactive BRs are used for experimental studies including physiological aspects name as brassinolide (BL), 28-homobrassinolide (28-HomoBL), and 24-epibrassinolide (24-EpiBL) (Vardhini and Anjum 2015). This hormone influences plant growth by affecting various physiological processes (Bajguz 2007). Under stress conditions, BRs play roles in activation of antioxidants (Kim et al. 2012; Lima and Lobato 2017; Tunc-Ozdemir and Jones 2017; Zou et al. 2018). Many researchers reported about the role of BRs in increasing plant tolerance under abiotic stresses through association with other phytohormones (Rajewska et al. 2016; Anwar et al. 2018; Banerjee and Roychoudhury 2018).

In this section, the relation of BRs with excessive ROS and respondent antioxidants in plants under salinity, drought, and waterlogging stress conditions has been accumulated and presented (Table 4). Different physiological and metabolic

Table 3 Regulation of antioxidants and related attributes by methyl jasmonate in plants under salinity and water stress

Phytohormone	Stress	Crop	Responses	References
Methyl jasmonate (MeJA)	Salinity	*Vigna unguiculata*	Improved stomatal conductance, photosynthesis, soluble proteins, and RWC.	Sadeghipour (2017)
MeJA	Salinity	*S. lycopersicum*	Increased photosynthetic rate, soluble sugar, proline, and free amino acid Elevated the activity of CAT and POD	Manan et al. (2016)
Jasmonate (JA)	Salinity	*Glycine max*	Increased K^+ and Ca^{2+}, RWC, proline, glycine betaine, soluble sugars, antioxidant enzymes, and leaf membrane stability	Farhangi-Abriz and Ghassemi-Golezani (2018)
JA	Salinity	*Hibiscus esculentus*	Reduced H_2O_2, MDA. Increased POD and APX activities	Azooz et al. (2015)
MeJA	Salinity	*Triticum aestivum*	Enhanced proline and dehydrins accumulation Decreased $O_2^{\bullet-}$ and MDA. Elevated APX and SOD activities	Avalbaev et al. (2020)
MeJA	Salinity	*Brassica napus*	Improved RWC. Decreased MDA and proline. Elevated activity of POD, CAT, and LOX	Ahmadi et al. (2018)
MeJA	Salinity	*Arabidopsis thaliana*	Increased SOD, CAT, guaiacol POD, proline, and anthocyanins	Yastreb et al. (2018)
JA	Drought	*Brassica* spp	Increased GR, MDHAR, DHAR, GPX activities Elevated Gly-I and Gly-II activities. Improved chl and RWC	Alam et al. (2014)
MeJA	Drought	*G. max*	Improved RWC, photosynthetic pigments, and POD activity	Mohamed and Latif (2017)
Jasmonic acid	Drought	*T. aestivum*	Lowered H_2O_2 and MDA contents Elevated SOD, POD, APC, and GR activities	Wang et al. (2021)
JA	Drought	*T. aestivum*	Enhanced proline and soluble sugar accumulation	Ilyas et al. (2017)
MeJA	Drought	*Verbascum nudicuale*	Increased H_2O_2, MDA, proline accumulation Reduced Chl *a* and Chl *b*, content Increased POD activity while SOD was reduced	Ghasemlou et al. (2019)

(continued)

Table 3 (continued)

Phytohormone	Stress	Crop	Responses	References
MeJA	Drought	*Zea mays*	Decreased MDA, H_2O_2, and LOX activity. Increased proline, soluble sugar, ABA accumulation Elevated the activity of CAT, POD, and SOD	Tayyab et al. (2020)
MeJA	Drought	*Hordeum vulgare*	Increased CAT, POD, and APX activities. Elevated total and reduced ascorbate.	Pazirandeh et al. (2013)
MeJA	Drought	*Z. mays*	Increased proline and IAA accumulation Stimulated the activity CAT, POD, and SOD	Abdelgawad et al. (2014)
JA	Drought, salinity	*Vitis* species	Reduced ROS generation Increased JA synthesis and the activity of SOD and POD	Fang et al. (2016)
MeJA	Drought	*Z. mays*	Elevated SOD, CAT activities with reduction of AsA content Reduced H_2O_2	Li et al. (1998)
MeJA	Oxidative stress	*Panax ginseng*	Increased APX and GST activities Elevated AsA level and its homeostasis Reduced the activity of SOD and CAT	Ali et al. (2006)

processes are significantly influenced by BRs and thus results in better photosynthesis, growth, and biomass accumulation of stressed plants (Ahammed et al. 2020). Exogenous application of BRs increased SOD, POD, CAT activities, content of proline and sugar in *Cucumis sativus* under saline condition (Yang et al. 2019). From several experiments, it has been clear that BRs are endogenous growth regulators by including the isolation of BR-insensitive and -deficient mutants of *Arabidopsis* (Clouse et al. 1996; Szekeres et al. 1996). This phytohormone plays its role in signal transduction pathways for regulating various biological responses (Kim and Wang 2010; Clouse 2011; Guo et al. 2013; Dejonghe et al. 2014; Nolan et al. 2017).

24-epiBL (5 μM) spraying to NaCl-exposed *P. vulgaris* had an effective role on enzymatic antioxidants responses and proline content (Rady 2011). Exogenous BRs application to salt-stressed rice showed better seed germination, root elongation, growth with increased Chl accumulation, and nitrate reductase activity (Anuradha and Rao 2001). Moreover, Xu et al. (2015) suggested about BRs-modulated nutrient and sugar accumulation with reports of increased soluble sugar contents in *Vitis vinifera* through regulation of invertases and sucrose synthase along with respective gene expression (Xu et al. 2015). Similarly, BRs application improved the stress marker protein expression in *A. thaliana* and *B. napus* as well as better germination

Table 4 Regulation of antioxidants and related attributes by brassinosteroid in plants under salinity and water stress

Phytohormone	Stress	Crop	Responses	References
Brassinosteroid with kinetin (Kn) and epi-brassinolide (EBL)	Salinity	*Solanum lycopersicum*	Increased the activity of SOD, CAT, GST, APX, and GR Reduced ROS (H_2O_2 and $O^{\bullet}$) production	Ahanger et al. (2020b)
Brassinosteroid (BR)	Drought	*Oryza sativa*	Increased CAT, SOD, and APX activities. Improved photosynthesis and intercellular CO_2 concentration. Declined malondialdehyde (MDA) and H_2O_2 production.	Farooq et al. (2009)
BR	Drought	*S. lycopersicum*	Increased SOD activity	Jangid and Dwivedi (2017)
BR	Drought	*Capsicum annuum*	Decreased H_2O_2 and MDA contents Enhanced level of ascorbate and glutathione	Kaya et al. (2019)
BR	Drought	*S. lycopersicum*	Increased SOD, GR, CAT, POD, and APX activities Improved proline and protein content	Behnamnia et al. (2009)
BR	Salinity	*O. sativa*	Enhanced SOD, APX, GPX, DHAR, and MDHAR activities	Sharma et al. (2013a, b)
BR	Drought	*Eucalyptus urphylla*	Elevated SOD, CAT, POX, and APX activities	Barros Junior et al. (2020)
BR	Salinity	*Zea mays*	Improved proline, glycine betaine, mannitol, and total sugar	Rattan et al. (2012)
BR	Drought	*Chorispora bungeana*	Increased SOD, CAT, APX, and GR activities	Li et al. (2012)
BR	Water stress	*Xanthoceras sorbifolia*	Improved RWC, proline, and soluble protein Stimulated SOD, POD, CAT, and APX activities	Li and Feng (2011)
BR	Salinity	*Z. mays*	Increased the activity of POD, CAT, DHAR, and MDHAR	Rattan et al. (2020)
BR	Salinity	*Phaseolus vulgaris*	Enhanced CAT, SOD, peroxidase, and GR activities	Rady (2011)

(continued)

Table 4 (continued)

Phytohormone	Stress	Crop	Responses	References
BR	Drought	*Glycine max*	Increased POD and SOD activities	Zhang et al. (2008)
BR	Drought	*Raphanus sativus*	Elevated the activity of SOD, CAT, and APX	Mahesh et al. (2013)

and growth germination and plant growth (Kagale et al. 2007). Leaf water status and CO_2 assimilation also increased in drought-stressed rice through BRs application (Farooq et al. 2009). Expression of oxidative stress marker genes was increased with lowered MDA in salt-affected rice by the application of EBL, while brassinosteroids genes like *OsBRI1* were increased with downregulation of *SalT* (Sharma et al. 2013a, b). The combined application of Kn and EBL upregulated the antioxidant metabolism and osmolyte accumulation during salt stress as well as ROS production reduced resulted in the protection of photosynthesis (Ahanger et al. 2020a).

The genomewide analysis of target genes *BZR1* and *BES1* discovered the role of BRs on plant growth regulation linked with light and hormone-signaling pathways *1* (Sun et al. 2010; Yu et al. 2011). During drought stress for plant endurance, BES1 degradation plays a valuable role to limit BR-regulated growth (Nolan et al. 2017). Similarly, overexpression of receptor gene *BRL3* increased plant survival capacity during drought stress without any damage in plant growth in BR mutant along with improvement of osmoprotective metabolites (Fàbregas et al. 2018). These BRs enhanced drought tolerance in *S. lycopersicum* and inhibited the capacity by overexpression BR receptor gene *SlBRI1* (Nie et al. 2019).

Drought-stressed *Eucalyptus urophylla* showed the improvement photosynthetic pigments, photosystem II efficiency, electron flux, and CO_2 fixation as well as enzymatic antioxidants activity when treated with EBR (Barros Junior et al. 2020). Mahesh et al. (2013) revealed that, exogenous application of BRs increased SOD, and CAT activities with elevated $O_2^{\bullet -}$ scavenging capacity as an indication of amelioration of oxidative damage under water stress conditions.

3 Crosstalk among Phytohormones to Regulate Plant Stress Responses

There is a cross-link among hormonal signal transduction cascades for plant development and responses against abiotic stresses. Signal transduction-mediated regulation of gene expression influences different metabolic processes including biosynthesis, hormonal activities, and developmental processes (Khan et al. 2020). Therefore, the crosstalk of auxins, cytokinin, and gibberellins with ABA, ET, SA, and jasmonates is potentially involved in plant defense responses (Nishiyama et al. 2013). Transcription of auxin-induced gene (*IAA19*) helps in auxin transportation and stimulates plant roots growth with DELLA proteins (Achard et al. 2003; Fu and Harberd 2003). Auxin-responsive genes (*IAA* genes) with transcriptional regulators, namely AUXIN RESPONSE FACTORS (ARFs) and AUXIN/INDOLE-3-ACETIC ACID (Aux/IAA) proteins help for plant growth directly with bioactive auxin (Depuydt and Hardtke 2011). Auxin acts as antagonistically in crosstalk with CK. Transcription factors of ARR1 and ARR12 are involved in auxin signaling-mediated regulation of root growth; both factors also work to activate SHY2 for the transition zone of root–apex (Müller and Sheen 2008).

Moreover, the nature of the interaction between JA and SA is either antagonistic or collaborative to cope with stress responses (Ju et al. 2012). Moreover, the interaction between ABA and JA was studied in *Arabidopsis* which is intensively regulated by MYC2 along with its homologs such as MYC3 and MYC4. This MYC2 orthodox, for example, JAM YC2 and JAMYC10 also regulated the ABA/JA wounding responsive gene expression (Fernández-Calvo et al. 2011).

Stomatal conductance, osmotic adjustment, leaf senescence as well as photoassimilate allocations are also regulated by interaction of ABA and SA; while an antagonistic interaction exists between ET and ABA for root formation upon waterlogging stress (Khan et al. 2020). An increase of ET activates negative regulators of ABA signaling naming ABII and ABI2 (Asselbergh et al. 2008). Furthermore, synergistic interaction exists between JA and GA for regulating metabolic functions including trichome initiation, sesquiterpenoid biosynthesis in the plant cell (Tanaka et al. 2006). Conversely, GA-dependent growth responses are antagonistically regulated by JA via JAZs-DELLAs. It was noted that *Arabidopsis* JAZ9 ties the DELLA protein RGA during the absence of JA and thus inhibition of growth-promoting TF PIF3 (phytochrome-interacting factor3) and JA-mediated JAZ breakdown was prevented as well as resulted in delays of GA-mediated DELLA degradation (Yang et al. 2012a, b, c).

In addition, JA and ET inhibited AUXs and thus repressed the expansion and cell cycle processes of the leaf (Shi et al. 2012). Conversely, AUXs were considered as repressors for JA biosynthesis and JA/Et-dependent nicotine responses.

ABA signaling is controlled by ABA receptor NtPHYL4, while JA acts in balancing plant growth with stress acclimatization (Lackman et al. 2011). Thereafter, ABA-induced stomatal closure is inhibited by ET concentration, which was confirmed by the result of slow stomatal closure in *Arabidopsis* ethylene overproducer 1 mutants (Tanaka et al. 2005).

The MeJA interacts with ABA-mediated stomatal closure within 10 minutes by CDPK production as well as increases the Ca^{2+} influx (Munemasa et al. 2007). The ABA signaling is also regulated by auxin response factors in *Arabidopsis* (Liu et al. 2008). Exogenously BR application increased the ET production in *Arabidopsis* (Hansen et al. 2009). Therefore, BR causes the upregulation of *1-aminocyclopropane-1-carboxylate synthase (ACS)* expression, which is the required main gene for ET biosynthesis (Polko and Kieber 2019). Again, it was reported that ethylene response factor protein (JERF3) raises the oxidative genes expression for lowering ROS accumulation tolerance as the indication for BR and ET interaction-mediated ROS sequestration (Wu et al. 2008).

An experiment was conducted to study the crosstalk between BRs and ethylene in tomato plants under salt stress (Zhu et al. 2016). Later, this study proposed that ethylene biosynthesis is affected by BRs and increases ACS (ethylene synthesis hormone) activity and stabilizing EILs (ethylene-insensitive3-like, ethylene transcription factor family). BRs uplift plant growth and development in association with auxins, cytokinins, gibberellins (Domagalska et al. 2010), ABA (Domagalska et al. 2010), ET (Manzano et al. 2011), SA (Divi et al. 2010), and JA (Creelman and Mullet 1997; Peng et al. 2011).

However, the interaction of the above hormones in ROS mitigation through upregulating antioxidants needs further more studies.

4 Conclusions and Recommendations

To combat excess ROS, phytohormones essentially play role in increasing the responses of both enzymatic and nonenzymatic antioxidants. Among the enzymatic antioxidants like SOD, CAT, POD, APX, GPX, GR, MDHAR, and DHAR, as well as nonenzymatic antioxidants like AsA and GSH are important to reduce oxidative damage for cellular survival under mentioned abiotic stresses. This chapter focused mainly on reviewing both endogenous and exogenous phytohormone-mediated responses of antioxidants to ameliorate oxidative stress. Their effects on related parameters like pigments, cellular stability, ion homeostasis, and so on are also mentioned. The information provided in this chapter can be helpful in understanding the role of phytohormones in improving stress tolerance, although there are still further works needed to access the exact mechanism of actions. Therefore, more attention is required on waterlogging or flooding stress amelioration by phytohormones, which is still in lack. Besides, methylglyoxal, a potential cytotoxin that needs two enzymes viz., glyoxalase-I and glyoxalase-II, but information on hormonal regulation on these enzymes is still needed to study. Therefore, the development of phytohormone-engineered plants showing more antioxidants and glyoxalases can be important to increase the adaptability of plants under abiotic stress conditions. Finding the crop-specific doses of phytohormones and application methods would be a further area of research.

References

Abdelaal KA, Attia KA, Alamery SF, El-Afry MM, Ghazy AI, Tantawy DS, Al-Doss AA, El-Shawy ESE, Abu-Elsaoud M, A. and Hafez, Y.M. (2020a) Exogenous application of proline and salicylic acid can mitigate the injurious impacts of drought stress on barley plants associated with physiological and histological characters. Sustainability 12(5):1736

Abdelaal KA, EL-Maghraby LM, Elansary H, Hafez YM, Ibrahim EI, El-Banna M, El-Esawi M, Elkelish A (2020b) Treatment of sweet pepper with stress tolerance-inducing compounds alleviates salinity stress oxidative damage by mediating the physio-biochemical activities and antioxidant systems. Agronomy 10(1):26

Abdelgawad ZA, Khalafaallah AA, Abdallah MM (2014) Impact of methyl jasmonate on antioxidant activity and some biochemical aspects of maize plant grown under water stress condition. Agric Sci 5(12):1077

Abido WAE, Allem A, Zsombic L (2019) Effect of gibberellic acid on germination of six wheat cultivars under salinity stress levels. Asian J Biol Sci 12(1):51–60

Achard P, Renou JP, Berthomé R, Harberd NP, Genschik P (2008) Plant DELLAs restrain growth and promote survival of adversity by reducing the levels of reactive oxygen species. Curr Biol 18(9):656–660

Achard P, Vriezen WH, Van Der Straeten D, Harberd NP (2003) Ethylene regulates Arabidopsis development via the modulation of DELLA protein growth repressor function. Plant Cell 15(12):2816–2825

Agarwal S, Sairam RK, Srivastava GC, Tyagi A, Meena RC (2005) Role of ABA, salicylic acid, calcium and hydrogen peroxide on antioxidant enzymes induction in wheat seedlings. Plant Sci 169(3):559–570

Ahammed GJ, Li X, Liu A, Chen S (2020) Brassinosteroids in plant tolerance to abiotic stress. J Plant Growth Regul 39:1451–1464

Ahanger MA, Alyemeni MN, Wijaya L, Alamri SA, Alam P, Ashraf M, Ahmad P (2018) Potential of exogenously sourced kinetin in protecting *Solanum lycopersicum* from NaCl-induced oxidative stress through up-regulation of the antioxidant system, ascorbate-glutathione cycle and glyoxalase system. PLoS One 13(9):e0202175

Ahanger MA, Aziz U, Alsahli AA, Alyemeni MN, Ahmad P (2020a) Influence of exogenous salicylic acid and nitric oxide on growth, photosynthesis, and ascorbate-glutathione cycle in salt stressed *Vigna angularis*. Biomol Ther 10(1):42

Ahanger MA, Mir RA, Alyemeni MN, Ahmad P (2020b) Combined effects of brassinosteroid and kinetin mitigates salinity stress in tomato through the modulation of antioxidant and osmolyte metabolism. Plant Physiol Biochem 147:31–42

Ahmad P, Alyemeni MN, Ahanger MA, Egamberdieva D, Wijaya L, Alam P (2018) Salicylic acid (SA) induced alterations in growth, biochemical attributes and antioxidant enzyme activity in faba bean (*Vicia faba* L.) seedlings under NaCl toxicity. Russ J Plant Physiol 65(1):104–114

Ahmad P, Nabi G, Ashraf M (2011) Cadmium-induced oxidative damage in mustard [*Brassica juncea* (L.) *Czern. & Coss.*] plants can be alleviated by salicylic acid. S Afr J Bot 77(1):36–44

Ahmadi FI, Karimi K, Struik PC (2018) Effect of exogenous application of methyl jasmonate on physiological and biochemical characteristics of *Brassica napus* L. *cv. Talaye* under salinity stress. S Afr J Bot 115:5–11

Al Mahmud J, Biswas PK, Nahar K, Fujita M, Hasanuzzaman M (2019) Exogenous application of gibberellic acid mitigates drought-induced damage in spring wheat. Acta Agrobot 72:2

Alam MM, Nahar K, Hasanuzzaman M, Fujita M (2014) Exogenous jasmonic acid modulates the physiology, antioxidant defense and glyoxalase systems in imparting drought stress tolerance in different brassica species. Plant Biotechnol Rep 8(3):279–293

Aldesuquy H, Ghanem H (2015) Exogenous salicylic acid and trehalose ameliorate short term drought stress in wheat cultivars by up-regulating membrane characteristics and antioxidant defense system. J Hortic

Al-Hakimi AMA, Hamada AM (2001) Counteraction of salinity stress on wheat plants by grain soaking in ascorbic acid, thiamin or sodium salicylate. Biol Plant 44(2):253–261

Ali MB, Yu KW, Hahn EJ, Paek KY (2006) Methyl jasmonate and salicylic acid elicitation induces ginsenosides accumulation, enzymatic and non-enzymatic antioxidant in suspension culture Panax ginseng roots in bioreactors. Plant Cell Rep 25(6):613–620

Alsahli A, Mohamed AK, Alaraidh I, Al-Ghamdi A, Al-Watban A, El-Zaidy M, Alzahrani SM (2019) Salicylic acid alleviates salinity stress through the modulation of biochemical attributes and some key antioxidants in wheat seedlings. Pak J Bot 51(5):1551–1559

Alscher RG, Donahue JL, Cramer CL (1997) Reactive oxygen species and antioxidants: relationships in green cells. Physiol Plant 100(2):224–233

AlTaey DKA (2017) Alleviation of salinity effects by poultry manure and gibberellin application on growth and peroxidase activity in pepper. Int J Environ Agric Biotechnol 2(4):e.238861

Anjum SA, Tanveer M, Hussain S, Tung SA, Samad RA, Wang L, Khan I, ur Rehman N, Shah AN, Shahzad B (2016) Exogenously applied methyl jasmonate improves the drought tolerance in wheat imposed at early and late developmental stages. Acta Physiol Plant 38(1):25

Anuradha S, Rao SSR (2001) Effect of brassinosteroids on salinity stress induced inhibition of seed germination and seedling growth of rice (*Oryza sativa* L.). Plant Growth Regul 33(2):151–153

Anwar A, Liu Y, Dong R, Bai L, Yu X, Li Y (2018) The physiological and molecular mechanism of brassinosteroid in response to stress: a review. Biol Res 51:46

Apel K, Hirt H (2004) Reactive oxygen species: metabolism, oxidative stress, and signal transduction. Ann Rev Plant Biol 55:373–399

Arraes FBM, Beneventi MA, de Sa MEL, Paixao JFR, Albuquerque EVS, Marin SRR, Purgatto E, Nepomuceno AL, Grossi-de-Sa MF (2015) Implications of ethylene biosynthesis and signaling in soybean drought stress tolerance. BMC Plant Biol 15(1):1–20

Artner C, Benkova E (2019) Ethylene and cytokinin: partners in root growth regulation. Mol Plant 12(10):1312–1314

Asselbergh B, De Vleesschauwer D, Höfte M (2008) Global switches and fine-tuning—ABA modulates plant pathogen defense. Mol Plant-Microbe Interact 21:719

Avalbaev A, Allagulova C, Maslennikova D, Fedorova K, Shakirova F (2020) Methyl Jasmonate and Cytokinin mitigate the salinity-induced oxidative injury in wheat seedlings. J Plant Growth Regul:1–12

Azmat A, Yasmin H, Hassan MN, Nosheen A, Naz R, Sajjad M, Ilyas N, Akhtar MN (2020) Co-application of bio-fertilizer and salicylic acid improves growth, photosynthetic pigments and stress tolerance in wheat under drought stress. PeerJ 8:e9960

Azooz MM, Metwally A, Abou-Elhamd MF (2015) Jasmonate-induced tolerance of Hassawi okra seedlings to salinity in brackish water. Acta Physiol Plant 37(4):77

Bajguz A (2007) Metabolism of brassinosteroids in plants. Plant Physiol Biochem 45:95–107

Bajguz A, Piotrowska-Niczyporuk A (2014) Interactive effect of brassinosteroids and cytokinins on growth, chlorophyll, monosaccharide and protein content in the green alga *Chlorella vulgaris* (Trebouxiophyceae). Plant Physiol Biochem 80:176–183

Bandurska H (2013) Salicylic acid: an update on biosynthesis and action in plant response to water deficit and performance under drought. In: Hayat S, Ahmad A, Alyemeni MN (eds) Salicylic acid: plant growth and development. Springer, Dordrecht, pp 1–14

Banerjee A, Roychoudhury A (2018) Interactions of brassinosteroids with major phytohormones: antagonistic effects. J Plant Growth Regul 37(4):1025–1032

Bano A, Ullah F, Nosheen A (2012) Role of abscisic acid and drought stress on the activities of antioxidant enzymes in wheat. Plant Soil Environ 58(4):181–185

Barros Junior UO, Lima MD, Alsahli AA, Lobato AK (2020) Unraveling the roles of brassinosteroids in alleviating drought stress in young *Eucalyptus urophylla* plants: implications on redox homeostasis and photosynthetic apparatus. Physiol Plant 172:748

Bashar KK (2018) Hormone dependent survival mechanisms of plants during post-waterlogging stress. Plant Signal Behav 13(10):e1529522

Behnamnia M, Kalantari KM, Ziaie J (2009) The effects of brassinosteroid on the induction of biochemical changes in *Lycopersicon esculentum* under drought stress. Turk J Bot 33(6): 417–428

Beltrano J, Montaldi E, Carbone A (1997) Emission of water stress ethylene in wheat (*Triticum aestivum* L.) ears: effects of rewatering. Plant Growth Regul 21(2):121–126

Benjamins R, Scheres B (2008) Auxin: the looping star in plant development. Ann Rev Plant Biol 59:443–465

Bielach A, Podlešáková K, Marhavý P, Duclercq J, Cuesta C, Müller B, Grunewald W, Tarkowski P, Benková E (2012) Spatiotemporal regulation of lateral root organogenesis in Arabidopsis by cytokinin. Plant Cell 24(10):3967–3981

Bishopp A, Help H, El-Showk S, Weijers D, Scheres B, Friml J, Benková E, Mähönen AP, Helariutta Y (2011) A mutually inhibitory interaction between auxin and cytokinin specifies vascular pattern in roots. Curr Biol 21(11):917–926

Bowling SA, Clarke JD, Liu Y, Klessig DF, Dong X (1997) The cpr5 mutant of Arabidopsis expresses both NPR1-dependent and NPR1-independent resistance. Plant Cell 9(9):1573–1584

Bray EA (2002) Abscisic acid regulation of gene expression during water-deficit stress in the era of the Arabidopsis genome. Plant Cell Environ 25(2):153–161

Brossa R, López-Carbonell M, Jubany-Marí T, Alegre L (2011) Interplay between abscisic acid and jasmonic acid and its role in water-oxidative stress in wild-type, ABA-deficient, JA-deficient, and ascorbate-deficient Arabidopsis plants. J Plant Growth Regul 30(3):322–333

Bukhat S, Manzoor H, Zafar ZU, Azeem F, Rasul S (2020) Salicylic acid induced photosynthetic adaptability of *raphanus sativus* to salt stress is associated with antioxidant capacity. J Plant Growth Regul 39(2):809–822

Cao S, Chen H, Zhang C, Tang Y, Liu J, Qi H (2016) Heterologous expression and biochemical characterization of two lipoxygenases in oriental melon, *Cucumis melo var. makuwa Makino*. PLoS One 11(4):e0153801

Catinot J, Buchala A, Abou-Mansour E, Métraux JP (2008) Salicylic acid production in response to biotic and abiotic stress depends on isochorismate in *Nicotiana benthamiana*. FEBS Lett 582(4): 473–478

Cha YR, Kim S, Lee JH, Shim IS (2020) Enhanced drought tolerance in Chinese cabbage (*Brassica campestris* L.) seedlings upon Pretreatment with exogenous salicylic acid. 원예과학기술지 38(1):9–20

Chakrabarti N, Mukherji S (2003) Alleviation of NaCl stress by pretreatment with phytohormones in *Vigna radiata*. Biol Plant 46(4):589–594

Chandrasekar V, Sairam K, R. and Srivastava, G.C. (2000) Physiological and biochemical responses of hexaploid and tetraploid wheat to drought stress. J Agron Crop Sci 185(4):219–227

Chang Z, Liu Y, Dong H, Teng K, Han L, Zhang X (2016) Effects of cytokinin and nitrogen on drought tolerance of creeping bentgrass. PLoS One 11(4):e0154005

Chen JH, Jiang HW, Hsieh EJ, Chen HY, Chien CT, Hsieh HL, Lin TP (2012) Drought and salt stress tolerance of an *Arabidopsis glutathione S-transferase U17* knockout mutant are attributed to the combined effect of glutathione and abscisic acid. Plant Physiol 158(1):340–351

Cheng MC, Ko K, Chang WL, Kuo WC, Chen GH, Lin TP (2015) Increased glutathione contributes to stress tolerance and global translational changes in Arabidopsis. Plant J 83(5):926–939

Chiang YH, Zubo YO, Tapken W, Kim HJ, Lavanway AM, Howard L, Pilon M, Kieber JJ, Schaller GE (2012) Functional characterization of the GATA transcription factors GNC and CGA1 reveals their key role in chloroplast development, growth, and division in Arabidopsis. Plant Physiol 160(1):332–348

Choi DS, Hwang BK (2011) Proteomics and functional analyses of pepper abscisic acid–responsive 1 (*ABR1*), which is involved in cell death and defense signaling. Plant Cell 23(2):823–842

Chojak-Koźniewska J, Linkiewicz A, Sowa S, Radzioch MA, Kuźniak E (2017) Interactive effects of salt stress and *pseudomonas syringae pv. Lachrymans* infection in cucumber: involvement of antioxidant enzymes, abscisic acid and salicylic acid. Environ Exp Bot 136:9–20

Clouse SD (2011) Brassinosteroid signal transduction: from receptor kinase activation to transcriptional networks regulating plant development. Plant Cell 23(4):1219–1230

Clouse SD, Langford M, McMorris TC (1996) A brassinosteroid-insensitive mutant in *Arabidopsis thaliana* exhibits multiple defects in growth and development. Plant Physiol 111(3):671–678

Cracker LE, Abeles FB (1969) Abscission: role of abscisic acid. Plant Physiol 44(8):1144–1149

Cramer GR, Urano K, Delrot S, Pezzotti M, Shinozaki K (2011) Effects of abiotic stress on plants: a systems biology perspective. BMC Plant Biol 11(1):163

Creelman RA, Mullet JE (1997) Oligosaccharins, brassinolides, and jasmonates: nontraditional regulators of plant growth, development, and gene expression. Plant Cell 9(7):1211–1223

Cutler SR, Rodriguez PL, Finkelstein RR, Abrams SR (2010) Abscisic acid: emergence of a core signaling network. Annu Rev Plant Biol 61:651–679

Dejonghe W, Mishev K, Russinova E (2014) The brassinosteroid chemical toolbox. Curr Opin Plant Biol 22:48–55

Depuydt S, Hardtke CS (2011) Hormone signalling crosstalk in plant growth regulation. Curr Biol 21(9):365–373

Desikan R, Last K, Harrett-Williams R, Tagliavia C, Harter K, Hooley R, Hancock JT, Neill SJ (2006) Ethylene-induced stomatal closure in Arabidopsis occurs via AtrbohF-mediated hydrogen peroxide synthesis. Plant J 47(6):907–916

Divi UK, Rahman T, Krishna P (2010) Brassinosteroid-mediated stress tolerance in Arabidopsis shows interactions with abscisic acid, ethylene and salicylic acid pathways. BMC Plant Biol 10(1):1–14

Domagalska MA, Sarnowska E, Nagy F, Davis SJ (2010) Genetic analyses of interactions among gibberellin, abscisic acid, and brassinosteroids in the control of flowering time in *Arabidopsis thaliana*. PLoS One 5(11):e14012

Ebrahimian E, Bybordi A (2012) Effect of salinity, salicylic acid, silicium and ascorbic acid on lipid peroxidation, antioxidant enzyme activity and fatty acid content of sunflower. Afr J Agric Res 7(25):3685–3694

Efroni I, Han SK, Kim HJ, Wu MF, Steiner E, Birnbaum KD, Hong JC, Eshed Y, Wagner D (2013) Regulation of leaf maturation by chromatin-mediated modulation of cytokinin responses. Dev Cell 24(4):438–445

Egamberdieva D (2009) Alleviation of salt stress by plant growth regulators and IAA producing bacteria in wheat. Acta Physiol Plant 31(4):861–864

El-Esawi MA, Elansary HO, El-Shanhorey NA, Abdel-Hamid AM, Ali HM, Elshikh MS (2017) Salicylic acid-regulated antioxidant mechanisms and gene expression enhance rosemary performance under saline conditions. Front Physiol 8:716

Elhakem A (2020) Salicylic acid ameliorates salinity tolerance in maize by regulation of phytohormones and osmolytes. Plant Soil Environ 66(10):533–541

El-Katony TM, El-Bastawisy ZM, El-Ghareeb SS (2019) Timing of salicylic acid application affects the response of maize (*Zea mays* L.) hybrids to salinity stress. Heliyon 5(4):e01547

Fàbregas N, Lozano-Elena F, Blasco-Escámez D, Tohge T, Martínez-Andújar C, Albacete A, Osorio S, Bustamante M, Riechmann JL, Nomura T, Yokota T (2018) Overexpression of the vascular brassinosteroid receptor BRL3 confers drought resistance without penalizing plant growth. Nat Commun 9(1):1–13

Fahad S, Bano A (2012) Effect of salicylic acid on physiological and biochemical characterization of maize grown in saline area. Pak J Bot 44(4):1433–1438

Fahad S, Hussain S, Bano A, Saud S, Hassan S, Shan D, Khan FA, Khan F, Chen Y, Wu C, Tabassum MA (2015) Potential role of phytohormones and plant growth-promoting rhizobacteria in abiotic stresses: consequences for changing environment. Environ Sci Pollut Res 22(7):4907–4921

Faiss M, Zalubìlová J, Strnad M, Schmülling T (1997) Conditional transgenic expression of the ipt gene indicates a function for cytokinins in paracrine signaling in whole tobacco plants. Plant J 12(2):401–415

Fang L, Su L, Sun X, Li X, Sun M, Karungo SK, Fang S, Chu J, Li S, Xin H (2016) Expression of Vitis amurensis NAC26 in Arabidopsis enhances drought tolerance by modulating jasmonic acid synthesis. J Exp Bot 67(9):2829–2845

Farber M, Attia Z, Weiss D (2016) Cytokinin activity increases stomatal density and transpiration rate in tomato. J Exp Bot:erw398

Fardus J, Matin MA, Hasanuzzaman M, Hossain MS, Nath SD, Hossain MA, Rohman MM, Hasanuzzaman M (2017) Exogenous salicylic acid-mediated physiological responses and improvement in yield by modulating antioxidant defense system of wheat under salinity. Notulae Scientia Biologicae 9(2):219–232

Farhangi-Abriz S, Ghassemi-Golezani K (2018) How can salicylic acid and jasmonic acid mitigate salt toxicity in soybean plants? Ecotoxicol Environ Saf 147:1010–1016

Farhangi-Abriz S, Ghassemi-Golezani K (2019) Jasmonates: mechanisms and functions in abiotic stress tolerance of plants. Biocatal Agric Biotechnol 20:101210

Faried HN, Ayyub CM, Amjad M, Ahmed R, Wattoo FM, Butt M, Bashir M, Shaheen MR, Waqas MA (2017) Salicylic acid confers salt tolerance in potato plants by improving water relations, gaseous exchange, antioxidant activities and osmoregulation. J Sci Food Agric 97(6): 1868–1875

Farooq M, Wahid A, Basra SMA (2009) Improving water relations and gas exchange with brassinosteroids in rice under drought stress. J Agron Crop Sci 195(4):262–269

Fayez KA, Bazaid SA (2014) Improving drought and salinity tolerance in barley by application of salicylic acid and potassium nitrate. J Saudi Soc Agric Sci 13(1):45–55

Fernández-Calvo P, Chini A, Fernández-Barbero G, Chico JM, Gimenez-Ibanez S, Geerinck J, Eeckhout D, Schweizer F, Godoy M, Franco-Zorrilla JM, Pauwels L (2011) The Arabidopsis bHLH transcription factors MYC3 and MYC4 are targets of JAZ repressors and act additively with MYC2 in the activation of jasmonate responses. Plant Cell 23(2):701–715

Fernando VD, Schroeder DF (2016) Role of ABA in Arabidopsis salt, drought, and desiccation tolerance. In: Abiotic and biotic stress in plants-recent advances and future perspectives. Intech Open, pp 507–524

Feussner I, Wasternack C (2002) The lipoxygenase pathway. Annu Rev Plant Biol 3(1):275–297

Freitas VS, de Souza Miranda R, Costa JH, de Oliveira DF, de Oliveira Paula S, de Castro Miguel E, Freire RS, Prisco JT, Gomes-Filho E (2018) Ethylene triggers salt tolerance in maize genotypes by modulating polyamine catabolism enzymes associated with H2O2 production. Environ Exp Bot 145:75–86

Frey A, Effroy D, Lefebvre V, Seo M, Perreau F, Berger A, Sechet J, To A, North HM, Marion-Poll A (2012) Epoxycarotenoid cleavage by NCED5 fine-tunes ABA accumulation and affects seed dormancy and drought tolerance with other NCED family members. Plant J 70(3):501–512

Fu X, Harberd NP (2003) Auxin promotes Arabidopsis root growth by modulating gibberellin response. Nature 421(6924):740–743

Gao SQ, Chen M, Xu ZS, Zhao CP, Li L, Xu HJ, Tang YM, Zhao X, Ma YZ (2011) The soybean GmbZIP1 transcription factor enhances multiple abiotic stress tolerances in transgenic plants. Plant Mol Biol 75(6):537–553

Garcion C, Lohmann A, Lamodière E, Catinot J, Buchala A, Doermann P, Métraux JP (2008) Characterization and biological function of the ISOCHORISMATE SYNTHASE2 gene of Arabidopsis. Plant Physiol 147(3):1279–1287

Gashaw A, Theerawitaya C, Samphumphuang T, Cha-um S, Supaibulwatana K (2014) CPPU elevates photosynthetic abilities, growth performances and yield traits in salt stressed rice (*Oryza sativa* L. *spp. indica*) via free proline and sugar accumulation. Pestic Biochem Physiol 108:27–33

Gautam S, Singh PK (2011) Effects of salicylate on growth and biochemical changes in maize seedlings under salt stress. IIOAB J 2:16–20

Gergoff G, Chaves A, Bartoli CG (2010) Ethylene regulates ascorbic acid content during dark-induced leaf senescence. Plant Sci 178(2):207–212

Gharbi E, Martínez JP, Benahmed H, Lepoint G, Vanpee B, Quinet M, Lutts S (2017) Inhibition of ethylene synthesis reduces salt-tolerance in tomato wild relative species *Solanum chilense*. J Plant Physiol 210:24–37

Ghasemlou F, Amiri H, Karamian R, Mirzaie-asl A (2019) Alleviation of the effects of on drought stress Verbascum nudicuale by methyl jasmonate and titanium dioxide nanoparticles. Iran J Plant Physiol 9(4):2911–2920

Ghassemi-Golezani K, Farhangi-Abriz S (2018) Changes in oil accumulation and fatty acid composition of soybean seeds under salt stress in response to salicylic acid and jasmonic acid. Russ J Plant Physiol 65(2):229–236

Ghodrat V, Rousta MJ (2012) Effect of priming with gibberellic acid (GA3) on germination and growth of corn (*Zea mays* L.) under saline conditions. Int J Agric Crop Sci (IJACS) 4(13): 882–885

Gill SS, Tuteja N (2010) Reactive oxygen species and antioxidant machinery in abiotic stress tolerance in crop plants. Plant Physiol Biochem 48(12):909–930

Giraudat J, Parcy F, Bertauche N, Gosti F, Leung J, Morris PC, Bouvier-Durand M, Vartanian N (1994) Current advances in abscisic acid action and signaling. Plant Mol Biol 26(5):1557–1577

Grove MD (1979) A unique plant growth promoting steroid from *Brassica napus* pollen. Nature 281:215–217

Grunewald W, Friml J (2010) The march of the PINs: developmental plasticity by dynamic polar targeting in plant cells. EMBO J 29(16):2700–2714

Gujjar RS, Banyen P, Chuekong W, Worakan P, Roytrakul S, Supaibulwatana K (2020) A synthetic Cytokinin improves photosynthesis in rice under drought stress by modulating the abundance of proteins related to stomatal conductance, chlorophyll contents, and rubisco activity. Plan Theory 9(9):1106

Guo H, Li L, Aluru M, Aluru S, Yin Y (2013) Mechanisms and networks for brassinosteroid regulated gene expression. Curr Opin Plant Biol 16(5):545–553

Guo Y, Jiang Q, Hu Z, Sun X, Fan S, Zhang H (2018) Function of the auxin-responsive gene *TaSAUR75* under salt and drought stress. Crop J 6(2):181–190

Hai NN, Chuong NN, Tu NHC, Kisiala A, Hoang XLT, Thao NP (2020) Role and regulation of cytokinins in plant response to drought stress. Plan Theory 9(4):422

Halo BA, Khan AL, Waqas M, Al-Harrasi A, Hussain J, Ali L, Adnan M, Lee IJ (2015) Endophytic bacteria (*Sphingomonas sp. LK11*) and gibberellin can improve Solanum lycopersicum growth and oxidative stress under salinity. J Plant Interact 10(1):117–125

Han Y, Mhamdi A, Chaouch S, Noctor G (2013) Regulation of basal and oxidative stress-triggered jasmonic acid-related gene expression by glutathione. Plant Cell Environ 36(6):1135–1146

Hansen M, Chae HS, Kieber JJ (2009) Regulation of ACS protein stability by cytokinin and brassinosteroid. Plant J 57(4):606–614

Hartung W (2010) The evolution of abscisic acid (ABA) and ABA function in lower plants, fungi and lichen. Funct Plant Biol 37(9):806–812

Hartung W, Wilkinson S, Davies WJ (1998) Factors that regulate abscisic acid concentrations at the primary site of action at the guard cell. J Exp Bot 49:361–367

Hasanuzzaman M, Bhuyan MHM, Zulfiqar F, Raza A, Mohsin SM, Mahmud JA, Fujita M, Fotopoulos V (2020) Reactive oxygen species and antioxidant defense in plants under abiotic stress: revisiting the crucial role of a universal defense regulator. Antioxidants 9(8):681

Hassanein RA, Amin AAE, Rashad ESM, Ali H (2015) Effect of thiourea and salicylic acid on antioxidant defense of wheat plants under drought stress. Int J ChemTech Res 7(01):346–354

Hedden P, Thomas SG (2012) Gibberellin biosynthesis and its regulation. Biochem J 444(1):11–25

Hernández I, Cela J, Alegre L, Munné-Bosch S (2012) Antioxidant defenses against drought stress. In: Plant responses to drought stress. Springer, Berlin, Heidelberg, pp 231–258

Hönig M, Plíhalová L, Husičková A, Nisler J, Doležal K (2018) Role of cytokinins in senescence, antioxidant defence and photosynthesis. Int J Mol Sci 19(12):4045

Hose E, Clarkson DT, Steudle E, Schreiber L, Hartung W (2001) The exodermis: a variable apoplastic barrier. J Exp Bot 52(365):2245–2264

Hose E, Steudle E, Hartung W (2000) Abscisic acid and hydraulic conductivity of maize roots: a study using cell-and root-pressure probes. Planta 211(6):874–882

Hou Y, Meng K, Han Y, Ban Q, Wang B, Suo J, Lv J, Rao J (2015) The persimmon 9-lipoxygenase gene DkLOX3 plays positive roles in both promoting senescence and enhancing tolerance to abiotic stress. Front Plant Sci 6:e.1073

Husen A, Iqbal M, Sohrab SS, Ansari MKA (2018) Salicylic acid alleviates salinity-caused damage to foliar functions, plant growth and antioxidant system in Ethiopian mustard (*Brassica carinata* A. Br.). Agric Food Secur 7(1):1–14

Hussain I, Rasheed R, Ashraf MA, Mohsin M, Shah SMA, Rashid DA, Akram M, Nisar J, Riaz M (2020) Foliar applied acetylsalicylic acid induced growth and key-biochemical changes in chickpea (*Cicer arietinum* L.) under drought stress. Dose-Response 18(4):1559325820956801

Ilyas N, Gull R, Mazhar R, Saeed M, Kanwal S, Shabir S, Bibi F (2017) Influence of salicylic acid and jasmonic acid on wheat under drought istress. Commun Soil Sci Plant Anal 48(22): 2715–2723

Jakab G, Ton J, Flors V, Zimmerli L, Métraux JP, Mauch-Mani B (2005) Enhancing Arabidopsis salt and drought stress tolerance by chemical priming for its abscisic acid responses. Plant Physiol 139(1):267–274

Jangid KK, Dwivedi P (2017) Physiological and biochemical changes by nitric oxide and brassinosteroid in tomato (*Lycopersicon esculentum* mill.) under drought stress. Acta Physiol Plant 39(3):73

Javid MG, Sorooshzadeh A, Moradi F, Modarres Sanavy SAM, Allahdadi I (2011) The role of phytohormones in alleviating salt stress in crop plants. Aust J Crop Sci 5(6):726–734

Jayasinghe T, Perera P and Wimalasekera R (2019) Effect of Foliar application of Gibberellin in mitigating salt stress in tomato (*Solanum lycopersicum*),'Thilina'Variety. In *Proceedings of the 6th International Conference on Multidisciplinary Approaches (iCMA)*

Jiang M, Zhang J (2001) Effect of abscisic acid on active oxygen species, antioxidative defence system and oxidative damage in leaves of maize seedlings. Plant Cell Physiol 42(11): 1265–1273

Jiang M, Zhang J (2002) Role of abscisic acid in water stress-induced antioxidant defense in leaves of maize seedlings. Free Radic Res 36(9):1001–1015

Ju C, Yoon GM, Shemansky JM, Lin DY, Ying ZI, Chang J, Garrett WM, Kessenbrock M, Groth G, Tucker ML, Cooper B (2012) CTR1 phosphorylates the central regulator EIN2 to control ethylene hormone signaling from the ER membrane to the nucleus in Arabidopsis. Proc Natl Acad Sci 109(47):19486–19491

Kagale S, Divi UK, Krochko JE, Keller WA, Krishna P (2007) Brassinosteroid confers tolerance in *Arabidopsis thaliana* and *Brassica napus* to a range of abiotic stresses. Planta 225(2):353–364

Kamal AHM, Komatsu S (2016) Jasmonic acid induced protein response to biophoton emissions and flooding stress in soybean. J Proteome 133:33–47

Kang SM, Khan AL, Waqas M, You YH, Kim JH, Kim JG, Hamayun M, Lee IJ (2014) Plant growth-promoting rhizobacteria reduce adverse effects of salinity and osmotic stress by regulating phytohormones and antioxidants in *Cucumis sativus*. J Plant Interact 9(1):673–682

Kasahara H, Hanada A, Kuzuyama T, Takagi M, Kamiya Y, Yamaguchi S (2002) Contribution of the mevalonate and methylerythritol phosphate pathways to the biosynthesis of gibberellins in Arabidopsis. J Biol Chem 277(47):45188–45194

Kaur M, Zhawar VK (2018) Analysis of ABA regulation of antioxidant metabolism under ABA and water stress in wheat cultivars. Acta Sci Agric 2(3):18–23

Kaya C, Ashraf M, Wijaya L, Ahmad P (2019) The putative role of endogenous nitric oxide in brassinosteroid-induced antioxidant defence system in pepper (*Capsicum annuum* L.) plants under water stress. Plant Physiol Biochem 143:119–128

Keshishian EA, Hallmark HT, Ramaraj T, Plačková L, Sundararajan A, Schilkey F, Novák O, Rashotte AM (2018) Salt and oxidative stresses uniquely regulate tomato cytokinin levels and transcriptomic response. Plant Direct 2(7):e00071

Khalvandi M, Siosemardeh A, Roohi E, Keramati S (2021) Salicylic acid alleviated the effect of drought stress on photosynthetic characteristics and leaf protein pattern in winter wheat. Heliyon 7(1):e05908

Khan AL, Hamayun M, Kang SM, Kim YH, Jung HY, Lee JH, Lee IJ (2012) Endophytic fungal association via gibberellins and indole acetic acid can improve plant growth under abiotic stress: an example of *Paecilomyces formosus* LHL10. BMC Microbiol 12(1):1–14

Khan MIR, Asgher M, Khan NA (2014) Alleviation of salt-induced photosynthesis and growth inhibition by salicylic acid involves glycinebetaine and ethylene in mungbean (*Vigna radiata* L.). Plant Physiol Biochem 80:67–74

Khan N, Bano A, Ali S, Babar MA (2020) Crosstalk amongst phytohormones from planta and PGPR under biotic and abiotic stresses. Plant Growth Regul 90(2):189–203

Kim BH, Kim SY, Nam KH (2012) Genes encoding plant-specific class III peroxidases are responsible for increased cold tolerance of the brassinosteroid-insensitive 1 mutant. Mol Cells 34(6):539–548

Kim EH, Kim YS, Park SH, Koo YJ, Do Choi Y, Chung YY, Lee IJ, Kim JK (2009) Methyl jasmonate reduces grain yield by mediating stress signals to alter spikelet development in rice. Plant Physiol 149(4):1751–1760

Kim TW, Wang ZY (2010) Brassinosteroid signal transduction from receptor kinases to transcription factors. Annu Rev Plant Biol 61:681–704

Kim Y, Seo CW, Khan AL, Mun BG, Shahzad R, Ko JW, Yun BW, Park SK, Lee IJ (2018) Exo-ethylene application mitigates waterlogging stress in soybean (*Glycine max* L.). BMC Plant Biol 18(1):1–16

Kim YH, Khan AL, Waqas M, Shim JK, Kim DH, Lee KY, Lee IJ (2014) Silicon application to rice root zone influenced the phytohormonal and antioxidant responses under salinity stress. J Plant Growth Regul 33(2):137–149

Kim YH, Kim MD, Choi YI, Park SC, Yun DJ, Noh EW, Lee HS, Kwak SS (2011) Transgenic poplar expressing Arabidopsis NDPK2 enhances growth as well as oxidative stress tolerance. Plant Biotechnol J 9(3):334–347

Kumari S, Kumar S, Prakash P (2018) Exogenous application of cytokinin (6-BAP) ameliorates the adverse effect of combined drought and high temperature stress in wheat seedling. J Pharmacog Phytochem 7(1):1176–1180

Kuromori T, Shinozaki K (2010) ABA transport factors found in Arabidopsis ABC transporters. Plant Signal Behav 5(9):1124–1126

Lackman P, González-Guzmán M, Tilleman S, Carqueijeiro I, Pérez AC, Moses T, Seo M, Kanno Y, Häkkinen ST, Van Montagu MC, Thevelein JM (2011) Jasmonate signaling involves the abscisic acid receptor PYL4 to regulate metabolic reprogramming in Arabidopsis and tobacco. Proc Natl Acad Sci 108(14):5891–5896

Larson RA (1988) The antioxidants of higher plants. Phytochemistry 27(4):969–978

Lee J, Nam J, Park HC, Na G, Miura K, Jin JB, Yoo CY, Baek D, Kim DH, Jeong JC, Kim D (2007) Salicylic acid-mediated innate immunity in Arabidopsis is regulated by SIZ1 SUMO E3 ligase. Plant J 49(1):79–90

Lee KH, Piao HL, Kim HY, Choi SM, Jiang F, Hartung W, Hwang I, Kwak JM, Lee IJ, Hwang I (2006) Activation of glucosidase via stress-induced polymerization rapidly increases active pools of abscisic acid. Cell 126(6):1109–1120

Lee S, Kim SG, Park CM (2010) Salicylic acid promotes seed germination under high salinity by modulating antioxidant activity in Arabidopsis. New Phytol 188(2):626–637

Lehmann J, Atzorn R, Brückner C, Reinbothe S, Leopold J, Wasternack C, Parthier B (1995) Accumulation of jasmonate, abscisic acid, specific transcripts and proteins in osmotically stressed barley leaf segments. Planta 197(1):156–162

Li CH, Wang G, Zhao JL, Zhang LQ, Ai LF, Han YF, Sun DY, Zhang SW, Sun Y (2014a) The receptor-like kinase SIT1 mediates salt sensitivity by activating MAPK3/6 and regulating ethylene homeostasis in rice. Plant Cell 26(6):2538–2553

Li KR, Feng CH (2011) Effects of brassinolide on drought resistance of *Xanthoceras sorbifolia* seedlings under water stress. Acta Physiol Plant 33(4):1293–1300

Li L, Staden JV, Jäger AK (1998) Effects of plant growth regulators on the antioxidant system in seedlings of two maize cultivars subjected to water stress. Plant Growth Regul 25(2):81–87

Li T, Hu Y, Du X, Tang H, Shen C, Wu J (2014b) Salicylic acid alleviates the adverse effects of salt stress in *Torreya grandis cv. Merrillii* seedlings by activating photosynthesis and enhancing antioxidant systems. PLoS One 9(10):e109492

Li YH, Liu YJ, Xu XL, Jin M, An LZ, Zhang H (2012) Effect of 24-epibrassinolide on drought stress-induced changes in *Chorispora bungeana*. Biol Plant 56(1):192–196

Lima JV, Lobato AKS (2017) Brassinosteroids improve photosystem II efficiency, gas exchange, antioxidant enzymes and growth of cowpea plants exposed to water deficit. Physiol Mol Biol Plants 23(1):59–72

Liu HH, Tian X, Li YJ, Wu CA, Zheng CC (2008) Microarray-based analysis of stress-regulated microRNAs in *Arabidopsis thaliana*. RNA 14(5):836–843

Lin Y, Yang L, Paul M, Zu Y, Tang Z (2013) Ethylene promotes germination of Arabidopsis seed under salinity by decreasing reactive oxygen species: evidence for the involvement of nitric oxide simulated by sodium nitroprusside. Plant Physiol Biochem 73:211–218

Liu J, Hasanuzzaman M, Wen H, Zhang J, Peng T, Sun H, Zhao Q (2019) High temperature and drought stress cause abscisic acid and reactive oxygen species accumulation and suppress seed germination growth in rice. Protoplasma 256(5):1217–1227

Liu SJ, Xu HH, Wang WQ, Li N, Wang WP, Møller IM, Song SQ (2015) A proteomic analysis of rice seed germination as affected by high temperature and ABA treatment. Physiol Plant 154(1): 142–161

Liu W, Zhang Y, Yuan X, Xuan Y, Gao Y, Yan Y (2016) Exogenous salicylic acid improves salinity tolerance of *Nitraria tangutorum*. Russ J Plant Physiol 63(1):132–142

Liu Y, Ye N, Liu R, Chen M, Zhang J (2010) H_2O_2 mediates the regulation of ABA catabolism and GA biosynthesis in Arabidopsis seed dormancy and germination. J Exp Bot 61(11):2979–2990

Loutfy N, Sakuma Y, Gupta DK, Inouhe M (2020) Modifications of water status, growth rate and antioxidant system in two wheat cultivars as affected by salinity stress and salicylic acid. J Plant Res 133(4):549–570

Lu S, Su W, Li H, Guo Z (2009) Abscisic acid improves drought tolerance of triploid bermudagrass and involves H2O2-and NO-induced antioxidant enzyme activities. Plant Physiol Biochem 47(2):132–138

Lubovská Z, Dobrá J, Štorchová H, Wilhelmová N, Vanková R (2014) Cytokinin oxidase/dehydrogenase overexpression modifies antioxidant defense against heat, drought and their combination in *Nicotiana tabacum* plants. J Plant Physiol 171(17):1625–1633

Ma R, Xu S, Chen Y, Guo F, Wu R, Okyere S, Wang F, Jing Y, Wang X (2019) Effects of exogenous application of salicylic acid on drought performance of medicinal plant, Fritillaria przewalskii maxim. Phytoprotection 99(1):27–35

Ma X, Zheng J, Zhang X, Hu Q, Qian R (2017) Salicylic acid alleviates the adverse effects of salt stress on *Dianthus superbus* (Caryophyllaceae) by activating photosynthesis, protecting morphological structure, and enhancing the antioxidant system. Front Plant Sci 8:600

Madhavan S, Chrominiski A, Smith BN (1983) Effect of ethylene on stomatal opening in tomato and carnation leaves. Plant Cell Physiol 24(3):569–572

Maggio A, Barbieri G, Raimondi G, De Pascale S (2010) Contrasting effects of GA 3 treatments on tomato plants exposed to increasing salinity. J Plant Growth Regul 29(1):63–72

Maghsoudi K, Emam Y, Ashraf M, Arvin MJ (2019) Alleviation of field water stress in wheat cultivars by using silicon and salicylic acid applied separately or in combination. Crop Pasture Sci 70(1):36–43

Mahesh K, Balaraju P, Ramakrishna B, Rao SSR (2013) Effect of brassinosteroids on germination and seedling growth of radish (*Raphanus sativus* L.) under PEG-6000 induced water stress. Am J Plant Sci 2013:9

Manan A, Ayyub CM, Pervez MA, Ahmad R (2016) Methyl jasmonate brings about resistance against salinity stressed tomato plants by altering biochemical and physiological processes. Pak J Agric Sci 53:1

Manzano S, Martínez C, Megías Z, Gómez P, Garrido D, Jamilena M (2011) The role of ethylene and brassinosteroids in the control of sex expression and flower development in *Cucurbita pepo*. Plant Growth Regul 65(2):213–221

Mehla N, Sindhi V, Josula D, Bisht P, Wani SH (2017) An introduction to antioxidants and their roles in plant stress tolerance. In: Reactive oxygen species and antioxidant Systems in Plants: role and regulation under abiotic stress. Springer, Singapore, pp 1–23

Mehrotra R, Bhalothia P, Bansal P, Basantani MK, Bharti V, Mehrotra S (2014) Abscisic acid and abiotic stress tolerance–different tiers of regulation. J Plant Physiol 171(7):486–496

Mercado-Blanco J, van der Drift KM, Olsson PE, Thomas-Oates JE, van Loon LC, Bakker PA (2001) Analysis of the pmsCEAB gene cluster involved in biosynthesis of salicylic acid and the siderophore pseudomonine in the biocontrol strain pseudomonas fluorescensWCS374. J Bacteriol 183(6):1909–1920

Mittler R (2006) Abiotic stress, the field environment and stress combination. Trends Plant Sci 11(1):15–19

Mohamed HI, Latif HH (2017) Improvement of drought tolerance of soybean plants by using methyl jasmonate. Physiol Mol Biol Plants 23(3):545–556

Mohammadi MHS, Etemadi N, Arab MM, Aalifar M, Arab M, Pessarakli M (2017) Molecular and physiological responses of Iranian perennial ryegrass as affected by Trinexapac ethyl, Paclobutrazol and abscisic acid under drought stress. Plant Physiol Biochem 111:129–143

Moravcová Š, Tůma J, Dučaiová ZK, Waligórski P, Kula M, Saja D, Słomka A, Bąba W, Libik-Konieczny M (2018) Influence of salicylic acid pretreatment on seeds germination and some defence mechanisms of *Zea mays* plants under copper stress. Plant Physiol Biochem 122:19–30

Müller B, Sheen J (2008) Cytokinin and auxin interaction in root stem-cell specification during early embryogenesis. Nature 453(7198):1094–1097

Munemasa S, Oda K, Watanabe-Sugimoto M, Nakamura Y, Shimoishi Y, Murata Y (2007) The coronatine-insensitive 1 mutation reveals the hormonal signaling interaction between abscisic acid and methyl jasmonate in Arabidopsis guard cells. Specific impairment of ion channel activation and second messenger production. Plant Physiol 143(3):1398–1407

Munns R (2011) Plant adaptations to salt and water stress: differences and commonalities. Adv Bot Res 57:1–32

Mýtinová Z, Motyka V, Haisel D, Lubovská Z, Trávníčková A, Dobrev P, Holík J, Wilhelmová N (2011) Antioxidant enzymatic protection during tobacco leaf ageing is affected by cytokinin depletion. Plant Growth Regul 65(1):23–34

Nakabayashi R, Yonekura-Sakakibara K, Urano K, Suzuki M, Yamada Y, Nishizawa T, Matsuda F, Kojima M, Sakakibara H, Shinozaki K, Michael AJ (2014) Enhancement of oxidative and drought tolerance in Arabidopsis by overaccumulation of antioxidant flavonoids. Plant J 77(3): 367–379

Nambara E, Marion-Poll A (2005) Abscisic acid biosynthesis and catabolism. Annu Rev Plant Biol 56:165–185

Naveed M, Sajid H, Mustafa A, Niamat B, Ahmad Z, Yaseen M, Kamran M, Rafique M, Ahmar S, Chen JT (2020) Alleviation of salinity-induced oxidative stress, improvement in growth, physiology and mineral nutrition of canola (*Brassica napus* L.) through calcium-fortified composted animal manure. Sustainability 12(3):846

Nazar R, Iqbal N, Syeed S, Khan NA (2011) Salicylic acid alleviates decreases in photosynthesis under salt stress by enhancing nitrogen and sulfur assimilation and antioxidant metabolism differentially in two mungbean cultivars. J Plant Physiol 168(8):807–815

Nie S, Huang S, Wang S, Mao Y, Liu J, Ma R, Wang X (2019) Enhanced brassinosteroid signaling intensity via SlBRI1 overexpression negatively regulates drought resistance in a manner opposite of that via exogenous BR application in tomato. Plant Physiol Biochem 138:36–47

Nishiyama R, Watanabe Y, Leyva-Gonzalez MA, Van Ha C, Fujita Y, Tanaka M, Seki M, Yamaguchi-Shinozaki K, Shinozaki K, Herrera-Estrella L, Tran LSP (2013) Arabidopsis AHP2, AHP3, and AHP5 histidine phosphotransfer proteins function as redundant negative regulators of drought stress response. Proc Natl Acad Sci 110(12):4840–4845

Nolan T, Chen J, Yin Y (2017) Cross-talk of Brassinosteroid signaling in controlling growth and stress responses. Biochem J 474(16):2641–2661

Okamoto M, Kuwahara A, Seo M, Kushiro T, Asami T, Hirai N, Kamiya Y, Koshiba T and Nambara E (2006) CYP707A1 and CYP707A2, which encode ABA 8′-hydroxylases, are indispensable for a proper control of seed dormancy and germination in Arabidopsis1. *Published on March, 16*, 106

Okamoto M, Tatematsu K, Matsui A, Morosawa T, Ishida J, Tanaka M, Endo TA, Mochizuki Y, Toyoda T, Kamiya Y, Shinozaki K (2010) Genome-wide analysis of endogenous abscisic acid-mediated transcription in dry and imbibed seeds of Arabidopsis using tiling arrays. Plant J 62(1): 39–51

Okuma E, Nozawa R, Murata Y, Miura K (2014) Accumulation of endogenous salicylic acid confers drought tolerance to Arabidopsis. Plant Signal Behav 9(3):e28085

Palma F, López-Gómez M, Tejera NA, Lluch C (2014) Involvement of abscisic acid in the response of Medicago sativa plants in symbiosis with *Sinorhizobium meliloti* to salinity. Plant Sci 223: 16–24

Park J-S, Okumura Y, Tachikawa H, Neiman AM (2013) SPO71 encodes a developmental stage-specific partner for Vps13 in Saccharomyces cerevisiae. Eukaryot Cell 12(11):1530–1537. https://doi.org/10.1128/EC.00239-13

Pavlů J, Novák J, Koukalová V, Luklová M, Brzobohatý B, Černý M (2018) Cytokinin at the crossroads of abiotic stress signalling pathways. Int J Mol Sci 19(8):2450

Pazirandeh MS, Hasanloo T, Niknam V, Shahbazi M, Mabood HE, Ghaffari A (2013) Effects of drought and methyl jasmonate on antioxidant activities of selected barley genotypes. J Agrobiol 30:71–82

Peng J, Li Z, Wen X, Li W, Shi H, Yang L, Zhu H, Guo H (2014) Salt-induced stabilization of EIN3/EIL1 confers salinity tolerance by deterring ROS accumulation in Arabidopsis. PLoS Genet 10(10):e1004664

Peng Z, Han C, Yuan L, Zhang K, Huang H, Ren C (2011) Brassinosteroid enhances jasmonate-induced anthocyanin accumulation in Arabidopsis seedlings. J Integr Plant Biol 53(8):632–640

Per TS, Khan MIR, Anjum NA, Masood A, Hussain SJ, Khan NA (2018) Jasmonates in plants under abiotic stresses: crosstalk with other phytohormones matters. Environ Exp Bot 145:104–120

Pereira A (2016) Plant abiotic stress challenges from the changing environment. Front Plant Sci 7: 1123

Polko JK, Kieber JJ (2019) 1-Aminocyclopropane 1-carboxylic acid and its emerging role as an ethylene-independent growth regulator. Front Plant Sci 10:1602

Qin X, Zeevaart JA (2002) Overexpression of a 9-cis-epoxycarotenoid dioxygenase gene in *Nicotiana plumbaginifolia* increases abscisic acid and phaseic acid levels and enhances drought tolerance. Plant Physiol 128(2):544–551

Rady MM (2011) Effect of 24-epibrassinolide on growth, yield, antioxidant system and cadmium content of bean (*Phaseolus vulgaris* L.) plants under salinity and cadmium stress. Sci Hortic 129(2):232–237

Rajewska I, Talarek M, Bajguz A (2016) Brassinosteroids and response of plants to heavy metals action. Front Plant Sci 7:629

Rao SSR, Vardhini BV, Sujatha E, Anuradha S (2002) Brassinosteroids–a new class of phytohormones. Curr Sci:1239–1245

Rate DN, Cuenca JV, Bowman GR, Guttman DS, Greenberg JT (1999) The gain-of-function Arabidopsis acd6 mutant reveals novel regulation and function of the salicylic acid signaling pathway in controlling cell death, defenses, and cell growth. Plant Cell 11(9):1695–1708

Rattan A, Kapoor D, Kapoor N, Bhardwaj R, Sharma A (2020) Brassinosteroids regulate functional components of antioxidative defense system in salt stressed maize seedlings. J Plant Growth Regul:1–11

Rattan A, Kapoor N, Bhardwaj R (2012) Role of brassinosteroids in osmolytes accumulation under salinity stress in *Zea mays* plants. Int J Sci Res 3(9):1822–1827

Raza A, Razzaq A, Mehmood SS, Zou X, Zhang X, Lv Y, Xu J (2019) Impact of climate change on crops adaptation and strategies to tackle its outcome: A review. Plan Theory 8(2):34

Razmi N, Ebadi A, Daneshian J, Jahanbakhsh S (2017) Salicylic acid induced changes on antioxidant capacity, pigments and grain yield of soybean genotypes in water deficit condition. J Plant Interact 12(1):457–464

Rhaman MS, Imran S, Rauf F, Khatun M, Baskin CC, Murata Y, Hasanuzzaman M (2021) Seed priming with phytohormones: An effective approach for the mitigation of abiotic stress. Plan Theory 10(1):37

Riyazuddin R, Verma R, Singh K, Nisha N, Keisham M, Bhati KK, Kim ST, Gupta R (2020) Ethylene: a master regulator of salinity stress tolerance in plants. Biomol Ther 10(6):959

Rohman MM, Islam MR, Naznin T, Omy SH, Begum S, Alam SS, Amiruzzaman M, Hasanuzzaman M (2019) Maize production under salinity and drought conditions: oxidative stress regulation by antioxidant defense and glyoxalase systems. In: Plant abiotic stress tolerance. Springer, Cham, pp 1–34

Ruan J, Zhou Y, Zhou M, Yan J, Khurshid M, Weng W, Cheng J, Zhang K (2019) Jasmonic acid signaling pathway in plants. Int J Mol Sci 20(10):2479

Ryu H, Cho YG (2015) Plant hormones in salt stress tolerance. J Plant Biol 58(3):147–155

Sadeghipour O (2017) Amelioration of salinity tolerance in cowpea plants by seed treatment with methyl jasmonate. Legume Res Int J 40:6

Saeidi-Sar S, Abbaspour H, Afshari H, Yaghoobi SR (2013) Effects of ascorbic acid and gibberellin A 3 on alleviation of salt stress in common bean (*Phaseolus vulgaris* L.) seedlings. Acta Physiol Plant 35(3):667–677

Sahay S, Khan E, Gupta M (2019) Nitric oxide and abscisic acid protects against PEG-induced drought stress differentially in brassica genotypes by combining the role of stress modulators, markers and antioxidants. Nitric Oxide 89:81–92

Salah A, Zhan M, Cao C, Han Y, Ling L, Liu Z, Li P, Ye M, Jiang Y (2019) γ-Aminobutyric acid promotes chloroplast ultrastructure, antioxidant capacity, and growth of waterlogged maize seedlings. Sci Rep 9(1):1–19

Samea-Andabjadid S, Ghassemi-Golezani K, Nasrollahzadeh S, Najafi N (2018) Exogenous salicylic acid and cytokinin alter sugar accumulation, antioxidants and membrane stability of faba bean. Acta Biol Hung 69(1):86–96

Saradhi PP, Suzuki I, Katoh A, Sakamoto A, Sharmila P, Shi DJ, Murata N (2000) Protection against the photo-induced inactivation of the photosystem II complex by abscisic acid. Plant Cell Environ 23(7):711–718

Sauer M, Balla J, Luschnig C, Wiśniewska J, Reinöhl V, Friml J, Benková E (2006) Canalization of auxin flow by Aux/IAA-ARF-dependent feedback regulation of PIN polarity. Genes Dev 20(20):2902–2911

Sauter A, Davies WJ, Hartung W (2001) The long-distance abscisic acid signal in the droughted plant: the fate of the hormone on its way from root to shoot. J Exp Bot 52(363):1991–1997

Schopfer P (2001) Hydroxyl radical-induced cell-wall loosening in vitro and in vivo: implications for the control of elongation growth. Plant J 28(6):679–688

Sedghi M, Seyed Sharifi R, Pirzad AR, Amanpour-Balaneji B (2012) Phytohormonal regulation of antioxidant systems in petals of drought stressed pot marigold (*Calendula officinalis* L.). J Agric Sci Technol 14(4):869–878

Seltmann MA, Stingl NE, Lautenschlaeger JK, Krischke M, Mueller MJ, Berger S (2010) Differential impact of lipoxygenase 2 and jasmonates on natural and stress-induced senescence in Arabidopsis. Plant Physiol 152(4):1940–1950

Senaratna T, Touchell D, Bunn E, Dixon K (2000) Acetyl salicylic acid (aspirin) and salicylic acid induce multiple stress tolerance in bean and tomato plants. Plant Growth Regul 30(2):157–161

Seo M, Koshiba T (2002) Complex regulation of ABA biosynthesis in plants. Trends Plant Sci 7(1): 41–48

Seo M, Koshiba T (2011) Transport of ABA from the site of biosynthesis to the site of action. J Plant Res 124:501–510

Seo PJ, Park CM (2010) MYB96-mediated abscisic acid signals induce pathogen resistance response by promoting salicylic acid biosynthesis in Arabidopsis. New Phytol 186(2):471–483

Seo PJ, Xiang F, Qiao M, Park JY, Lee YN, Kim SG, Lee YH, Park WJ, Park CM (2009) The MYB96 transcription factor mediates abscisic acid signaling during drought stress response in Arabidopsis. Plant Physiol 151(1):275–289

Shakar M, Yaseen M, Mahmood R, Ahmad I (2016) Calcium carbide induced ethylene modulate biochemical profile of *Cucumis sativus* at seed germination stage to alleviate salt stress. Sci Hortic 213:179–185

Sharma A, Shahzad B, Kumar V, Kohli SK, Sidhu GPS, Bali AS, Handa N, Kapoor D, Bhardwaj R, Zheng B (2019) Phytohormones regulate accumulation of osmolytes under abiotic stress. Biomol Ther 9(7):285

Sharma A, Sidhu GPS, Araniti F, Bali AS, Shahzad B, Tripathi DK, Brestic M, Skalicky M, Landi M (2020) The role of salicylic acid in plants exposed to heavy metals. Molecules 25(3):540

Sharma E, Sharma R, Borah P, Jain M, Khurana JP (2015) Emerging roles of auxin in abiotic stress responses. In: Elucidation of abiotic stress signaling in plants. Springer, New York, NY, pp 299–328

Sharma I, Ching E, Saini S, Bhardwaj R, Pati PK (2013a) Exogenous application of brassinosteroid offers tolerance to salinity by altering stress responses in rice variety Pusa Basmati-1. Plant Physiol Biochem 69:17–26

Sharma L, Dalal M, Verma RK, Kumar SV, Yadav SK, Pushkar S, Kushwaha SR, Bhowmik A, Chinnusamy V (2018) Auxin protects spikelet fertility and grain yield under drought and heat stresses in rice. Environ Exp Bot 150:9–24

Sharma R, Priya P, Jain M (2013b) Modified expression of an auxin-responsive rice CC-type glutaredoxin gene affects multiple abiotic stress responses. Planta 238(5):871–884

Shemi R, Wang R, Gheith ESM, Hussain HA, Hussain S, Irfan M, Cholidah L, Zhang K, Zhang S, Wang L (2021) Effects of salicylic acid, zinc and glycine betaine on morpho-physiological growth and yield of maize under drought stress. Sci Rep 11(1):1–14

Shen QJ, Casaretto JA, Zhang P, Ho THD (2004) Functional definition of ABA-response complexes: the promoter units necessary and sufficient for ABA induction of gene expression in barley (*Hordeum vulgare* L.). Plant Mol Biol 54(1):111–124

Sheng XF, Xia JJ (2006) Improvement of rape (*Brassica napus*) plant growth and cadmium uptake by cadmium-resistant bacteria. Chemosphere 64(6):1036–1042

Shi H, Chen L, Ye T, Liu X, Ding K, Chan Z (2014) Modulation of auxin content in Arabidopsis confers improved drought stress resistance. Plant Physiol Biochem 82:209–217

Shi Y, Tian S, Hou L, Huang X, Zhang X, Guo H, Yang S (2012) Ethylene signaling negatively regulates freezing tolerance by repressing expression of CBF and type-A ARR genes in Arabidopsis. Plant Cell 24(6):2578–2595

Shinozaki K, Yamaguchi-Shinozaki K (2000) Molecular responses to dehydration and low temperature: differences and cross-talk between two stress signaling pathways. Curr Opin Plant Biol 3(3):217–723

Siddiqi KS, Husen A (2019) Plant response to jasmonates: current developments and their role in changing environment. Bull Natl Res Centre 43(1):1–11

Siddiqui MH, Alamri S, Alsubaie QD, Ali HM (2020) Melatonin and gibberellic acid promote growth and chlorophyll biosynthesis by regulating antioxidant and methylglyoxal detoxification system in tomato seedlings under salinity. J Plant Growth Regul 39(4):1488–1502

Singh SK, Singh AK, Dwivedi P (2017) Modulating effect of salicylic acid in tomato plants in response to waterlogging stress. Int J Agric Env Biotechnol 10(1):1

Soares C, de Sousa A, Pinto A, Azenha M, Teixeira J, Azevedo RA, Fidalgo F (2016) Effect of 24-epibrassinolide on ROS content, antioxidant system, lipid peroxidation and Ni uptake in *Solanum nigrum* L. under Ni stress. Environ Exp Bot 122:115–125

Sohag AAM, Tahjib-Ul-Arif M, Brestic M, Afrin S, Sakil MA, Hossain MT, Hossain MA, Hossain MA (2020) Exogenous salicylic acid and hydrogen peroxide attenuate drought stress in rice. Plant Soil Environ 66(1):7–13

Souri MK, Tohidloo G (2019) Effectiveness of different methods of salicylic acid application on growth characteristics of tomato seedlings under salinity. Chem Biol Technol Agric 6(1):1–7

Suhaib M, Ahmad I, Munir M, Iqbal MB, Abuzar MK, Ali S (2018) Salicylic acid induced physiological and ionic efficiency in wheat under salt stress. Pak J Agric Res 31(1):79–85

Sun Y, Fan XY, Cao DM, Tang W, He K, Zhu JY, He JX, Bai MY, Zhu S, Oh E, Patil S (2010) Integration of brassinosteroid signal transduction with the transcription network for plant growth regulation in Arabidopsis. Dev Cell 19(5):765–777

Szekeres M, Németh K, Koncz-Kálmán Z, Mathur J, Kauschmann A, Altmann T, Rédei GP, Nagy F, Schell J, Koncz C (1996) Brassinosteroids rescue the deficiency of CYP90, a cytochrome P450, controlling cell elongation and de-etiolation in Arabidopsis. Cell 85(2):171–182

Tanaka Y, Sano T, Tamaoki M, Nakajima N, Kondo N, Hasezawa S (2005) Ethylene inhibits abscisic acid-induced stomatal closure in Arabidopsis. Plant Physiol 138(4):2337–2343
Tanaka Y, Sano T, Tamaoki M, Nakajima N, Kondo N, Hasezawa S (2006) Cytokinin and auxin inhibit abscisic acid-induced stomatal closure by enhancing ethylene production in Arabidopsis. J Exp Bot 57:2259–2266
Tayyab N, Naz R, Yasmin H, Nosheen A, Keyani R, Sajjad M, Hassan MN, Roberts TH (2020) Combined seed and foliar pre-treatments with exogenous methyl jasmonate and salicylic acid mitigate drought-induced stress in maize. PLoS One 15(5):e0232269
Tewari S, Mishra A (2018) Flooding stress in plants and approaches to overcome. In: Plant metabolites and regulation under environmental stress. Academic Press, pp 355–366
Thu NBA, Hoang XLT, Truc MT, Sulieman S, Thao NP, Tran LSP, Pandey G (2017) Cytokinin signaling in plant response to abiotic stresses. Mech Plant Hormone Signal Under Stress 1(chpt. 4):71–100
Tognetti VB, Van Aken O, Morreel K, Vandenbroucke K, Van De Cotte B, De Clercq I, Chiwocha S, Fenske R, Prinsen E, Boerjan W, Genty B (2010) Perturbation of indole-3-butyric acid homeostasis by the UDP-glucosyltransferase UGT74E2 modulates Arabidopsis architecture and water stress tolerance. Plant Cell 22(8):2660–2679
Toh S, Imamura A, Watanabe A, Nakabayashi K, Okamoto M, Jikumaru Y, Hanada A, Aso Y, Ishiyama K, Tamura N, Iuchi S (2008) High temperature-induced abscisic acid biosynthesis and its role in the inhibition of gibberellin action in Arabidopsis seeds. Plant Physiol 146(3): 1368–1385
Torun H, Novák O, Mikulík J, Pěnčík A, Strnad M, Ayaz FA (2020) Timing-dependent effects of salicylic acid treatment on phytohormonal changes, ROS regulation, and antioxidant defense in salinized barley (*Hordeum vulgare* L.). Sci Rep 10(1):1–17
Tunc-Ozdemir M, Jones AM (2017) BRL3 and AtRGS1 cooperate to fine tune growth inhibition and ROS activation. PLoS One 12(5):e0177400
Vardhini BV (2012a) Application of brassinolide mitigates saline stress of certain metabolites of sorghum grown in Karaikal. J Phytology 4(4):1–3
Vardhini BV (2012b) Effect of brassinolide on certain enzymes of sorghum grown in saline soils of Karaikal. J Phytology 4(2):30–33
Vardhini BV, Anjum NA (2015) Brassinosteroids make plant life easier under abiotic stresses mainly by modulating major components of antioxidant defense system. Front Environ Sci 2:67
Verma V, Ravindran P, Kumar PP (2016) Plant hormone-mediated regulation of stress responses. BMC Plant Biol 16(1):1–10
Vujicic MM, Milošević SM, Sabovljević MS, Sabovljević AD (2017) Effect of ABA treatment on activities of antioxidative enzymes in selected bryophyte species. Botanica Serbica 41(1):11–15
Vwioko E, Adinkwu O, El-Esawi MA (2017) Comparative physiological, biochemical, and genetic responses to prolonged waterlogging stress in okra and maize given exogenous ethylene priming. Front Physiol 8:632
Wan D, Li R, Zou B, Zhang X, Cong J, Wang R, Xia Y, Li G (2012) Calmodulin-binding protein CBP60g is a positive regulator of both disease resistance and drought tolerance in Arabidopsis. Plant Cell Rep 31(7):1269–1281
Wang J, Sommerfeld M, Hu Q (2009) Occurrence and environmental stress responses of two plastid terminal oxidases in *Haematococcus pluvialis* (Chlorophyceae). Planta 230(1):191–203
Wang L, Qin L, Liu W, Zhang D, Wang Y (2014a) A novel ethylene-responsive factor from *Tamarix hispida*, ThERF1, is a GCC-box-and DRE-motif binding protein that negatively modulates abiotic stress tolerance in Arabidopsis. Physiol Plant 152(1):84–97
Wang X, Li Q, Xie J, Huang M, Cai J, Zhou Q, Dai T, Jiang D (2021) Abscisic acid and jasmonic acid are involved in drought priming-induced tolerance to drought in wheat. Crop J 9(1): 120–132
Wang Y, Liu S, Zhang H, Zhao Y, Zhao H, Liu H (2014b) Glycine betaine application in grain filling wheat plants alleviates heat and high light-induced photoinhibition by enhancing the psbA transcription and stomatal conductance. Acta Physiol Plant 36(8):2195–2202

Wani SH, Kumar V, Shriram V, Sah SK (2016) Phytohormones and their metabolic engineering for abiotic stress tolerance in crop plants. Crop J 4(3):162–176

Wasilewska A, Vlad F, Sirichandra C, Redko Y, Jammes F, Valon C, dit Frey NF, Leung J (2008) An update on abscisic acid signaling in plants and more.... Mol Plant 1(2):198–217

Wen B (2015) Effects of high temperature and water stress on seed germination of the invasive species Mexican sunflower. PLoS One 10(10):e0141567

Wen B, Liu M, Tan Y, Liu Q (2016) Sensitivity to high temperature and water stress in recalcitrant *Baccaurea ramiflora* seeds. J Plant Res 129(4):637–645

Wu L, Zhang Z, Zhang H, Wang XC, Huang R (2008) Transcriptional modulation of ethylene response factor protein JERF3 in the oxidative stress response enhances tolerance of tobacco seedlings to salt, drought, and freezing. Plant Physiol 148(4):1953–1963

Xiao S, Yuan LB, Dai YS, Xie LJ, Yu LJ, Zhou Y, Lai YX, Yang YC, Xu L, Chen QF (2017a) Jasmonate regulates plant responses to reoxygenation through activation of antioxidant synthesis. Plant Physiol

Xiao XO, Zeng YM, Cao BH, Lei JJ, Chen QH, Meng CM, Cheng YJ (2017b) PSAG12-IPT overexpression in eggplant delays leaf senescence and induces abiotic stress tolerance. J Hortic Sci Biotechnol 92(4):349–357

Xing Q, Liao J, Cao S, Li M, Lv T, Qi H (2020) CmLOX10 positively regulates drought tolerance through jasmonic acid-mediated stomatal closure in oriental melon (*Cucumis melo* var. *makuwa Makino*). Sci Rep 10(1):1–14

Xu F, Xi ZM, Zhang H, Zhang CJ, Zhang ZW (2015) Brassinosteroids are involved in controlling sugar unloading in *Vitis vinifera* 'cabernet Sauvignon'berries during véraison. Plant Physiol Biochem 94:197–208

Xu L, Xiang G, Sun Q, Ni Y, Jin Z, Gao S, Yao Y (2019) Melatonin enhances salt tolerance by promoting MYB108A-mediated ethylene biosynthesis in grapevines. Hortic Res 6(1):1–14

Xu Y, Huang B (2017) Transcriptional factors for stress signaling, oxidative protection, and protein modification in ipt-transgenic creeping bentgrass exposed to drought stress. Environ Exp Bot 144:49–60

Yalpani N, León J, Lawton MA, Raskin I (1993) Pathway of salicylic acid biosynthesis in healthy and virus-inoculated tobacco. Plant Physiol 103(2):315–321

Yamaguchi S (2008) Gibberellin metabolism and its regulation. Annu Rev Plant Biol 59:225–251

Yang DL, Yao J, Mei CS, Tong XH, Zeng LJ, Li Q, Xiao LT, Sun TP, Li J, Deng XW, Lee CM (2012a) Plant hormone jasmonate prioritizes defense over growth by interfering with gibberellin signaling cascade. Proc Natl Acad Sci 109(19):E1192–E1200

Yang L, Zu YG, Tang ZH (2013) Ethylene improves Arabidopsis salt tolerance mainly via retaining K+ in shoots and roots rather than decreasing tissue Na+ content. Environ Exp Bot 86:60–69

Yang XY, Jiang WJ, Yu HJ (2012b) The expression profiling of the lipoxygenase (LOX) family genes during fruit development, abiotic stress and hormonal treatments in cucumber (*Cucumis sativus* L.). Int J Mol Sci 13(2):2481–2500

Yang Z, Yu J, Merewitz E, Huang B (2012c) Differential effects of abscisic acid and glycine betaine on physiological responses to drought and salinity stress for two perennial grass species. J Am Soc Hortic Sci 137(2):96–106

Yang P, Azher Nawaz M, Li F, Bai L, Li J (2019) Brassinosteroids regulate antioxidant system and protect chloroplast ultrastructure of autotoxicity-stressed cucumber (*Cucumis sativus* L.) seedlings. Agronomy 9(5):265

Yastreb TO, Kolupaev YE, Shvidenko NV, Dmitriev AP (2018) Action of methyl jasmonate and salt stress on antioxidant system of Arabidopsis plants defective in jasmonate signaling genes. Ukrainian Biochem J 90(5):50–59

Ye N, Zhu G, Liu Y, Zhang A, Li Y, Liu R, Shi L, Jia L, Zhang J (2012) Ascorbic acid and reactive oxygen species are involved in the inhibition of seed germination by abscisic acid in rice seeds. J Exp Bot 63(5):1809–1822

Yoshida T, Fujita Y, Sayama H, Kidokoro S, Maruyama K, Mizoi J, Shinozaki K, Yamaguchi-Shinozaki K (2010) AREB1, AREB2, and ABF3 are master transcription factors that cooperatively regulate ABRE-dependent ABA signaling involved in drought stress tolerance and require ABA for full activation. Plant J 61(4):672–685

Younis M, El-Shahaby O, Alla MMN, El-Bastawisy Z (2003) Kinetin alleviates the influence of waterlogging and salinity on growth and affects the production of plant growth regulators in *Vigna sinensis* and *Zea mays*. Agronomie 23(4):277–285

Yu LL, Liu Y, Zhu F, Geng XX, Yang Y, He ZQ, Xu F (2020) The enhancement of salt stress tolerance by salicylic acid pretreatment in Arabidopsis thaliana. Biol Plant 64:150–158

Yu X, Li L, Zola J, Aluru M, Ye H, Foudree A, Guo H, Anderson S, Aluru S, Liu P, Rodermel S (2011) A brassinosteroid transcriptional network revealed by genome-wide identification of BESI target genes in *Arabidopsis thaliana*. Plant J 65(4):634–646

Zayed M, El-Kafafi ES, El Hafnawy SF, El-Araby HG (2017) Effect of auxin treatment on growth and physiological traits in two sunflower cultivars under saline conditions. J Plant Prod 8(2): 335–345

Zhang M, Zhai Z, Tian X, Duan L, Li Z (2008) Brassinolide alleviated the adverse effect of water deficits on photosynthesis and the antioxidant of soybean (Glycine max L.). Plant Growth Regul 56(3):257–264

Zhang Y, Li Y, Hassan MJ, Li Z, Peng Y (2020) Indole-3-acetic acid improves drought tolerance of white clover via activating auxin, abscisic acid and jasmonic acid related genes and inhibiting senescence genes. BMC Plant Biol 20:1–12

Zhang Z, Li F, Li D, Zhang H, Huang R (2010) Expression of ethylene response factor JERF1 in rice improves tolerance to drought. Planta 232(3):765–774

Zhou L, Xu H, Mischke S, Meinhardt LW, Zhang D, Zhu X, Li X, Fang W (2014) Exogenous abscisic acid significantly affects proteome in tea plant (*Camellia sinensis*) exposed to drought stress. Hortic Res 1(1):1–9

Zhu T, Deng X, Zhou X, Zhu L, Zou L, Li P, Zhang D, Lin H (2016) Ethylene and hydrogen peroxide are involved in brassinosteroid-induced salt tolerance in tomato. Sci Rep 6(1):1–15

Zou LJ, Deng XG, Zhang LE, Zhu T, Tan WR, Muhammad A, Zhu LJ, Zhang C, Zhang DW, Lin HH (2018) Nitric oxide as a signaling molecule in brassinosteroid-mediated virus resistance to cucumber mosaic virus in *Arabidopsis thaliana*. Physiol Plant 163(2):96–210

Zwack PJ, Rashotte AM (2015) Interactions between cytokinin signalling and abiotic stress responses. J Exp Bot 66(16):4863–4871

Plant Phenolic Compounds for Abiotic Stress Tolerance

Khursheda Parvin, Kamrun Nahar, Sayed Mohammad Mohsin, Jubayer Al Mahmud, Masayuki Fujita, and Mirza Hasanuzzaman

Abstract Abiotic stresses (drought, salinity, metal toxicity, heat, cold, extreme light, nutrient deficiency, UV radiation) are causing adverse climatic situations for plants survival. Eventually different physiological and metabolic alterations are induced in plants to loss their potentiality to survive and even cause death. However, plants are well equipped with coordinated and organized defense systems against abiotic stresses. As phytoprotectant, phenolic compounds (PCs) are promising group for inducing tolerance in plants against abiotic stresses. Plants synthesize these potential metabolites to modulate their defense mechanism when exposed to stresses. But their active actions including physiological, and signaling responses

K. Parvin
Laboratory of Plant Stress Responses, Faculty of Agriculture, Kagawa University, Miki-cho, Kagawa, Japan

Department of Horticulture, Faculty of Agriculture, Sher-e-Bangla Agricultural University, Dhaka, Bangladesh

K. Nahar
Department of Agricultural Botany, Faculty of Agriculture, Sher-e-Bangla Agricultural University, Dhaka, Bangladesh

S. M. Mohsin
Laboratory of Plant Stress Responses, Faculty of Agriculture, Kagawa University, Miki-cho, Kagawa, Japan

Department of Plant Pathology, Faculty of Agriculture, Sher-e-Bangla Agricultural University, Dhaka, Bangladesh

J. Al Mahmud
Department of Agroforestry and Environmental Science, Faculty of Agriculture, Sher-e-Bangla Agricultural University, Dhaka, Bangladesh

M. Fujita (✉)
Laboratory of Plant Stress Responses, Faculty of Agriculture, Kagawa University, Miki-cho, Kagawa, Japan
e-mail: fujita@ag.kagawa-u.ac.jp

M. Hasanuzzaman
Department of Agronomy, Faculty of Agriculture, Sher-e-Bangla Agricultural University, Dhaka, Bangladesh

M. Hasanuzzaman et al. (eds.), *Managing Plant Production Under Changing Environment*, https://doi.org/10.1007/978-981-16-5059-8_8

in stressed plants still scattered. Antioxidative nature of these compounds is also promising properties for gaining plant tolerance to adverse climatic conditions. The present review is an attempt to coordinate the significant informations about PCs and their contributions to sustain plants. We have summarized and critically evaluated the available literatures of PCs to strengthen the plant tolerance by their antioxidant natures. Finally, there is a vast hope for protecting plants by using this group as phytoprotectant under abiotic stress.

Keywords Abiotic stress · Phenolic compound · Reactive oxygen species · Defense system

1 Introduction

Phenolic compounds (PCs) are the majority recurrent and extensive groups of substances of plant kingdom. More than 8,000 identified phenolic structures have been conferred in plants distributed in almost all the organs (Dai and Mumper 2010; Chen et al. 2015). In plant cells, PCs are found mainly in cell wall, vacuole, chloroplasts, and nucleus (Khlestkina 2013). Plants PCs perform multiple functions such as PCs dissociate oxidative phosphorylation, enhance cell division, modulate seed germination, they act as precursors of lignin, and structural component of cell wall (Hura et al. 2012; Ha et al. 2016); they have allelopathic, antibacterial, antifungicidal, and antipathogenic activities (Nduche and Otaka 2019; Santra and Banerjee 2020). Accumulation of PCs is determined by the activity and ultrastructural organization of chloroplasts (Zaprometov and Nikolaeva 2003). Phenolic compounds also improve photosynthetic performances by increasing photosynthetic pigments (chlorophyll, Chl; and carotenoid) contents (Mekawy et al. 2018; Parvin et al. 2019; Zhang et al. 2020), modulating net photosynthetic rate (Pn), and stomatal conductance, photosystem II (PS II) activity, and improving Chl fluorescence parameters (Mekawy et al. 2018; Zhang et al. 2020). Phenolic compounds are able in osmoregulation in plants which give multiple stress protection. Exogenous PCs supplementation caused higher accumulation of proline (Pro), glycine betaine (GB), trehalose (Tre), and various soluble sugars and were associated with osmotic stress protection (El-Soud et al. 2013; Parvin et al. 2019). Modulation of membrane potential and H^+-ATPase activity (Ghassemi-Golezani and Farhangi-Abriz 2018), regulation of Na^+ and K^+ uptake, and maintainance of ion and nutrient homeostasis were conferred by PCs addition under different stresses including salinity (Parvin et al. 2019), drought (Chavoushi et al. 2020), and heavy metal (Shahid et al. 2018). Lipid peroxidation inhibition due to the alkoxyl-radicals trapping as well as suppressing oxidizing enzymes activities was reported due to putative effect of PCs (Shahid et al. 2018). Phenolic compound has been proved itself as an efficient reactive oxygen species (ROS) scavenger not only by its unique structure but also by enhancing the antioxidant defense system (Parvin et al. 2019). Phenolic compounds-induced stimulation in antioxidants responses including both nonenzymatic and enzymatic components and glyoxalase enzymes (glyoxalase I, Gly I, and glyoxalase II, Gly II) have been reported to decrease ROS and methylglyoxal (MG) production, respectively to confer oxidative stress tolerance (Saleh and Madany 2015; Parvin et al. 2019). In recent research findings, exogenous application of PCs improved plants performance under different stress conditions including salt (Parvin et al. 2019; Ahanger et al. 2020), drought (Chavoushi et al. 2020), waterlogging (Xuan and Khang 2018), high temperature (Zhang et al. 2020), low temperature (Ozfidan-Konakci et al. 2019), metal (Yildiztugay et al. 2019), and excess light (Yang et al. 2019). The improved stress tolerance was discernible from improved growth, development, and physiological performance, which was further conferred by upregulated gene expression related to biosynthesis of endogenous PCs (Singh et al. 2017) and antioxidant genes and other defense-related genes (Li et al. 2013; Cheng et al. 2018; Ignatenko et al. 2019). Phenolic compounds can regulate auxin biosynthesis which has effect on plant growth, and development (Kovaleva et al. 2007; Tanase et al. 2019).

Previously published reports emphasized mainly the health benefit or antimicrobial properties of PCs. But these plant derivatives itself either are beneficial to enhance its self-defense system against different abiotic stresses that were not profoundly explored. This is a review to look behind and forward to understand the functioning of PCs in various aspects which will be eye opening to exploit PCs more efficiently in plant science.

2 Plant Phenolic Compounds: Types and Nature

Phenolic compounds are the second most abundant organic compounds group after cellulose, which contains an aromatic ring with one or more hydroxyl groups (de la Rosa et al. 2019). The structural variability is the main reason behind the diversification of PCs in nature (Waśkiewicz et al. 2013). Due to having great diversification in their carbon skeleton such as from consisting of single aromatic ring to complex polymeric substances, PCs are classified into further various subgroups.

Classification can be expressed in following way presented in a flow diagram (Fig. 1) by following previous scientists (Jaganath and Crozier 2010; Waśkiewicz et al. 2013; de la Rosa et al. 2019).

2.1 *Simple Phenol*

Phenol is the structure of benzene-based group. The phenols remain in many ways similar to alcohols of aliphatic structures where the hydroxyl group is attached to a chain of carbons. The presence of the aromatic ring influences the phenolic hydroxyl group. Due to the aromatic ring, the hydrogen of the phenolic hydroxyl is labile, which makes the phenols acidic weak (Vermerris and Nicholson 2008).

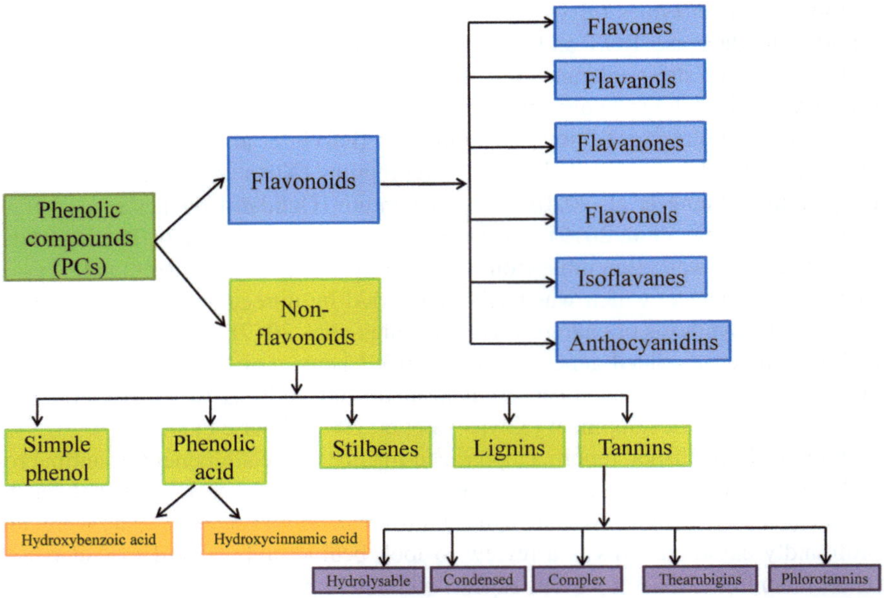

Fig. 1 Illustration on the classification of phenolic compound

Simple phenols are low to medium molecular weight compounds which are soluble and not bound to membranes compounds. Examples of simple phenols (C6) include catechol and phloroglucinol. Although most of the complex plant polyphenols contain these two simple phenols as their structural parts, but exist as uncommon in plant tissues (Giada 2013). Other examples included orcinol, 4-methylresorcinol, 2- methylresorcinol, resorcinol, pyrogallol, and hydroquinone.

2.2 *Phenolic Acid*

Most important and diverse group among non-flavonoids is phenolic acid (C6-Cn) consisting of single phenyl group substituted by one carboxylic group and one or more OH groups. Although most of the members of this group are simpler and smaller than flavonoids but some are in complex structure with higher molecular weight. Depending on the length of chain attaching with the carboxylic group, phenolic acids can be divided mainly into hydroxybenzoic and hydroxycinnamic acids, while hydroxyphenyl acids are also phenolic acids such as acetic, propanoic, and pentaenoic acid (de la Rosa et al. 2019).

Hydroxybenzoic acids (C6–C1) derived from benzoic acid are hardly ever found as free form, but commonly found as glycosylated linked with small organic acids like quinic, maleic, tartaric, or bound to structural cells components such as cellulose, proteins, lignin of plants (de la Rosa et al. 2019). Structural variations of hydroxybenzoic acids are due to the hydroxylations and methoxylations of the aromatic cycle. The *p*-hydroxybenzoic acid, vanillic, syringic, and protocatechuic acids are four leading families of this group, while first three are released from lignin by alkaline hydrolysis (Murkovic 2003). Salicylic acid is also hydroxybenzoic acids. Hydroxybenzoic aldehydes are common derivatives of hydroxybenzoic acids like vanillic acid–derived vanillin and syringic acid–derived syringealdehide (de la Rosa et al. 2019).

As like hydroxybenzoic, hydroxycinnamic acids are also rarely found as free form but widely distributed as bound with small or large molecules. They are derived from cinnamic acid and these derivatives are most popular due to their abundance and diversified nature. The *p*-coumaric acid, caffeic acid, and the methylated forms of ferulic and sinapic acids are the major four hydroxycinnamic acids. Consequently, most of the derivatives are essentially present as combined forms of these four basic molecules (Bravo 1998; Murkovic 2003; de la Rosa et al. 2019). There are two main types of soluble derivatives that have been identified. First groups are consisting of an ester bond between the carboxylic function of phenolic acid and the alcoholic group of an organic compound such as chlorogenic acid. Other groups are linking by a glycosidic bond with one of the phenolic groups of the molecule as, for example, *p*-coumaric acid *O*-glucoside. Thereafter, nature of the bonds and the involved molecules are the main reasons behind the diversity of the hydroxycinnamic acids in plants (Murkovic 2003). However, caffeic acid is predominant in many fruits except citrus and pineapple because of constituting over 75% of total hydroxycinnamic acids, where *p*-coumaric acid is dominant (Murkovic 2003). It has been accounted that phenolic acids occupied almost one-third of dietary PCs, then phenolic acids and flavonoids comprise the majority of PCs in edible plant products.

2.3 Tannins

Those polyphenols are bind with protein and precipitate known as tannins. Thereafter, tannins can be further divided into hydrolyzable, condensed, complex tannins, thearubigins, and phlorotannins. Hydrolyzable tannins consist of a sugar molecule (glucose or polysaccharide) and much hydroxyls group esterified with gallic acid (galotannins) and ellagic acid (ellagitannins). Under acidic condition, hydrolyzable tannins are able to cleave with their monomers (Tsimogiannis and Oreopoulou 2019). Pentagalloyl glucose, lambertianin C, sorgum procyanidin, and fucodiphlorethol are the characteristic examples of gallotannins, ellagitannins, condensed tannin, and phlorotannin, respectively. Catechin monomer is the structural component of the condensed tannins, which form complex tannins by coupling with hydrolyzable tannins as, for example, acutissimin (Navarro et al. 2017). On the other hand, phlorotannins are a very rare category with its higher hydrophilicity, which are oligomers of phloroglucinol (Catarino et al. 2017). Fucodiphlorethol is the characteristic example of phlorotannins.

Gallic acid is well-known component of hydrobenzoic acid, which forms complex structures called hydrolyzable tannins with carbohydrates and further divided into gallotannins, which provided sugar and gallic acid on hydrolysis, and ellagitannins—produce ellagic acid (Murkovic 2003; de la Rosa et al. 2019). Punicalagin is a kind of ellagitannin.

2.4 *Lignin*

Based on phenylpropanoid units, lignins (C6-C3)n are the complex phenolic heteropolymers produced from the oxidative polymerization of hydroxycinnamoyl alcohol derivatives (Vogt 2010). Complicated monolignol (coumaryl, coniferyl, and sinapyl alcohol) and lignan (dimmers of monolignols) units randomly coupled to form a three-dimensional polymer called lignins (Vanholme et al. 2010). As a second most abundant polymer lignins provide structural supports to plant while its hydrophobicity facilitates water transport through vascular tissue.

Biochemically similar to lignins, lignans are another major kind of phenolic polymer formed from the two units of phenylpropanoid units (C6–C3)2 (Katerova et al. 2012). To yield monolignol-derived dimmers and oligomers from phenylalanine via dimerization of substituted cinnamic alcohols, lignans are synthesized. The most common lignans include the secoisolariciresinol, matairesinol, lariciresinol, and pinoresinol.

2.5 *Stilbenes*

Stilbenes (C6-C2-C6) contain a trans (or cis) ethene bond substituted by a phenyl group on both sides of its C−C double bond. Resveratrol and lunularic acid are the most abundant natural stilbenes (Katerova et al. 2012; de la Rosa et al. 2019).

2.6 *Flavonoids Compounds*

The most abundant and bioactive group of PCs is flavonoids and hold about two-thirds of dietary PCs (de la Rosa et al. 2019). Overall, 10,000 members have been enlisted under flavonoids which stand for third biggest group of natural products following the alkaloids (12,000) and terpenoids (30,000) (Dixon and Pasinetti 2010; Santos et al. 2017).

These C15 compounds (C6-C3-C6) structurally possess a phenyl benzopyran skeleton with two phenyl rings (A and B) connected by a three-carbon bridge called heterocyclic pyran ring (C ring). The basic skeleton of flavonoid can able to link with many substitutions including hydroxyl groups, methyl groups, sugars (e.g., glucose, galactose, rhamnose), etc. Thus an additional hydroxyl group at *o*-position in the B ring of flavonoids enhances their antioxidant competence (Edreva et al. 2008). Plants use these flavonoids during their growth and defense mechanism (Havsteen 2002). Moreover, sugars and hydroxyl groups are responsible to increase the water solubility of flavonoids, while methyl and isopentyl groups make them lipophilic. On the basis of the position of pyran ring, flavonoids compounds can be classified into further six groups where individual components have a distinction pattern in hydroxylation and methylation of rings A and B (de la Rosa et al. 2019). These are flavones, flavanols, flavanones, flavonols, isoflavones, and anthocyanidins.

Flavones are an important subgroup of flavonoids where substituents are linked on 4′, 5, and 7 C positions with the absence of –OH group in 3-C position where a ketone group exists in 4-C position. Most of them have –OH group at 5-C of A position (Harborne 1980; Katerova et al. 2012). The basic structural difference of flavanols is from flavones having –OH group at 3-C position. Unlike flavones, flavanones have no double bond in between 2-C and 3-C positions in structure but hold the ketone group (Katerova et al. 2012; Panche et al. 2016). Flavonols contain a ketone group in 4-C position as like flavones but unlikely there is a double bond in between 2-C and 3-C position with –OH group at 3-C (Katerova et al. 2012; Tsimogiannis and Oreopoulou 2019).

Isoflavones are mostly similar to flavones except the attachment of B ring at 3-C instead of 2-C of C ring (Panche et al. 2016). They act as a precursor for the phytoalexins development during plant–microbe interactions (Aoki et al. 2000; Panche et al. 2016).

Anthocyanidins structurally contain pyrylium (+ charge) in the C ring included two double bonds. Mostly, C-ring holds additional 3-OH substitution (Tsimogiannis and Oreopoulou 2019). Up to present, identified anthocyanidins in nature are roughly 300 together with their glycosides which are water-soluble and called anthocyanins (Katerova et al. 2012; Panche et al. 2016). These renowned anthocyanins are color pigments responsible for making colors in leaves, flowers, and fruits even the roots of plants.

3 Synthesis of Phenolic Compounds in Plants under Optimal and Stressful Environments

All PCs in plant are usually synthesized by similar biosynthetic pathways; shikimate pathway, phenylpropanoid pathway, and flavonoid pathway (Jiang et al. 2013; Naikoo et al. 2019). However, polyketide malonate/acetate pathways synthesize PCs are mostly of simple phenols (Cheynier et al. 2013). The pathways of PCs biosynthesis in plant (Figs. 2–4) have been illustrated in the following section.

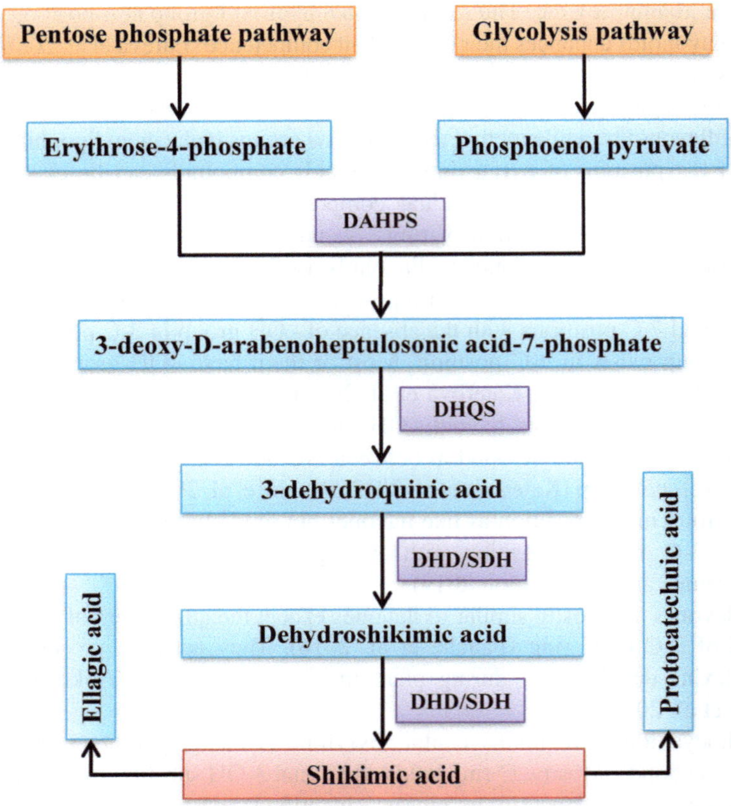

Fig. 2 Shikimic acid pathway of PC biosynthesis. *DAHPS* 3-deoxy-D-arabinoheptulosonate 7-phosphate synthase, *DHQS* 3-dehydroquinate synthase, *DHD* 3-dehydroquinate dehydratase, *SDH* shikimate dehydrogenase

3.1 Shikimate Pathway

The biosynthesis of PC is initiated by shikimate pathway by using intermediate metabolites phosphoenol pyruvate (PEP) and erythose-4-phosphate to produce phenylalanine, while erythose-4-phosphate and PEP are the products of pentose phosphate pathway and glycolysis pathway, respectively (Cheynier et al. 2013; Naikoo et al. 2019). The combination of erythose-4-phosphate and PEP produces 3-deoxy-D-arabinoheptulosonic acid-7-phosphate (DAHP) with the help of DAHP synthase, which turns into shikimic acid after reduction (Dias et al. 2016). Some PCs such as protocatechuic acid and ellagic acid can synthesize directly from shikimic acid (Fig. 2) and also lead to the formation of phenylalanine (Cohen and Kennedy 2010; Dias et al. 2016).

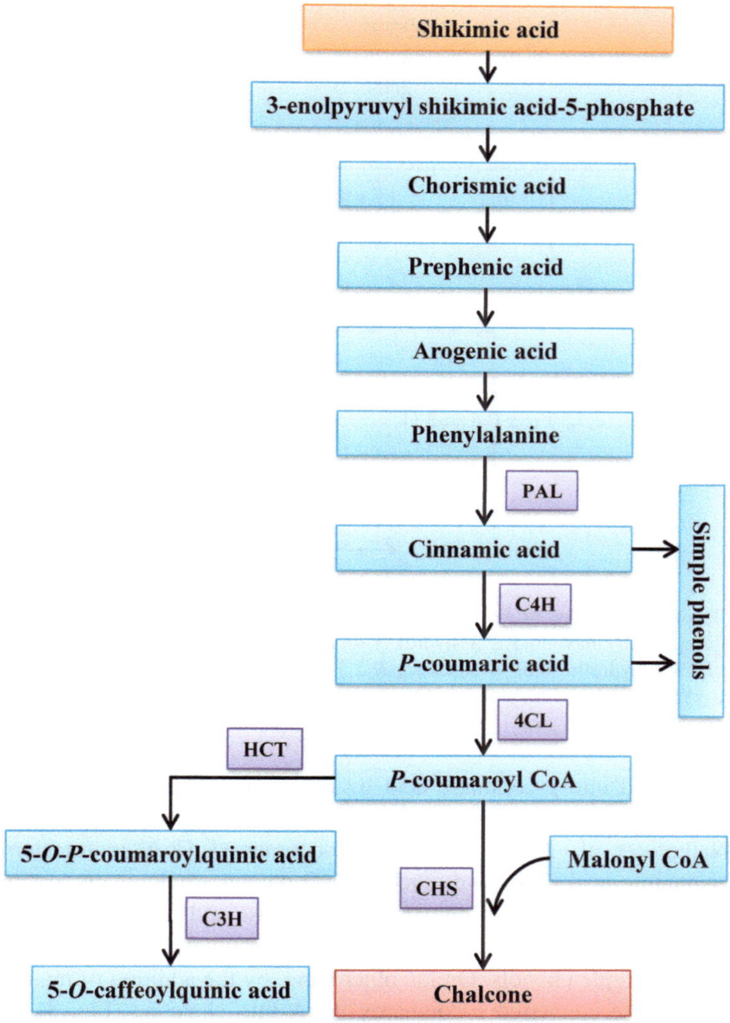

Fig. 3 Phenylpropanoid pathway of PC biosynthesis. *PAL* Phenylalanine ammonia-lyase, *C4H* cinnamate-4-hydroxylase, *4CL* 4-coumaroyl CoA-ligase, *HCT* hydroxycinnamoyl transferase, *C3H* p-coumarate-3-hydroxylase, *CHS* chalcone synthase

3.2 Phenylpropanoid Pathway

In the beginning of phenylpropanoid pathway, shikimic acid is converted into phenylalanine through the production of some intermediate compounds such as 3-enolpyruvyl shikimic acid-5-phosphate, chorismic acid, prephenic acid, and arogenic acid (Dias et al. 2016). By the influence of phenylalanine ammonia-lyase, phenylalanine produces the PC cinnamic acid. The complex compound cinnamic

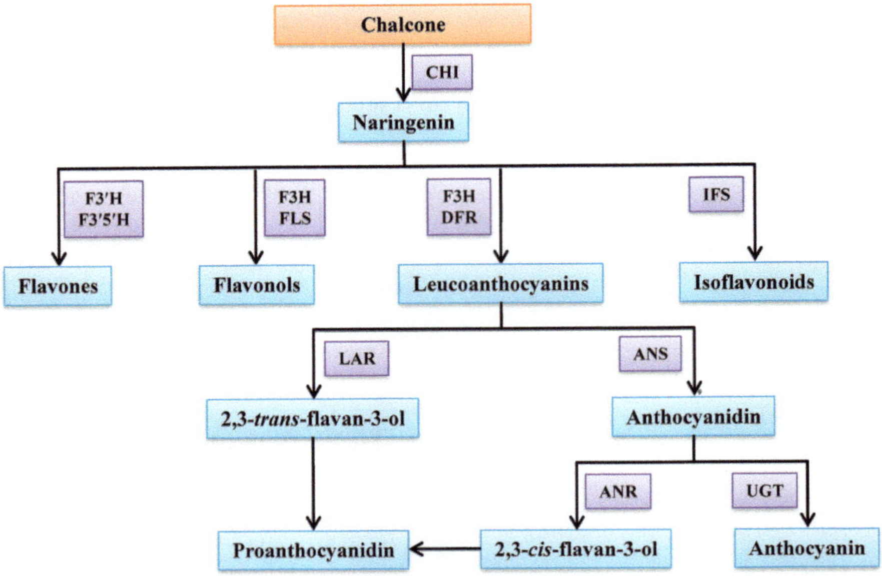

Fig. 4 Flavonoid pathway of PC biosynthesis. *CHI* chalcone isomerase, *F3′H* flavonoid 3-′-hydroxylase, *F3′5′H* flavonoid 3′,5′-hydroxylase, *F3H* flavanone 3-hydroxylase, *FLS* flavonol synthase, *DFR* dihydroflavonol 4-reductase, *IFS* isoflavone synthase, *LAR* leucoanthocyanidin reductase, *ANS* anthocyanidin synthase, *ANR* anthocyanidin reductase, *UGT* uridine diphosphate glycosyltransferase

acid is converted into *p*-coumaric acid by cinnamate-4-hydroxylase and leads to production of *p*-coumaroyl CoA by 4-coumaroyl CoA-ligase activity (Jiang et al. 2013; Naikoo et al. 2019). Therefore, *p*-coumaroyl CoA is transferred into hydroxycinnamic acid derivatives such as 5-*O*-*p*-coumaroylquinate acid and 5-*O*-caffeoylquinic acid by hydroxycinnamoyl transferase and *p*-coumarate-3-hydroxylase, respectively. In addition, cinnamic acid and *p*-coumaric acid are responsible to produce some simple phenols. On the other hand, by the addition of malonyl CoA, chalcone synthase converted *p*-coumaroyl CoA into chalcone (Fig. 3) (Cheynier et al. 2013; Naikoo et al. 2019).

3.3 *Flavonoid Pathway*

Chalcone is the primary compound for flavonoid pathway, where chalcone is converted into naringenin by chalcones isomerase activity. Therefore, naringenin is converted into various forms of PCs such as flavones, flavonols, anthocyanidin, anthocyanin, proanthocyanidins, and so on by the activity of different enzymes and among them anthocyanidin synthase, dihydroflavonol reductase, flavone synthase, flavonol synthase, flavanone 3-hydroxylase, isoflavone synthase, anthocyanidin reductase, and leucoanthocyanidin reductase are the most important (Fig. 4) (Cheynier et al. 2013; Jiang et al. 2013).

4 Metabolism of the Phenolic Compound under Abiotic Stresses

Although PCs synthesis in plants is endogenously controlled process, it is also regulated by external environmental factors including light, temperature, and wounding (Crozier et al. 2006). These compounds act as signaling molecules and protect plants from environmental stress-induced ROS generation and thus stimulating stress tolerance. Higher biosynthesis of polyphenols is one of the plants responds toward abiotic stresses to enhance the evolutionary fitness. Indeed, these accumulated phenolics give plant higher tolerance against abiotic stresses including salinity, drought, heavy metal, temperature, pesticides, and UV radiations (Fig. 5; Naikoo et al. 2019; Sharma et al. 2019). Thereafter, plants have the ability to enhance PCs biosynthesis upon adverse environmental growing conditions (Selmar 2008). As PCs are nonenzymatic antioxidants and that is why capable of reducing cell membrane peroxidation by detoxifying ROS as well as protects plants from oxidative stress (Sharma et al. 2019).

Activities of phenylalanine ammonia lyase (PAL), chalcone synthase (CHS), and other enzymes are responsible to PCs accumulation (Naikoo et al. 2019). Increased enzymatic activities are accompanied with the upregulation of the transcript levels of genes of key biosynthetic enzymes such as PAL, C4H (cinnamate 4-hydroxylase), 4CL (4-coumarate: CoA ligase), CHS, CHI (chalcone isomerase), F3H (flavanone3-hydroxylase), F30H (flavonoid 30-hydroxylase), F3050H (flavonoid 3050-hydroxylase), DFR (dihydroflavonol 4-reductase), FLS (flavonol synthase), IFS (isoflavone synthase), IFR (isoflavone reductase), and UFGT (UDP flavonoid glycosyltransferase) under various abiotic stresses (Fig. 6) (Ma et al. 2014; Leng et al. 2015; Zhou et al. 2018; Chen et al. 2019c; Gharibi et al. 2019; Sharma et al. 2019).

4.1 *Salinity*

Upon salt stress, plants showed higher accumulation of PCs as a part of antioxidant defense and thus scavenged stress-induced toxic ROS (Bistgani et al. 2019; Chen et al. 2019a). This is supported by the salt-induced stimulation in phenylpropanoid biosynthetic pathway (Rossi et al. 2016; Al-Ghamdi and Elansary 2018; Bistgani et al. 2019). It is reported that genes of flavonoid biosynthetic pathways including *VvbHLH1* are regulated to enhance flavonoid production as well as for higher tolerance (Wang et al. 2016; Golkar and Taghizadeh 2018). Therefore, salt stress

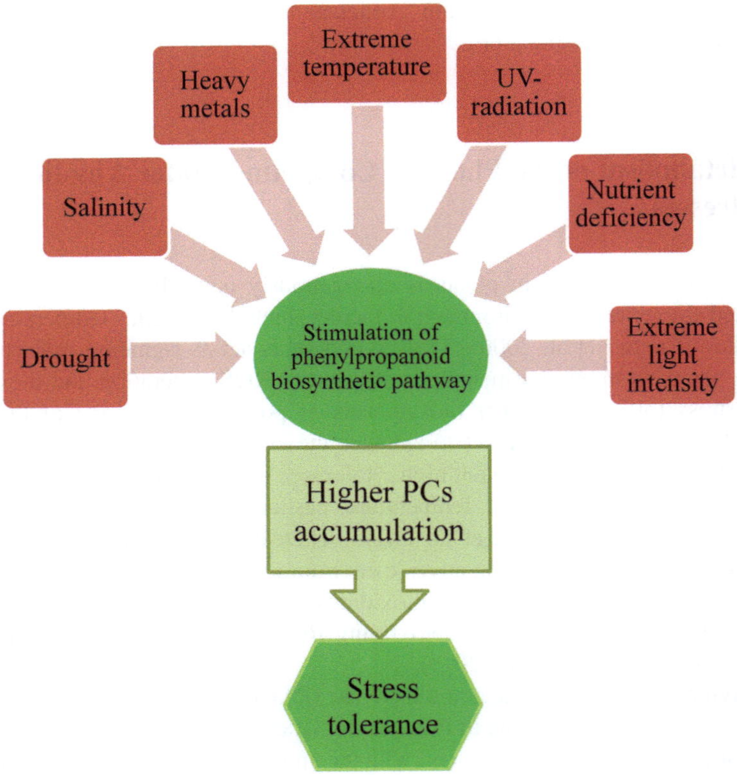

Fig. 5 Abiotic stress-induced responses in the status of PCs production. Various environmental stresses cause the stimulation in phenylpropanoid biosynthetic pathway due to higher activities of responsible enzymes toward accumulation of higher PCs for attuning higher tolerance

showed the overexpression of *NtCHS1* genes for flavonoids accumulation in tobacco (Chen et al. 2019b). Upon salinity, flavone biosynthesis also improved in *Glycine max* along with upstimulation of flavones synthase genes expression including *GmFNSII-1* and *GmFNSII-2* (Yan et al. 2014). *Fragaria ananassa* showed the increment in the transcript levels of *PAL, C4H, F3H, DFR,* and *FLS* under salt stress (Perin et al. 2019). Consequently, various PC accumulated in *Thymus* spp. under saline conditions including gallic acid, rosmarinic acid, cinnamic acid, chlorogenic acid, rutin, and quercetin (Bistgani et al. 2019) while anthocyanins in *Abelmoschus esculentus* (Dkhil and Denden 2012). The contents of total phenolics (as hydroxybenzoic acids—gallic acid, vanillic acid, syringic acid, ellagic acid, *p*-hydroxybenzoic acid, and as hydroxycinnamic acids—caffeic acid, chlorogenic acid, *p*-coumaric acid, m-coumaric acid, ferulic acid, sinapic acid, trans-cinnamic acid) and flavonoids—iso-quercetin, hyperoside, rutin were increased in *Amaranthus tricolor* under saline conditions (Sarker and Oba 2018a). Similarly, salinity increased different kinds of PCs in different crop species including *Asparagus aethiopicus* (Al-Ghamdi and Elansary 2018), *Chenopodium quinoa* (Aloisi

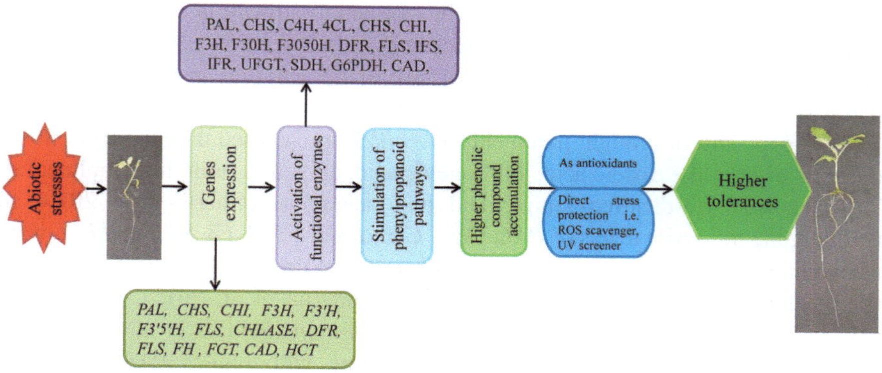

Fig. 6 Plant PCs metabolism under abiotic stresses conditions. Environmental stresses cause upstimulation of key genes and associated enzymes of phenolic biosynthetic pathways which resulted in higher accumulation of different PCs. These PCs later act as antioxidant and/or direct stress scavenger (*CHS* chalcone synthase, *CHI* chalcone isomerase, *F3′H* flavonoid 3′-hydroxylase, *F3′5′H* flavonoid 3′,5′-hydroxylase, *F3H* flavanone 3-hydroxylase, *FLS* flavonol synthase, *IFS* isoflavone synthase, *PAL* phenylalanine ammonia-lyase, *C4H* cinnamate-4-hydroxylase, *4CL* 4-coumaroyl CoA-ligase, *HCT* hydroxycinnamoyl transferase, *F30H* flavonoid 30-hydroxylase, *F3050H* flavonoid 3050-hydroxylase, *DFR* dihydroflavonol 4-reductase, *IFR* isoflavone reductase, *UFGT* UDP flavonoid glycosyltransferase, *HCT* hydroxycinnamoyl transferase, *SDH* shikimate dehydrogenase, *CAD* cinnamyl alcohol dehydrogenase, *G6PDH* glucose-6-phosphate dehydrogenase, *CHLASE* forcholrophyllase, *HCT* hydroxycinnamoyl transferase, *FGT* flavonoid glycosyltransferases)

et al. 2016), *Hordeum vulgare* (Ma et al. 2019), *Solanum lycopersicon* (Martinez et al. 2016), *Triticum aestivum* (Kaur and Zhawar 2015).

Salt-induced excessive ROS in plants needs active participation of antioxidants to be scavenged for gaining higher tolerance. Consequently, the total phenolic and flavonoid contents reduced in salt exposed *Oryza sativa* because of elevated ROS and showed their antioxidants properties (Rahman et al. 2016).

4.2 Drought

Accumulation of PCs in plants under drought stress provides tolerance. *Arabidopsis* showed higher drought tolerance by higher flavonoid accumulation (Nakabayashi et al. 2014). Upon drought stress, stimulation of the flavonols biosynthesis was occurred along with higher accumulation in *Trifolium repens* which was correlated with better resistance (Ballizany et al. 2012). Drought-induced upregulation of PCs synthesis plays crucial roles in antioxidant defense system as these are nonenzymatic in nature and thus protect plants growth (Nichols et al. 2015; Rezayian et al. 2018). Drought-stressed *S. lycopersicon* tolerance was regulated by higher flavonoids content including quercetin and kaempferol (Sanchez-Rodriguez et al. 2011).

Modulation of the phenylpropanoid biosynthetic pathway is considered as the main reason for drought-mediated enhancement in PCs accumulation (Sharma et al. 2019). Where main responsible genes of enzymatic activities in phenylpropanoid pathway were also regulated to result in the stimulation of PCs biosynthesis. About 21 days exposure of drought treatment to *Achillea* spp. caused the increment in the contents of PCs like caffeic acid, chlorogenic acid, apigenin, rutin, luteolin luteolin-7-O-glycoside, 1,3-dicaffeoylquinic acid, and kaempferol (Gharibi et al. 2019). In addition, the transcript levels of *PAL, CHS, CHI, F3H, F3'H, F3'5'H, and FLS* were also enhanced there. Similar results were observed in drought-stressed *F. ananassa* (Perin et al. 2019), *Lotus japonicas* (Garcia-Calderon et al. 2015), *T. aestivum* (Ma et al. 2014), and in *Chrysanthemum morifolium* (Hodaei et al. 2018). The increased activity of PAL enzyme activity along with higher *PAL* expression enhanced the drought tolerance in *Brassica napus* (Rezayian et al. 2018). Drought stress caused the increment in the polyphenols contents like flavonoids, proanthocyanidins, and flavonols in *Larrea* spp. (Varela et al. 2016).

4.3 Heavy Metals

Heavy metals regulate the activities of PAL, shikimate dehydrogenase (SDH), glucose-6-phosphate dehydrogenase (G6PDH), and cinnamyl alcohol dehydrogenase (CAD), which results in the motivation of the phenylpropanoid biosynthetic pathway in plants depending on transcript levels of responsible genes of these enzymes (Chen et al. 2019b; Mishra and Sangwan 2019). Thus, metal-induced higher phenolics contents enhance the plants tolerances by protecting them from oxidative damage (Kohli et al. 2018; Handa et al. 2019). Among PCs, flavonoids act as metals chelator and thereby able to decrease the hydroxyl radical in plant cells as their characteristics defense responses (Mira et al. 2002; Williams et al. 2004). Consequently, excessive metal exposure caused a higher accumulation of flavonoids in plants (Kaur et al. 2017; Handa et al. 2019). The metals stress like copper (Cu), lead (Pb), and cadmium (Cd) enhanced the PCs including chlorogenic acid and rutin accumulation in *Zea mays* as defense tactics (Kısa et al. 2016). Likely, *B. juncea* showed the increment in total phenol, flavonoids, and anthocyanins levels by 44%, 45%, and 42%, respectively supported by elevated chlorophyllase (*CHLASE*), *PAL*, and *CHS* expressions under chromium (Cr) toxicity (Handa et al. 2019). Upon exposure to aluminum (Al, 50 μM) stress, 27% higher total phenolic content was recorded in *Fagopyrum esculentum* while flavonoid and anthocyanin contents were also increased with threefold higher PAL activity (Smirnov et al. 2015).

4.4 Extreme Temperature

Temperature stress including both high and low enhances the production of PCs like phenolic acids, flavonoids, flavonols, anthocyanins, and so on to protect plants (Ancillotti et al. 2015; Wang et al. 2019b). Accumulation of PCs in *Festuca trachyphylla* under long heat exposure of 38/33 °C was evaluated (Wang et al. 2019a). In this study, phenolic acid, including caffeic acid, ferulic acid, salicylic acid, 4-hydroxybenzoic acid, 3,4-dyhydroxybenzoic acid, benzoic acid, cinnamic acid, gallic acid, coumaric acid, homovanillic acid, and vanillic acid, was significantly increased. In addition, about 118% and 117% increased content of homovanillic and caffeic acid, respectively, was observed under short-term heat stress of 7 hours, while 3,4-dyhydroxybenzoic acid accumulated greatly upto 214% under 21 days of heat exposure. These increments of PCs in *F. trachyphylla* were associated with higher tolerances. Similarly, heat stress (44 °C) enhanced the caffeic acid, coumaric acid, and anthocyanins biosynthesis in *Daucus carota* for attaining oxidative stress tolerance (Commisso et al. 2016).

Similarly, exposure to low-temperature stress (15/10 °C) of *Capsicum annuum* plants caused the higher accumulation of phenolic acid named chlorogenic acid and flavonoids including luteolin-7-O-glucoside and apigenin-7-O-glucoside (León-Chan et al. 2017). On the other hand, chilling stress enhanced the accumulation of PCs in cell wall like suberin and lignin which act in increasing plant resistance (Griffith and Yaish 2004). Recently, Naikoo et al. (2019) supported this by disclosing the nature of PC in increasing cell wall thickness for protecting plant from chilling injury and consequent cell collapse. Therefore, low temperature-induced stimulation in phenolic biosynthesis through increasing the expression of *PAL, CAD,* and hydroxycinnamoyl transferase (*HCT*) is associated with higher PCs accumulation in plants for better tolerances (Zhou et al. 2018).

4.5 UV Radiation

Plants counteract the UV toxicity and protect cellular machineries by cellular phenolics-induced shield just below the epidermal layer (Sharma et al. 2019). In addition, as flavonoids possess the ability to absorb UV radiations then hence act as light screen under harmful UV radiation and thus defending plants growth (Lattanzio 2013; Sharma et al. 2019).

It is reported that the mechanism of plants to increase flavonoids biosynthesis under UV radiations for attaining higher tolerances by acting as potential antioxidant (Agati et al. 2012). This is might be one of the reasons for higher production of phenolics specially flavonoids in plants of higher altitude than temperate zones. Upon excessive UV condition, some key genes such as *PAL*, *CHS*, *CHI*, *FLS*, *DFR*, *FHT* (flavanone 3-hydroxylase), *FGT* (flavonoid glycosyltransferases) are up-stimulated, which later causes the activation of phenolics biosynthesis pathways

(Goyal et al. 2014; Ghasemi et al. 2019). Thus, higher flavonoids accumulations in plants under UV stress are the cumulative responses of corresponding genes regulations along with the stimulation of flavonoids production pathways.

5 Phenolic Compounds as Stress Biomarker

Abiotic stresses causes difficulties in plant survival by causing extreme ROS accumulation followed by inhibition in growth and physiological metabolisms. Moreover, acute ROS is also potential stress marker to evaluate stress intensity. In addition, the role of PCs as nonenzymatic antioxidants in plant defense system bestows a rationale to be considered as a stress marker. Different plant species under different kinds of stresses showed the higher or lower accumulation of phenolics, which might be due to stress intensity and hence showed scavenging of ROS (Rahman et al. 2016).

Salinity reduced the PCs including phenol and total flavonoids in rice owing to overproduced ROS where again accumulated in higher upon manganese-mediated lowered ROS contents (Rahman et al. 2016). Drought-affected plants showed higher accumulation of total flavonoids and phenylpropanoids in correspondence with higher ROS production (Sarker and Oba 2018b). This increment was dependent on water stress severity and varietal tolerance levels and maximum increase was recorded from 25% field capacity (FC)-induced drought than 80% and 50% FC. The flavonoids contents significantly increased as protector against UV-radiation stress to alleviate the damage (Huang et al. 2016). Phenolic compound production is the oxidative action of toxic metal including Cr, Pb, Ni, Cd, and acts in alleviation of metal-mediated oxidative damage indicated by lipid peroxidation inhibition due to the alkoxyl radicals trapping as well as suppressing oxidizing enzymes activities (Shahid et al. 2018). Plants outfit with higher phenolics accumulation as secondary metabolites to cope with Cr-induced excessive generation of ROS (Shahid et al. 2018). The study of Shahid et al. (2018) reported that the production of phenolic acids enlisted as *p*-hydroxybenzoic acid (PHBA), ferulic acid (FA), *p*-coumaric acid (PCA), gallic acid (GLA), and vanillic acid (VA) increased in Kinnow mandarin plant as one of the mechanism of stress defense and proposed as an indicator of stress tolerance. High temperature of 39/30 °C increased the endogenous FA production in *Vaccinium corymbosum*.

6 Phenolic Compound-Induced Regulation of Plant Stress Tolerances

There is an interlink between abiotic stress and endogenous PCs accumulation in plants toward tolerance and survival. Upon various environmental stresses, plants enhance PCs contents, which results in higher tolerances by stimulating various mechanisms. Therefore, exogenous application of PCs to attain higher plant stress tolerance is getting priority among plant scientists for increasing plants growth under stress. Here, we have summarized the role of exogenous PCs-mediated plant tolerances to different abiotic stresses based on available latest literatures (Table 1).

Different kinds of phenolic acid are very potential to increase the plant antioxidants capacity as well as growth under stress conditions. Salicylic acid (SA) is one of the most popular hydroxybenzoic types of phenolic acid by showing its potential role in plant growth and developments like photosynthesis, enzymatic activity, minerals contents, stress tolerance (Ahanger et al. 2020). It is established that SA potentially strengthens the plant immune systems and tolerance under environmental stresses including salinity (Nazar et al. 2011), drought (Chavoushi et al. 2020), metal (Ahmad et al. 2011; Moravcová et al. 2018), extreme temperature, and so on. Upon salt stress, SA caused higher osmolyte accumulation, inhibition in toxic Na^+ and Cl^- uptake with higher essential minerals accumulations, stimulation in antioxidants potentiality, and reduction of ROS generation (Khan et al. 2014; Ahanger et al. 2020). In addition, SA increased photosynthetic pigments contents and thus improved photosynthesis rate, which later resulted in higher growth and biomass production in salt-stressed *Vigna angularis* (Ahanger et al. 2020). Exogenous application of 200 ppm SA enhanced growth, Chl contents, water status, and tolerance index in drought-stressed *Ocimum basilicum,* which is sensitive to drought (Damalas 2019). In this study, SA alleviated the adverse effects of osmotic stress under 40% FC of soil moisture by increasing Pro and Chl synthesis which might be the reasons for increasing tolerance index and growth indicated as plant height, biomass production. Salicylic acid also showed higher plant tolerance to metal toxicity including Ni (Zaid et al. 2019), Cd (Ahmad et al. 2011), Cu (Moravcová et al. 2018). Upon metal stresses, SA reduced the metal uptake, increased essential minerals accumulation, while increased the photosynthetic pigments content with higher stomatal conductance, intercellular CO_2, and water use efficiency (Zaid et al. 2019). Moreover, SA showed the higher growth and tolerance of *Arabidopsis thaliana, Capsicum annuum,* and *T. aestivum* upon extreme light (Yang et al. 2019), heat and cold stress (Ignatenko et al. 2019), respectively. These tolerances were evaluated by higher Pro, soluble sugar accumulation, and cellular water status with lower cell death.

Foliar application of *p*-hydroxybenzoic acids alleviated the damaging effects of water shortage in *Oryza sativa* (Quan and Xuan 2018) by increasing Chl *a*, Chl *b*, Car contents with lowered leaf rolling, and drying damage. Here, endogenous phenolic and flavonoid contents were also increased in *p*-hydroxybenzoic acids treated plants under drought stress. In another study, 24 h pretreated *O. sativa* by

Table 1 Role of phenolic compounds as a potential stress marker

Crops	Abiotic stress	Phenolic compounds	Status in plant cells	References
Glycine max L.	Salinity	Anthocyanins, flavonoid	Decreased	Simaei et al. (2012)
Oryza sativa L.	Salinity	Phenol, flavonoid	Decreased	Rahman et al. (2016)
Brassica oleracea L. *var. acephala*	Salinity	Ferulic acid	Increased	Linić et al. (2019)
B. rapa ssp. *Pekinensis; B. oleracea* var. *capitata*	Salinity	Caffeic, salicylic, and 4-coumaric acid	Decreased	Linić et al. (2019)
O. sativa L.	Salinity	Total flavonoids	Decreased by 22% in shoots and 41% increased in roots	Mekawy et al. (2018)
Amaranthus tricolor L.	Drought	Flavonoids and phenylpropanoids	Increased up to 42% and 169%, respectively	Sarker and Oba (2018b)
Fagopyrum tataricum L.	UV radiation	Rutin and quercetin	Increased up to 9%	Huang et al. (2016)
Triticum aestivum L.	UV radiation	Total phenolic content including ferulic acid, *p*-coumaric acid, vanillic acid	Increased	Chen et al. (2019c)
Kinnow mandarin (*Citrus nobilis* Lour × *Citrus deliciosa* ten)	Cr toxicity	*p*-hydroxybenzoic acid, ferulic acid, *p*-coumaric acid, gallic acid, and vanillic acid	Increased	Shahid et al. (2018)
Kandelia obovata	Cd, Zn toxicity	Total phenolic compound	Increased upto 45% in roots and 127% in shoots	Chen et al. (2019b)
T. aestivum L.	Cu, Cd, Hg, Pb toxicity	Total phenolic acid including protocatechuic, *P*-hydroxybenzoic, caffeic, vanillic, syringic, *P*-coumaric, sinapic, and ferulic acid	Increased	Colak et al. (2019)
Vaccinium corymbosum L.	Drought	Vanillic acid	Increased	An et al. (2019)
Carthamus tinctorius L.	Drought	Anthocyanin, phenol, rutin, quercetin, luteloin, apigenin	Increased about 1.2–2.3 times	Chavoushi et al. (2020)

(continued)

Table 1 (continued)

Crops	Abiotic stress	Phenolic compounds	Status in plant cells	References
Deschampsia antarctica	Waterlogging	Total phenolic	Decreased by 20%	Park and Lee (2019)
Zea mays L.	Waterlogging	Phenol	Decreased	Jaiswal and Srivastava (2016)
Coffea arabica L.	Cold	Total phenolic content and anthocyanin content	Increased by 8% and 21%, respectively	Acidri et al. (2020)
Festuca trachyphylla (Hack.) Krajina	High temperature	Ferulic acid, caffeic acid, syringic acid, coumaric acid, vanillic acid, 3,4-dyhydroxybenzoic acid gallic acid, cinnamic acid, benzoic acid, and, 4-hydroxybenzoic acid	Increased about 31–2% individually	Wang et al. (2019a)
V. corymbosum L.	High temperature	Ferulic acid	Increased about 2%	Cheng et al. (2018)

protocatechuic acid boosted up the survival capacity upto 6 days of submergence conditions (Xuan and Khang 2018). Protocatechuic acid enhanced the endogenous phenolic, flavonoid, and protocatechuic acid accumulations which were correlated with higher plant growth and photosynthetic pigment levels (Table 2).

However, tolerance of *O. sativa* increased to 15 days of drought (Quan and Xuan 2018) and 6 days of submergence stress (Xuan and Khang 2018) by VA at 50 μM and 1 mM, respectively. In both of these cases, endogenous phenolic acid and flavonoids contents were increased in VA-treated plants with higher Chl and car synthesis which resulted in higher plant growth and survival under stress conditions. Again, 2 days pretreatment of VA prompted *Vaccinium corymbosum* tolerance to drought (An et al. 2019). Application of VA in this study showed the higher water content, osmotic potential with improved Pro, and soluble sugars in drought-treated plants. It was suggested that exogenous VA-mediated higher endogenous VA accumulation caused the higher plant tolerances in *V. corymbosum.*

Ferulic acid—hydroxycinnamic type of phenolic acid protected the *Cucumis sativus* plants from dehydration toxicity imposed by 10% of polyethylene glycol (PEG) (Li et al. 2013). Ferulic acid was applied as pretreatments for 2 days of duration to plants prior to drought stress which showed the tolerances of stressed *C. sativus* plants by increasing endogenous FA content. In addition, FA pretreatment increased the water status, Pro, and soluble sugar contents in stressed plants, which resulted in higher plant growth. Recently, FA was evaluated in mitigating heat stress in *V. corymbosum* (Cheng et al. 2018). Supplementation of FA caused higher water

Table 2 Phenolic compound-induced plant responses under abiotic stresses

Exogenous phenolic compounds application	Plant species	Abiotic stresses	Plant tolerance responses	References
Rutin; 2, 4 mM, as pretreatment	*Agrostis stolonifera* L.	35 °C; 49 days	• Improved turf quality • Increased chlorophyll (Chl) content	Merewitz and Liu (2019)
Ferulic acid (FA); 0.6 mM, as pretreatment (1 day)	*Vaccinium corymbosum* L.	35/30 °C; 3 days	• Increased relative water content (RWC) and decreased osmotic potential • Enhanced proline (Pro) and soluble sugar contents • Improved endogenous FA content	Cheng et al. (2018)
FA: 0.5 mM, as pretreatment (2 days)	*Cucumis sativus* L.	10% PEG, 2 days	• Increased fresh weight (FW) and dry weight (DW) of leaf, shoot, and root • Improved RWC to control level enhanced endogenous ferulic acid, Pro, and soluble sugar contents	Li et al. (2013)
Salicylic acid (SA), 250 μM	*Carthamus tinctorius* L.	25% of field capacity	• Increased shoot and root length with FW and DW of roots • Lowered free Pro content	Chavoushi et al. (2019)
SA, 0.3 mM	*Arabidopsis thaliana* L.	High light, 3 h	• Increased Pro and soluble sugar at 17% and 46%, respectively • Lowered cell death with less cellulose deposition • Regulated stomatal closure and increased stomatal conductance (gs)	Yang et al. (2019)
SA,0.1 mM, as foliar spray	*Capsicum annuum* L.	35 °C, 5 days	• Enhanced survival and growth rate with reduction of oxidative damage • Lessened the area and degree of chlorosis • Reduced leaf electrical conductivity and increased RWC and root vigority • Increased Chl and Pro contents • Enhanced soluble sugar content	Zhang et al. (2020)

(continued)

Table 2 (continued)

Exogenous phenolic compounds application	Plant species	Abiotic stresses	Plant tolerance responses	References
SA, 100 μM, as pretreatment (1 days)	*Triticum aestivum* L.	4 °C, 7 days	• Increased cold tolerance during whole period of stress • Elevated Pro content	Ignatenko et al. (2019)
SA, 1 mM as foliage spray	*Vigna angularis*	100 mM NaCl, 20 days	• Enhanced shoot length and plant dry weight • Increased total Chl and carotenoid (Car) contents • Improved net photosynthesis (Pn), stomatal conductance (gs), intercellular CO_2 (Ci), transpiration (E), and PSII activity (Fv/Fm) • Augmented Pro, glycine betaine (GB), sugar, and RWC significantly • Lowered Na^+ and Cl^- contents with higher N, K, and Ca contents	Ahanger et al. (2020)
Vanillic acid (VA), 40 μM, as pretreatment (2 days)	*V. corymbosum* L.	10% (w/v) polyethylene glycol (PEG), 2 days	• Increased RWC and osmotic potential significantly • Improved endogenous VA content • Enhanced soluble sugar and Pro contents	An et al. (2019)
VA, 25 and 50 μM, as pretreatment (2 days)	*Oryza sativa* L.	Withholding water; 15 days	• Enhanced Chl *a*, *b*, and carotenoids (Car) contents • Promoted phenolic and flavonoids contents including vanillin and *p*-hydroxybenzoic acid	Quan and Xuan (2018)
VA, 1 mM, as pretreatment (24 h)	*O. sativa* L.	Submergence; 3 and 6 days	• Increased shoot height and survival percentage • Enhanced Chl *a* and *b* contents • Improved total phenolics and flavonoid contents with higher endogenous VA but reduced the protocatechuic acid content	Xuan and Khang (2018)

(continued)

Table 2 (continued)

Exogenous phenolic compounds application	Plant species	Abiotic stresses	Plant tolerance responses	References
p-hydroxybenzoic acids, 25 and 50 μM, as pretreatment (2 days)	*O. sativa* L.	Withholding water; 15 days	• Protected leaf rolling and drying • Enhanced Chl *a*, *b*, and Car contents • Promoted phenolic and flavonoids contents	Quan and Xuan (2018)
Quercetin (Qu), 25 μM as co-treatment	*Solanum lycopersicum* L.	150 mM NaCl; 5 days	• Increased length of shoot and root and their FW and DW • Improved Chl and Car contents • Reduced Na^+ content with higher K^+ accumulation • Enhanced Ca^{2+} and Mg^{2+} contents • Improved RWC content with higher pro accumulation	Parvin et al. (2019)
Qu, 100 μM as co-treatment	*A. thaliana* L.	Paraquat; 0.15 and 0.30 μM	• Increased plant growth indicated by higher relative FW • Enhanced the expansion of leaf	Kurepa et al. (2016)
Qu, 100 μM as co-treatment	*Nicotiana tabacum* L.	Paraquat; 0.5 μM	• Increased plant growth leaf surface • Improved Chl content	Kurepa et al. (2016)
Qu, 100 μM as co-treatment	*Lemna gibba* L.	Paraquat 1 μM	• Increased Chl level	Kurepa et al. (2016)
Coumarine (COU), 50 ppm, as seed priming (4 h)	*T. aestivum* L.	100 and 150 mM NaCl; 14 days	• Enhanced shoot length, FW of shoot and root, DW of shoot and shoot moisture content • Decreased Na^+ accumulation by shoot while increased in roots • Improved K^+ content in both shoot and root tissue with higher K^+/Na^+ • Caused higher accumulation of soluble sugar • Increased Pro accumulation significantly	Saleh and Madany (2015)

(continued)

Table 2 (continued)

Exogenous phenolic compounds application	Plant species	Abiotic stresses	Plant tolerance responses	References
			• Enhanced total phenolics and flavonoids contents along with higher PAL activity • Increased endogenous COU, syringic acid, and VA contents	
Gallic acid, 1, 2 mM, as co-treatment	*Glycine max* L.	5 °C; 72 h	• Increased relative growth rate (RGR) significantly • Enhanced RWC • Increased Pro content by 51%	Ozfidan-Konakci et al. (2019)
Gallic acid, 1, 2 mM, as co-treatment	*G. max* L.	10 °C; 72 h	• Increased RGR significantly • Increased 20% RWC • Improved Pro accumulation	Ozfidan-Konakci et al. (2019)
Caffeic acid, 25 μM, as pretreatment (2 days)	*C. sativus* L.	15/8 °C; 1 day	• Increased fresh and dry weight of leaf, shoot, and roots • Enhanced RWC • Augmented endogenous caffeic and FA contents • Increased Pro and soluble sugar contents	Wan et al. (2015)
Protocatechuic acid, 1 mM, as pretreatment (24 h)	*O. sativa* L.	Submergence; 3 and 6 days	• Improved shoot length and survival percentage • Increased Chl *b* content by 40% • Augmented total phenolics and flavonoid contents with reduction of endogenous protocatechuic acid	Xuan and Khang (2018)
Ellagic acid, 50 ppm, as seed treatment	*Cicer arietinum* L.	PEG-induced osmotic stress (−0.2, −0.4, −0.6 and − 0.8 MPa)	• Improved germination rate • Increased radicle length, FW, and DW of seedlings • Enhanced the level of GB and Pro • Amplified the accumulation of total phenolics and flavonoids	El-Soud et al. (2013)

(continued)

Table 2 (continued)

Exogenous phenolic compounds application	Plant species	Abiotic stresses	Plant tolerance responses	References
			• Enhanced the endogenous ellagic acid, phenylalanine, and shikimic acid contents • Increased phenylalanine ammonia lyase (PAL) and chalcone synthase (CHS) activities	
Apigenin 10 ppm, asseed treatment (24 h)	*O. sativa* L.	50 mM NaCl;14 days	• Enhanced shoot and root length by 12% and 10%, respectively • Enhanced total Chl and Car contents by 12% and 21%, respectively • Increased total flavonoids contents by 73% • Decreased Na^+ accumulation with higher K^+ • Reduced Na^+/K^+	Mekawy et al. (2018)

content by increasing endogenous FA, Pro, and soluble sugar accumulation under 35 °C of high temperature.

Caffeic acid (CA) application significantly reduced the chilling stress-induced injury in *C. sativus* (Wan et al. 2015). Exogenous CA improved leaf, shoot, and root fresh weight and their dry weight of 15 °C treated plants. Suffering of *C. sativus* by reducing water status under chilling stress was also inhibited by CA regulated higher Pro and soluble sugar accumulations. Moreover, exogenous CA-mediated higher endogenous caffeic and FA generation were closely linked with higher chilling stress tolerances (Wan et al. 2015).

It was reported that GLA is potential to improve plants growth and tolerance under chilling stress (Ozfidan-Konakci et al. 2019). Hydroponically grown 21 days old *G. max* plants were co-treated by GLA and chilling stress of 5 and 10 °C for the next 72 h. Therefore, GLA-treated plants showed higher growth rate and water status with augmented Pro content under stress condition. Endogenous phenolic and flavonoids contents were also increased by CA treatment in stressed plants which were strongly correlated with alleviation of stressed damage.

Quercetin (Qu) is flavonol type of flavonoids possesses strong antioxidant properties in controlling physiological and biochemical attributes in plants under adverse growth condition. Exogenous application of Qu inhibited osmotic and ionic stresses by promoting Pro and checking Na^+ uptake with higher K^+ and Ca^{2+}, respectively in *S. lycopersicum* at saline condition (Parvin et al. 2019). Therefore, Qu enhanced the growth of *S. lycopersicum* by increasing plant height, root length, biomass accumulation, Chl and Car contents, and thus showed its role in making plants more tolerant to stress. Moreover, Qu also protected *A. thaliana*, *Nicotiana tabacum*, and *Lemna gibba* from paraquat toxicity and improved their growth and photosynthetic pigments accumulation (Kurepa et al. 2016).

Coumarin (COU) effectively enhanced *T. aestivum* tolerance to salt stress when applied as seed-priming agent (Saleh and Madany 2015). Plants from coumarin treatment increased PAL activity resulted in higher total phenolic and flavonoids including syringic acid, VA, and COU contents in salt-stressed plants. Therefore, COU-treated *T. aestivum* plants showed higher Pro and soluble sugar accumulation with better moisture contents under stress condition. Again, salt-derived ionic toxicity from higher Na^+ accumulation was also inhibited by COU treatment through improving K^+ accumulation and thus finally improved plant growth and biomass contents (Saleh and Madany 2015). Another flavonoid—apigenin recommended for salt stress mitigation in *O. sativa* growth under 50 mM NaCl exposure for 14 days (Mekawy et al. 2018). Apigenin caused higher endogenous flavonoids accumulation in stressed plants which showed better tolerance by increasing plant growth and photosynthetic pigments level. Inhibitory effect of apigenin on Na^+ accumulation with higher K^+ uptakes thus disclosed its mitigating role to salt stress.

Ellagic acid increased seed germination with seedling emergence rate and growth rate of *Cicer arietinum* under water crisis (El-Soud et al. 2013). Drought-induced plants suffering from lower moisture contents were controlled by ellagic acid by improving osmolyte accumulation including Pro and GB. Thus, ellagic acid–mediated drought tolerance symptoms were supported from higher endogenous phenolic compound synthesis.

7 Antioxidant Properties of Phenolic Compounds

Since the discovery of PCs, scientists revealed diverse beneficial effects derived from this group (Fig. 7). Phenolic compounds are well recognized as major determinant of antioxidant potentials of foods and a natural source of antioxidants for human being since long time (Parr and Bolwell 2000). But this antioxidant potential of PC, how does act for plants under stress condition has been being explored recently. In this section, we will arrange and illustrate information regarding the antioxidant properties of PCs and their mode of action in different plant species under stresses (Table 3).

Before going to the experimental evidence let's look through the structural uniqueness of PC which gives it the antioxidant nature. Two main things in structural organization of PC made it an effective antioxidant. First one is the number and position of the hydroxyl groups and second one is the pattern of substitutions on the aromatic rings. Phenolic compounds are concerned in oxidative-reduction reactions and act as machinery of electron transport chains in mitochondria and chloroplasts (Minatel et al. 2017; Babenko et al. 2019). The antioxidant activity of PCs includes

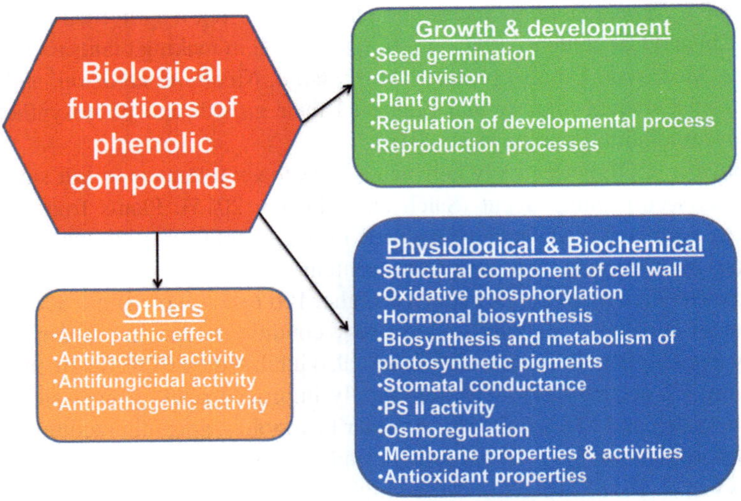

Fig. 7 Biological functions of PCs. Phenolics enhance plant growth and developmental processes as well as involve in diversified physiological and biochemical processes. Antipathogenic and allopathic activities are other notable biological functions of PCs

the capacity to scavenge free radicals, donating hydrogen atoms, electrons, or chelate metal cations (Afanas'ev et al. 1989; Minatel et al. 2017).

The hydroxyl and carboxyl groups of PCs possess high tendency to bind metals specially Fe and Cu and thus may inactivate these metal ions by chelating and later causes the suppression in the superoxide-driven Fenton reaction one of the vital source for ROS (Rice-evans et al. 1997; Arora et al. 1998). In addition, as antioxidant, PCs also directly trapping the lipid alkoxyl radical which resulted in the inhibition in lipid peroxidation depending on the number and position of the hydroxyl groups in their structural phenomenon (Millic et al. 1998). Thereafter, PCs most specifically flavonoids have the capability to stabilize membrane and inhibit the reaction of peroxidation through lowering membrane fluidity and hindering free radical's diffusion (Arora et al. 2000; Blokhina et al. 2003).

They are also able to accumulate at the membrane surface due to their interactivity with membrane phospholipids through hydrogen bonding. Thus, flavonoids significantly prevent the access of deleterious molecules, which are responsible for oxidative damage to the membrane component and consequently maintain the membranes' integrity (Verstraeten et al. 2003).

It is also established that they can directly scavenge ROS by donating electrons or hydrogen atoms (Moran-Palacio et al. 2014) which shows their strong antioxidant action. Naturally, plant composed with antioxidant defense system, comprising with prominent antioxidants which are involved in ROS detoxification as a disciplined manner for keeping this under nontoxic level. Subsequently, it is also mandatory to explore the involvement of PCs in strengthening this system.

Table 3 Phenolic compounds-induced antioxidative responses to alleviate oxidative stress in plants

Stress	Plant species	Exogenously applied phenolic compound	Antioxidative response induced by phenolic compound	References
NaCl; 70 mM	*Glycine max* L.	Caffeic acid, 100 μM	• Improved ROS scavenging ability • Soybean root nodule $O_2^{\bullet-}$ content upregulated • Increased SOD activity • Cu/Zn SOD isoform expression augmented	Klein et al. (2013)
NaCl; 100 mM	*Vigna angularis*	Salicylic acid (SA, 1 mM)	• Increased antioxidant system enzymes' activities: SOD, CAT, APX, DHAR, and GR • AsA, GSH, and tocopherol contents increased • Decreased H_2O_2, $O_2^{\bullet-}$, and LOX activity • Lipid peroxidation and electrolyte leakage decreased	Ahanger et al. (2020)
NaCl; 50, 100, 150 mM	*Triticum aestivum* L.cv. Misr 1	Coumarin (COU, 50 ppm)	• Peroxidase (POD) and total antioxidant capacity (TAC) were upregulated	Saleh and Madany (2015)
NaCl; 50 mM	*Oryza sativa* L. cv. Koshihikari	Apigenin,10 ppm	• Decreased H_2O_2 and lipid peroxidation • Increased CAT and APX activity • Higher accumulation of nonenzymatic antioxidants	Mekawy et al. (2018)
NaCl, 50 mM	Strawberry (*Fragaria* × *ananassa* Duch. cv. 'Gaviota')	SA, 100 and 500 μM	• Diminished lipid peroxidation as well as membrane damage • Elevated SOD, CAT, and APX activity	Samadi et al. (2019)

(continued)

Table 3 (continued)

Stress	Plant species	Exogenously applied phenolic compound	Antioxidative response induced by phenolic compound	References
NaCl, 150 mM NaCl	*Solanum lycopersicum* L. cv. Pusa Ruby	Quercetin (Qu, 15 and 25 μM)	• Higher AsA and GSH content, AsA: DHA and GSH:GSSG ratio resulted • Higher GST, MDHAR, and GR activity was demonstrated • Histochemical detection of oxidative stress markers ($O_2^{\bullet -}$ and H_2O_2) decreased, compared to salt stress • Reduction of H_2O_2 generation and LOX activity • Reduced lipid peroxidation/MDA level and electrolyte leakage • Decreased MG production by increasing Gly I and Gly II activities	Parvin et al. (2019)
Drought,85% soil water content	*O. sativa* L.; varieties: Nep nanhngua Hai phong (Q8) and Re nuoc (Q2),	Vanillic acid (VA) and *p*-hydroxybenzoic acid (PHBA); 25 and 50 μM VA, 25 and 50 μM PHBA	• Increased antioxidant activity or DPPH radical scavenging	Quan and Xuan (2018)
Drought (reduced field capacity water by 25%)	*Carthamus tinctorius* L.	SA, 250 μM	• Increased SOD and CAT activities • Gene expression of two subunits of Fe- and Cu-SOD • Reduced free radicals and lipid peroxidation	Chavoushi et al. (2019)
Dehydration stress (10% (w/v) PEG 6000, 2 days	*Vaccinium corymbosum* L.	VA, 40 μM	• Augmented the transcript levels of genes encoding eight antioxidant enzymes in leaves, including Fe SOD, chloroplast Cu/Zn SOD, cytoplasmic Cu/Zn SOD, CAT, guaiacol peroxidase (GPOX), GPX, GR, and DHAR • Improved activities of SOD, GPX	An et al. (2019)

			and elevated contents of reduced GSH and AsA • Decreased H_2O_2, $O_2^{\bullet-}$, and MDA contents	
Dehydration stress induced by 10% PEG	*Cucumis sativus* L. cv. Yuexiu no. 3	Ferulic acid (FA); 0.5 mM, 2 days	• Relative expression levels of *Cu/Zn-SOD*, *Mn-SOD*, *CAT*, and *GPOX* enhanced • Activities of SOD, CAT, and *GPOX* increased • Content of MDA, $O_2^{\bullet-}$ and H_2O_2 decreased	Li et al. (2013)
Osmotic stress (0, −0.2, −0.4, −0.6 and − 0.8 MPa created by PEG)	*Cicer arietinum* L. (var. Giza 195)	Ellagic acid, 0.001 M	• Enhanced the total antioxidantcapacity (FRAP). • Lessened lipid peroxidation levels (MDA), electrolyte leakage (EL), and H_2O_2. • Increased GSH contents • Raised the activity CAT, SOD, GR	El-Soud et al. (2013)
Cadmium toxicity (Cd; 1.0 and 2.0 mM; 72 h)	*O. sativa* L. cv. BRRI dhan54	VA; 50 μM)	• Increased phytochelatin, biological accumulation factor (BCF), biological accumulation co-efficient (BAC), Cd translocation factor (TF) • Suppressed H_2O_2 content (27%), LOX activity (19%), MDA content, and EL (leaf and root; 44 and 38%, respectively) • Caused the increment in AsA:DHA and GSH:GSSG balance. • Increased APX, MDHAR, DHAR and GR activities by 24%, 46%, 90%, and 26%, respectively • Augmented SOD, CAT, GPX, GST activity by 14%, 39%, 22%, and 70%, respectively	Bhuyan et al. (2020)

(continued)

Table 3 (continued)

Stress	Plant species	Exogenously applied phenolic compound	Antioxidative response induced by phenolic compound	References
Submergence stress	*O. sativa* L. cv. Koshihikari	Protocatechuic acid (PA) and VA, 0.1–1.0 mM	• APX and SOD activities were enhanced • Expression of genes encoding APX and SOD enzymes was favored	Xuan and Khang (2018)
High temperature stress; 40 °C/25 °C	*Capsicum annuum* L.	SA, 0.01 and 0.1 mM	• Increase of POD activity • Upredulated AsA and GSH contents. • Reduced $O_2^{\bullet-}$ scavenging lowered MDA content.	Zhang et al. (2020)
High temperature stress, 39/30 °C, 3 days	*V. corymbosum* L. cv. Bluecrop	FA; 0.3, 0.4, 0.5, 0.6, and 0.7 mM	• Transcriptions of *Fe-SOD*, *Cyt Cu/Zn-SOD*, *Chl Cu/Zn-SOD*, *CAT*, *GPX*, GPOX, *APX* (*G*), *MDHAR* (*H*), *DHAR* (*I*), and *GR* were upregulated • Activities of SOD and GPX increased • Augmented the content of AsA, GSH	Cheng et al. (2018)
High temperature stress, 35 °C	*Agrostis stolonifera* L.	Rutin, 2 mM	• Reduced EL	Merewitz and Liu (2019)
Cold stress, 5 and 10 °C	*G. max* L.	Gallic acid, 1 and 2 mM)	• Reduction of H_2O_2 and TBARS accumulation, thus reduced lipid peroxidation • Increase in the activities of SOD, CAT, and • Decline of MDHAR, DHAR activities and contents of, total ascorbate (tAsA), and GSH.	Ozfidan-Konakci et al. (2019)
Low temperature stress, 4 °C	*T.aestivum* L. cv. Moskovskaya 39	SA, 100 µM	• Boosted activity of SOD, CAT, and GPOX	Ignatenko et al. (2019)

			• Diminished MDA and H_2O_2 accumulation • Improved the levels of *TaMnSOD*, *TaFeSOD,* and *TaCAT* gene transcripts were increased	
Cold stress (15/8 °C)	*C. sativus* L. cv. Jinchun no. 4	Caffeic acid, 25 μM	• Enhanced SOD, GPOX, CAT, GPX, APX, MDHAR, DHAR, and GR activities • Increased transcript levels of Cu/Zn SOD, GPX, and *MnSOD* genes	Wan et al. (2015)
High light stress, 3 h	*Arabidopsis thaliana* L.	SA, 0.3 mM	• Increased the activities of POD, SOD and CAT, APX, GPX, and GR • Increased AsA and GSH, content with reduction of DHA and GSSG contents • Reduced $O_2^{\bullet-}$ and H_2O_2 contentby 18% and 23%, respectively • Decreased MDA and EL by 24% and 18%, respectively	Yang et al. (2019)
Excess boron (B), 4 and 8 mM	*T. aestivum* L.	FA, 25 and 75 μM	• Increased SOD activity. • $O_2^{\bullet-}$ and H_2O_2 increased noticeably. • Increased the activities of CAT and POD. • APX, MDHAR, DHAR, and GR; increased level of tAsA or DHA • REducedlipid peroxidation (TBARS) • Increased radical scavenging activity	Yildiztugay et al. (2019)

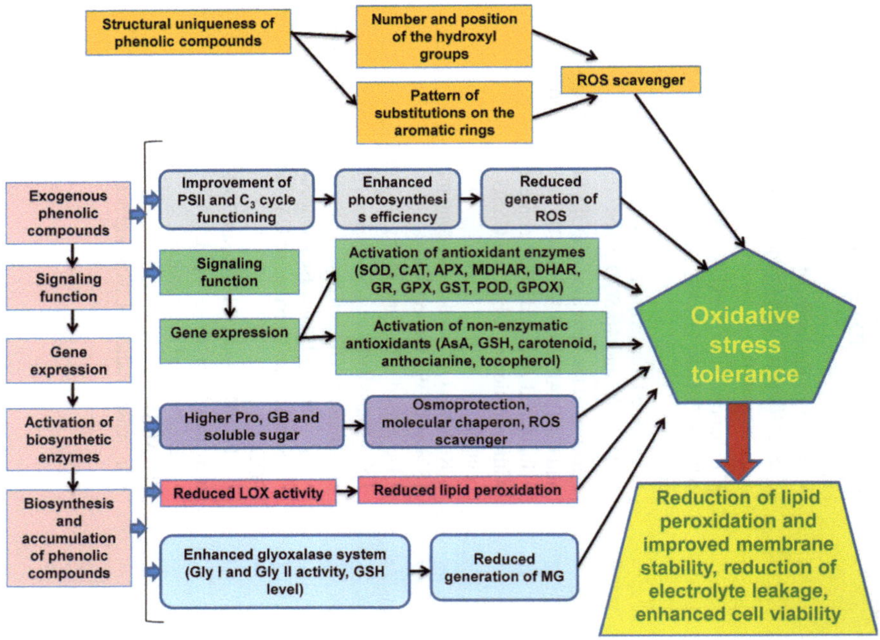

Fig. 8 Antioxidant activities of PCs and other mechanisms to enhance oxidative stress tolerance

The frontline defense comes from superoxide dismutase (SOD) which dismutase $O_2^{\bullet-}$ and turn into more toxic, then can be detoxified directly by catalase (CAT) or through AsA-GSH pool starting from ascorbate peroxidase (APX) activity (Parvin et al. 2019). Hence, instead of APX, H_2O_2 can be scavenged by using phenolics with AsA (Michalak 2006). Therefore, PCs also strengthen the plant antioxidants activities subsequently reduced the ROS generation upon abiotic stress conditions (Parvin et al. 2019).

From the available published articles, the PCs have been presented as ROS detoxifier and antioxidant defense contributor under stress condition (Fig. 8).

Exogenous SA activated the antioxidant defense system in *V. angularis* under salinity (100 mM NaCl) which alleviated oxidative damage by reducing H_2O_2, as a result, malondialdehyde (MDA) and electrolyte leakage (EL) decreased (Ahanger et al. 2020). Foliar application of SA decreased oxidative stress significantly, which was correlated with higher activities of SOD, CAT, APX, dehydroascorbate reductase (DHAR), and glutathione reductase (GR), as well as the increased content of nonenzymatic antioxidant components such as AsA and GSH. The impending function of COU to recover wheat plants' tolerance to salinity (50, 100, and 150 mM NaCl, 2 weeks) was explored. Increased activity of peroxidase (POD) and elevated total antioxidant capacity (TAC) were acquired at COU-primed plants under stress as indicator of inhibition of oxidative stress (Saleh and Madany 2015). Upon salt-induced elevated ROS generation, exogenous PCs including VA and Qu caused the suppression of ROS with the strengthening of antioxidant activities in tomato (Parvin et al. 2019, 2020).

Quercetin and VA alleviated salt-induced higher generation, including H_2O_2, $O_2^{\bullet-}$, MDA content, and lipoxygenase (LOX) activity. Higher AsA and GSH content, increased AsA:DHA, and GSH:GSSG were attributed by both Qu and VA under salinity. Where both Qu and VA-induced higher glutathione *S*-transferase (GST), monodehydroascorbate reductase (MDHAR), and GR activity were demonstrated, which were in corroboration with decreased oxidative stress. In addition, VA caused the higher activities of SOD and CAT, which was reduced in case of Qu supplementation under salinity might be due to the flavonoid-mediated direct ROS scavenging.

Mechanism of apigenin-induced stimulation in antioxidants activities with declined H_2O_2 and lipid peroxidation was ascertained in salt (50 mM NaCl) stressed *O. sativa* (Mekawy et al. 2018). Higher accumulation of nonenzymatic antioxidants such as reduced GSH and AsA; amplified transcript levels of genes encoding eight antioxidant enzymes in leaves, including Fe SOD, chloroplast Cu/Zn SOD, cytoplasmic Cu/Zn SOD, CAT, guaiacol peroxidase (GPOX), GPX, GR, and DHAR; higher activities of SOD, GPX, CAT, and APX are the strong evidence of an effective antioxidant defense system as induced by apigenin.

Caffeic acid (CA, 100 μM) application strengthened the protection mechanism in *G. max* under NaCl (70 mM) stress, which was confirmed from lessened $O_2^{\bullet-}$ content. The reason behind was found not only the increased SOD activity but also the augmented Cu/Zn SOD isoform expression (Klein et al. 2013).

Safflower, *C. tinctorius* was examined for SA-induced modulation of antioxidants responses for drought tolerance at 25% and 100% of FC. Salicylic acid promoted the SOD and CAT activities and gene expression of two subunits of Fe- and Cu-SOD under drought stress. Consequently, reduced free radicals production and lipid peroxidation was confirmed by SA as a sign of drought tolerance (Chavoushi et al. 2020). Drought tolerance in rice was enhanced by VA and *p*-hydroxybenzoic acid (PHBA) by enhancing antioxidant activity and thus increasing DPPH radical scavenging phenomenon (Quan and Xuan 2018). Ferulic acid pretreated seedlings demonstrated higher CAT, SOD, and GPOX activities and upregulated relative expression levels of *Cu/Zn-SOD*, *Mn-SOD,* and *CAT* under dehydration stress (induced by 10% PEG 6000) (Li et al. 2013). Therefore, reduction of $O_2^{\bullet-}$, H_2O_2, and MDA was obtained as an outcome of exogenous FA treatment under dehydration stress. Vanillic acid showed its role in alleviating the detrimental effects of dehydration stress in *V. corymbosum* by upregulating and improving the antioxidant defense system as well as suppression in H_2O_2, $O_2^{\bullet-}$ and MDA contents (An et al. 2019). Augmented the transcript levels of genes encoding eight antioxidant enzymes in leaves, including Fe SOD, chloroplast Cu/Zn SOD, cytoplasmic Cu/Zn SOD, CAT, GPOX, GPX, GR, and DHAR with the improved activities of SOD, GPX, and elevated contents of GSH and AsA acted behind to decrease the oxidative damage.

Exogenous PCs also can significantly improve submergence tolerance, which was reported from the study of Xuan and Khang (2018). Exogenously applied

0.1–1.0 mM PA and VA increased the expression of genes encoding APX and SOD antioxidant enzymes.

Metal/metalloid is another vital abiotic stress in this industrial era by which plants suffer from oxidative stress along with insufficient antioxidative responses. This PCs group is also able to check this metal-induced higher ROS production through metal chelation and strengthening the antioxidants defense. Exogenous VA caused the increment in phytochelatin content as well as biological accumulation factor (BAF), biological accumulation coefficient (BAC), and translocation factor (TF) in rice under Cd toxicity (Bhuyan et al. 2020). In this recent study, VA showed its promising roles in inhibition of LOX activity, enhancement of AsA-GSH pool along with activities of SOD, CAT, APX, MDHAR, DHAR, GR, GPX, and GST, which caused the reduction in ROS production, MDA, and EL.

Gallic acid is another PC, which also resulted in the reduction in H_2O_2 production and lipid peroxidation along with higher $OH^{\bullet}$ scavenging activity in wheat under Cd stress (Konakci 2019). The supplementation also showed the increment in SOD and CAT activities by which ROS accumulation was suppressed.

High temperature (40 °C/25 °C; day/night) induced oxidative stress indicated by high rise in $O_2^{\bullet-}$ and lipid peroxidation in *C. annuum* was inhibited by SA (Zhang et al. 2020). Salicylic acid application increased the content of AsA and GSH and POD activity. More in depth of PCs mediated heat stress was studied by Cheng et al. (2018) by applying FA against heat (39/30 °C for 3 days) affected *V. corymbosum* plants focusing on antioxidants actions. The cycle of AsA-GSH was boost up by involving its enzymatic components naming APX, MDHAR, DHAR, and GR due to FA addition under heat stress. In addition, transcriptions levels of genes encoding Fe-SOD, cytoplasmic Cu/Zn SOD, chloroplast Cu/Zn SOD were also increased in this condition along with elevated activity of POD, CAT, and GPX. Thus, FA resulted in the reduction in the production of $O_2^{\bullet-}$, H_2O_2, and MDA (Cheng et al. 2018).

As like heat, cold is also responsible for oxidative stress for plant and thus restricted plant survival. Therefore, exogenous CA and GLA showed their ability in contributing in cold stress tolerance in *C. sativus* and *G. max*, respectively (Wan et al. 2015; Ozfidan-Konakci et al. 2019). The CA application upregulated the activity of SOD, GPOX, CAT, GPX, MDHAR, APX, GR, and DHAR, which was correlated with the increased transcript levels of Cu/Zn SOD, GPX, and Mn-SOD genes in cold stress plant. Therefore, CA mediated higher AsA and GSH levels are also supported by above results under cold stress for acquiring the tolerance. But, GLA caused the reduction of H_2O_2 and lipid peroxidation by increasing SOD, CAT, and peroxidases (POX) activities rather than the antioxidants of AsA-GSH pools. Again, SA is also having the capability to increase SOD, CAT, and GPOX invented for effective decrease of MDA, H_2O_2 content under cold stress (Ignatenko et al. 2019). Levels of *TaMnSOD*, *TaFeSOD,* and *TaCAT* gene transcripts were also amplified by SA application under cold stress condition.

Extreme light induced elevated $O_2^{\bullet-}$ and H_2O_2 generation along with higher MDA and EL in *A. thaliana* can be controlled by PCs which was checked with SA application (Yang et al. 2019). Here, SA (0.3 mM) caused the upregulation of antioxidant activity of POD, SOD, CAT, APX, GPX, and GR as well as AsA and GSH redox balance which resulted in the suppression of above mentioned oxidative stress marker significantly. Plants also suffered similarly like abiotic stressed under excessive nutrient-induced oxidative damage. Therefore, PCs also showed their ability to check such toxicity. The supplementation of FA caused the reduction of $O_2^{\bullet -}$, H_2O_2, lipid peroxidation, and increased the activity of $OH^{\bullet}$ scavenging in *T. aestivum* under excess boron (B) stress (Yildiztugay et al. 2019). Higher AsA redox balance and the enhanced activities of SOD, CAT, POX, APX, MDHAR, DHAR, and GR were observed in FA-added plants grown under B stress.

It is interesting that PCs also can stimulate the antioxidants activity for lowering the ROS production in plants under normal growth condition; which would have prospect for increasing the plant yield potential. It was reported that, different concentration of rutin and GLA were applied in *O. sativa* for demonstrating their roles. Enhanced the free radical scavenging activity and total antioxidant capacity resulted in decreased H_2O_2 and $O_2^{\bullet -}$; and thus contributed in better growth performance (Singh et al. 2017).

Exogenous COU addition caused higher SOD activity in *Salvia hispanica* (chia) when grown under normal growth condition (Nkomo et al. 2019). As a result, COU decreased $O_2^{\bullet -}$ and H_2O_2 generation as well as oxidative damage (decreased MDA). The growth was also promoted.

Role of the secondary metabolite trans-cinnamic acid on *Z. mays* leaves resulted in a higher level of AsA and enhanced radical scavenging capacity to ensure oxidative stress tolerance (Araniti et al. 2018).

From the above discussion, it is apparent that PCs bear antioxidant capacity both under stress condition and non-stress condition. In some cases, excess accumulation of PCs demonstrated detrimental effects on plants. Some research findings demonstrated the enhanced ROS scavenging capacity without any changes of antioxidant system components. So, further research is needed to unravel the dilemma.

8 Methylglyoxal Detoxification by Phenolic Compounds

Upon stress condition, plants also suffer from higher MG generation which is also responsible for further ROS production. In addition, the role of glyoxalase system is by which toxic MG is detoxified. Our studies also find out that PCs also can suppress this MG level through strengthening the glyoxalase enzymes activities and other components of glyoxalase system. Among PCs, VA is a very promising member under the subgroup of phenolic acid; while Qu is another member under flavonoids, both of these had been studied in MG detoxification under abiotic stress. Upon salt stress, *S. lycopersicum* was suffered from elevated MG production, which was inhibited by supplementation of VA and Qu separately (Parvin et al. 2019, 2020). Both VA and Qu caused the upregulation of Gly I and Gly II activities significantly which were correlated with the suppression of MG at notable rate. This VA also showed its capability in MG detoxification where increased its Gly I and Gly II

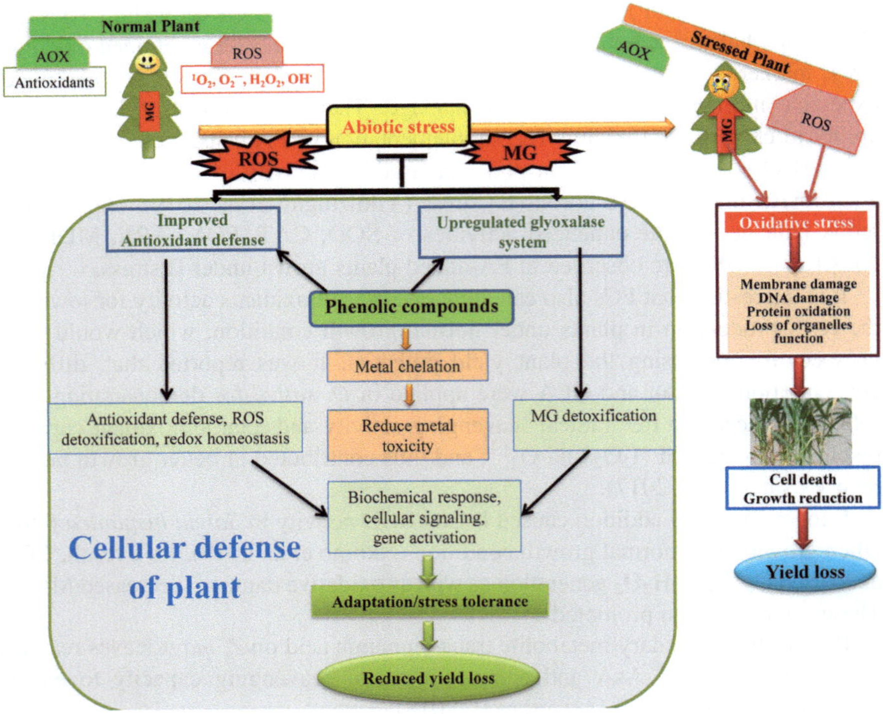

Fig. 9 Relationship between abiotic stresses and plant responses. Antioxidant components including PCs are potentially involved in cellular defense against stress-induced ROS. While plants suffer due to improper regulation of antioxidants and glyoxalases defense systems

activity by 213% and 43%, respectively in *O. sativa* under Cd toxicity (Bhuyan et al. 2020). In all these cases, a higher level of GSH was observed which is needed to use in MG detoxification by glyoxalase enzymes; GSH is recycled back after the MG detoxification process. Therefore, like many other phytoprotectants, PCs can maintain MG levels for plants during normal metabolism through coordinating antioxidants defense system by making bridge via using GSH.

So, finally if we summarized the properties of PCs in regulating ROS and MG balance there will be a systematic scenario about PCs-modulated plant responses upon oxidative stress (Fig. 9). Therefore, as nonenzymatic antioxidants, PCs can work together with other antioxidants in a system and chelate the diverse range of ions and scavenge ROS consequently which hinder oxidation of lipid and DNA damage (Rani et al. 2018). Most importantly, both GST and GSH contribute to accumulation of some flavonoids which act as metal binder and to be accumulated into the vacuole as inert form (Petrussa et al. 2013; Landi et al. 2015).

9 Conclusion

Plants are obligated to face various unfavorable environmental factors entitled by abiotic stresses throughout their life cycle due to its sessile nature that is responsible for significant yield loss. Phenolic compounds are an important group of secondary metabolites intregated into plant for playing vital roles to improve plant growth and morphology by mitigating abiotic stresses. In plants, PCs are commonly synthesized by shikimate, phenylpropanoid, and flavonoid pathways and consist of diversified structural variations including simple form such as phenol, phenolic acid (vanillic acid, caffeic acid), and also polyphenols (e.g., stilbenes and flavonoids). Phenolic compounds play key roles in plant development mechanisms including cell division, photosynthesis, nutrient balance, regulation of hormone, and reproduction. Most of the PCs have antioxidant properties to regulate oxidative stress, while some PCs act as signaling molecules, structural polymers, and UV screeners. Besides, PCs are also involved to plant responses under abiotic stresses such as salinity, drought, extreme temperature, flooding, and some have ability to chelate metal ions. Under abiotic stress conditions, plant synthesizes various PCs that regulate abiotic stresses-induced negative effects by scavenging ROS and harmful toxic compounds. Exogenous application of PCs also shows positive role to enhance stress tolerance in crop plants. Therefore, plant growth status improved by the influence of PCs independently or the interaction with growth-regulating substances. However, many unknown approaches also can be included in stress tolerance, which is not identified yet. Nowadays, scientific communities are showing immense interest to reveal the molecular approaches of PCs in alleviating the role of oxidative stresses. Therefore, further comprehensive studies are required to explore as well as compile all possible mechanisms of PCs, including genes response about mitigation of oxidative damages under abiotic stresses and plant tolerance.

Acknowledgment We acknowledge Dr. Iqbal R. Khan, Department of Botany, School of Chemical and Life Sciences, Jamia Hamdard, New Delhi, India for reviewing of the draft. We also thankful to Md. Rakib Hossain Raihan, Department of Agronomy, Sher-e-Bangla Agricultural University, Dhaka, Bangladesh for his critical reading and formatting of the manuscript.

References

Acidri R, Sawai Y, Sugimoto Y, Handa T, Sasagawa D, Masunaga T, Yamamoto S, Nishihara E (2020) Exogenous kinetin promotes the nonenzymatic antioxidant system and photosynthetic activity of coffee (*Coffea arabica* L.) plants under cold stress conditions. Plants 9(2):281. https://doi.org/10.3390/plants9020281

Afanas'ev IB, Dcrozhko AI, BrodskiiA V, Kostyuk VA, Potapovitch AI (1989) Chelating and free radical scavenging mechanisms of inhibitory action of rutin and quercetin in lipid peroxidation. Biochem Pharmacol 38:1763–1769

Agati G, Azzarello E, Pollastri S, Tattini M (2012) Flavonoids as antioxidants in plants: location and functional significance. Plant Sci 196:67–76

Ahanger MA, Aziz U, Alsahli AA, Alyemeni MN, Ahmad P (2020) Influence of exogenous salicylic acid and nitric oxide on growth, photosynthesis, and ascorbate-glutathione cycle in salt stressed *Vigna angularis*. Biomol Ther 10:42. https://doi.org/10.3390/biom10010042

Ahmad P, Nabi G, Ashraf M (2011) Cadmium-induced oxidative damage in mustard [*Brassica juncea* (L.) Czern. & Coss.] plants can be alleviated by salicylic acid. S Afr J Bot 77(1):36–44

Al-Ghamdi AA, Elansary HO (2018) Synergetic effects of 5-aminolevulinic acid and *Ascophyllum nodosum* seaweed extracts on *Asparagus* phenolics and stress related genes under saline irrigation. Plant Physiol Biochem 129:273–284

Aloisi I, Parrotta L, Ruiz KB, Landi C, Bini L, Cai G, Biondi S, Del Duca S (2016) New insight into quinoa seed quality under salinity: changes in proteomic and amino acid profiles, phenolic content, and antioxidant activity of protein extracts. Front Plant Sci 7:656. https://doi.org/10.3389/fpls.2016.00656

An YQ, Sun L, Wang XJ, Sun R, Cheng ZY, Zhu ZK, Yan GG, Li YX, Bai JG (2019) Vanillic acid mitigates dehydration stress responses in blueberry plants. Russ J Plant Physiol 66:806–817

Ancillotti C, Bogani P, Biricolti S, Calistri E, Checchini L, Ciofi L, Gonnelli C, Bubba MD (2015) Changes in polyphenol and sugar concentrations in wild type and genetically modified *Nicotiana langsdori* Weinmann in response to water and heat stress. Plant Physiol Biochem 97:52–61

Aoki T, Akashi T, Ayabe S (2000) Flavonoids of leguminous plants: structure, biological activity, and biosynthesis. J Plant Res 113:475–488

Araniti F, Lupini A, Mauceri A, Zumbo A, Sunseri F, Abenavoli MR (2018) The allelochemical trans-cinnamic acid stimulates salicylic acid production and galactose pathway in maize leaves: a potential mechanism of stress tolerance. Plant Physiol Biochem 128:32–40

Arora A, Byrem TM, Nari MG, Strasburg GM (2000) Modulation of liposomal membranes fluidity by flavonoids and isoflavonoids. Arch Biochem Biophys 373:102–109

Arora A, Nair MG, Strasburg GM (1998) Structure-activity relationships for antioxidant activities of a series of plavonoids in a liposomal system. Free Radic Biol Med 24:1355–1363

Babenko LM, Smirnov OE, Romanenko KO, Trunova OK, Kosakivsk IV (2019) Phenolic compounds in plants: biogenesis and functions. Ukr Biochem J 91:41635484. https://doi.org/10.15407/ubj91.03.005

Ballizany WL, Hofmann RW, Jahufer MZZ, Barrett BA (2012) Multivariate associations of flavonoid and biomass accumulation in white clover (*Trifolium repens*) under drought. Funct Plant Biol 39:167–177

Bhuyan MHMB, Parvin K, Mohsin SM, Mahmud JA, Hasanuzzaman M, Fujita M (2020) Modulation of cadmium tolerance in rice: insight into vanillic acid-induced upregulation of antioxidant defense and glyoxalase systems. Plants 9:188. https://doi.org/10.3390/plants9020188

Bistgani ZE, Hashemi M, DaCosta M, Craker L, Maggi F, Morshedloo MR (2019) Effect of salinity stress on the physiological characteristics, phenolic compounds and antioxidant activity of *Thymus vulgaris* L. and *Thymus daenensis* Celak. Ind Crop Prod 135:311–320

Blokhina O, Virolainen E, Fagerstedt KV (2003) Antioxidants, oxidative damage and oxygen deprivation stress: a review. Ann Bot 91:179–194

Bravo L (1998) Polyphenols: chemistry, dietary sources, metabolism, and nutritional significance. Nutr Rev 56:317–333

Catarino MD, Silva AMS, Cardoso SM (2017) Fucaceae: a source of bioactive phlorotannins. Int J Mol Sci 18:1327. https://doi.org/10.3390/ijms18061327

Chavoushi M, Najafi F, Salimi A, Angaji SA (2019) Improvement in drought stress tolerance of safflower during vegetative growth by exogenous application of salicylic acid and sodium nitroprusside. Ind Crop Prod 134:168–176

Chavoushi M, Najafi F, Salimi A, Angaji SA (2020) Effect of salicylic acid and sodium nitroprusside on growth parameters, photosynthetic pigments and secondary metabolites of safflower under drought stress. Sci Hortic 259:108823. https://doi.org/10.1016/j.scienta.2019.108823

Chen S, Wang Q, Lu H, Li J, Yang D, Liu J, Yan C (2019a) Phenolic metabolism and related heavy metal tolerance mechanism in *Kandelia obovata* under Cd and Zn stress. Ecotoxicol Environ Saf 169:134–143

Chen S, Wu F, Li Y, Qian Y, Pan X, Li F, Wang Y, Wu Z, Fu C, Lin H, Yang A (2019b) NtMYB4 and NtCHS1 are critical factors in the regulation of flavonoid biosynthesis and are involved in salinity responsiveness. Front Plant Sci 10:178. https://doi.org/10.3389/fpls.2019.00178

Chen Y, Xiao H, Zheng J, Liang G (2015) Structure-thermodynamics-antioxidant activity relationships of selected natural phenolic acids and derivatives: an experimental and theoretical evaluation. PLoS One 10:e0121276. https://doi.org/10.1371/journal.pone.0121276

Chen Z, Ma Y, Yang R, Gu Z, Wang P (2019c) Effects of exogenous Ca^{2+} on phenolic accumulation and physiological changes in germinated wheat (*Triticum aestivum* L.) under UV-B radiation. Food Chem 288:368–376

Cheng ZY, Sun L, Wang XJ, Sun R, An YQ, An BL, Zhu MX, Zhao CF, Bai JG (2018) Ferulic acid pretreatment allèviates heat stress in blueberry seedlings by inducing antioxidant enzymes, proline, and soluble sugars. Biol Plant 62:534–542

Cheynier V, Comte G, Davies KM, Lattanzio V, Martens S (2013) Plant phenolics: recent advances on their biosynthesis, genetics, and ecophysiology. Plant Physiol Biochem 72:1–20

Cohen SD, Kennedy JA (2010) Plant metabolism and the environment: implications for managing phenolics. Crit Rev Food Sci Nutr 50:620–643

Colak N, Torun H, Gruz J, Strnad M, Ayaz FA (2019) Exogenous N-acetylcysteine alleviates heavy metal stress by promoting phenolic acids to support antioxidant defence systems in wheat roots. Ecotoxicol Environ Saf 181:49–59

Commisso M, Toffali K, Strazzer P, Stocchero M, Ceoldo S, Baldan B, Levi M, Guzzo F (2016) Impact of phenylpropanoid compounds on heat stress tolerance in carrot cell cultures. Front Plant Sci 7:1439. https://doi.org/10.3389/fpls.2016.01439

Crozier A, Jaganath IB, Clifford MN (2006) Phenols, polyphenols and tannins: an overview. In: Crozier A, Clifford MN, Ashihara H (eds) Plant secondary metabolites: occurrence, structure and role in the human diet, 1. Blackwell, London

Dai J, Mumper RJ (2010) Plant phenolics: extraction, analysis and their antioxidant and anticancer properties. Molecules 15:7313–7352

Damalas CA (2019) Improving drought tolerance in sweet basil (*Ocimum basilicum*) with salicylic acid. Sci Hortic 246:360–365

de la Rosa LA, Moreno-Escamilla JO, Rodrigo-García J, Alvarez-Parrilla E (2019) Phenolic compounds. In: Elhadi MY (ed) Postharvest physiology and biochemistry of fruits and vegetables. Elsevier, Cambridge, pp 253–271

Dias MI, Sousa MJ, Alves RC, Ferreira IC (2016) Exploring plant tissue culture to improve the production of phenolic compounds: a review. Ind Crop Prod 82:9–22

Dixon RA, Pasinetti GM (2010) Flavonoids and isoflavonoids: from plant biology to agriculture and neuroscience. Plant Physiol 154:453–457

Dkhil BB, Denden M (2012) Effect of salt stress on growth, anthocyanins, membrane permeability and chlorophyll fluorescence of okra (*Abelmoschus esculentus* L.) seedlings. Am J Plant Physiol 7:174–183

Edreva A, Velikova V, Tsonev T, Dagnon S, Gürel A, Aktaş L, Gesheva E (2008) Stress-protective role of secondary metabolites: diversity of functions and mechanisms. Stress protection by secondary metabolites. Gen Appl Plant Physiol 34(1–2):67–78

El-Soud WA, Hegab MM, AbdElgawad H, Zinta G, Asard H (2013) Ability of ellagic acid to alleviate osmotic stress on chickpea seedlings. Plant Physiol Biochem 71:173–183

Garcia-Calderon M, Pons-Ferrer T, Mrazova A, Pal'ove-Balang P, Vilkova M, Perez-Delgado CM, Vega JM, Eliasova A, Repcak M, Marquez AJ, Betti M (2015) Modulation of phenolic metabolism under stress conditions in a *Lotus japonicus* mutant lacking plastidic glutamine synthetase. Front Plant Sci 6:760. https://doi.org/10.3389/fpls.2015.00760

Gharibi S, Tabatabaei BES, Saeidi G, Talebi M, Matkowski A (2019) The effect of drought stress on polyphenolic compounds and expression of flavonoid biosynthesis related genes in *Achillea pachycephala* Rech.F. Phytochemistry 162:90–98

Ghasemi S, Kumleh HH, Kordrostami M (2019) Changes in the expression of some genes involved in the biosynthesis of secondary metabolites in *Cuminum cyminum* L. under UV stress. Protoplasma 256:279–290

Ghassemi-Golezani K, Farhangi-Abriz S (2018) Foliar sprays of salicylic acid and jasmonic acid stimulate H+-ATPase activity of tonoplast, nutrient uptake and salt tolerance of soybean. Ecotoxicol Environ Saf 166:18–25

Giada MDLR (2013) Food phenolic compounds: Main classes, sources and their antioxidant power. In: Morales-Gonzalez JA (ed) Oxidative stress and chronic degenerative diseases—a role for antioxidants. InTech, Rijeka, pp 87–112

Golkar P, Taghizadeh M (2018) In vitro evaluation of phenolic and osmolite compounds, ionic content, and antioxidant activity in saower (*Carthamus tinctorius* L.) under salinity stress. Plant Cell Tissue Org Cult 134:357–368

Goyal A, Siddiqui S, Upadhyay N, Soni J (2014) Effects of ultraviolet irradiation, pulsed electric field, hot water and ethanol vapours treatment on functional properties of mung bean sprouts. J Food Sci Technol 51:708–714

Griffith M, Yaish MW (2004) Antifreeze proteins in overwintering plants: a tale of two activities. Trends Plant Sci 9:399–405

Ha K, Jo S, Mannam V, Kwon YI, Apostolisdis E (2016) Stimulation of phenolics, antioxidant and α-glucosidase inhibitory activities during barley (*Hordeum vulgare* L.) seed germination. Plant Foods Hum Nutr 71:211–217

Handa N, Kohli SK, Sharma A, Thukral AK, Bhardwaj R, Abd Allah EF, Alqarawi AA, Ahmad P (2019) Selenium modulates dynamics of antioxidative defence expression, photosynthetic attributes and secondary metabolites to mitigate chromium toxicity in *Brassica juncea* L. plants. Environ Exp Bot 161:180–192

Harborne JB (1980) Plant phenolics. In: Pirson A, Zimmermann MH (eds) Encyclopedia of plant physiology, vol 8. Springer, Berlin, New York, pp 329–402

Havsteen B (2002) The biochemistry and medical significance of the flavonoids. Pharmacol Ther 96:67–202

Hodaei M, Rahimmalek M, Arzani A, Talebi M (2018) The effect of water stress on phytochemical accumulation, bioactive compounds and expression of key genes involved in flavonoid biosynthesis in *Chrysanthemum morifolium* L. Ind Crop Prod 120:295–304

Huang X, Yao J, Zhao Y, Xie D, Jiang X, Xu Z (2016) Efficient rutin and quercetin biosynthesis through flavonoids-related gene expression in *Fagopyrum tataricum* Gaertn. Hairy root cultures with UV-B irradiation. Front Plant Sci 7:63. https://doi.org/10.3389/fpls.2016.00063

Hura T, Hura K, Dziurka K, Ostrowska A, Bączek-Kwinta R, Grzesiak M (2012) An increase in the content of cell wall-bound phenolics correlates with the productivity of triticale under soil drought. J Plant Physiol 169:1728–1736

Ignatenko A, Talanova V, Repkina N, Titov A (2019) Exogenous salicylic acid treatment induces cold tolerance in wheat through promotion of antioxidant enzyme activity and proline accumulation. Acta Physiol Plant 41(6):80. https://doi.org/10.1007/s11738-019-2872-3

Jaganath IB, Crozier A (2010) Dietary flavonoids and phenolic compounds. In: Fraga CG (ed) Plant phenolics and human health. Wiley, New Jersey, Canada, pp 1–50

Jaiswal A, Srivastava JP (2016) Nitric oxide mitigates waterlogging stress by regulating antioxidative defense mechanism in maize (*Zea mays* L.) roots. Bangladesh J Bot 45:517–524

Jiang X, Liu Y, Li W, Zhao L, Meng F, Wang Y, Tan H, Yang H, Wei C, Wan X, Gao L (2013) Tissue-specific, development-dependent phenolic compounds accumulation profile and gene expression pattern in tea plant [*Camellia sinensis*]. PLoS One 8(4):e62315. https://doi.org/10.1371/journal.pone.0062315

Katerova Z, Todorova D, Tasheva K, Sergiev I (2012) Influence of ultraviolet radiation on plant secondary metabolite production. Genet Plant Physiol 2(3–4):113–144

Kaur L, Zhawar VK (2015) Phenolic parameters under exogenous ABA, water stress, salt stress in two wheat cultivars varying in drought tolerance. Ind J Plant Physiol 20:151–156

Kaur R, Yadav P, Sharma A, Thukral AK, Kumar V, Kohli SK, Bhardwaj R (2017) Castasterone and citric acid treatment restores photosynthetic attributes in *Brassica juncea* L. under Cd (II) toxicity. Ecotoxicol Environ Saf 145:466–475

Khan MIR, Asgher M, Khan NA (2014) Alleviation of salt-induced photosynthesis and growth inhibition by salicylic acid involves glycinebetaine and ethylene in mungbean (*Vigna radiata* L.). Plant Physiol Biochem 80:67–74

Khlestkina E (2013) The adaptive role of flavonoids: emphasis on cereals. Cereal Res Commun 41: 185–198

Kısa D, Elmastas M, Öztürk L, Kayır Ö (2016) Responses of the phenolic compounds of *Zea mays* under heavy metal stress. Appl Biol Chem 59:813–820

Klein A, Keyster M, Ludidi N (2013) Caffeic acid decreases salinity-induced root nodule superoxide radical accumulation and limits salinity-induced biomass reduction in soybean. Acta Physiol Plant 35:3059–3066

Kohli SK, Handa N, Sharma A, Gautam V, Arora S, Bhardwaj R, Wijaya L, Alyemeni MN, Ahmad P (2018) Interaction of 24-epibrassinolide and salicylic acid regulates pigment contents, antioxidative defense responses, and gene expression in *Brassica juncea* L. seedlings under Pb stress. Environ Sci Pollut Res 25:15159–15173

Konakci CO (2019) Does exogenously applied gallic acid regulate the enzymatic and non-enzymatic antioxidants in wheat roots exposed to cadmium stress? Celal Bayar Üniversitesi Fen Bilimleri Dergisi 15(3):279–285

Kovaleva LV, Zakharova EV, Minkina YV (2007) Auxin and flavonoids in the progame phase of fertilization in petunia. Russ J Plant Physiol 54(3):396–401

Kurepa J, Shull TE, Smalle JA (2016) Quercetin feeding protects plants against oxidative stress. F1000 Res 5:2430. https://doi.org/10.12688/f1000research.9659.1

Landi M, Tattini M, Gould KS (2015) Multiple functional roles of anthocyanins in plant-environment interactions. Environ Exp Bot 119:4–17

Lattanzio V (2013) Phenolic compounds: introduction. In: Ramawat KG, Merillon JM (eds) Natural products: phytochemistry, botany and metabolism of alkaloids, phenolics and terpenes. Springer, Heidelberg, pp 1543–1580

Leng X, Jia H, Sun X, Shangguan L, Mu Q, Wang B, Fang J (2015) Comparative transcriptome analysis of grapevine in response to copper stress. Sci Rep 5:17749. https://doi.org/10.1038/srep17749

León-Chan RG, López-Meyer M, Osuna-Enciso T, Sañudo-Barajas JA, Heredia JB, León-Félix J (2017) Low temperature and ultraviolet-B radiation affect chlorophyll content and induce the accumulation of UV-B-absorbing and antioxidant compounds in bell pepper (*Capsicum annuum*) plants. Environ Exp Bot 139:143–151

Li DM, Nie YX, Zhang J, Yin JS, Li Q, Wang XJ, Bai JG (2013) Ferulic acid pretreatment enhances dehydration-stress tolerance of cucumber seedlings. Biol Plant 57(4):711–717

Linić I, Šamec D, Grúz J, Bok VV, Strnad M, Salopek-Sondi B (2019) Involvement of phenolic acids in short-term adaptation to salinity stress is species-specific among brassicaceae. Plan Theory 8(6):155. https://doi.org/10.3390/plants8060155

Ma D, Sun D, Wang C, Li Y, Guo T (2014) Expression of flavonoid biosynthesis genes and accumulation of flavonoid in wheat leaves in response to drought stress. Plant Physiol Biochem 80:60–66

Ma Y, Wang P, Gu Z, Tao Y, Shen C, Zhou Y, Han Y, Yang R (2019) Ca^{2+} involved in GABA signal transduction for phenolics accumulation in germinated hulless barley under NaCl stress. Food Chem:X 2:100023. https://doi.org/10.1016/j.fochx.2019.100023

Martinez V, Mestre TC, Rubio F, Girones-Vilaplana A, Moreno DA, Mittler R, Rivero RM (2016) Accumulation of flavonols over hydroxycinnamic acids favors oxidative damage protection under abiotic stress. Front Plant Sci 7:838. https://doi.org/10.3389/fpls.2016.00838

Mekawy AMM, Abdelaziz MN, Ueda A (2018) Apigenin pretreatment enhances growth and salinity tolerance of rice seedlings. Plant Physiol Biochem 130:94–104

Merewitz EB, Liu S (2019) Improvement in heat tolerance of creeping bentgrass with melatonin, rutin, and silicon. J Am Soc Hort Sci 144(2):141–148

Michalak A (2006) Phenolic compounds and their antioxidant activity in plants growing under heavy metal stress. Polish J Environ Stud 15(4):523–530

Millic BL, Djilas SM, Canadanovic-Brunet JM (1998) Antioxidative activity of phenolic compounds on the metal-ion breakdown of lipid peroxidation system. Food Chem 61:443–447

Minatel IO, Borges CV, Ferreira MI, Gomez HAG, Chen CYO, Lima GPP (2017) Phenolic compounds: functional properties, impact of processing and bioavailability. In: Soto-Hernandez M, Palma-Tenango M, del Rosario G-MM (eds) Phenolic compounds—biological activity. Intech, Rijeka

Mira L, Fernandez MT, Santos M, Rocha R, Florencio MH, Jennings KR (2002) Interactions of flavonoids with iron and copper ions: a mechanism for their antioxidant activity. Free Radic Res 36:1199–1208

Mishra B, Sangwan NS (2019) Amelioration of cadmium stress in *Withania somnifera* by ROS management: active participation of primary and secondary metabolism. Plant Growth Regul 87:403–412

Moran-Palacio EF, Zamora-Álvarez LA, Stephens-Camacho NA, Yáñez-Farías GA, Virgen-Ortiz-A, Martínez-Cruz O, Rosas-Rodríguez JA (2014) Antioxidant capacity, radical scavenging kinetics and phenolic profile of methanol extracts of wild plants of southern Sonora, Mexico. Trop J Pharm Res 13(9):1487–1493

Moravcová Š, Tůma J, Dučaiová ZK, Waligórski P, Kula M, Saja D, Słomka A, Bąba W, Libik-Konieczny M (2018) Influence of salicylic acid pretreatment on seeds germination and some defence mechanisms of *Zea mays* plants under copper stress. Plant Physiol Biochem 122:19–30

Murkovic M (2003) Phenolic compounds. In: Caballero B, Trugo C, Finglas PM (eds) Encyclopedia of food sciences and nutrition, 2nd edn. Academic Press, Amsterdam, pp 4507–4514

Naikoo MI, Dar MI, Raghib F, Jaleel H, Ahmad B, Raina A, Khan FA, Naushin F (2019) Role and regulation of plants phenolics in abiotic stress tolerance: An overview. In: Khan MIR, Ferrante A, Reddy PS, Khan NA (eds) Plant signaling molecules: role and regulation under stressful environments. Elsevier, Amsterdam, pp 157–168

Nakabayashi R, Yonekura-Sakakibara K, Urano K, Suzuki M, Yamada Y, Nishizawa T, Matsuda F, Kojima M, Sakakibara H, Shinozaki K, Michael AJ (2014) Enhancement of oxidative and drought tolerance in *Arabidopsis* by overaccumulation of antioxidant flavonoids. Plant J 77: 367–379

Navarro M, Kontoudakis N, Canals JM, Garcıa-Romero E, Gomez-Alonso S, Zamora F, Hermosın-Gutierrez I (2017) Improved method for the extraction and chromatographic analysis on a fused-core column of ellagitannins found in oak-aged wine. Food Chem 226:23–31

Nazar R, Iqbal N, Syeed S, Khan NA (2011) Salicylic acid alleviates decreases in photosynthesis under salt stress by enhancing nitrogen and sulfur assimilation and antioxidant metabolism differentially in two mungbean cultivars. J Plant Physiol 168(8):807–815

Nduche MU, Otaka CL (2019) Phytochemical screening and antimicrobial activity of *Talinium triangulare* (JACQ) Willd, *Ocimum gratissimum* L., *Chromoleana odorata* L., and *Aloe vera* (L.) Burm. F. Int J Res Pharm Biosci 6:1–12

Nichols SN, Hofmann RW, Williams WM (2015) Physiological drought resistance and accumulation of leaf phenolics in white clover interspecific hybrids. Environ Exp Bot 119:40–47

Nkomo M, Gokul A, Keyster M, Klein A (2019) Exogenous p-Coumaric acid improves *Salvia hispanica* L. seedling shoot growth. Plan Theory 8:546. https://doi.org/10.3390/plants8120546

Ozfidan-Konakci C, Yildiztugay E, Yildiztugay A, Kucukoduk M (2019) Cold stress in soybean (*Glycine max* L.) roots: exogenous gallic acid promotes water status and increases antioxidant activities. Bot Serb 43:59–71

Panche AN, Diwan AD, Chandra SR (2016) Flavonoids: an overview. J Nutr Sci 5:e47. https://doi.org/10.1017/jns.2016.41

Park JS, Lee EJ (2019) Waterlogging induced oxidative stress and the mortality of the Antarctic plant, *Deschampsia antarctica*. J Ecol Environ 43(1):29. https://doi.org/10.1186/s41610-019-0127-2

Parr AJ, Bolwell GP (2000) Phenols in the plant and in man. The potential for possible nutritional enhancement of the diet by modifying the phenols content or profile. J Sci Food Agric 80:985–1012

Parvin K, Hasanuzzaman M, Bhuyan MHM, Mohsin SM, Fujita M (2019) Quercetin mediated salt tolerance in tomato through the enhancement of plant antioxidant defense and glyoxalase systems. Plan Theory 8:247. https://doi.org/10.3390/plants8080247

Parvin K, Nahar K, Hasanuzzaman M, Bhuyan MHMB, Mohsin SM, Fujita M (2020) Exogenous vanillic acid enhances salt tolerance of tomato: insight into plant antioxidant defense and glyoxalase systems. Plant Physiol Biochem 150:109–120

Perin EC, Da Silva MR, Borowski JM, Crizel RL, Schott IB, Carvalho IR, Rombaldi CV, Galli V (2019) ABA-dependent salt and drought stress improve strawberry fruit quality. Food Chem 271:516–526

Petrussa E, Braidot E, Zancani M, Peresson C, Bertolini A, Patui S, Vianello A (2013) Plant flavonoids-biosynthesis, transport and involvement in stress responses. Int J Mol Sci 14:14950–14973

Quan NT, Xuan TD (2018) Foliar application of vanillic and *p*-hydroxybenzoic acids enhanced drought tolerance and formation of phytoalexin momilactones in rice. Arch Agron Soil Sci 64: 1831–1846

Rahman A, Hossain MS, Mahmud JA, Nahar K, Hasanuzzaman M, Fujita M (2016) Manganese-induced salt stress tolerance in rice seedlings: regulation of ion homeostasis, antioxidant defense and glyoxalase systems. Physiol Mol Biol Plants 22(3):291–306

Rani R, Arora S, Kaur J, Manhas RK (2018) Phenolic compounds as antioxidants and chemopreventive drugs from Streptomyces cellulosae strain TES17 isolated from rhizosphere of *Camellia sinensis*. BMC Complement Altern Med 18:82. https://doi.org/10.1186/s12906-018-2154-4

Rezayian M, Niknam V, Ebrahimzadeh H (2018) Differential responses of phenolic compounds of *Brassica napus* under drought stress. Iran J Plant Physiol 8:2417–2425

Rice-Evans CA, Miller NJ, Paganga G (1997) Antioxidant properties of phenolic compounds. Trends Plant Sci 2:152–159

Rossi L, Borghi M, Francini A, Lin X, Xie DY, Sebastiani L (2016) Salt stress induces dierential regulation of the phenylpropanoid pathway in *Olea europaea* cultivars Frantoio (salt-tolerant) and Leccino (salt-sensitive). J Plant Physiol 204:8–15

Saleh AM, Madany MMY (2015) Coumarin pretreatment alleviates salinity stress in wheat seedlings. Plant Physiol Biochem 88:27–35

Samadi S, Habibi G, Vaziri A (2019) Effects of exogenous salicylic acid on antioxidative responses, phenolic metabolism and photochemical activity of strawberry under salt stress. Iran J Plant Physiol 9(2):2685–2694

Sanchez-Rodriguez E, Moreno DA, Ferreres F, Rubio-WilhelmiMdel M, Ruiz JM (2011) Differential responses of five cherry tomato varieties to water stress: changes on phenolic metabolites and related enzymes. Phytochemistry 72:723–729

Santos EL, Maia BHLNS, Ferriani AP, Teixeira SD (2017) Flavonoids: classification, biosynthesis and chemical ecology. In: Justino GC (ed) Flavonoids from biosynthesis to human health, vol 6. InTech, Rijeka, p 482

Santra HK, Banerjee D (2020) Natural products as fungicide and their role in crop protection. In: Singh J, Yadav A (eds) Natural bioactive products in sustainable agriculture. Springer, Singapore

Sarker U, Oba S (2018a) Augmentation of leaf color parameters, pigments, vitamins, phenolic acids, flavonoids and antioxidant activity in selected *Amaranthus tricolor* under salinity stress. Sci Rep 8:12349. https://doi.org/10.1038/s41598-018-30897-6

Sarker U, Oba S (2018b) Drought stress effects on growth, ROS markers, compatible solutes, phenolics, flavonoids, and antioxidant activity in *Amaranthus tricolor*. Appl Biochem Biotechnol 186(4):999–1016

Selmar D (2008) Potential of salt and drought stress to increase pharmaceutical significant secondary compounds in plants. Landbauforschung Volkenrode 58:139–144

Shahid MA, Balal RM, Khan N, Rossi L, Rathinasabapathi B, Liu G, Khan J, Cámara-Zapata JM, Martínez-Nicolas JJ, Garcia-Sanchez F (2018) Polyamines provide new insights into the biochemical basis of Cr-tolerance in Kinnow mandarin grafted on diploid and double-diploid rootstocks. Environ Exp Bot 156:248–260

Sharma A, Shahzad B, Rehman A, Bhardwaj R, Landi M, Zheng B (2019) Response of phenylpropanoid pathway and the role of polyphenols in plants under abiotic stress. Molecules 24(13):2452. https://doi.org/10.3390/molecules24132452

Simaei M, Khavari-Nejad RA, Bernard F (2012) Exogenous application of salicylic acid and nitric oxide on the ionic contents and enzymatic activities in NaCl-stressed soybean plants. Am J Plant Sci 3:1495–1503

Singh B, Singh JP, Kaur A, Singh N (2017) Phenolic composition and antioxidant potential of grain legume seeds: a review. Food Res Int 101:1–16

Smirnov OE, Kosyan AM, Kosyk OI, Taran NY (2015) Response of phenolic metabolism induced by aluminium toxicity in *Fagopyrum esculentum* moench. Plants. Ukr Biochem J 87:129–135

Tanase C, Coşarcă S, Muntean DL (2019) A critical review of phenolic compounds extracted from the bark of woody vascular plants and their potential biological activity. Molecules 24:1182. https://doi.org/10.3390/molecules24061182

Tsimogiannis D, Oreopoulou V (2019) Classification of phenolic compounds in plants. In: Watson RR (ed) Polyphenols in plants, 2nd edn. Elsevier, Amsterdam, pp 263–284

Vanholme R, Demedts B, Morreel K, Ralph J, Boerjan W (2010) Lignin biosynthesis and structure. Plant Physiol 153:895–905

Varela MC, Arslan I, Reginato MA, Cenzano AM, Luna MV (2016) Phenolic compounds as indicators of drought resistance in shrubs from *Patagonian shrublands* (Argentina). Plant Physiol Biochem 104:81–91

Vermerris W, Nicholson R (2008) Families of phenolic compounds and means of classification. Phenolic compound biochemistry. Springer, Dordrecht

Verstraeten SV, Keen CL, Schmitz HH, Fraga CG, Oteiza PI (2003) Flavan-3-ols and procyanidins protect liposomes against lipid oxidation and disruption of the bilayer structure. Free Radic Biol Med 34:84–92

Vogt T (2010) Phenylpropanoid biosynthesis. Mol Plant 3(1):2–20

Wan YY, Zhang Y, Zhang L, Zhou ZQ, Li X, Shi Q, Wang XJ, Bai JG (2015) Caffeic acid protects cucumber against chilling stress by regulating antioxidant enzyme activity and proline and soluble sugar contents. Acta Physiol Plant 37(1):1706. https://doi.org/10.1007/s11738-014-1706-6

Wang F, Zhu H, Chen D, Li Z, Peng R, Yao Q (2016) A grape bHLH transcription factor gene, *VvbHLH1*, increases the accumulation of flavonoids and enhances salt and drought tolerance in transgenic *Arabidopsis thaliana*. Plant Cell Tissue Organ Cult 125(2):387–398

Wang J, Yuan B, Huang B (2019a) Differential heat-induced changes in phenolic acids associated with genotypic variations in heat tolerance for hard fescue. Crop Sci 59:667–674

Wang L, Shan T, Xie B, Ling C, Shao S, Jin P, Zheng Y (2019b) Glycine betaine reduces chilling injury in peachfruit by enhancing phenolic and sugar metabolisms. Food Chem 272:530–538

Waśkiewicz A, Muzolf-Panek M, Goliński P (2013) Phenolic content changes in plants under salt stress. In: Ahmad P, Azooz MM, Prasad MNV (eds) Ecophysiology and responses of plants under salt stress. Springer, New York, pp 283–314

Williams RJ, Spencer JP, Rice-Evans C (2004) Flavonoids: antioxidants or signalling molecules? Free Radic Biol Med 36:838–849

Xuan T, Khang D (2018) Effects of exogenous application of protocatechuic acid and vanillic acid to chlorophylls, phenolics and antioxidant enzymes of rice (*Oryza sativa* L.) in submergence. Molecules 23(3):620. https://doi.org/10.3390/molecules23030620

Yan J, Wang B, Jiang Y, Cheng L, Wu T (2014) GmFNSII-controlled soybean flavone metabolism responds to abiotic stresses and regulates plant salt tolerance. Plant Cell Physiol 55:74–86

Yang ZC, Wu N, Tang L, Yan XH, Yuan M, Zhang ZW, Yuan S, Zhang HY, Chen YE (2019) Exogenous salicylic acid alleviates the oxidative damage of *Arabidopsis thaliana* by enhancing antioxidant defense systems under high light. Biol Plant 63:474–483

Yildiztugay E, Ozfidan-Konakci C, Karahan H, Kucukoduk M, Turkan I (2019) Ferulic acid confers tolerance against excess boron by regulating ROS levels and inducing antioxidant system in wheat leaves (*Triticum aestivum*). Environ Exp Bot 161:193–202

Zaid A, Mohammad F, Wani SH, Siddique KM (2019) Salicylic acid enhances nickel stress tolerance by up-regulating antioxidant defense and glyoxalase systems in mustard plants. Ecotoxicol Environ Saf 180:575–587

Zaprometov MN, Nikolaeva TN (2003) Chloroplasts isolated from kidney bean leaves are capable of phenolic compound biosynthesis. Russ J Plant Physiol 50(5):623–626

Zhang Z, Lan M, Han X, Wu J, Wang-Pruski G (2020) Response of ornamental pepper to high-temperature stress and role of exogenous salicylic acid in mitigating high temperature. J Plant Growth Regul 39:133–146. https://doi.org/10.1007/s00344-019-09969-y

Zhou P, Li Q, Liu G, Xu N, Yang Y, Zeng W, Chen A, Wang S (2018) Integrated analysis of transcriptomic and metabolomic data reveals critical metabolic pathways involved in polyphenol biosynthesis in *Nicotiana tabacum* under chilling stress. Funct Plant Biol 46:30–43

Efficacy of Various Amendments for the Phytomanagement of Heavy Metal Contaminated Sites and Sustainable Agriculture. A Review

Sanobar Iqbal, Mujahid Farid, Muhammad Zubair, Zaki Ul Zaman Asam, Shafaqat Ali, Muhammad Abubakar, Sheharyaar Farid, and Muhammad Rizwan

Abstract Soil has the ability to persist contaminants for a longer duration which lowers the quality of soil for agricultural use. Soils augmented with a load of heavy and toxic metals and organic pollutants of industrial sources show the severe effect on the crops. This effort will gather information about the effect of environmental pollution on the crops and remediation of the contaminated soil using various amendments. Emphasize is focused on the physiochemical and biological mechanisms along with advanced phytoremediation techniques. New developments of research are included on the morpho-physiological and biochemical responses of important food crops that grow under heavy metals stress. Among the applied practices the best application recently working are the organic acids amendments having a chelating ability. Significant facts of plants growth-promoting bacteria which favor the phytoremediation because of converting rhizospheric situation, increase the plant biomass, and bioaccumulation of heavy metals are reported. A

S. Iqbal · M. Farid (✉) · Z. U. Z. Asam · M. Abubakar
Department of Environmental Sciences, University of Gujrat,Hafiz Hayat Campus, Gujrat, Pakistan

M. Zubair
Department of Chemistry, University of Gujrat, Hafiz Hayat Campus, Gujrat, Pakistan

S. Ali
Department of Environmental Sciences and Engineering, Government College University, Faisalabad, Pakistan

Department of Biological Sciences and Technology, China Medical University, Taichung, Taiwan

S. Farid
Institute of Biomedical Sciences ABEL SALAZAR, University of Porto, Porto, Portugal

Department of Biology, Ecology, and Evolution, University of Liege, Liege, Belgium

M. Rizwan
Department of Environmental Sciences and Engineering, Government College University, Faisalabad, Pakistan

M. Hasanuzzaman et al. (eds.), *Managing Plant Production Under Changing Environment*, https://doi.org/10.1007/978-981-16-5059-8_9

broader approach is used to best summarize the literature regarding the efficacy of soil amendments that can play a positive role in enhancing the efficiency of phytomanagement.

Keywords Soil · Heavy metal · Phytomanagement · Chelators · Bacteria · Wastewater

1 Introduction

In the present condition environmental pollution is the most critical and burning problem throughout the world and many preventive measures are being adopted but these are not enough. The word heavy metal includes few metallic elements that are strongly toxic and are very fatal even in a very minute quantity (Lenntech 2004; Farid et al. 2020a, b). About 20% of the land is contaminated in Europe due to land pollution and also their toxicity traditional preventive techniques are present, but these are not sufficient for eliminating or minimizing the land degradation due to hazardous materials the report gives to the European union in 2002 elaborated charge for removing the polluted sites which are affected by the human activities such as industries is about to 100 million dollars the file issued in the United States approximately in 1990s to remove the thousands of polluted sites almost 500,000 (Saier et al. 2010). Conversion of fertile land into barren land has become an emerging issue worldwide so, it is need of the hour to develop methods for decreasing or minimizing lethal materials (Ji et al. 2011; Bech et al. 2012; Choi et al. 2013). Robustness is higher than the water that's why it has negative reactions on the health of human beings such as the sliver, arsenic, mercury, lead, chromium, copper, cadmium, and iron, the second definition of one of the known groups is platinum which covered a species by the components of it and their mixture (Farlex 2005). Numerous modes by the body to take the metals into food and air, and also from the skin adjustment in the human system. It's a very fast process than detoxification and its symptoms are not shown earlier due to more time because the more ratio in the immune system causes bad impacts on it and useless by using lethal chemicals (Suruchi et al. 2012; Rizwan et al. 2017a, b). From an environmental perspective, all lethal metals are equally dangerous and toxic and can stay in the soil for long periods of time and cause degradation (Abrahams 2002). Extended periods and small-span contact cause dangerous effects (Abrahams 2002; Schröder et al. 2003; Schrøder et al. 2004). Awareness needs to be provided to humans about undesirable influences when it is present into the soil (Thuy et al. 2000). Contamination in the aquatic region is a water quality problem, in developing countries, due to overpopulation and excessive use of resources, and bad sanitation in cities (Mintz and Baier 2000). Heavy metal in the soil put adverse impacts on the soil and the nature of plants (Gichner et al. 2006). Entrance anthropogenically created example particulate matter in water, naturally exists and is not destroyed easily because it deposits in the body of a human (Olayinka and Alo 2004; Lenntech 2004).

Significant compounds with their permissible limits cause injurious impacts on the environment (Lone et al. 2008). Chromium is not commonly effective but few compounds which enter into human beings cause lung, respiratory, skin issues allowable amount of 0.05 mg/L. PCRWR by research in 23 megacities of Pakistan, 1% lower water samples harmless boundaries of chromium (Tariq et al. 2008). Nickel a metal that is present in the atmosphere, lithosphere and hydrosphere is required in smaller and minute quantities for normal functioning and used for the folic acid and vitamin B_{12} functionality as obtained from the study taken on the animal that is the reason for the heart problem and lung disorder (Midrar-Ul-Haq et al. 2005). Lead is a very poisonous material in the soil (Raja et al. 2008). Throughout from domestic and industrial areas also from traffic diseases are mostly the kidney and nervous damages. Scientifically developed methods to remove metals becoming older and non-economical as excavation and solidification, which put the poor impact on the fertility of the soil (Pulford and Watson 2003). Technical development and industrial revolt had increased heavy metals pollution in the environment, the persistent nature of heavy metal is lethal to the environment in the soil. Human activities such as excavating of essential resources and overconsumption of are the sources of dense metals (Marwa et al. 2012; Farid et al. 2015). Persistent in the soil has been occurred for thousands of years because it is not remediated by the microorganisms (Tangahu et al. 2011). The capacity of plants to remediate metals in soil is called phytoremediation and is more beneficial than other methods because its biological control is combined with the agronomic traits, environmentally friendly, require less disposal hence used to minimize the soil pollution by using green plants (Ali et al. 2013). One of the best treatments which is given by the scientist is the phytoextraction which uses the high accumulating plants that contain high biomass and help to remove metals from the soil. Afterward, the metals that move through the root system to the above plants among them some metals have less bioavailability into the soil even they are removed from the soil using chelates which are used for lowering the limitation of the phytoextraction to increase the removal of the metals (Evangelou et al. 2007). Following plants have the more accumulating power like *Brassica napus* L. it is used in the containing pollutant soil and grows for removal of heavy metals and their oil used as biodiesel (Park et al. 2008). 40% to 44% lubricant spores. *Helianthus annuus* L. has the capability due to high biomass extraction of Cd and Pb is the best for the phytoremediation because it works more efficiently (Lasat 2002). Another example is Dhanyaka and their existence in every country such as China, Netherland, Bangladesh, Eastern, and central Europe because of easy growth in the severe conditions and frequently changing of weather doesn't affect the best period for the cultivation of the coriander is October to February the *Coriandrum sativum* L. is the herb which has the high amount of the best food qualities. Coriander is a tropical plant, which grows easily in harsh and constantly changing environment growth coriander seeds is second, the tenth month contain all food quality use for the decrease of the heavy metals such as lead, copper, zinc (Khan et al. 2010). *C. sativum* L remarkable as a point of the dietary product as cabbage has many health effects on the human that which proved that the quantity of the elements in the plant which are affected by the amount of harmful quantity into the soil from the result obtained manifest that roots accumulate the Mn, Fe, Cu, and Zn in soil. Zinc absorbed by the roots is monitored by *C. sativum* exposed attraction Ni with reasonable to great points noticed in the root, leaves. Based on that accumulated data obtained in this study, this plant can be more effective in the phytoremediation of metals from the field (Ximénez-Embún et al. 2001). This technique is used for the last 20 years and many scientists have used it to remove lead, chromium, mercury, and arsenic with sunflower to accumulate these metals. It has the quality of high accumulation and uptake the elements which are present in trace amount. It is estimated that the sunflower plant has the capacity to reduce the Pb, Cr, Zn, Cd. Similarly, *C. sativum* is also used for the extraction of metals for phytoremediation (Nehnevajova et al. 2005).

2 Environmental Pollution

Due to population eruption in the twentieth-century contamination of heavy metal in water, soil, and air affected badly on both environment stability and as well as living things on the earth, heavy metal deposition is not degradable due to the biological action of microorganisms, unlike organic contaminants (Amir et al. 2020). The polluted environment dominated the harmful effects of heavy metal ions (Ajmal et al. 2003). The main sources of pollution in the environment are the anthropogenic source, the revolution of industries, and the expansion of population (Martín-González et al. 2006). The persistent nature of heavy metal poses a significant threat to human health, biomagnification, and containments food chain (Rajendran et al. 2003; Oelofse et al. 2007). Dominant up-to-date towns by hurried expansion trade advance (Liu et al. 2016). The pollution of the metals approximately two decades is become the emerging issue due to the high population and industries development the experiment honk which tells the more amount of Cu and Zn, urban topsoil matched with rural dust and the concentration of the lead and their size is more than countryside soil (Fatima et al. 2020). Effects of metals on humans are skin cancer because of its contact from hand to mouth and majorly effect on the children (Saeedi et al. 2012).

Heavy metal pollution is the most important issue of the environment. Human negligence and cross limits of metals which effect on human beings cause the whole world people into trouble like soil and health also have bad impacts on the terrestrial and aquatic system by the excessive transmission to air to water and soil (Wang and Qin 2007; Driscoll et al. 2013; Zhao et al. 2014; Nezhad et al. 2014). A large decomposition like the wet and dry form of metals from factories such as the electronic and also the fuel incineration and sorting of the waste and fertilizers and also home waste is playing chief part contribution to metropolitan soil pollution heavy metal (Pastor and Hernández 2012; Wang et al. 2012). Soil has unique quality against the deposition in soil and other toxic substance anthropogenic activities such as cadmium, chromium sixth is danger Hg, Pb, Zn known major metals (Wuana and Okieimen 2011). Dominate reasons for the dispersion of metals are the features of geography and worst impact on surroundings by high population more use of natural resources causes environmental damage. Its transfer by inhaled and ingest in humans (Ji et al. 2013; Varrica et al. 2014). It is a different quality of a very long life not degrade easily and has more impacts on children (Varrica et al. 2014; Ji et al. 2013). Transfer metals are influenced by the following variables: temperature, pH, and also the adsorption and soil nature and the natural things which are also involved in the transfer of the metals such as the physiological and biological alteration (Zhao et al. 2014). Soil unable to grow the healthy plants the soil quality changes the whole water cycle change which results in floods and regularly transfer of metals cause the mutagenesis and carcinogenesis effects. Environment filthiness, general matter. Utmost latest three decades intensifying natural contamination universal (Kimani-Murage 2007). Water and soil contamination affects sanitation as well as result in expanded well-being dangers, and has been involved in the ascent of "malignant growth towns." The term "malignant growth town" ordinarily indicates a town where the horribleness pace of disease is fundamentally higher than the normal level, most likely brought about by natural contamination.

2.1 *Types of Pollution*

2.1.1 Soil Pollution

The unpleasant changes in the chemical and physical nature of the soil, which destroyed the characteristic of soil which for dwelling and cultivation is known as soil pollution. Now it's the component of the universe the balance and biochemical and geochemical cycles of some essential HMs are disturbed due to human activities (Kabata-Pendias and Mukherjee 2007). Rapid urbanization and industrialization are one of the major causes of the HMs pollution of soil in the environment (Gowd et al. 2010). The HMs pollution of soil is dangerous for humans due to its persistent nature, bioaccumulation, biomagnification in the food chain along its toxicity. However, soil pollution concerns agriculture and non-agriculture areas. HMs soil pollution is caused due to human activities present at each stage of urban soil formation. In recent decades, soil pollution has drawn worldly attention (Solgi et al. 2012). HMs soil pollution is one of the major issues in China, China has only 7% farmland with 22% of the world (Zhao et al. 2015). The two agencies which conduct the survey are the protection of the environment and also land and resources 2005 to 2013 and again state by the people of China in 2014 about 82.4% of soil pollution due to Heavy metals (including Cd, Cr, Hg, Pb, As, Cu, Zn, and Ni through the report explored about the trend of heavy metal caused soil pollution but does not describe concentration and distribution of metals for this purpose every year dozen of paper was published but most of the research in a small region or limited areas and then analyzed it (Wong et al. 2002).

2.1.2 Sources of Soil Pollution

Two sources of soil pollution of heavy metals are originated from natural and anthropogenic. Heavy metals may be inherited into the soil from parent materials. The concentration of HMs relatively high in mafic and ultramafic rock as compared with felsic rock and they are responsible for the contamination of agricultural soil, crops, and vegetables (Khair et al. 2020). Industrial activities, traffic emission, inadequate waste disposal, fuel combustion, and mining activities are responsible for soil contamination in urban areas (Bullock and Gregory 2009). However, the environment also plays an important role in the transformation of HMs to a longer distance. Pollutants are absorbed by the dust due to larger surface areas and move to many miles with air current. Urban dust acts as both a sink and the source of heavy metals and contributes to the accumulation and distribution of HMs in the urban atmosphere.

2.2 *Major Contaminants of Soil*

2.2.1 Organic Pollutant

There are thousands of POPS chemicals that come from the series. POPS are insistent in nature having prolonged half-lives in air, soil, and sediments or biota have the tendency to change in gas, in environmental temperature thus they have ability to volatile from flora, water bodies, and soils in air and due to their ability of resist to breakdown reactions in the air, they travel miles before being replaced. Polycyclic aromatic hydrocarbons are formed by the burning and combustion of organic compounds and are categorized as POPs. Their existence is caused by man-made activities and pollution of PAHs especially in high-concentration industrial areas the possible sources of organic pollutants contain. Industrial waste leakage, leakage of petrochemical tanks, inappropriate usage of pesticide improper use of cleansers anti-free and oil petroleum products most common pollutants that pollute underground water quality as a result of underground storage tanks and during the transportation of petroleum's (Bakht et al. 2020). These chemicals stay in the environment for a long time, they travel from warm regions to colder regions and are called global transport. They accumulate in the food chain due to biomagnification.

2.2.2 Inorganic Pollutants

The soil in mining areas is not good for agricultural practices because its contaminant by heavy metals (Baldantoni et al. 2016). Rich in nature and by activities of human its cause bad impacts on them (Wilson and Pyatt 2007). These damaged soils are polluted with heavy metals and contain low concentrations of necessary nutrients and organic matter the performance of crops and vegetables is reduced due to high toxicity and low overall quality (Zhuang et al. 2009; Munir et al. 2020). Nature of these substances such as the heavy metals discharged into that water and later used for the useful purpose is a major source for damaging the health of the human. Moreover, its discharge into the environment causes more and more complications in the environment and food system. It is absorbed into the soil, which is farmed by the process of irrigation by wastewater, this does not cause pollution of soil but also a reason of high accumulation of heavy metals into the plants and the food badly effects. Pollutants which are very deadly enter into the human by the food because it's an only way of transfer from one side to human side it also depends on the plant type gathering of heavy metals their efficiency of metal absorption (Maalik et al. 2020; Rattan et al. 2005).

2.3 *Sources of Waste Water*

2.3.1 Textile Industry

Cotton was an important sector of industrialization in the nineteenth century during the Industrial Revolution, the industry became extraordinary due to technological advancements and efficiency improvements. Its raw materials and finished products are very popular in the international market in 1800, the United Kingdom owned 95% of the world's pins, 69% in 1850, and 58% in 1900. In 1831, the cotton industry accounted for 22% of the United Kingdom's value-added industries and 50% of UK merchandise exports. From 1780 to 1860, while promoting environmental protection, it maintained an average annual TFP growth rate of 2.6%, accounting for a quarter of the total TFP growth in the economy. Industrial pollution is one of the problems currently facing Nigeria in order to ensure, living hygienic atmosphere, the country makes various efforts to fight various industries in country one of the sources of pollution is industrial wastewater the pollution of air, soil, and water by industrial sewage is associated with a high burden of disease (WHO 2002).which may be one of the reasons for the current decline in life expectancy in the country compared to developed countries (WHO 2003). It has been shown that some of the heavy metals contained in these wastewaters (free form in wastewater or adsorbed in suspended solids) are carcinogens (Tamburlini et al. 2002). And other chemicals shown here are also toxic, depending on the dosage and contact time. These chemicals are not only toxic to humans, but also aquatic organisms and may cause food contamination. Ammonia is harmful to fish or other aquatic organisms, with a free (non-ionized) concentration of 10–50 μg/L or higher pH, and sulfur in wastewater is an environmental problem (WHO 2000). Because it causes low air quality if not maintained properly, it will threaten humans, vegetation and materials the same is true of pH, and if water for human use does not meet the required level, it is considered to cause health problems. The textile industry is the main source of such wastewater (Ghoreishi and Haghighi 2003). This is due to the nature of its activities and requires a large amount of water, which ultimately leads to a large amount of wastewater they are one of the largest users of water and pollution. The LC50 of textile wastewater is more closed and has extremely high toxicity at an acute toxicity unit (ATU) level between 22 and 960. Dyes were tested for overall toxicity throughout the process. In addition, dye baths may have higher BOD/COD levels, color, toxicity, surfactants, fibers, and turbidity, and may contain heavy metals.

2.3.2 Tanning Industry

There are a large number of tanneries with a number of 800 plus in Pakistan whereas another 237 units are situated in Kasur District. Effluents of tanning ventures containing Cr and different metals are released into the nearby water bodies which are utilized for agriculture. An amount of chromium past as far as possible (<2 ppm) makes water unacceptable for crop development (Farid et al. 2017a, b). The significant level of Cr and supplement substance in the gushing has been accounted for to hinder the seed germination and seedling development, which maybe because of the nearness of inordinate measure of disintegrated solids, chlorides, sulfides, chromium, high biological oxygen demand (BOD), and chemical oxygen demand (COD) estimations of the emanating (Mishra and Bera 1995). Wastewater-loaded water system water influences plant development and yield (Barman and Lal 1994). And the gathering of harmful overwhelming metals is biomagnified at various trophic levels through the evolved ways of life. The aggregation, be that as it may, relies upon the plant species, the components, its bioavailability, redox, pH, cations trade limit, broke down oxygen, temperature, and discharge of roots stamped harmfulness of chromium has been accounted for regarding photosynthetic colors in green growth and other higher plants. Cr harmfulness produces chlorosis and rots in plants (Farid et al. 2018a, b; Cervantes et al. 2001). The diminishing in chlorophyll substance is because of chromium, going after iron at the utilitarian site which may be meddling with the useful metal (Mg^{2+}) in the porphyrin ring (Mengel and Kirkby 2001). Chromium introduction results in complete loss of development in sidelong roots while lesser focus begins harming root top, stomata, and cotyledon hair appear to be fell and plasma layer has all the earmarks of being confined from the cell divider under cytological examinations.

2.3.3 Surgical Industry

Cobalt and chromium more in quantity by the usage of the ions and elements free by the wear and/or erosion of the embed initiating hip replacements, the metallic particles are created prevalently at the verbalization through wear of the restricting surfaces wear from the femoral stem and acetabula segments waterway so occur co, Cr metal wear particles are commonly under 50 nm in size, at least when disengaged from hip test system investigations of metal-on-metal enunciations (Williams et al. 2006). Although larger particles can be framed from the agglomeration of littler particles. The arrival of metal from joint substitutions is by a wide margin the most portrayed inside metal presentation to emerge from a careful gadget this is expected partially to the acknowledgment of neighborhood tissue responses in light of garbage when the inserts were first presented during the. Expanded way of life is making an incredible hazard to the earth by creating an enormous amount of waste contingent on the nature and wellspring of waste, it might be delegated clinical, mechanical, agrarian, metropolitan, vehicular waste and so on the quantity of waste is 90% to 70% (Allsopp et al. 2001; Grochowalski 1998). It is a well-presenting process that heats up by the deep oxidation at peal temperature at about 900°C for degradation of the waste. And is viewed as one of the four essential approaches to oversee strong squanders, related to source decrease and reuse, reusing treating the soil, and landfilling (Lee et al. 2000). This technique is issued the policy for the society safety as the biomedical waste in India.

2.3.4 Impacts of Waste Water on Food

Water pollution is a big problem and also the soil which has bad impacts on the food well-being condition when all is said in done. Developing degrees of contamination and over-utilization of assets request a type of arrangement. Anthropogenic effect on regular habitats and particularly on amphibian environments is right now a subject of expanding concern decay outward liquid (Nobukawa and Sanukida 2002). The possible reasons for circumstance, different known and unknown routes. In addition, the enormous amount of wastewater released by these plants is causing many further natural and ecological effects. Assessment of such a lot of wastewaters and its effects is wanted to help improve water use effectiveness, limit natural and ecological effects, just as encourage strategy detailing various invested individuals would thus be able to profit by the pertinent dynamic for supporting natural insurance and biological system preservation If sewage is just incompletely rewarded before it is discarded, it can pollute water and mischief colossal measures of untamed life. On the other hand, spilling or flooding can make totally untreated sewage enter waterways and other water sources, making them become contaminated (Yang and Carlson 2004). Lessen the natural effects related with giving medicinal services yet the present status of social insurance treatment over the U.S. also, internationally, offers a huge open door for productivity enhancements, possibly prompting decreases in costs, asset use and squander, and ecological effects.

2.4 Major Contaminants of Waste Water

2.4.1 Heavy Metals (HMs)

Heavy metal discusses here metal component is a moderately great thickness or toxic even at low concentrations (Lenntech 2004). Heavy metals contain cadmium, silver, chromium, zinc, arsenic, lead, copper, mercury, iron, and platinum groups heavy metals are involved in several household waste industrial processes, vehicle emissions, and agricultural activities due to their environmental stability, bioaccumulation, and high toxicity, they are considered to be one of the most serious pollutants and heavy metals are not very soluble in water they tend to adsorb on particles suspended in the sea and affect marine life (Latif et al. 2020). Toxins usually do not show toxic effects immediately after entering the environment and organisms. They are usually only visible after a few years. Heavy metal pollution from surface water and groundwater sources can cause serious soil pollution. When the extracted minerals are poured onto the soil surface for artificial coating, the pollution will increase. The great system impact of the metals on the surrounding and also on the plants this is one of worst situations faced by all over the people it destroys the natural beauty of soil by filling its poisonous molecules Food consumption is low due to poor soil quality by metals big problem nowadays end all the feature of crops by stressing on it (Farid et al. 2015; Marwa et al. 2012). Then they invent more reliable techniques and also cost-effective to save us from this exposure stress becomes low due to usage of the chelator for enhancement of phytoremediation process the huge studies performed on polluted sites growing of plants on toxic places to check their absorption power (Khalid et al. 2020). Existing consequences the ratio of metals is more in vegetable by the test due to the Seepage from dirty water that emits from the factories cause health controversy.

2.4.2 Impacts of Heavy Metals on Animals

Many xenobiotics, excluding heavy metals, store in human beings. Some of them can quickly detoxify, but the vast majority are stored in tissues and organs bioconcentration and factors and nutrient transfer factors can measure the level of metal retained in the body. Bioconcentration is a process that causes the concentration of pollutants in the body to be higher than the environmental concentration taking into account the health of humans and animals.

2.4.3 Impacts of Heavy Metals on Human Well-being

The best CSOIL exposure model contains exposure calculations for the health risk assessment of contaminated soil, which sets the critical limit of soil based on a given acceptable daily intake (ADI) value. The model covers many human exposure pathways, such as crops, meat, the share of drinking water, and air and soil consumption. Using the CSOIL model to determine the critical soil limits associated with ADI depends to a large extent on many assumptions about food supply (Lijzen et al. 2001). Introducing heavy metals in groundwater and surface water through human activities (nonstandard industrial processes, urban waste, excessive, and sometimes unnecessary chemical substances used in agricultural production) (Midrar-Ul-Haq et al. 2005). In most parts of Pakistan, the content of heavy metals exceeds the limits allowed by WHO, summarizing the results of many reports. Some heavy metals are essential for health, but when the concentration is limited, high concentrations can have harmful effects on health. Zinc (Zn) and copper (Cu) are important for health, but the concentration is limited (Solomons and Ruz 1998).

2.4.4 Chromium (Cr)

Chromium itself shows a vital part in the body, but it is not toxic itself, but some of the compounds are toxic. In the human body, chromium metal is an indispensable element in the metabolic process it can regulate blood sugar levels and help transport glucose into cells, which can be used to produce the energy our body needs a very small amount of chromium due to its contribution to the metabolism of fats, proteins, carbohydrates, carbohydrates, and other nutrients, chromium also plays an important role in the prevention of cardiovascular diseases chromium deficiency includes symptoms such as irregular blood sugar, fatigue, high cholesterol, and anxiety (Ashfaq et al. 2020). Compared to trivalent chromium (+3), hexavalent chromium (+6) is more toxic, and this form is caused by industrial pollution and excessive consumption of this form can cause skin irritation, digestive problems, and lung cancer according to the World Health Organization, the maximum allowable concentration of chromium in water is 0.05 mg/L. Pakistan Council of Research in Water Resources (PCRWR) conducted a study in 23 major cities in Pakistan and found that only 1% of groundwater samples exceeded the safety limit for chromium. Many researchers report that the concentration of chromium exceeds the WHO safety limit (0.05 mg/L) a chromium content of 9.80 mg/L (2.12 mg/L) was found in drinking water samples collected in Kasur residential areas in Punjab (Tariq et al. 2008). For comparison, a 25% sample was taken in Karachi (Sindh Province), which 75% of the samples were from various sources in Khyber Pakhtunkhwa Province, used for drinking water analysis, and found that the chromium concentration exceeded the maximum allowable value of drinking water (Midrar-Ul-Haq et al. 2005). If industrial activities are discharged without treatment, the concentration of heavy metals will also be affected Lahore and Sialkot have several branches of the leather and tanning industry, which can have adverse health effects due to more focus and release of chromium in the atmosphere (Ullah et al. 2009).

2.4.5 Nickel (Ni)

The metal which is known as nickel it has a very high concentration in the soil, water, and atmosphere, although our demand for nickel is not large, it is still at a tracking level it is not clear why we need this mineral nickel application research plays a role in the use of folic acid and vitamin B12 in the human body comprehensive studies of hormone compounds have also shown that nickel concentration occurs in the hormone group (Kasprzak et al. 2003). Absorbing too much nickel can cause adverse health effects, such as lung cancer, nose cancer, throat cancer and prostate cancer, diseases, dizziness, asthma, and heart problems; people are exposed to nickel by inhaling nickel-containing air, drinking water, and food; the WHO value of a nickel is 0.02 mg/L, which is harmless to our body (McGregor and Peake 2000). In Pakistan, the nickel concentration in groundwater and surface water is between 0 and 3.66 mg/L, and between 0 and 1.52 mg, the study shows that the nickel content in Khyber Pakhtunkhwa province has a higher nickel content (0.0023.66 mg/L) in Karachi (Sindh) lawsuit, which ranges from (0.01–2).

2.4.6 Arsenic (As)

This is metalloid that is ubiquitous, terrestrial in this region, and is measured as a universal well-being risk factor essentially, arsenic is concentrated in the earths crust, rocky sediments, and gradually leached into drinking water (Vahter 2008). Exposure to inorganic arsenic through consumption of contaminated food, water, air, and occupational exposure, but does not contain arsenic organic products (mainly seafood, such as fish, oysters, shrimp, mussels, etc.) have a serious impact on human health. Arsenic has a low and long-term occurrence in many medical complications called "arsenic poisoning" (McCarty et al. 2011). Arsenic exposure is the result of natural or man-made sources ingestion through the skin, inhalation and absorption are one of the basic methods for arsenic to enter the body pentavalent and trivalent arsenic compounds can be quickly and densely absorbed from strong foods in addition, the absorption of sodium arsenate is high, and the absorption of inorganic tetravalent arsenic is poor, however, arsenic trisulfide and lead scales and their useful materials such as hair and nails all contain coating systems it is usually described as being available in the body as needed the skin is considered to be more sensitive to arsenic poisoning and higher initial high symptoms (Rahman et al. 2009). Abnormal skin is a long-term feature of adult exposure to arsenic, in addition, men and women work together to cause skin-induced symptoms due to arsenic. As can be seen from the main features, skin lesions caused by exposure to arsenic are melanosis, keratosis, and pigmentation (Rahman et al. 2009). Those who absorb the highest load after oral administration should consider the Senate. Health issues are faced by the nations Nepal, Bangladesh, Myanmar, India, Vietnam, and China (Islam-Ul-Haq et al. 2007).

2.4.7 Lead (Pb)

This is venomous heavy metal on earth (Raja et al. 2008). Too much data depict delineates, such as industrial waste, household paint, and automobile exhaust. Lead has no well-known functions in the human body (Raja et al. 2008). However, excessive amounts of lead can adversely affect health and may damage major organs and body systems failed research, diseases of the hematopoietic system, cardiovascular diseases, neurological diseases, and effects on the immune system are the most common diseases related to lead interaction. (Riess and Halm 2007). In pregnant women, even low lead levels can affect newborns, low birth weight, and miscarriage.

2.4.8 Effects of Heavy Metals on Plants

Overwhelming metals are conceivably harmful to plants phytotoxicity brings about chlorosis, frail plant development, yield gloom, and may even be joined by decreased supplement take-up, messes in plant digestion, and in leguminous plants, a diminished capacity to focus atomic nitrogen. In substantial metal-dirtied soils, plant development can be repressed by metal retention. Nonetheless, some plant species can aggregate genuinely a lot of substantial metals without demonstrating pressure, which speaks to a likely hazard for creatures and people (Oliver 1997). Substantial metal take-up by crops developing in polluted soil is an expected peril to human well-being in view of transmission in the natural way of life (Ginocchio et al. 2002).

2.4.9 Effects of Heavy Metals on Soil Microorganisms

Overwhelming metals are constant in soils, prompting a major issue in a biological system and making dangers human well-being through bioaccumulation in plants and creatures or bio fixation in the natural pecking order (Nabula et al. 2010; Adrees et al. 2015). Many investigations on the e-squander pollution have been done and uncover the huge effects of overwhelming metal defilement on natural quality and general well-being. More critically, the co-event of substantial metals and natural poisons at e-squander destinations displays progressively entangled collaborations in compound procedures, adsorption practices, and organic procedures. The conjunction of various poisons brings about contention for the coupling destinations of adsorbents and chemicals restraining microbial digestion systems and along these lines lessening the corruption effectiveness of natural contaminations. Hence, a more profound investigation of the environmental impacts of e-squander removal at present draws expanding considerations, which may significantly profit land the board and reclamation in e-squander reusing districts. Overwhelming metals are poisonous to all the microorganisms practically by influencing the development, morphology, and digestion and repressing basic cell capacities, for example, protein blend and the trustworthiness of cell layers prompting the adjustments in capacity, movement, and a decent variety of soil microbial network. Because of the high affectability to natural changes in their living spaces, the structure and a decent variety of soil microbial networks are progressively examined noteworthy negative connections be tween's dirt chemical movement, microbial wealth/assorted variety, and substantial metal tainting inclinations have been broadly talked about. For instance, the raised substantial metal fixation antagonistically influences the all-out populace of microscopic organisms and action oomycetes, and enzymatic exercises in soil biological systems. An examination of soil microbial ordered structure at an e-squander site shows fundamentally modified soil smaller scale biotas between the tainted and reference soils (Liu et al. 2015). As of late, numerous investigations are concentrating on the biological impacts of substantial metals on microbial network structure and assorted variety with the improvement of high-throughput sequencing.

2.5 *Treatments Strategies*

2.5.1 Thermal Treatments

Then, wastes are transformed into naturally constant waste forms such as glass and ceramic (Quina et al. 2008). As a result, heavy metals are isolated and their discharge is reserved. Sintering is a process in which solid wastes containing heavy metals (SWCHM) are heated until wastes are changed to a compact ceramic material. Heating temperatures of this process are 900–1000°C, below the melting point of wastes (Quina et al. 2008; Zacco et al. 2014). The chemical phases of wastes are reconfigured through this process (Chandler et al. 1997). Thermal treatments are research hotspots among numerous control methods. The immobilization mechanisms of these methods are vital to reveal the flexibility decline of heavy metals. In order to augment the immobilization effects of each technique, it is essential to recognize the immobilization methods (Ucaroglu and Talinli 2012). Normally, vitrification is accomplished by mixing the SWCHM with glass-forming forerunners, warming mixtures until they liquidize, and obtain a formless similar glass after the liquid cools down (Chandler et al. 1997).

2.5.2 Physiochemical Treatment

Few of the best treatments which are used for the removing of the dirt such as the metals by the procedure of the redox reaction that is chemical, give and take of ions, precipitator, recovery electrolyte, solidification, a number of physiochemical claim to these ways controlled due to useful boundaries departure is done by alkali precipitation buildup solid slurry (Afshan et al. 2015). When the quantity of the metals in the series 10–100 mg/l, the interchange of ions and adsorption is wasteful (Eccles 1999). Bio sorbents of heavy metals, yeasts, microbes are applied, depend on the chemistry of ions to check these sources to handle metals is good, maybe not precise superficial belongings organisms, major composition cell and chemical–physical, environmental impression like pH, heat metal attentiveness matching ions metals gaze added rehabilitated tools, transformed bacteria Streptococcus is microbes, Saccharomyces Aspergillums, yeast and fungus extract the Cr, Fe six and ready of permanganate potassium till turned into pink lightly (Sağ and Kutsal 1996). Advanced micro devices for the conjoint interpretation bitterness, based on the sulfate-reducing bacteria, succeed in shrinkage sulfates on hydrogen sulfide due to gradual corrosion of carbon-based substrates. More metals ions in this way so bioreactor is sulfide hydrogen not soluble is has low pH (Tokunaga and Hakuta 2002).

2.5.3 Biological Treatments

Bioremediation

In in situ practices bioremediation applied to topsoil and groundwater at the site with trifling trouble. These techniques are usually the most desired options due to lower cost and less disturbances since they offer treatment in place evading excavation and conveyance of pollutants treatment is partial by the depth of the soil that can be efficiently treated. These techniques include the mine or elimination of polluted soil. Composting is a process in which organic wastes are blackened by microbes, at raised temperatures. The hotness of the manure is about 55–65°C, this is one high in the micro-organic material where the removal of solid organic material is a simple way to conduct this by soil quarried and separated to eliminate big pillars and rubbishes (Bouwer and Zehnder 1993).

Phytoremediation

As progressively manufacturing responsibile for the rise in earth heavy metals are the main issue which are very dangerous and destroying for the ecological system and also bad for the health. Soil damaged cautions is taken to remediate the soil by the different methods such as chemically, physically, and biologically (Vara Prasad and de Oliveira Freitas 2003). Get move out from soil called excavation of soil, filling of the plot known as the entombment deep, acids and chelator which are very high amounted chemical ways cause soil quality low. Phytoremediation, using of high biomass plants this one is the new all of them to remove metals. Stressed stakes plants used for this, no determined plants of less expansion and biomass. Frequently unfinished for soil stages unapproachable system of roots is hyperactive collectors (Keller et al. 2003). That is natural kind reasonably occasional taxa, present in the geographically out-of-the-way zones through extremely minor populations environmentally friendly circumstances cause a success for this technique (Baker et al. 1994). Several studies are taken by the scientists in which they tell harmful metals become the crucial point in the soil one of them Ni increasing day by day in the soil, which is used for the purpose of cultivation.Heavy metals showed the bad reactions on the plants, it has been reported that *Brassica napus* L depicted damaged cell membrane when exposed to radiation. Another low applied radiation is more in development. Citric acid plays an essential role in reducing the negative effect of Ni toxins in both plants cured, non-treated speaks microwave treatment cause positive outcomes, boost the phytoremediation (Ehsan et al. 2014).

Phyto-Desalination

Arundo donax and the cattail are the plants altered fit to water clean system known as the abbreviated form of CWs also done to decrease of the effectiveness of matter organic, pathogens, nutrients were applied in the constructed swamplands, helophytic as reciprocal reed (Carballeira et al. 2016; Vymazal 2013). Halophytes that have the quality of bear of salts done for phytoremediation salts in the clean water project (Shelef et al. 2012, 2013). A new way of management salinity, brings good outcomes better yields. Skill to gather metals and salts (Manousaki and Kalogerakis 2011). biggest achievements for water by using this technique very and better (Manousaki and Kalogerakis 2011; Suaire et al. 2016).

Phyto Volatilization

The method by plants used to get rid of pollutants then throw into the atmosphere by leaves (Ghosh and Singh 2005). The procedure of phytovolatilization depends deeply on the physical features of the pollutant itself in order to get into the plant, the contaminant must have the proper chemistry to permit through root tissue once inside the plant it can be phytovolatilized into the atmosphere through the apertures. This was completed for conversion into the vaporizations of Se and Hg, was a feast the pollutants thinned range of volatilization did not presence danger (Meagher 2000). Brassicaceae relate plants for example broccoli, cabbage (*Brassica oleracea* var. *capitata*) promising to volatilize Se >10 g/ha/d. when phytovolatilization is combined that the phytoextraction eventual, experience, phytoremediation, saving Se-danger soil can be amplified about 2–3 times (Zayed and Terry 1994).

Rhizo-Degradation

It exists in act plant roots, related microflora, emission yields to abolish chemical part of the root Specific study tells the rhizosphere impacts happening remediation that is hydrocarbon poly aromatic this containment organic. However, rhizodegradation has not been adequately advanced for the extensive claim. Fine root progress possibilities of enhanced rhizodegradation accomplishment of Italian ryegrass in the greater boundary of diesel growth of the plant (Karim et al. 2006).

Phytodegradation

Phytodegradation consequences when pollutants are degraded within plants concluded metabolic methods or through deprivation exterior to the plant through properties of enzymes produced by plant alteration, acceptance and metabolism methods that occur within a specific use account for phytodegradation mechanisms uptake and conversion are reliant on contaminant hydrophobicity, solubility, and separation the ability to reduce the big collection of pollutant which are organic due to the involvement of root plant favorable effect of willow on the outcome of organic contaminant Plain Flow willow results showed by on the oil spill soil because the talent, remove oil since, soil due to its microbes presence with roots. Now it's used in the Siberian taiga for removing the oil making the green area (Chralovich 2000). Results showed a 57% reduction in the oil organic in those plots which are planted with willow and this is against the 15% on the plots which are fallow (Vervaeke et al. 2003). Good things transport oxygen to root zone study of planting of willows on landfill. Has shown that those sites have advanced methane oxidation rates associated with plots without plants, trees with their widespread root organizations may offer a better environment for methane-oxidizing bacteria thus permitting its release into the atmosphere. Likely results in the remediation of shallow aquifer sites polluted with ethanol-blended gasoline spills using willow and its acceptance to enlarged levels of ethanol, has been reported (Corseuil and Moreno 2001).

Phytostabilization

Phytostabilization can bring about the obsession with portable or harmful contaminants inside the core of plant either due to restriction, assimilation and collection by origins of plants Willows' capacity seize substantial pollutants such as metals and different toxins, root frameworks, stopping their flow inside nature, can be of extraordinary down to earth use the work of plants in this strategy is, most importantly, diminishing the measure of water permeating through the dirt framework, which makes it conceivable to stay away from the development of harmful and hazardous leachates, furthermore invigorating sorption procedures of overwhelming metals in soil because of the impacts of plants on the dirt condition, bioavailable types of overwhelming metals can change over to less effectively accessible ones. Mixes found in the roots (phytosiderophores, natural acids, phenol mixes) respond with metal particles and accelerate them as insoluble salts just as collecting metals in roots. Next, they tie them in the cell divider and collect just as in vacuoles. Different types of grass, for example, red fescue (*Festuca rubra* L.) is the most helpful during the time spent the supported phytostabilization of substantial metals in soils (Touceda-González et al. 2017). Phyto stabilization might be supported by the use of different soil added substances (helped phytostabilization) the impact of added substances must be broke down in a wide range, on account of which it will be conceivable to survey the impacts of individual added substances on physical and concoction soil properties the versatility of vast metals, the plant, the viability of the phytostabilization procedure (Sylvain et al. 2016; Radziemska et al. 2017).

2.6 *Phytoextraction*

Best ever treatment name as the phytoextraction that using the big and more biomass plants which collects metals from the polluted regions badly affected by the industries from soil and water and then fold from lower ground to upper through plant tissues. Depend on its bioavailability significant metals have low in soils, then we use this also chelates for better results (Evangelou et al. 2007).

2.6.1 *Brassica napus* (L.)

Phytoremediation is a very new and interesting technique that uses natural plants which are green in nature to pull-down the effect of metals from the society. It is cost effective (Boonyapookana et al. 2005). And the plant which has biofuel nature is used as the phytoextractor of heavy metals. *Brassica napus* L. ability to extract natural gift and there. Good quality plant (Park et al. 2008). Contain most oil in the seeds.

2.6.2 *Eichhornia crassipes*

Native of the south, Asia, America region and its preferable growing rate 17.5 metrics has been reported in everyday record. Amphibian plant. Used for phytoremediation (Zhu et al. 1999). Water hyacinth industrialized above panel, flood plan soils stream the Yamuna in Delhi. Reasons are for gathering countless metals, exclusion, Co, Al, that operated perfection of water (Zhu et al. 1999; Mishra and Tripathi 2008) Declared phytoremediation metals by this plant for, Ni wastewater cleansing due to this (Youngchul et al. 2006). Systems employing active wetland plants for metals extraction (Lone et al. 2008).

2.6.3 *Zea mays* (L.)

Oil usage by the industry as in the burning process, traffic batteries also contain the cadmium convert into the soil, air, and living beings. Unimportant poisonous overwhelming metal and moderately portable, in this way effectively moved to natural pecking order from the dirtied rural plot (Ahmad et al. 2011). Take some seeds of it (*Zea mays* L.) collections from Sahiwal-2002 they set experiment the research company held into the pot of plastic with a set of sand enclosed 300 mg kg-1 controls of the cdcl2 extract of citrus outcomes prediction cause more the level of phytoextraction 0, 0.25, 0.5, 1 and 2 g kg^{-1} (Turgut et al. 2004). Generally, tolerate distinct Cd in soil. Then using citrus acid by the method maize as taken by the example after collection about 10 days of plants, roots clean by the 5 mm super cold $CaCl_2$ eject extracellular Cd. Satisfied deionized water then two roots and shoots were removed, tarnished by channel paper, and gauged broiler dried at 70 °C handled solution of H_2SO_4 and H_2O_2 instrument Atomic Absorption Spectrophotometer volume of the plant to translocate irresistible metals and (Anamika et al. 2009).

2.6.4 *Coriandrum sativum* (L.)

Dhanyaka is popular by the name of it and mostly from the eastern and central European countries, Bangladesh, China. The controls of the substantial metal content in restorative and fragrant plants speak to one of the components for the assessment of their quality, the high overwhelming metal substance in some therapeutic plants emerges from their capacity to amass specific metals. Be that as it may, high substantial metal take-up due to specific properties of this dirt, for example, acridity and/or the nearness of metal-bearing minerals, which favors the versatility of overwhelming metals in dirt and their high accessibility to plants. The plant requires at any rate 17 components to finish their life cycles, including the overwhelming metals Cu, Zn, and Ni. Plants additionally aggregate unimportant metals, for example, Pb and Cd, when each is available on the earth. Some metal particles such as cadmium, lead, and mercury have harmful jobs in biochemical responses in our body, it is a solid connection between miniaturized scale supplement, nourishment of plants, creatures, and people. Overwhelming metal take-up, translocation, and sequestration are key parts of a plant's capacity to amass and adapt to high groupings of substantial metals, for example, Cr, Mn, Fe, Co, Ni, Cu, Zn, As, and Pb are an expanding ecological issue around the world *C. sativum* L. was chosen for the substantial metal accumulation the various pieces of *C. sativum* L. many plants may have the potential for lead phytoremediation that have not yet been tried on the off chance that remediation properties were recognized in another plant, the plant could help in the expulsion of lead from the earth *C. sativum* L. has been appeared to diminish lead harmfulness in the blood due to its chelating properties. On the off chance that *C. sativum* can chelate lead when completely developed, it is conceivable that it might likewise retain lead from its condition during development this starter and proof of idea study is intended to look at the phytoremediative properties of *C. sativum* L. so as to decide if it can possibly work as a modest yet compelling approach to expel lead from nature (Tangahu et al. 2011). Different mentor talks about the various plants, lethal of metals *C. sativum* L. composition of plant absorbed more things from the soil into roots on the test takes by the guider, preparing solution outcomes minor Ni to become aware of by this plant (Yang and Carlson 2004). After a lot of reading on *C. sativum* information delivered is best in phytoremediation then in this study was taken in solution hydroponic which made by different chemicals for the growth of coriander due to its favorable ability for lessening the metals from the world (Zaheer et al. 2015). Imaginary circumstances set up hydroponic scheme seeds sow cured with radiation microwave previous knowledge identified when seeds show more progress is definitely use of radiation at low level. Citric acid is considered as one of the best phytoextraction amendment and results in higher accumulation by *C. sativum* (Ali et al. 2013).

2.6.5 *Helianthus annuus* (L.)

Cadmium and lead are taken by the sunflower higher biomass than others in grams 6.32 in the phytoremediation should be used of very extraordinary biomass for the damage soil recovery which is full of metals in the lead 50 treatment (Papoyan and Kochian 2004). Assembly that is differed with species obsessions, orderings extensive metals, in height biomass that be able to mass dangerous metals efficiently. Each characteristics of contagions, conditions of plants variations in rhizosphere capacity of bioaccumulation extremely, with pH, Eh rules, must improve the phytoremediation settings infected of metals (Lasat 2002).

2.6.6 *Spinacia oleracea* (L.)

Vegetables are a significant piece of our day-by-day diet and on a normal 130 g vegetables are devoured by a grown-up every day in Bangladesh. Spinach (*Spinacia oleracea*) is a yearly plant and may get by overwinter in calm areas it is one among the most well-known vegetables in winter in Bangladesh and leaves are the palatable piece of it spinach contains shallow root framework and supplements take-up by it shifts with soil and climatic conditions transportation and collection of substantial metals in plants additionally relies upon sorts of soil, soil pH, soil natural issue content, nearness of other compound and kind of plant species two significant effects brought about by substantial metal gathering, one is its passage into the human eating routine and another is declining crop creation because of restraint of metabolic procedures (Singh and Agrawal 2008). Substantial metal sullying of the food things is one of the most significant evaluation boundaries of food quality confirmation (Khan et al. 2008). Accordingly, global and national guidelines on food quality have brought down the most extreme admissible degrees of harmful metals in food things because of expanded attention to the hazard (Radwan and Salama 2006). The spinach is an edible plant investigates by the people who are experts in research the quantity of the metals in leaves of this plant more and more in both soil factories contaminated and farm polluted because spinach established in both soils results in chromium more health risks for male and female. *T. officinale* metals stayed r in shoots more than roots according to the performing action by the researcher has the very effective nature for subsiding the risk conveys deleterious special effects on plant normal mechanisms such as the leaf fatness is cheap, mitochondria organization badly vanished.

2.7 *Amendments*

2.7.1 Biochar Absorbent

Biochar absorbent and carbonaceous material are obtained from the biological pyrolysis. Many quantities that are used as feedstock, mud, plant ingredients, compost the wood biochar is the old usage about thousands years this is a new step for the invention of it which contain high cations because its very helpful to change their ability and its convert into the alkaline biochar cause many advantages to the plant and also good reason for the soil to high the activity which are biological (Lehmann et al. 2011; Paz-Ferrero et al. 2012). Following benefits of this contribute to decreasing the level of greenhouse gas from the forming points and high the carbon of soil appropriation contrary kind of carbon. Modifications of soil class (Paz-Ferreiro et al. 2012). Green economy can be high by this (Jeffery et al. 2011; Liu et al. 2013). A lot of good services of it explained by authors Manufacturing of biochar is the good idea for resizing the carbon biomass for becoming stable. This is done for decreasing greenhouse releasing and turned into biofuel (Inyang et al. 2012). Research on biochar predicts very best for adaptation of the soil its cause the soil good (Hossain et al. 2010). Remove both contaminants which are inorganic and organic (Uchimiya et al. 2010). Composition is too soft and has many functional groups good for remediate the heavy metals when in the marine region. But less settlements on the effect of alteration organic that is biochar on the transfer of metals (Beesley et al. 2010).

2.7.2 Organic Acids

The citric acid has a low molecular weight and has less leaching and high degradation power known as chelating agents. Second research again on it every metal compact the plant stamina in this author use Cu give the lesser EDTA makes the task for power metals destroy green pigment *a*, *b* totally to some extent vanished (Farid et al. 2013). Stress oxidative due to excessive aggregate of metals in soil more in before statistics (Sing and Agrawal 2010). If enzyme activity extend predicts plant tolerant acids brings produced of more biomass components of photosynthetic, and also transfer gas features One more turn on this plant by the consultants of science applied the acid citric output is brought good stem, roots, and leaves production and protect the electrolyte leakage. Transportation of Cd maybe others by tissues (Kasha et al. 2001). Phytoremediation phenomena best all over the globe should be implemented.

Citric Acid

Immobilization of generous metals over and done with the development of lime, phosphate (Ebbs et al. 1998), and calcium carbonate is a good procedure. This remediation propels have the upsides of instantly diminishing the risk factors creating from metal contamination anyway simply be seen as fleeting choice because these metals have been ousted from the earth condition considering a growing need to address regular tarnishing, various remediation progress have been made to treat soil, leachate, wastewater and groundwater corrupted by various poisons recalling for situ and ex-situ systems (Aboulroos et al. 2006). Standard procedures to remediate metal-contaminated soils including soil flushing, change, vitrification, warm desorption, epitome. It can be used at the incredibly tarnished site anyway not material to tremendous zones, Thats the explanation all the while, they crush soil structure and reduces soil creation (Evangelou et al. 2007). In addition, various masters have gotten the usage of built chelators, for instance, ethylene diamine tetra acidic destructive, ethylene glycol tetra acidic destructive, citrus remove the other fabricated chelators to overhaul bio-availability of overpowering metals (Evangelou et al. 2006; Awokunmi et al. 2015).

Oxalic Acid

A very lethal, has no color, but very powerful and also a part of carboxylic acids due to same formula ($H_2C_2O_4$), it is molecular in nature, weight is 90.03488. Logically taking place, plants and vegetables have acid concentrations ascorbic and oxalic which reduce the metals from the polluted soil also for the purpose of cleaning (Kos and Leštan 2004). Oxalic acid also removes metals very early and its reliable not too much for soil cleansing drive.

Ethylenediaminetetraacetic Acid EDTA

EDTA is crystal in nature and has no color and also a weak acid linkages with the amino carbon that belongs to poly-family and its weight and formula is (292.24), ($C_{10}H_{16}N_2O_8$). The reaction between organic and inorganic both are happened by this by giving ions of hydrogen know the phenomena of the neutralization its work with metal by the way of binds of four carboxylates and two amines for making of multiplexes of metals by it costly and more use in the fabric factory to avoid the alteration of colors food factory also apply for their purposes as a tool of conserving, impounding in cosmetics use in hospital medicine as therapy purposes for the bank of blood it is the anticoagulant very effectual (Meers et al. 2005). Stabilization is strong but degradability has less level (Kos and Lestan 2004). The ability of the plants produced many acids which are organic in nature, oxalic, acetic, citric, malonic, succinic, formic, lactic acids as root exudates (Labanowski et al. 2008). These plants are very thrilling for the assimilate t metal unclean soil due to up the phytoavailability of metals by organic acids develop the process of Classification of the chelators artificial, usual. The attraction of certain metals, that why the phytoextraction may not apply for all it's for the copper and zinc, and lead, cadmium. Good for removal (Meers et al. 2005).

2.7.3 Microbes

Overall social change has caused broad ecological anthropological medical issues extensive assortment of synthetics, overwhelming metals, pesticides, chlorinated solvents, distinguished, various normal assets, such as soil, air, water (Mansour and Gad 2010). Because of their significance for down-to-earth applications, metal-open minded plant–organism affiliations have been the target of specific consideration because of the capability of microorganisms for bio amassing metals from the contaminated condition and its impacts on metal activation/immobilization, and subsequently, it additionally upgrading metal take-up and plant development. Synergistic utilization of organisms and also plant, tidy up metal-polluted soils. Microorganisms are in the circle of rhizo that involved with of plant infrastructures through defiled metal milieu soil proceeding organisms really microscopic relate some qualities for the production of plant, for example, PGPB these plants connected organisms walk from the point of the soil mass to rhizosphere of that plant which is convincingly settled rhizosphere fundamental basics plants. Supposed rhizobacteria, giving helpful assistants of plant progress (Kapulnik and Kushnir 1991). Name is given in a way Rhizobacteria viz., Achromobacter, Arthrobacter, Azotobacter, Azospirillum, Bacillus, Enterobacter, Pseudomonas, and Serratia. Give the plants more benefits for their growth in worst situations such as metals soil (Tokala et al. 2002). Correct the cell of plants breakdown pressure the metals in way lots of extraction of the metals from it (Welbaum et al. 2004). The best PGPB lessening the phytotoxicity of metals by the way of biosorption by the method of absorption 1.0–1.5 mm size very rapidly accumulate the metals from the soil inorganic (Khan et al. 2007). Microorganisms are very helpful in the procedure of phytoremediation. Most known plant growth-promoting bacteria (PGPB) play important role in the degradation of metals provided with additions of organic acids. Soil vaccination by the surfactant providing (*Bacillus* sp. J119) for increasing the tomato biomass by cd extraction by theses test because to protecting the seeds also activation of metals for upgrading the transfer of metals. Believably expands phytoextraction, amending affluence, approachability, carriage the metal enhancements, shrinking soil pH, coming chelators, P solubilization, redox ups and downs by unalike metabolites conveyed, PGPB (Dimkpa et al. 2009).

2.8 *Radiations*

2.8.1 Beta Radiation

Fission products are more by the activity which are generated by the humans these are Radionuclides has physical life half takes many years continuing pollution in the earth isotopes throw out by the alpha and beta that early drop radiation (Ezaz et al. 2020). The photosynthetic reaction is prompted due to extraordinary 90Sr coverage ranks by the study which are taken by scientists. A previous investigation on the effects of beta irradiation on field crops has shown a wide range of sensitivity between the three crops studied while wheat and lettuce were sensitive to beta-radiation, corn was not but pea crop was grown under field conditions and exposed to beta radiation arising from a fallout simulant Specifically, the effects on yield, the sensitivity of reproductive vs. vegetative tissue, and possible relative importance of beta vs. gamma radiation arising from fallout, based on theoretical calculations, are examined. Use of the radiation that is ionizing on plants results will be injurious to some extents depending on buildup localization effect of this first on DNA bind with it cause depressed its function destroy many processes which are metabolic, such as enzyme and production, hormonal work blend bad impacts on amino acids then go into the cells, radionuclide (Rabie et al. 1996). *Pharsalus vulgaris* seeds, is one of the best studies taken on by applying the low radiation of beta rays, resulted in detruction of ontogenic periods, when exposed to increase doses of radiation, it cause bad effects on the plant such as less shoot and root, branches, leaves production.

2.8.2 Alpha Radiation

Those alpha-particle radiations which are naturally occurring are ubiquitous in the environment (Kennedy et al. 2002). The alpha radioactivity in cigarettes/tobacco had been a great role in causing lungs cancer among smokers. The radionuclides uptakes into the roots of tobacco from the soil due to the heavy use of phosphate fertilizer from the root it moved upward leaves of tobacco and accumulate there. Accumulation of uranium and their decaying products into the plant and their foodstuffs are harmful to humans but the concentration of accumulation depends on the age and nature of the plants. For radiological protection, it is important to known the way of radionuclides elements from fertilizers to tobacco plants and human body many foodstuffs elements which are radioactive that is grip by nuts, fruits and leafy plants of tobacco and vegetables from air soil also (Ekdal et al. 2006).

2.8.3 Microwave Radiation

Microwaves are a piece, range of electromagnetic, viewed as heat running and repetition of 300 million series every second (300 MHz) to 300 billion (300 GHz), Relate toward frequency scope of 1 m to 1 mm dejected this nonionizing retained magnetic which is electro atomic side by side shows as deviations in vibrational life, particles, warmth (Banik et al. 2003). The feeble power of microwaves didnt influence plant development however expanded dosages eased back seed germination (Oprica 2008). Vegetable yield depends on the assortment as well as feature of seeds virtue, growth strength as indicated by the quality necessities this for many types of seed classes, sprouting of radish should attain, 80%, carrots—70%, and tomatoes—at any rate 75%. Detailed are microwaves, diminish incubation of seeds. The impact of microwave additionally relied upon its source, for example, the cucumber developed better under microwave-fueled sulfur lights than under metal halide lights. Microwave influenced photosynthetic shade substance in rewarded rye seedlings, particularly when the examples were additionally rewarded with the fluid arrangement of KNO3. Use of MW illumination procedure for the warm treatment of waste slop that the possible uses of MW vitality as healing options for different kinds of squanders (defiled soils, ooze, or wastewater) have pulled in consistently developing examination endeavors during the most recent decade. Its balance out overwhelming metals and the significant main impetuses that have caused this ascent in the utilization of MW innovation are that use of MW vitality is better than ordinary warming as a result of its capacity to warm quickly, quicken response rates, give moment on/off control and increment vitality effectiveness, which results from its capacity to specifically initiate or stifle response pathways or specifically heat substances (Jothiramalingam et al. 2010).

2.8.4 Gamma Radiation

The radiation of the gamma best form of electromagnetic has 10-kV of electrons and peak level inflowing as compared to others just as alpha, beta. Uses process that is atomic for the agriculture plant upgrading, radiance by seeds result genetic, which cause more authorize plant grain produce, worth. The main motive behind decreasing the pathogens as a safety measure. Policies are reliant on physiological fluctuations, the difficulty of production of seeds, roots, shoots which is estimated for the grains of oat prolongation by irradiated comprehensive light of seeds wheat moderated shoot and root extents on germination. Radiation of gamma is very used in characters that are physiological. Radicals change substantial parts, plant cells work differently on the morphology, some of the things which depend on the radiation natural attraction, the composition of plants. ultra-auxiliary observations plants on which use of radiations visible simple fluctuations in chloroplasts but with 50 Gyexposed chloroplasts were gradually sensitive at extraordinary percentage this, organelles comparable consequences accounted to be activated due to ecological stress UV, considerable metals, acidic torrent, great light (Quaggiotti et al. 2004). Little bit of light does not affect the modifications in ultrastructure chloroplasts high beams cause the letdowns, blend of protein, hormone balance, leaf gas trade, water skill, and multiple drives when apply on seeds. Rapid oxidative burden for overproduction requires responsive oxygen species (ROS, e.g., superoxide radicals ($O_2^{\bullet -}$), hydroxyl radicals (OH^-), and H_2O_2 (Apel and Hirt 2004)).

3 Conclusions

Literature suggests that various amendments like chelating agents, biochar, and early treatment of plant seeds with a range of radiation can enhance the efficiency of phytoremediation and could be used for the treatment of industrial wastewater and heavy metal contaminated sites. Furthermore, it is worth mentioning that the efficacy of these amendments is directly linked with the plant species and variety. The selection of a suitable amendment to treat a contaminated site for a specific plant species and variety might be different as compared to the other species. For this purpose, detailed studies are required to evaluate the efficacy of these amendments and the suitability of plant species.

Acknowledgments The authors are thankful to the University of Gujrat, Gujrat, Pakistan, and the Higher Education Commission (HEC) of Pakistan for providing technical and library support. The present study is part of HEC funded NRPU funded project number 2017/HEC/R&D/NRPU/8996.

References

Aboulroos SA, Helal MID, Kamel MM (2006) Remediation of Pb and Cd polluted soils using in situ immobilization and phytoextraction techniques. Soil Sedim Contam Int J 15(2):199–215

Abrahams PW (2002) Soils: their implications to human health. Sci Total Environ 291(1–3):1–32

Adrees M, Ali S, Rizwan M, Zia-ur-Rehman M, Ibrahim M, Abbas F, Farid M, Qayyum MF, Irshad MK (2015) Mechanisms of silicon-mediated alleviation of heavy metal toxicity in plants: a review. Ecotoxicol Environ Saf 119:186–197

Afshan S, Ali S, Bharwana SA, Rizwan M, Farid M, Abbas F, Abbasi GH (2015) Citric acid enhances the phytoextraction of chromium, plant growth, and photosynthesis by alleviating the oxidative damages in *Brassica napus* L. Environ Sci Pollut Res 22(15):11679–11689

Ahmad MN, Mokhtar MN, Baharuddin AS, Hock LS, Ali SRA, Abd-Aziz S, Hassan MA (2011) Changes in physicochemical and microbial community during co-composting of oil palm frond with palm oil mill effluent anaerobic sludge. BioRes 6(4):4762–4780

Ajmal M, Rao RAK, Anwar S, Ahmad J, Ahmad R (2003) Adsorption studies on rice husk: removal and recovery of Cd (II) from wastewater. Bioresour Technol 86(2):147–149

Ali H, Khan E, Sajad MA (2013) Phytoremediation of heavy metals—concepts and applications. Chemosphere 91(7):869–881

Allsopp M, Costner P, Johnston P (2001) Incineration and human health. Environ Sci Pollut Res 8(2):141–145

Amir W, Farid M, Ishaq HK, Farid S, Zubair M, Rizwan M, Raza N, Ali S (2020) Accumulation potential and tolerance response of *Typha latifolia* L. under citric acid assisted phytoextraction of lead and mercury. Chemosphere 257:127247

Anamika S, Eapen S, Fulekar MH (2009) Phytoremediation of cadmium, lead and zinc by Brassica juncea L. Czern and Coss. J Appl Biosci 13:726–736

Apel K, Hirt H (2004) Reactive oxygen species: metabolism, oxidative stress, and signal transduction. Annu Rev Plant Biol 55:373–399

Ashfaq H, Abubakar M, Ghulzar H, Farid M, Yaqoob S, Komal N, Azam Z, Hamza A, Ali S, Adrees M (2020) Phytoremediation potential of oilseed crops for lead- and nickel-contaminated soil. In: Plant ecophysiology and adaptation under climate change: mechanisms and perspectives II, mechanisms of adaptation and stress amelioration. Springer, Singapore. https://doi.org/10.1007/978-981-15-2172-0_31

Awokunmi EE, Asaolu SS, Adefemi SO, Gbolagade AY (2015) Contributions of municipal solid waste to heavy metal concentration in soil near okeese dumpsite, Ilesha, Osun State, Nigeria. Inter J Environ Protect 5(1):44–51

Baker J, Grewal D, Parasuraman A (1994) The influence of store environment on quality inferences and store image. J Acad Mark Sci 22(4):328–339

Bakht S, Safdar K, Khair KU, Fatima A, Fayyaz A, Ali SM, Munir H, Farid M (2020) Response of major food crops under drought stress; physiological and biochemical response. In: Agronomic crops - volume 3: stress responses and tolerance. Springer, Singapore

Baldantoni D, Morra L, Zaccardelli M, Alfani A (2016) Cadmium accumulation in leaves of leafy vegetables. Ecotoxic Environ Safety 123:89–94

Banik SBASGS, Bandyopadhyay S, Ganguly S (2003) Bioeffects of microwave—a brief review. Bioresour Technol 87(2):155–159

Barman SC, Lal MM (1994) Accumulation of heavy metals (Zn, Cu, Cd and Pb) in soil and cultivated vegetables and weeds grown in industrially polluted fields. J Environ Biol 15(2): 107–115

Bech J, Duran P, Roca N, Poma W, Sánchez I, Roca-Pérez L, Poschenrieder C (2012) Accumulation of Pb and Zn in Bidens triplinervia and Senecio sp. spontaneous species from mine spoils in Peru and their potential use in phytoremediation. J Geochem Explor 123:109–113

Beesley L, Moreno-Jiménez E, Gomez-Eyles JL (2010) Effects of biochar and greenwaste compost amendments on mobility, bioavailability and toxicity of inorganic and organic contaminants in a multi-element polluted soil. Environ Pollut 158(6):2282–2287

Boonyapookana B, Parkpian P, Techapinyawat S, DeLaune RD, Jugsujinda A (2005) Phytoaccumulation of lead by sunflower (Helianthus annuus), tobacco (Nicotiana tabacum), and vetiver (Vetiveria zizanioides). J Environ Sci Health 40(1):117–137

Bouwer EJ, Zehnder AJ (1993) Bioremediation of organic compounds—putting microbial metabolism to work. Trends Biotech 11(8):360–367

Bullock P, Gregory PJ (eds) (2009) Soils in the urban environment. John Wiley & Sons

Carballeira T, Ruiz I, Soto M (2016) Effect of plants and surface loading rate on the treatment efficiency of shallow subsurface constructed wetlands. Ecologi Engine 90:203–214

Cervantes C, Campos-García J, Devars S, Gutiérrez-Corona F, Loza-Tavera H, Torres-Guzmán JC, Moreno-Sánchez R (2001) Interactions of chromium with microorganisms and plants. FEMS Microbiol Rev 25(3):335–347

Chandler AJ, Eighmy TT, Hjelmar O, Kosson DS, Sawell SE, Vehlow J, Hartlén J (1997) Municipal solid waste incinerator residues. Elsevier

Choi AM, Ryter SW, Levine B (2013) Autophagy in human health and disease. New England J Medicin 368(7):651–662

Chralovich E (2000) Vtoraya zizn "mertvoy semli". Lesnoe chozaystvo 6:26

Corseuil HX, Moreno FN (2001) Phytoremediation potential of willow trees for aquifers contaminated with ethanol-blended gasoline. Water Res 35(12):3013–3017

Dimkpa C, Weinand T, Asch F (2009) Plant–rhizobacteria interactions alleviate abiotic stress conditions. Plant Cell Environ 32(12):1682–1694

Driscoll CT, Mason RP, Chan HM, Jacob DJ, Pirrone N (2013) Mercury as a global pollutant: sources, pathways, and effects. Environ Sci Techn 47(10):4967–4983

Ebbs SD, Norvell WA, Kochian LV (1998) The effect of acidification and chelating agents on the solubilization of uranium from contaminated soil, vol 27, no 6. American Society of Agronomy, Crop Science Society of America, and Soil Science Society of America, pp 1486–1494

Eccles H (1999) Treatment of metal-contaminated wastes: why select a biological process. Trends Biotech 17(12):462–465

Ehsan S, Ali S, Noureen S, Mahmood K, Farid M, Ishaque W, Rizwan M (2014) Citric acid assisted phytoremediation of cadmium by Brassica napus L. Ecotoxic Environ Safety 106:164–172

Ekdal ELÇİN, Karali TURGAY, Sac MM (2006) 210Po and 210Pb in soils and vegetables in Kucuk Menderes basin of Turkey. Radiat Meas 41(1):72–77

Evangelou MW, Ebel M, Schaeffer A (2006) Evaluation of the effect of small organic acids on phytoextraction of Cu and Pb from soil with tobacco Nicotiana tabacum. Chemosphere 63(6): 996–1004

Evangelou MW, Ebel M, Schaeffer A (2007) Chelate assisted phytoextraction of heavy metals from soil. Effect, mechanism, toxicity, and fate of chelating agents. Chemosphere 68(6):989–1003

Ezaz Z, Azhar R, Rana A, Ashraf S, Farid M, Mansha A, Naqvi SAR, Zahoor FA, Rasool N (2020) Current trends of phytoremediation in wetlands: mechanisms and applications. In: Plant eco-physiology and adaptation under climate change: mechanisms and perspectives II, mechanisms of adaptation and stress amelioration. Springer, Singapore

Farid M, Ali S, Ishaque W, Shakoor MB, Niazi NK, Bibi I, Dawood M, Gill RA, Abbas F (2015) Exogenous application of EDTA enhanced phytoremediation of cadmium by Brassica napus L. Int J Environ Sci Technol 12(12):3981–3992

Farid M, Ali S, Rizwan M, Ali Q, Abbas F, Bukhari SAH, Saeed R, Wu L (2017a) Citric acid assisted phytoextraction of chromium by sunflower; morpho-physiological and biochemical alterations in plants. Ecotoxicol Environ Saf 145:90–102

Farid M, Ali S, Rizwan M, Ali Q, Saeed R, Nasir T, Abbasi GH, Rehmani MIA, Ata-Ul-Karim ST, Bukhari SAH (2018b) Phytomanagement of chromium contaminated soils through sunflower under exogenously applied 5-aminolevulinic acid. Ecotoxicol Environ Saf 151(30):255–265

Farid M, Ali S, Rizwan M, Saeed R, Tauqeer HM, Sallah-Ud-Din R, Azam A, Raza N (2017b) Microwave irradiation and citric acid assisted seed germination and phytoextraction of nickel (Ni) by Brassica napus L.; morpho-physiological and biochemical alterations under Ni stress. Environ Sci Pollut Res 24(25):2150–2164

Farid M, Ali S, Rizwan M, Yasmeen T, Arif MS, Riaz M, Saqib M, Zia ur Rehman M, Ayub MA (2020a) Combined effects of citric acid and 5-aminolevulinic acid in mitigating chromium toxicity in sunflower (Helianthus annuus l.) grown in cr spiked soil. Pak J Agri Sci

Farid M, Ali S, Zubair M, Saeed R, Rizwan M, Sallah-Ud-Din R, Azam A, Ashraf R, Ashraf W (2018a) Glutamic acid assisted phytomanagement of silver contaminated soils through sunflower; physiological and biochemical response. Environ Sci Pollut Res 25(25):25390–25400

Farid M, Farid S, Zubair M, Rizwan M, Ishaq HK, Ali S, Ashraf U, Alhaithloul HAS, Gowayed S, Soliman MH (2020b) Efficacy of Zea mays L. for the management of marble effluent contaminated soil under citric acid amendment; morpho-physiological and biochemical response. Chemosphere 240:124930

Farid M, Shakoor MB, Ehsan S, Ali S, Zubair M, Hanif MS (2013) Morphological, physiological and biochemical responses of different plant species to cd stress. Int J Chem Biochem Sci 3:53–60

Farlex I (2005) Definition: environment, the free dictionary. Farlex Inc. Publishing, USA

Fatima A, Farid M, Alharby HF, Bamagoos AA, Rizwan M, Ali S (2020) Efficacy of fenugreek plant for ascorbic acid assisted phytoextraction of copper (Cu); A detailed study of Cu induced morpho-physiological and biochemical alterations. Chemosphere 251:126424

Ghoreishi SM, Haghighi R (2003) Chemical catalytic reaction and biological oxidation for treatment of non-biodegradable textile effluent. Chem Eng J 95(1–3):163–169

Ghosh M, Singh SP (2005) A review on phytoremediation of heavy metals and utilization of it's by products. Asian J Energy Environ 6(4):18

Gichner T, Patková Z, Száková J, Demnerová K (2006) Toxicity and DNA damage in tobacco and potato plants growing on soil polluted with heavy metals. Ecotoxic Environ Safety 65(3): 420–426

Ginocchio R, Rodríguez PH, Badilla-Ohlbaum R, Allen HE, Lagos GE (2002) Effect of soil copper content and pH on copper uptake of selected vegetables grown under controlled conditions. Environ Toxicol Chem Int J 21(8):1736–1744

Gowd SS, Reddy MR, Govil PK (2010) Assessment of heavy metal contamination in soils at Jajmau (Kanpur) and Unnao industrial areas of the Ganga Plain, Uttar Pradesh, India. *J Hazard Mater* 174(1–3):113–121

Grochowalski A (1998) PCDDs and PCDFs concentration in combustion gases and bottom ash from incineration of hospital wastes in Poland. Chemosphere 37(9–12):2279–2291

Hossain MA, Hasanuzzaman M, Fujita M (2010) Up-regulation of antioxidant and glyoxalase systems by exogenous glycinebetaine and proline in mung bean confer tolerance to cadmium stress. Physiol Mol Biol Plant 16(3):259–272

Inyang M, Gao B, Yao Y, Xue Y, Zimmerman AR, Pullammanappallil P, Cao X (2012) Removal of heavy metals from aqueous solution by biochars derived from anaerobically digested biomass. Bioresour Technol 110:50–56

Islam-Ul-Haq M, Deedar N, Wajid H (2007, December) Groundwater arsenic contamination–a multi directional emerging threat to water scarce areas of Pakistan. In: 6th Int IAHS Groundwater Quality Conference, held in Fremantle, Western Australia, pp 2–7

Jeffery S, Verheijen FG, van der Velde M, Bastos AC (2011) A quantitative review of the effects of biochar application to soils on crop productivity using meta-analysis. Agric Ecosystem Environ 144(1):175–187

Ji L, Lin Z, Alcoutlabi M, Zhang X (2011) Recent developments in nanostructured anode materials for rechargeable lithium-ion batteries. Energy Environ Sci 4(8):2682–2699

Ji K, Kim J, Lee M, Park S, Kwon HJ, Cheong HK, Choi K (2013) Assessment of exposure to heavy metals and health risks among residents near abandoned metal mines in Goseong, Korea. Environ Pollut 178:322–328

Jothiramalingam R, Lo SL, Chen CL (2010) Effects of different additives with assistance of microwave heating for heavy metal stabilization in electronic industry sludge. Chemos 78(5): 609–613

Kabata-Pendias A, Mukherjee AB (2007) Trace elements from soil to human. Springer Science & Business Media

Kapulnik Y, Kushnir U (1991) Growth dependency of wild, primitive and modern cultivated wheat lines on vesicular-arbuscular mycorrhiza fungi. Euphytica 56(1):27–36

Karim S, Toimil-Molares ME, Balogh AG, Ensinger W, Cornelius TW, Khan EU, Neumann R (2006) Morphological evolution of Au nanowires controlled by Rayleigh instability. Nanotech 17(24):5954

Kasha KJ, Hu TC, Oro R, Simion E, Shim YS (2001) Nuclear fusion leads to chromosome doubling during mannitol pretreatment of barley (Hordeum vulgare L.) microspores. J Exp Bot 52(359): 1227–1238

Kasprzak KS, Sunderman FW Jr, Salnikow K (2003) Nickel carcinogenesis. Mutat Res Fundam Mol Mech Mutagen 533(1–2):67–97

Keller C, Hammer D, Kayser A, Richner W, Brodbeck M, Sennhauser M (2003) Root development and heavy metal phytoextraction efficiency: comparison of different plant species in the field. Plant Soil 249(1):67–81

Kennedy TA, Naeem S, Howe KM, Knops JM, Tilman D, Reich P (2002) Biodiversity as a barrier to ecological invasion. Nature 417(6889):636–638

Khair KU, Farid M, Ashraf U, Zubair M, Rizwan M, Farid S, Ishaq HK, Iftikhar U, Ali S (2020) Citric acid enhanced phytoextraction of nickel (Ni) and alleviate *Mentha piperita* (L.) from Ni induced physiological and biochemical damages. Environ Sci Pollut Res 27:270110–227022

Khalid A, Farid M, Zubair M, Rizwan M, Iftikhar U, Ishaq HK, Farid S, Latif U, Hina K, Ali S (2020) Efficacy of Alternanthera bettzickiana to remediate copper and cobalt contaminated soil physiological and biochemical alterations. Int J Environ Res 14:243–255

Khan MA, Lee HJ, Lee WS, Kim HS, Ki KS, Hur TY, Choi Y (2007) Structural growth, rumen development, and metabolic and immune responses of Holstein male calves fed milk through step-down and conventional methods. J Dairy Sci 90(7):3376–3387

Khan S, Cao Q, Zheng YM, Huang YZ, Zhu YG (2008) Health risks of heavy metals in contaminated soils and food crops irrigated with wastewater in Beijing, China. Environ Pollut 152(3):686–692

Khan A, Ahmad A, Akhtar F, Yousuf S, Xess I, Khan LA, Manzoor N (2010) Ocimum sanctum essential oil and its active principles exert their antifungal activity by disrupting ergosterol biosynthesis and membrane integrity. Res Microbiol 161(10):816–823

Kimani-Murage EW, Ngindu AM (2007) Quality of water the slum dwellers use: the case of a Kenyan slum. J Urban Health 84(6):829–838

Kos B, Leštan D (2004) Chelator induced phytoextraction and in situ soil washing of Cu. Environ Pollut 132(2):333–339

Labanowski J, Monna F, Bermond A, Cambier P, Fernandez C, Lamy I, Van Oort F (2008) Kinetic extractions to assess mobilization of Zn, Pb, Cu, and Cd in a metal-contaminated soil: EDTA vs. citrate. Environ Pollut 152(3):693–701

Lasat MM (2002) Phytoextraction of toxic metals: a review of biological mechanisms. J Environ Quality 31(1):109–120

Latif U, Farid M, Rizwan M, Ishaq HK, Farid S, Ali S, El-Sheikh MA, Alyemeni MN, Wijaya L (2020) Physiological and biochemical response of *Alternanthera bettzickiana* (regel) G. Nicholson under acetic acid assisted phytoextraction of lead. Plan Theory 9(9):1084

Lee JY, Cole TB, Palmiter RD, Koh JY (2000) Accumulation of zinc in degenerating hippocampal neurons of ZnT3-null mice after seizures: evidence against synaptic vesicle origin. J Neurosci 20(11):RC79-RC79

Lehmann J, Rillig MC, Thies J, Masiello CA, Hockaday WC, Crowley D (2011) Biochar effects on soil biota–a review. Soil Biol Biochem 43(9):1812–1836

Lenntech K (2004) Water treatment and air purification. Rotter Dam Seweg, Netherlands

Lijzen J P A, Baars A J, Otte P F, Rikken M, Swartjes F A, Verbruggen E M J, Van Wezel A P (2001). Technical evaluation of the Intervention Values for Soil/sediment and Groundwater. Human and ecotoxicological risk assessment and derivation of risk limits for soil, aquatic sediment and groundwater

Liu L, Oza S, Hogan D, Perin J, Rudan I, Lawn JE, Black RE (2015) Global, regional, and national causes of child mortality in 2000–13, with projections to inform post-2015 priorities: an updated systematic analysis. Lancet 385(9966):430–440

Liu R, Men C, Liu Y, Yu W, Xu F, Shen Z (2016) Spatial distribution and pollution evaluation of heavy metals in Yangtze estuary sediment. Mar Pollut Bull 110(1):564–571

Liu X, Zhang Y, Han W, Tang A, Shen J, Cui Z, Zhang F (2013) Enhanced nitrogen deposition over China. Nature 494(7438):459–462

Lone MI, He ZL, Stoffella PJ, Yang X (2008) Phytoremediation of heavy metal polluted soils and water: progresses and perspectives. J Zhejiang University Sci B 9(3):210–220

Maalik U, Farid M, Zubair M, Ali S, Rizwan M, Shafqat M, Ishaq HK (2020) Rice production, augmentation, escalation and yield under water stress. In: Agronomic crops—Volume 3: Stress responses and tolerance. Springer, Singapore, pp 117–128. https://doi.org/10.1007/978-981-15-0025-1

Manousaki E, Kalogerakis N (2011) Halophytes present new opportunities in phytoremediation of heavy metals and saline soils. Ind Eng Chem Res 50(2):656–660

Mansour SA, Gad MF (2010) Risk assessment of pesticides and heavy metals contaminants in vegetables: a novel bioassay method using Daphnia magna Straus. Food Chemical Toxic 48(1): 377–389

Martín-González A, Díaz S, Borniquel S, Gallego A, Gutiérrez JC (2006) Cytotoxicity and bioaccumulation of heavy metals by ciliated protozoa isolated from urban wastewater treatment plants. Res Microb 157(2):108–118

Marwa EM, Meharg AA, Rice CM (2012) Risk assessment of potentially toxic elements in agricultural soils and maize tissues from selected districts in Tanzania. Sci Total Environ 416: 180–186

McCarty PL, Bae J, Kim J (2011) Domestic wastewater treatment as a net energy producer–can this be achieved

McGregor PK, Peake TM (2000) Communication networks: social environments for receiving and signalling behaviour. Acta Ethologica 2(2):71–81

Meagher RB (2000) Phytoremediation of toxic elemental and organic pollutants. Curr Opin Plant Biol 3(2):153–162

Meers E, Ruttens A, Hopgood MJ, Samson D, Tack FMG (2005) Comparison of EDTA and EDDS as potential soil amendments for enhanced phytoextraction of heavy metals. Chemosphere 58(8):1011–1022

Mengel K, Kirkby EA, Kosegarten H, Appel T (2001) Potassium. In: Principles of plant nutrition. Springer, Dordrecht, pp 481–511

Midrar-ul-Haq RA, Khattak HK, Puno MS, Saif KS (2005) Memon and NB Sial. Asian J Plant Sci 4:132

Mintz E, Baier K (2000) A simple system for water purification in developing countries. Centre for Disease Control and Prevention Bulletin, Atlanta Georgia

Mishra P, Bera AK (1995) Effect of tannery effluent on seed germination and early seedling growth in wheat. Seed Res 23:129–131

Mishra VK, Tripathi BD (2008) Concurrent removal and accumulation of heavy metals by the three aquatic macrophytes. Bioresour Technol 99(15):7091–7097

Munir MAM, Liu G, Yousaf B, Ali MU, Abbas Q, Ullah H (2020) Synergistic effects of biochar and processed fly ash on bioavailability, transformation and accumulation of heavy metals by maize (Zea mays L.) in coal-mining contaminated soil. Chemosphere 240:124845

Nabula G, Young SD, Black CR (2010) Assessing risk to human health from tropical leafy vegetables grown on soil amended with urban sewage sludge. Environ Pollut 159:368–376

Nehnevajova E, Herzig R, Federer G, Erismann KH, Schwitzguébel JP (2005) Screening of sunflower cultivars for metal phytoextraction in a contaminated field prior to mutagenesis. Int J Phytoremed 7(4):337–349

Nezhad MTK, Mohammadi K, Gholami A, Hani A, Shariati MS (2014) Cadmium and mercury in topsoils of Babagorogor watershed, western Iran: distribution, relationship with soil characteristics and multivariate analysis of contamination sources. Geoderma 219:177–185

Nobukawa T, Sanukida S (2002) Contributions of genotoxic precursors from tributary rivers and sewage effluents to the Yodo River in Japan. Water Res 36(4):989–995

Oelofse SHH, Hobbs PJ, Rascher J, Cobbing JE (2007, December) The pollution and destruction threat of gold mining waste on the Witwatersrand: a West Rand case study. In: 10th Int Symposium on Environ Issues and Waste manag in Energy and Mineral Production (SWEMP, 2007) Bangkok, pp 11–13

Olayinka KO, Alo BI (2004) Studies on industrial pollution in Nigeria: the effect of textile effluents on the quality of groundwater in some parts of Lagos. Nigerian J Health Biomed Sci 3(1):44–50

Oliver C (1997) Sustainable competitive advantage: combining institutional and resource-based views. Strateg Manage J 18(9):697–713

Oprica L (2008) Effect of microwave on the dynamics of some oxidoreductase enzymes in Brassica napus germination seeds. J Exp Mol Biol 9(3)

Papoyan ASHOT, Kochian LV (2004) Characterization of a heavy metal transporting p-type ATPase from Thlaspi caerulescens: does it play a role in heavy metal hyperaccumulation. Plant Physiol 136:3814–3823

Park J, Park J, Jang S, Kim S, Kong S, Choi J, Lee YH (2008) FTFD: an informatics pipeline supporting phylogenomic analysis of fungal transcription factors. Bioinformatics 24(7):1024–1025

Pastor J, Hernández AJ (2012) Heavy metals, salts and organic residues in old solid urban waste landfills and surface waters in their discharge areas: determinants for restoring their impact. J Environ Manag 95:S42–S49

Paz-Ferreiro J, Gasco G, Gutiérrez B, Mendez A (2012) Soil biochemical activities and the geometric mean of enzyme activities after application of sewage sludge and sewage sludge biochar to soil. Biol Fertil Soils 48(5):511–517

Pulford ID, Watson C (2003) Phytoremediation of heavy metal-contaminated land by trees—a review. Environ Int 29(4):529–540

Quaggiotti S, Trentin AR, Dalla Vecchia F, Ghisi R (2004) Response of maize (Zea mays L.) nitrate reductase to UV-B radiation. Plant Sci 167(1):107–116

Quina MJ, Bordado JC, Quinta-Ferreira RM (2008) Treatment and use of air pollution control residues from MSW incineration: an overview. Waste Manag 28(11):2097–2121

Rabie ABM, Dan Z, Samman N (1996) Ultrastructural identification of cells involved in the healing of intramembranous and endochondral bones. Int J Oral Maxillofacial Surgery 25(5):383–388

Radwan MA, Salama AK (2006) Market basket survey for some heavy metals in Egyptian fruits and vegetables. Food Chem Toxic 44(8):1273–1278

Radziemska M, Gusiatin ZM, Bilgin A (2017) Potential of using immobilizing agents in aided phytostabilization on simulated contamination of soil with lead. Ecologi Engin 102:490–500

Rahman MR, Shi ZH, Chongfa C (2009) Soil erosion hazard evaluation—an integrated use of remote sensing, GIS and statistical approaches with biophysical parameters towards management strategies. Ecol Model 220(13–14):1724–1734

Raja AR, Babu GV, Menezes G, Venkatesh T (2008) Lead toxicity as a result of herbal medication. Ind J Clini Biochem 23(2):200–203

Rajendran P, Muthukrishnan J, Gunasekaran P (2003) Microbes in heavy metal remediation

Rattan RK, Datta SP, Chhonkar PK, Suribabu K, Singh AK (2005) Long-term impact of irrigation with sewage effluents on heavy metal content in soils, crops and groundwater—a case study. Agri Eco Environ 109(3–4):310–322

Riess ML, Halm JK (2007) Lead poisoning in an adult: lead mobilization by pregnancy. J Gen Intern Med 22(8):1212–1215

Rizwan M, Ali S, Hussain A, Ali Q, Shakoor MB, Zia-ur-Rehman M, Farid M, Asma M (2017a) Effect of zinc-lysine on growth, yield and cadmium uptake in wheat (Triticum aestivum L.) and health risk assessment. Chemosphere 187:35–42

Rizwan M, Ali S, Qayyum MF, Ok YS, Adress M, Ibrahim M, Zia-ur-Reham M, Farid M, Abbas F (2017b) Effect of metal and metal oxide nanoparticles on growth and physiology of globally important food crops: a critical review. J Hazard Mater 322:2–16. https://doi.org/10.1016/j.jhazmat.2016.05.061

Saeedi M, Li LY, Salmanzadeh M (2012) Heavy metals and polycyclic aromatic hydrocarbons: pollution and ecological risk assessment in street dust of Tehran. J Hazard Mater 227:9–17

Sağ Y, Kutsal T (1996) The selective biosorption of chromium (VI) and copper (II) ions from binary metal mixtures by R. arrhizus. Process Biochem 31(6):561–572

Saier MH, Trevors JT (2010) Phytoremediation. Water Air Soil Pollut 205(1):61–63

Schröder P, Fischer C, Debus R, Wenzel A (2003) Reaction of detoxification mechanisms in suspension cultured spruce cells (Picea abies L. Karst.) to heavy metals in pure mixture and in soil eluates. Environ Sci Pollut Res 10(4):225–234

Schrøder N, Højlund Pedersen L, Juel Bitsch R (2004) 10,000 Years of climate change and human impact on the environment in the area surrounding Lejre. J Transdiscip Environ Stud 3(1):1–27

Shelef O, Gross A, Rachmilevitch S (2012) The use of Bassia indica for salt phytoremediation in constructed wetlands. Water Res 46(13):3967–3976

Shelef O, Gross A, Rachmilevitch S (2013) Role of plants in a constructed wetland: current and new perspectives. Water 5(2):405–419

Singh RP, Agrawal M (2008) Potential benefits and risks of land application of sewage sludge. Waste Manag 28(2):347–358

Solgi E, Esmaili-Sari A, Riyahi-Bakhtiari A, Hadipour M (2012) Soil contamination of metals in the three industrial estates, Arak, Iran. Bull Environ Contam Toxic 88(4):634–638

Solomons NW, Ruz M (1998) Trace element requirements in humans: an update. J Trace Elem Exp Med Off Publ Int Soc Trace Elem Res Hum 11(2–3):177–195

Suaire R, Durickovic I, Framont-Terrasse L, Leblain JY, De Rouck AC, Simonnot MO (2016) Phytoextraction of Na+ and Cl− by Atriplex halimus L. and Atriplex hortensis L.: A promising solution for remediation of road runoff contaminated with deicing salts. Ecologi Engin 94: 182–189

Suruchi GD, Anjali AC, Ashok MS (2012) Survey of methods for character recognition. Int J Engine Innov Techn (IJEIT) 1(5)

Sylvain B, Mikael MH, Florie M, Emmanuel J, Marilyne S, Sylvain B, Domenico M (2016) Phytostabilization of As, Sb and Pb by two willow species (S. viminalis and S. purpurea) on former mine technosols. Catena 136:44–52

Tamburlini G, Ehrenstein OSV, Bertollini R, World Health Organization (2002) Children's health and environment: a review of evidence: a joint report from the European Environment Agency and the WHO Regional Office for Europe

Tangahu BV, Sheikh Abdullah SR, Basri H, Idris M, Anuar N, Mukhlisin M (2011) A review on heavy metals (As, Pb, and Hg) uptake by plants through phytoremediation. Int J Chem Engin 2011

Tariq SR, Shah MH, Shaheen N, Jaffar M, Khalique A (2008) Statistical source identification of metals in groundwater exposed to industrial contamination. Environ Monit Assess 138(1): 159–165

Thuy HTT, Tobschall HJ, An PV (2000) Distribution of heavy metals in urban soils–a case study of Danang-Hoian Area (Vietnam). Environ Geo 39(6):603–610

Tokala RK, Strap JL, Jung CM, Crawford DL, Salove MH, Deobald LA, Morra MJ (2002) Novel plant-microbe rhizosphere interaction involving Streptomyces lydicus WYEC108 and the pea plant (Pisum sativum). Appl Environ Microbiol 68(5):2161–2171

Tokunaga S, Hakuta T (2002) Acid washing and stabilization of an artificial arsenic-contaminated soil. Chemosphere 46(1):31–38

Touceda-González M, Álvarez-López V, Prieto-Fernández Á, Rodríguez-Garrido B, Trasar-Cepeda C, Mench M, Kidd PS (2017) Aided phytostabilisation reduces metal toxicity, improves soil fertility and enhances microbial activity in Cu-rich mine tailings. J Environ Manag 186: 301–313

Turgut C, Pepe MK, Cutright TJ (2004) The effect of EDTA and citric acid on phytoremediation of Cd, Cr, and Ni from soil using Helianthus annuus. Environ Pollut 131(1):147–154

Ucaroglu S, Talinli I (2012) Recovery and safer disposal of phosphate coating sludge by solidification/stabilization. J Environ Manag 105:131–137

Uchimiya M, Lima IM, Klasson KT, Wartelle LH (2010) Contaminant immobilization and nutrient release by biochar soil amendment: roles of natural organic matter. Chemo 80(8):935–940

Ullah R, Malik RN, Qadir A (2009) Assessment of groundwater contamination in an industrial city, Sialkot, Pakistan. African J Environ Sci Techn 3(12)

Vahter M (2008) Health effects of early life exposure to arsenic. Basic Clin Pharmacol Toxicol 102(2):204–211

Vara Prasad MN, de Oliveira FHM (2003) Metal hyper accumulation in plants: biodiversity prospecting for phytoremediation technology. Electron J Biotechnol 6(3):285–321

Varrica D, Tamburo E, Milia N, Vallascas E, Cortimiglia V, De Giudici G, Losno R (2014) Metals and metalloids in hair samples of children living near the abandoned mine sites of Sulcis-Inglesiente (Sardinia, Italy). Environ Res 134:366–374

Vervaeke P, Luyssaert S, Mertens J, Meers E, Tack FMG, Lust N (2003) Phytoremediation prospects of willow stands on contaminated sediment: a field trial. Environ Pollut 126(2): 275–282

Vymazal J (2013) The use of hybrid constructed wetlands for wastewater treatment with special attention to nitrogen removal: a review of a recent development. Water Res 47(14):4795–4811

Wang X, Yang G, Feng Y, Ren G, Han X (2012) Optimizing feeding composition and carbon–nitrogen ratios for improved methane yield during anaerobic co-digestion of dairy chicken manure and wheat straw. Bioresour Technol 120:78–83

Wang XS, Qin Y (2007) Some characteristics of the distribution of heavy metals in urban topsoil of Xuzhou, China. Environ Geochem Health 29(1):11–19

Williams PL, Mishin Y, Hamilton JC (2006) An embedded-atom potential for the Cu–Ag system. Model Simul Mater Sci Eng 14(5):817

Wilson B, Pyatt FB (2007) Heavy metal dispersion, persistance, and bioccumulation around an ancient copper mine situated in Anglesey, UK. *Ecotoxi Environ Safety* 66(2):224–231

Wong SC, Li XD, Zhang G, Qi SH, Min YS (2002) Heavy metals in agricultural soils of the Pearl River Delta, South China. Environ Pollut 119(1):33–44

World Health Organization (2000) The world health report 2000: health systems: improving performance. World Health Organization

World Health Organization (2002) The world health report 2002: reducing risks, promoting healthy life. World Health Organization

World Health Organization, & International Society of Hypertension Writing Group (2003) 2003 World Health Organization (WHO)/International Society of Hypertension (ISH) statement on management of hypertension. J Hyperten 21(11):1983–1992

Wuana RA, Okieimen FE (2011) Heavy metals in contaminated soils: a review of sources, chemistry, risks and best available strategies for remediation. Int Scholarly Res Not 2011

Ximénez-Embún P, Madrid-Albarrán Y, Cámara C, Cuadrado C, Burbano C, Múzquiz M (2001) Evaluation of lupinus species to accumulate heavy metals from waste waters. Int J Phytoremed 3(4):369–379

Yang S, Carlson KH (2004) Solid-phase extraction–high-performance liquid chromatography–ion trap mass spectrometry for analysis of trace concentrations of macrolide antibiotics in natural and waste water matrices. J Chromatogr A 1038(1–2):141–155

Youngchul K, Gilson H, Jin-Woo L, Je-Chul P, Dong-Sup K, Min-Gi K, In-Soung C (2006) Experiences with constructed wetland systems in Korea. J Ocean University China 5(4): 345–350

Zacco A, Borgese L, Gianoncelli A, Struis RP, Depero LE, Bontempi E (2014) Review of fly ash inertisation treatments and recycling. Environ Chem Letters 12(1):153–175

Zaheer IE, Ali S, Rizwan M, Farid M, Shakoor MB, Gill RA, Ahmad R (2015) Citric acid assisted phytoremediation of copper by Brassica napus L. Ecotoxic Environ Safety 120:310–317

Zayed AM, Terry N (1994) Selenium volatilization in roots and shoots: effects of shoot removal and sulfate level. J Plant Physiol 143(1):8–14

Zhao Y, Yan Z, Qin J, Xiao Z (2014) Effects of long-term cattle manure application on soil properties and soil heavy metals in corn seed production in Northwest China. Environ Sci Pollut Res 21(12):7586–7595

Zhao S, Zhu L, Li D (2015) Microplastic in three urban estuaries, China. Environ Pollut 206: 597–604

Zhu YL, Pilon-Smits EA, Tarun AS, Weber SU, Jouanin L, Terry N (1999) Cadmium tolerance and accumulation in Indian mustard is enhanced by overexpressing γ-glutamylcysteine synthetase. Plant Physio 121(4):1169–1177

Zhuang P, McBride MB, Xia H, Li N, Li Z (2009) Health risk from heavy metals via consumption of food crops in the vicinity of Dabaoshan mine, South China. Sci Total Environ 407(5): 1551–1561

Exploring Plant Responses to Salinity and Implications of Halophytes as a Model for Salinity Improvement

Nicolle Louise Ferreira Barros, Deyvid Novaes Marques, and Cláudia Regina Batista de Souza

Abstract Inappropriate agricultural practices and environmental impacts are worsening soil salinity. This affects crop yield and, consequently, the dynamics of the international market and food security. According to the stage of development of the plant, the duration of exposure, and the intensity of stress, different responses are triggered to maintain vital metabolic reactions and the integrity of cellular components. The most consumed crops in the world, in general, are glycophytes, and the efforts to find salt-tolerant cultivars have not yet resulted in wide practical application in the field. Since halophytic plants can complete their life cycle under highly saline conditions, they can provide clues about pathways to be explored to improve glycophytes' response to salinity. In this context, the search for differences between glycophytes and halophytes has contributed to the identification of promising traits of the latter that can enable the achievement of the mentioned aim. Among them, the existence of transcripts unique to halophytes and unannotated, therefore, with unknown functions. Furthermore, although responses to salt are generally common between these two groups of plants, halophytes succeed, for example, regarding the balancing of the Na^+/K^+ ratio. It can occur through the ability to compartmentalize higher levels of Na^+ in vacuoles and to maintain or distribute K^+ more efficiently. Moreover, other highlights that can be explored include the ions usage for osmotic adjustment as a metabolically cheaper alternative and more powerful antioxidant system and stress signaling pathways.

Keywords Climate change · Ion homeostasis · Osmotic stress · Salt tolerance · Haloculture · Biosaline agriculture

N. L. F. Barros
Departamento de Genética, Universidade Federal do Rio Grande do Sul (UFRGS), Porto Alegre, RS, Brazil

D. N. Marques (✉)
Departamento de Genética, Universidade de São Paulo (USP), Escola Superior de Agricultura "Luiz de Queiroz" (ESALQ), Piracicaba, SP, Brazil

C. R. B. de Souza
Instituto de Ciências Biológicas, Universidade Federal do Pará (UFPA), Belém, Brazil

M. Hasanuzzaman et al. (eds.), *Managing Plant Production Under Changing Environment*, https://doi.org/10.1007/978-981-16-5059-8_10

1 Introduction

Imbalance in the salinity of soils is considered a drawback for world agriculture. It is caused by inappropriate cultivation practices such as flawed irrigation systems or excess minerals and ineffective drainage (Maathuis et al. 2014). In addition, another anthropogenic interference commonly associated with secondary salinization of soils is deforestation, which increases soil salt levels due to erosion and change in rainfall (Köster et al. 2019).

Moreover, induced salinity is a hindrance to the economy since a projection indicates that expenses with losses in world crop yield due to this abiotic stress will reach $ 27.3 billion. Salinity stress can also affect the environment by interfering in biogeochemical cycles like that of nitrogen, due to the modification of soil microorganisms (Joint Research Center, European Commission 2018).

In light of this, approaches that accelerate the accumulation of data for the production of salt-tolerant crops are urgently needed. Efforts are being made not only because of the economic losses that salinity stress has been causing but also because it represents another factor that threatens food security, besides population expansion and resulting greater demand for food.

One way to achieve this goal is to use halophytic plants as models since they face salinity efficiently and share response pathways with glycophytes. Thus, new insights about the mechanisms triggered by salinity can be obtained, which can guide analysis and accelerate the process of increasing tolerance in glycophytes. Here, an overview is presented about that plants use under salinity. Subsequently, some singularities of the responses of halophytes are pointed out, so that these differences, if involved with the success of these "salt-loving" plants, become targets for studies with glycophytes.

2 Plant Responses to Salinity Stress

2.1 Perceive to React: the Early Responses Triggered by Excess NaCl in Plant Cells

Root cells are the first to capture increasing variations in salinity by NaCl since Na^+ associates with cell wall components. Thus, this salt interacts with pectin, affecting the structure's rigidity (Proseus and Boyer 2012), and with wall-associated kinases (WAK), both calcium binders under normal conditions (Decreux and Messiaen 2005). Therefore, by destabilizing these pectin-Ca^{2+} and WAK-Ca^{2+} links, salinity stress can be perceived from the boost in free Ca^{2+} (Byrt et al. 2018). Increases in salt concentrations can also produce changes in the pH of the apoplast that cause the reinforcement of the wall, making cell expansion difficult (Geilfus 2017) and inhibiting root growth, one of the main effects of this stress (Byrt et al. 2018).

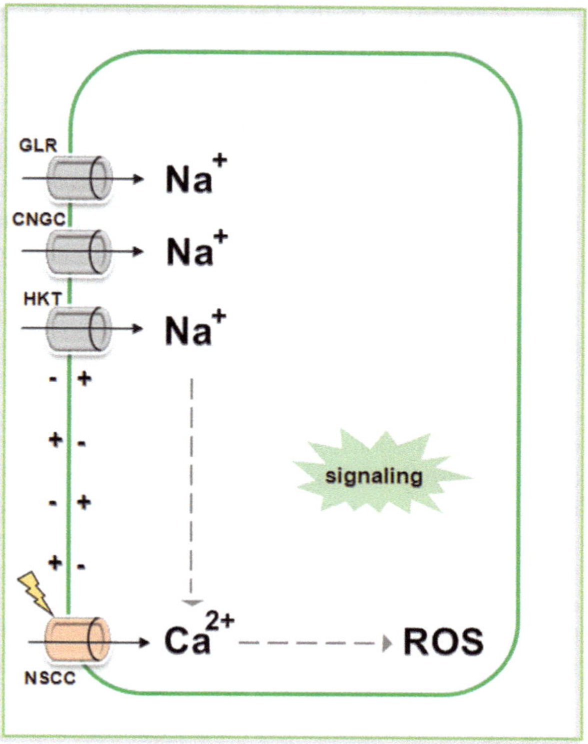

Fig. 1 Schematic illustration depicting the plant cell's sensing of salinity stress caused by NaCl. The influx of sodium is promoted by glutamate receptors (GLRs), cyclic nucleotide-gated channels (CNGCs), and high-affinity potassium transporters (HKT). Consequently, there is depolarization of the plasma membrane, which activates non-selective cation channels (NSCCs) for the entry of Ca^{2+} into the cytosol, stimulating the production of reactive oxygen species (ROS). Ca^{2+} and ROS initiate signaling cascades to respond to environmental disturbance

In this context, this abiotic stress also stimulates changes in the disposition and concentration of cellulose microfibrils (Koyro 1997) and callose, the latter being a likely negative regulator of Na^+ transport via plasmodesma to prevent the ion from accumulating in the shoot of the plant (Cui and Lee 2016).

Thus, Na^+ entry is promoted by ion channels in the plasma membrane (Fig. 1), such as glutamate receptors (GLRs) and cyclic nucleotide-gated channels (CNGCs), and by high-affinity potassium transporters (HKT) (Davenport 2002; Assmann 1995; Horie et al. 2001). This ion drives the increase in the concentration of cytosolic Ca^{2+} by influx via non-selective cation channels (NSCCs) (Demidchik and Maathuis 2007). Among other functions, Ca^{2+} induces the synthesis of ROS through NAPDH oxidases (Marino et al. 2012). In this context, it is possible that NaCl also causes an increase in the concentration of extracellular ATP, which is also related to the increase in Ca^{2+} and ROS contents (Sun et al. 2010).

Both Ca^{2+} and ROS act as signaling components for salinity stress. However, the scientific community has not so far identified specific Na^+ sensors in the plasma membrane (Yang and Guo 2018) nor how the cell interprets the signaling of these molecules as specific for salinity stress because Ca^{2+} and ROS are also involved in responses to other stresses such as drought, osmotic and cold (Maathuis et al. 2014).

2.2 Responses to the Developments of Salinity Stress: How Plants Deal with Osmotic and Ionic Stresses

Salinity stress causes osmotic stress, which involves the closure of stomata and damage to photosynthetic elements; and ionic stress, which results in a disproportion between Na^+ and K^+ concentrations, causing loss of homeostasis, a vital equilibrium for cellular functioning (Munns and Tester 2008; Köster et al. 2019).

The response to osmotic stress is regulated by hormones like abscisic acid (ABA), gibberellins (GA), jasmonic acid (JA), and ethylene. Of these, the first promotes the formation of ROS to indicate the closure of stomata (Zhang et al. 2001). It also leads to remodeling the roots in an attempt to prevent water deficit in plant tissues (Chen et al. 2006) and activate transcription factors such as bZIP, NAC, and MYB (Fang et al. 2017; Marques et al. 2017; Zhu et al. 2018; He et al. 2019), which induce the expression of response genes. Moreover, a reduction in GA and an increase in ethylene cause limitation of root growth (Belmecheri-Cherifi et al. 2019), while JA increases the antioxidant activity of the plant (Vatanparast et al. 2012).

It is worth mentioning that there is an interconnection between these hormones maintained by the DELLA (Asp–Glu–Leu–Leu–Ala) protein (Fig. 2). This biomolecule is only found in high concentrations when the opposite occurs with GA since the interaction between GA and a GA INSENSITIVE DWARF1 (GID1) receptor induces a conformational change in DELLA, followed by degradation by the

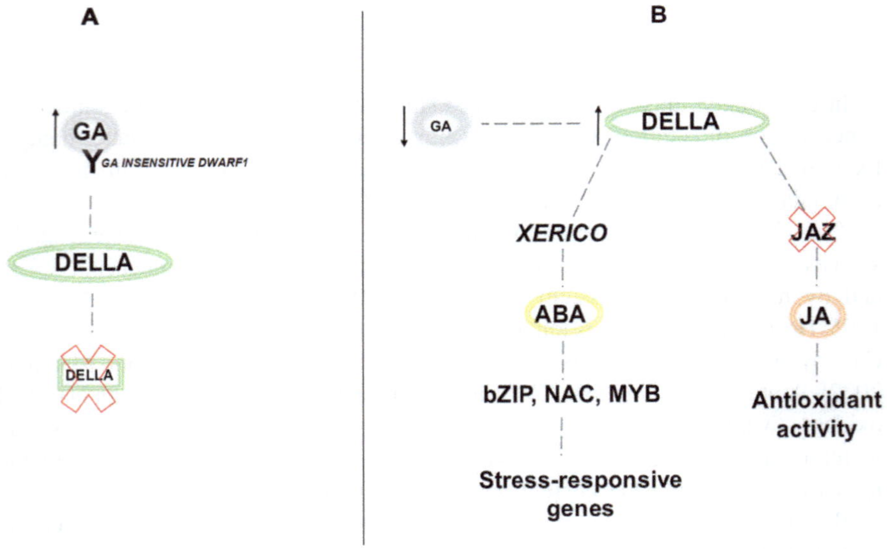

Fig. 2 Schematic illustration depicting the hormonal interconnection associated with the DELLA (Asp–Glu–Leu–Leu–Ala) protein. (a) High concentration of GA are detected by the GID1 receptor, and this interaction causes conformational changes in DELLA, followed by degradation of this protein. (b) Low levels of GA lead to high levels of DELLA, favoring the biosynthesis of ABA and JA and generating stress tolerance responses

ubiquitin-proteasome-26S system (Fu et al. 2002). Therefore, in the face of abiotic stresses, the concentration of GA decreases and DELLA levels become high, so that they cause induction of the *XERICO* gene for ABA synthesis (Ariizumi et al. 2013; Zeng et al. 2015) and impair the action of JAZ, a negative regulator of JA biosynthesis genes (Hou et al. 2010).

It is reported that ethylene contributes to salt tolerance by activating *Atrboh* genes (NADPH oxidase). Thus, this hormone favors the production of ROS in the *Arabidopsis* vascular tissue, limiting Na^+ to the roots and maintaining K^+ cellular (Jiang et al. 2013). Under such adverse conditions, ethylene signaling enables the scavenging of these reactive molecules (Peng et al. 2014), and the balance between these mechanisms ensures the role of ROS only as signaling agents, without toxic accumulation for the cell. Despite these results, there are still discrepancies between the positive or negative role of ethylene in dealing with such abiotic stress (Tao et al. 2015).

In contrast, the response to ionic stress comprises well-established metabolic pathways, among them SOS for apoplastic extrusion of Na^+, compartmentalization of Na^+ into vacuoles, xylem loading, and recycling of Na^+.

The SOS pathway is composed of molecules such as CBL4 (SOS3). This protein is activated by Ca^{2+}, which enters the cells as a result of Na^+ in abundance by the mechanisms already mentioned above. Subsequently, SOS3 activates CIPK24 (SOS2), which phosphorylates the S1138 residue and activates SOS1, a Na^+/H^+ antiporter (exchanger), whose function is to remove excess cytosolic Na^+ for the apoplast (Luan 2009; Quintero et al. 2011; Maathuis 2014). However, under homeostatic conditions, SOS2 has its kinase activity blocked by 14-3-3 proteins (Zhou et al. 2014) and is also negatively regulated by the GIGANTEA (GI) protein (Kim et al. 2013) (Fig. 3).

It is important to note that SOS1 is also related to xylem Na^+ loading (Katschnig et al. 2015), while SOS2 participates in the compartmentalization of Na^+ into vacuoles and regulates V-ATPase activity (Qiu et al. 2004; Batelli et al. 2007).

Regarding vacuolar sequestration, considering that this compartment does not experience damage of the same magnitude as the cytoplasm, Na^+ is stored in the vacuole to control cytoplasmic toxicity and reduce the cellular water potential to prevent water loss (Maathuis et al. 2014). Then, the sodium is assimilated by the vacuole via NHX1 exchanger/antiporter after extrusion of the H^+ purposely incorporated in excess by V-PPases and V-H^+-ATPases (Isayenkov et al. 2010; Assaha et al. 2017) (Fig. 3).

Another strategy applied by plants to deal with salinity stress is the flow of Na^+ between the roots and the shoot to stimulate an osmotic adjustment and save the energy that would eventually be used for synthesizing osmolytes (Bose et al. 2014). In this way, Na^+ can be loaded into the xylem by proteins such as SOS1 (Feki et al. 2014) and cation antiporters (CHX) (Guan et al. 2014). After performing this role, it is removed by HKT (Horie et al. 2009), so that there is no accumulation of this ion in leaves and no damage to photosynthesis (Almeida et al. 2017).

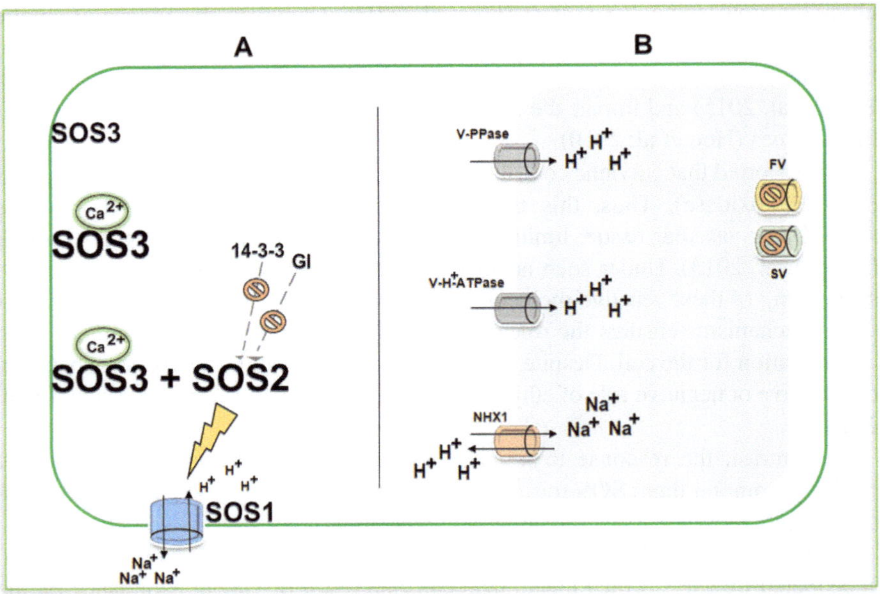

Fig. 3 Schematic illustration depicting the extrusion process and sodium vacuolar compartmentalization under salinity stress. (a) Ca^{2+} binds to SOS3 (CBL4), which, added to SOS2 (CIPK24), composes the SOS3–SOS2 complex, formed to activate SOS1 to remove sodium from the cytoplasm to the apoplast. Moreover, 14–3-3 and GI (GIGANTEA) proteins block SOS2 activity under normal conditions. (b) V-PPase and V-H^+-ATPase proton pumps establish an exaggerated influx of H^+ into the vacuole. This ion is eliminated into the cytosol in exchange for Na^+ entry via NHX1

2.3 *Involvement of Non-Coding Sequences in the Response to Salinity*

A cotton *CHR* zinc finger transcription factor was induced when its negative regulator miRNVL5 was inhibited. This gene in *Arabidopsis* alleviated ionic stress and promoted the growth and development of the primary root in this species (Gao et al. 2016). After treatment with exogenous ABA, which can simulate environmental stresses (Finkelstein 2013), miRNA477a-5p levels increased in roots of *Populus euphratica*, leading to root growth through the repression of GA. The levels of miR530b decreased, which could induce the expression of its target, the transcription factor *bHLH*, which is associated with the increase in endogenous ABA and promotion of root growth (Liu et al. 2014; Lian et al. 2018).

3 Halophytes as a Model for the Establishment of Salt-Tolerant Glycophytes

The enhancement of soil salinity is an expanding event globally. The probable incidence of this abiotic stress must be considered in advance, even in a land that is still suitable for cultivation. It can favor the production of crops tolerant to salinity stress, given that it is a long-term cumulative process and that the worsening of environmental impacts is likely. In this regard, a combination of increasingly diversified and precise biotechnology tools and plants with biological efficiency under salinity is vital to strategic approaches.

While most crops are glycophytes (species with a lower threshold of tolerance to salinity) (Yuan et al. 2019a), halophytic plants represent 1%–2% (Rozema and Flowers 2008) of the species previously described, and their economic potential is under discussion (Garza-Torres et al. 2020). It is worth mentioning that the low numerical representativeness of the latter group does not mean low diversity, since they are widespread among families such as Chenopodiaceae (Wang et al. 2020), Amaranthaceae (Cai and Gao 2020), Brassicaceae (Hajiboland et al. 2020), and Poaceae (Guo et al. 2020). This group of plants occupies coastal zones, mangrove forests, salt marshes, mudflats, steppes, and inland deserts (Shabala 2013). Among halophytes, as also occurs in glycophytes, there are varying degrees of salt tolerance (Bueno et al. 2020), although the strict definition assumes that such plants complete their life cycle successfully (Flowers et al. 1986). Thus, the same concentration of salt can affect the growth of one halophytic species and enhance the development of another. It is estimated that the optimal concentration for maintaining the metabolic reactions of halophytes varies from 100 mM–200 mM (Flowers and Colmer 2008) or 200 mM–300 mM (Katschnig et al. 2013).

Regarding strategies, variation occurs even between monocotyledonous and dicotyledonous halophytes. A salt exclusion tendency is attributed to the former, like most glycophytes, but not commonly observed in the second taxonomic group (Ben Hamed et al. 2018).

Based on the above, shared and particular features of the main strategies against salinity used by glycophytes and halophytes (see articles reviewed by Barros et al. 2021) (Fig. 4) will be presented. These topics emphasize the use of ions as convenient tools.

3.1 *Compartmentalization—Beneficial Accumulation Against Toxicity*

The imbalance in the concentration of saline ions in the soil drives the passive influx of Na^+ by HKT1, KT, KUP/HAK/KT-type transporters, AKT1-type channels, and NSCCs (Yuan et al. 2019b) into the cytoplasm, followed by the entry of Cl^-, although more slowly, due to the disparity between extra and intracellular charges

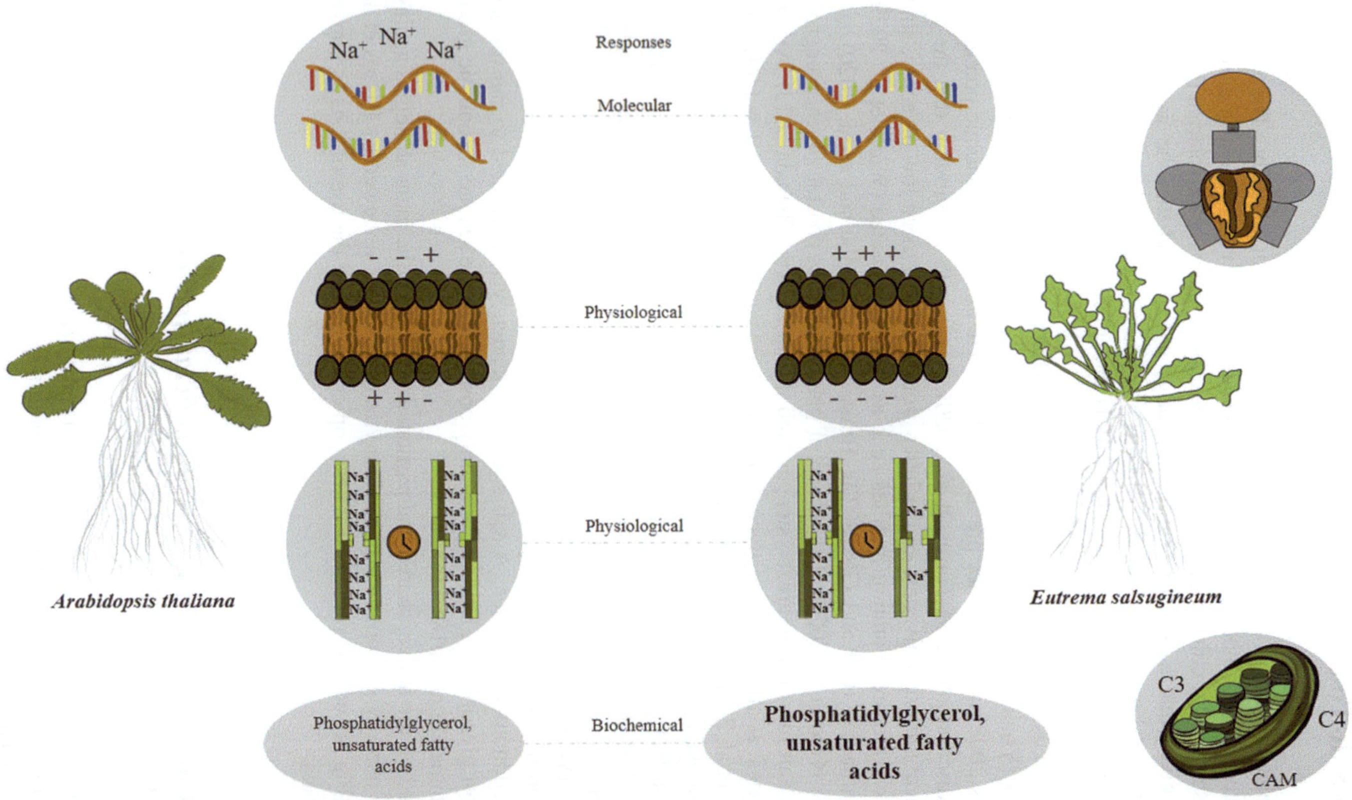

Fig. 4 Differences between the responses of glycophytes and halophytes to high salt concentrations. The center of this schematic illustration highlights some mechanisms against salt employed by glycophytes (represented here by *Arabidopsis thaliana*) and halophytes (represented here by *E. salsugineum*). At the molecular level, glycophytic salt-responsive genes are inducible by NaCl, while halophytes display a steady-state gene expression. At the physiological level, halophyte root cells can reduce the frequency of membrane depolarization by the improved activity of H^+ ATPases and the retention of K^+. On the other hand,

glycophytes are more prone to these events. Also, halophytes can regulate xylem loading efficiently since sodium accumulation is abrupt but transient in these plants, and it is performed only for osmotic adjustment purposes. In glycophytes, the Na^+ build-up is slow but continuous so that high levels of this ion persist during several days of treatment with salt. Finally, at the biochemical level, the accumulation of lipid compounds in halophytes may be associated with the protection of photosystems I and II. On the right, potentially exclusive halophyte strategies are shown, such as the removal of salt via glands and epidermal bladder cells, besides the ability to switch to C4 or CAM (Crassulacean acid metabolism) metabolism to overcome C3 limiting factors. It is noteworthy that the resources used by halophytes under high salt concentrations are diverse and that *E. salsugineum* was selected only as a representative of the group, which does not mean that it necessarily applies all these mechanisms

(Niu et al. 1995). Subsequent events include depolarization of the plasma membrane. This activates the channels that enable the influx of Ca^{2+} (Roberts and Tester 1997), the ion responsible for interrupting Na^+ access into the cell by blocking NSCCs (Demidchik and Tester 2002). Regarding Ca^{2+}, salt-tolerant species contain more NSCCs sensitive to this divalent cation (Chen et al. 2007), a characteristic that also prevents K^+ evasion by outward-rectifying channels (Shabala et al. 2007).

Since salinity disturbs the nutrition of plants, the preservation of levels of K^+ represents a primordial tolerance trait. It can be displayed either by retaining or relocating the previously available content (Zarei et al. 2020). Halophytes also show a considerable ability to uptake K^+ (Sun et al. 2017), which is crucial since plant cells require higher concentrations of K^+ in comparison to Na^+ (Tester and Davenport 2003). Halophytes can retain K^+ without high-energy consumption, which can be reallocated for growth (Percey et al. 2016b), while glycophytes such as pea (*Pisum sativum*) lose K^+ abundantly under salinity (Sun et al. 2017).

The increase in the Na^+/K^+ ratio in the cytoplasm commonly occurs as part of the situation imposed by salinity, which is harmful even for halophytes (Bartels and Dinakar 2013). Thus, to face this adverse condition, both halophytes and glycophytes apply vacuolar compartmentalization of Na^+. This protects cellular constituents and prevents, for example, the impairment of enzyme activity as a result of the absence of K^+ as a cofactor (Hasanuzzaman et al. 2018), lipid peroxidation, high production of reactive carbonyl species (RCS) (Yalcinkaya et al. 2019), changes in thylakoid membranes and Photosystem II (PSII) functioning (Pan et al. 2020).

This process depends on a proton gradient and a difference in transmembrane electric potential. Both of these components are part of the Proton Motive Force (PMF) (Kramer et al. 2003), which is necessary to address the H^+ to the vacuole under the control of the proton pumps (Shabala 2013). It is worth mentioning that, in halophytes such as *Puccinellia tenuiflora* and *Salicornia europaea*, this activity is superior to that of glycophyte antiporters (Liu et al. 2017). Questions remain concerning disparities between the maximum Na^+ concentrations that the cytoplasm of glycophytes and halophytes can tolerate and the threshold of accumulation of this ion in the chloroplasts of these two types of plants.

Another advantage is that some halophytes possibly demonstrate a steady-state via constitutive expression, which can affect the success of the compartmentalization of Na^+. Besides, these plants keep Na^+ in vacuoles. This can be achieved by decreasing both the number of channels (Aslam et al. 2011) and the permeability by deposition of lipids (Glenn et al. 1999), and the management of the duration of opening or channel activity in older leaves, which are targets for compartmentalization in quinoa (*Chenopodium quinoa*) (Bonales-Alatorre et al. 2013). Besides, the ability of halophytes to accumulate these ions in vacuoles increases the succulence of the shoot (Ben Hamed et al. 2018).

On the other hand, halophyte chloroplasts can store Na^+ and Cl^- without severe structural or physiological damage (Flowers et al. 2015). This is due to the improved balance of charges in the PSI (Percey et al. 2016a), intensification of grana formation in response to salt (Redondo-Gómez et al. 2010; Trotta et al. 2012) and greater resilience of enzymes such as fructose-1,6-bisphosphatase to salinity (Ghosh et al. 2001). Therefore, these characteristics of halophytes and vacuolar participation to maintain ionic homeostasis can work towards counteracting salinity stress.

3.2 *Flow Against Toxicity: Preventing Accumulation in Key Tissues*

After the influx of Na^+ to the cytoplasm, another alternative is long-distance movement through xylem loading carried out by SOS1, a type of NHX protein (Yadav et al. 2012). The transfer of ions to the shoot minimizes ionic toxicity since the distribution allows them to be compartmentalized in senescent leaves, which can later defoliate (Gorham 1995). Moreover, the ion can be excreted by salt glands (Yuan and Wang 2020) and used for osmotic adjustment (Haque et al. 2017). It is interesting how, even under normal growth conditions, the halophyte *Atriplex lentiformis* already maintains high levels of Na^+ in the xylem (Zarei et al. 2020), which corroborates the possibility of the steady-state already mentioned. Other evidence associated with the greater effectiveness of xylem loading in halophytes suggests that the root ion channels are less sensitive to the H_2O_2 signaling molecule, so there is less chance of Na^+ returning from the xylem (Zarei et al. 2020).

Considering the competence of halophytes to perform massive Na^+ uptake without harming the tissues (Yepes et al. 2018), the timing and duration of xylem loading must be critical. For example, there was a substantial and temporary increase in Na^+ in the xylem of *A. lentiformis* exposed to salt and a subsequent drop to levels comparable to those of the controls, in contrast to the slow but continuous increase in the xylem of bean plants (*Vicia faba* L.) (Zarei et al. 2020). Similarly, Na^+ accumulation was transitory in the quinoa protoplast compared to that of the pea, a glycophyte (Sun et al. 2017).

Following Na^+ circulation, the rest of the ions can return to the roots via unloading through the phloem, carried out by HKT transporters located on the plasma membrane of the xylem parenchyma cells (Ali et al. 2019). In halophytes, HKT1 proteins do not follow the relationship between amino acid sequence and expected function, which means that they display functional differences and, during high salinity, exhibit greater affinity for K^+ instead of Na^+, as in glycophytes (Ali et al. 2016).

Once in the roots, Na^+ can be sent to the apoplast by SOS1 and then out of the cell. Moreover, SOS1 is part of one of the main mechanisms of salt exclusion that is actively developed. Therefore, it requires the action of proton pumps on the plasma membrane (Shi et al. 2000). From this perspective, in the halophyte quinoa, the activities of the H^+-ATPase of the plasma membrane and SOS1 were prominent in comparison to those of pea, probably due to the considerable increase in the pH of the halophytes' cytosol, even in control plants, and to the subsequent influx of H^+ following the treatment with NaCl and abscisic acid (ABA) (Sun et al. 2017).

Although the description of the SOS pathway was established based on glycophytes such as *Arabidopsis* (Quintero et al. 2002), it also occurs in halophytes (Zhang et al. 2020).

3.3 Osmotic Adjustment: Halophytes Use Ions to Surpass the Trend Created by Na^+ and K^+ Imbalance

The main goal of this adjustment is to create a hyperosmotic condition so that the water continues to move into the cells to maintain the turgor and proper distribution of minerals and fixed carbon among the tissues (Flowers and Colmer 2015) since one of the most striking effects of salinity is dehydration. It is remarkable how, in general, halophytes choose ions as tools to reverse the unfavorable situation (Haque et al. 2017), varying in their preference for Na^+ (Slama et al. 2015), Cl^- (Bazihizina et al. 2019), or K^+, as observed in the families Gramineae and Cyperaceae (Kumar et al. 2019) and quinoa, concerning potassium (Sun et al. 2017). In contrast, glycophytes use compatible solutes such as proline, polyols, sucrose, and trehalose (Nikalje et al. 2018).

Thus, it is proposed that the use of each of these strategies has different ramifications regarding energy. The production and accumulation of organic compounds are more expensive (Haque et al. 2017) as they consume C and N in abundance, a characteristic that can suppress the development of plant tissues (Gharbi et al. 2017). Halophytes can apply one of the two options (Kumar et al. 2019), while glycophytes have limitations regarding ion usage for this purpose since these plants deal poorly with saline ions.

In this regard, the halophytic wild relative (*Solanum chilense*) of the cultivated tomato (*S. lycopersicum*) produces proline. However, the proline contribution to osmoregulation has been seen to be 12 times less than that applied by Na^+ in this plant. Nevertheless, the glycophytic *S. lycopersicum* exhibited higher levels of proline and reduced plant growth (Gharbi et al. 2017). Therefore, it is possible that the increase of proline contents in halophytes, if not crucial for osmotic adjustment, is important for other functions, such as protection of protein conformation (Abdallah et al. 2020) and improvement of the antioxidant system (Docimo et al. 2020).

3.4 The Regulation Performed by MicroRNAs May Also Be Responsible for the Differences Between Halophytes and Glycophytes Under Salinity

Research on regulatory aspects has also been directed towards miRNA, since these small non-coding RNAs may be crucial for the better performance of salt-tolerant plants. For example, the miRNA profile of *Mesembryanthemum crystallinum*, added to the expression levels of these non-coding sequences and the target salt-responsive genes, indicates that there is an inversely proportional relationship between them. Therefore, saline conditions resulted in inhibition of miRNAs that possibly regulate stress signaling and ionic homeostasis and, thus, enhanced the expression of genes that promote root growth, Na^+ transport to the shoot, and glycolysis (Chiang et al. 2016).

Similarly, in *S. europaea* the predominant putative targets participate in signaling and metabolism (Feng et al. 2015). In this context, the predicted target of miR156 is the same as in *A. thaliana* (Gou et al. 2011), *M. crystallinum* (Chiang et al. 2016), and *S. europaea* (Feng et al. 2015), but in those two halophytes this non-coding sequence was inhibited, and the SPL target was induced in *M. crystallinum*. This gene is related to root development and response to salinity (Feng et al. 2015). Furthermore, in *A. thaliana* (Liu et al. 2008) miR169 was induced by salt, while the opposite was observed in *S. europaea* (Feng et al. 2015) and *M. crystallinum* (Chiang et al. 2016). Consequently, the shared target NF-YA was upregulated in the latter species (Chiang et al. 2016), and it is involved in the same previously cited functions (Feng et al. 2015).

Accordingly, the expression patterns of miR171 also showed contrasts between Arabidopsis (Liu et al. 2008) and *E. salsugineum* (Wu et al. 2016), as well as between the miR319 of this glycophyte and *E. salsugineum* (Wu et al. 2016), whose targets are transcription factors. In the halophyte *Oryza coarctata*, most of the probable genes regulated by miRNAs are responsible for the response to the stimulus and metabolism of starch, sucrose, and phenylpropanoid, while others carry out K^+ transport, antioxidant activity, and sucrose transport (Mondal et al. 2015). These biological processes are efficiently performed by halophytes, and, therefore, these data encourage further research into these regulatory elements.

4 Conclusions

The use of halophytes for agricultural purposes, as already suggested, is a less viable alternative compared to the continuity of investments in salt-tolerant glycophytic plants. This is because crops are predominantly glycophytes, which are well-established and widely consumed commodities, whose features and cultivation practices are already standardized and well known. Developing comparative analyses between these two groups of plants can generate a greater understanding of the mechanisms of response to salinity, besides helping to highlight biological aspects that differentiate halophytes. Thus, it is possible to explore these unique characteristics and apply them in favor of glycophytes under this abiotic stress. Some issues must be taken into account for the optimization of investigations on the performance of halophytes against salt to diminish the disparities between the number of scientific results produced so far and the field applications. Thus, it is crucial to evaluate the reproductive aspects of these plants, as this stage of the life cycle is one of the most affected in crops under stress. Furthermore, epigenome analysis is pivotal for a comprehensive understanding of salt tolerance since epigenetic modifications constitute a way to achieving the phenotypic plasticity required under adverse conditions. Accordingly, ongoing studies about the influence of maternal habitat on the tolerance of halophytic offspring and the methylation patterns responsible for the C3-CAM transition have been filling these gaps. On the other hand, non-model wild species related to domesticated glycophytic crops can also be sources of molecular strategies for coping with salt stress. However, for this purpose, the phenotyping methods for the establishment of salt-sensitive and salt-tolerant species must be improved. It can facilitate the proper differentiation between salt-tolerant glycophytes and halophytes, the definition of concentrations considered stressful for each of these plants, and the maximum concentrations that each one tolerates. Finally, the molecular strategies that could be carried out in ongoing and future research include the use of tissue-specific and stress-induced halophytic promoters and the transformation of glycophytes with multiple halophytic tolerance genes simultaneously.

Acknowledgments Nicolle Louise Ferreira Barros thanks Conselho Nacional de Desenvolvimento Científico e Tecnológico (CNPq; grant number 140586/2021-0) and Coordenação de Aperfeiçoamento de Pessoal de Nível Superior (CAPES - Finance Code 001). The corresponding author (Deyvid Novaes Marques) thanks Fundação de Amparo à Pesquisa do Estado de São Paulo (FAPESP; grant numbers 2017/05544-0, 2018/20706-9, and 2020/12666-7) and CNPq (grant number 153102/2016-0).

References

Abdallah MM-S, El Sebai TN, Ramadan AAE-M, El-Bassiouny HMS (2020) Physiological and biochemical role of proline, trehalose, and compost on enhancing salinity tolerance of quinoa plant. Bull Natl Res Cent 44:96

Ali A, Maggio A, Bressan R, Yun D-J (2019) Role and functional differences of HKT1-type transporters in plants under salt stress. Int J Mol Sci 20:1059

Ali A, Raddatz N, Aman R, Kim S, Park HC, Jan M, Baek D, Khan IU, Oh D-H, Lee SY, Bressan RA, Lee KW, Maggio A, Pardo JM, Bohnert HJ, Yun D-J (2016) A single amino-acid substitution in the sodium transporter HKT1 associated with plant salt tolerance. Plant Physiol 171:2112–2126

Almeida DM, Oliveira MM, Saibo NJM (2017) Regulation of Na^+ and K^+ homeostasis in plants: towards improved salt stress tolerance in crop plants. Genet Mol Biol 40:326–345

Ariizumi T, Hauvermale AL, Nelson SK, Hanada A, Yamaguchi S, Steber CM (2013) Lifting DELLA repression of Arabidopsis seed germination by nonproteolytic gibberellin signaling. Plant Physiol 162:2125–2139

Aslam R, Bostan N, Nabgha-e-Amen MM, Safdar W (2011) A critical review on halophytes: salt tolerant plants. J Med Plant Res 5:7108–7118

Assaha DVM, Ueda A, Saneoka H, Al-Yahyai R, Yaish MW (2017) The role of Na^+ and K^+ transporters in salt stress adaptation in glycophytes. Front Physiol 8:5

Assmann SM (1995) Cyclic AMP as a second messenger in higher plants (status and future prospects). Plant Physiol 108:885–889

Barros NLF, Marques DN, Tadaiesky LBA, de Souza CRB (2021) Halophytes and other molecular strategies for the generation of salt-tolerant crops. Plant Physiol Biochem 162:581–591

Bartels D, Dinakar C (2013) Balancing salinity stress responses in halophytes and non-halophytes: a comparison between *Thellungiella* and *Arabidopsis thaliana*. Funct Plant Biol 40:819

Batelli G, Verslues PE, Agius F, Qiu Q, Fujii H, Pan S, Schumaker KS, Grillo S, Zhu K (2007) SOS2 promotes salt tolerance in part by interacting with the vacuolar H^{+}-TPase and upregulating its transport activity. Mol Cell Biol 27:7781–7790

Bazihizina N, Colmer TD, Cuin TA, Mancuso S, Shabala S (2019) Friend or foe? Chloride patterning in halophytes. Trends Plant Sci 24:142–151

Belmecheri-Cherifi H, Albacete A, Martínez-Andújar C, Pérez-Alfocea F, Abrous-Belbachir O (2019) The growth impairment of salinized fenugreek (*Trigonella foenum-graecum* L.) plants is associated to changes in the hormonal balance. J Plant Physiol 232:311–319

Ben Hamed K, Dabbous A, El Shaer H, Abdely C (2018) Salinity responses and adaptive mechanisms in halophytes and their exploitation for producing salinity tolerant crops. In: Kumar V, Wani S, Suprasanna P, Tran LS (eds) Salinity responses and tolerance in plants. Springer, Cham, pp 1–19

Bonales-Alatorre E, Shabala S, Chen Z-H, Pottosin I (2013) Reduced tonoplast fast-activating and slow-activating channel activity is essential for conferring salinity tolerance in a facultative halophyte, quinoa. Plant Physiol 162:940–952

Bose J, Rodrigo-Moreno A, Shabala S (2014) ROS homeostasis in halophytes in the context of salinity stress tolerance. J Exp Bot 65:1241–1257

Bueno M, Lendínez ML, Calero J, del Pilar CM (2020) Salinity responses of three halophytes from inland saltmarshes of Jaén (southern Spain). Flora 266:151589

Byrt CS, Munns R, Burton RA, Gilliham M, Wege S (2018) Root cell wall solutions for crop plants in saline soils. Plant Sci 269:47–55

Cai Z-Q, Gao Q (2020) Comparative physiological and biochemical mechanisms of salt tolerance in five contrasting highland quinoa cultivars. BMC Plant Biol 20:70

Chen C-W, Yang Y-W, Lur H-S, Tsai Y-G, Chang M-C (2006) A novel function of abscisic acid in the regulation of rice (*Oryza sativa* L.) root growth and development. Plant Cell Physiol 47:1–13

Chen Z, Pottosin II, Cuin TA, Fuglsang AT, Tester M, Jha D, Zepeda-Jazo I, Zhou M, Palmgren MG, Newman IA, Shabala S (2007) Root plasma membrane transporters controlling K^{+}/Na^{+} homeostasis in salt-stressed barley. Plant Physiol 145:1714–1725

Chiang C-P, Yim WC, Sun Y-H, Ohnishi M, Mimura T, Cushman JC, Yen HE (2016) Identification of ice plant (*Mesembryanthemum crystallinum* L.) microRNAs using RNA-seq and their putative roles in high salinity responses in seedlings. Front Plant Sci 2016:7

Cui W, Lee J-Y (2016) Arabidopsis callose synthases CalS1/8 regulate plasmodesmal permeability during stress. Nat Plants 2:16034

Davenport R (2002) Glutamate receptors in plants. Ann Bot 90:549–557

Decreux A, Messiaen J (2005) Wall-associated kinase WAK1 interacts with cell wall pectins in a calcium-induced conformation. Plant Cell Physiol 46:268–278

Demidchik V, Maathuis FJM (2007) Physiological roles of nonselective cation channels in plants: from salt stress to signalling and development. New Phytol 175:387–404

Demidchik V, Tester M (2002) Sodium fluxes through nonselective cation channels in the plasma membrane of protoplasts from Arabidopsis roots. Plant Physiol 128:379–387

Docimo T, De Stefano R, Cappetta E, Piccinelli AL, Celano R, De Palma M, Tucci M (2020) Physiological, biochemical, and metabolic responses to short and prolonged saline stress in two cultivated cardoon genotypes. Plant Theory 9:554

Fang Q, Jiang T, Xu L, Liu H, Mao H, Wang X, Jiao B, Duan Y, Wang Q, Dong Q, Yang L, Tian G, Zhang C, Zhou Y, Liu X, Wang H, Fan D, Wang B, Luo K (2017) A salt-stress-regulator from the poplar R2R3 MYB family integrates the regulation of lateral root emergence and ABA signaling to mediate salt stress tolerance in Arabidopsis. Plant Physiol Biochem 114:100–110

Feki K, Quintero FJ, Khoudi H, Leidi EO, Masmoudi K, Pardo JM, Brini F (2014) A constitutively active form of a durum wheat Na^+/H^+ antiporter SOS1 confers high salt tolerance to transgenic *Arabidopsis*. Plant Cell Rep 33:277–288

Feng J, Wang J, Fan P, Jia W, Nie L, Jiang P, Chen X, Lv S, Wan L, Chang S, Li S, Li Y (2015) High-throughput deep sequencing reveals that microRNAs play important roles in salt tolerance of euhalophyte *Salicornia europaea*. BMC Plant Biol 15:63

Finkelstein R (2013) Abscisic acid synthesis and response. Arab B 11:e0166

Flowers TJ, Colmer TD (2008) Salinity tolerance in halophytes. New Phytol 179:945–963

Flowers TJ, Colmer TD (2015) Plant salt tolerance: adaptations in halophytes. Ann Bot 115:327–331

Flowers TJ, Hajibagheri MA, Clipson NJW (1986) Halophytes. Q Rev Biol 61:313–337

Flowers TJ, Munns R, Colmer TD (2015) Sodium chloride toxicity and the cellular basis of salt tolerance in halophytes. Ann Bot 115:419–431

Fu X, Richards DE, Ait-ali T, Hynes LW, Ougham H, Peng J, Harberd NP (2002) Gibberellin-mediated proteasome-dependent degradation of the barley DELLA protein SLN1 repressor. Plant Cell 14:3191–3200

Gao S, Yang L, Zeng HQ, Zhou ZS, Yang ZM, Li H, Sun D, Xie F, Zhang B (2016) A cotton miRNA is involved in regulation of plant response to salt stress. Sci Rep 6:19736

Garza-Torres R, Troyo-Diéguez E, Nieto-Garibay A, Lucero-Vega G, Magallón-Barajas FJ, García-Galindo E, Fimbres-Acedo Y, Murillo-Amador B (2020) Environmental and management considerations for adopting the halophyte *Salicornia bigelovii* Torr. As a sustainable seawater-irrigated crop. Sustainability 12:707

Geilfus C-M (2017) The pH of the apoplast: dynamic factor with functional impact under stress. Mol Plant 10:1371–1386

Gharbi E, Martínez J-P, Benahmed H, Hichri I, Dobrev PI, Motyka V, Quinet M, Lutts S (2017) Phytohormone profiling in relation to osmotic adjustment in NaCl-treated plants of the halophyte tomato wild relative species *Solanum chilense* comparatively to the cultivated glycophyte *Solanum lycopersicum*. Plant Sci 258:77–89

Ghosh S, Bagchi S, Lahiri Majumder A (2001) Chloroplast fructose-1,6-bisphosphatase from *Oryza* differs in salt tolerance property from the *Porteresia* enzyme and is protected by osmolytes. Plant Sci 160:1171–1181

Glenn EP, Brown JJ, Blumwald E (1999) Salt tolerance and crop potential of halophytes. CRC Crit Rev Plant Sci 18:227–255

Gorham J (1995) Mechanism of salt tolerance of halophytes. In: Choukr-Allah R, Malcolm CV, Hamdy A (eds) Halophytes and biosaline agriculture. CRC Press, Boca Raton, pp 31–53

Gou J-Y, Felippes FF, Liu C-J, Weigel D, Wang J-W (2011) Negative regulation of anthocyanin biosynthesis in *Arabidopsis* by a miR156-targeted SPL transcription factor. Plant Cell 23:1512–1522

Guan R, Qu Y, Guo Y, Yu L, Liu Y, Jiang J, Chen J, Ren Y, Liu G, Tian L, Jin L, Liu Z, Hong H, Chang R, Gilliham M, Qiu L (2014) Salinity tolerance in soybean is modulated by natural variation in *GmSALT3*. Plant J 80:937–950

Guo R, Zhao L, Zhang K, Gao D, Yang C (2020) Genome of extreme halophyte *Puccinellia tenuiflora*. BMC Genomics 21:311

Hajiboland R, Bahrami-Rad S, Zeinalzade N, Atazadeh E, Akhani H, Poschenrieder C (2020) Differential functional traits underlying the contrasting salt tolerance in *Lepidium* species. Plant and Soil 448:315–334

Haque MI, Rathore MS, Gupta H, Jha B (2017) Inorganic solutes contribute more than organic solutes to the osmotic adjustment in *Salicornia brachiata* (Roxb.) under natural saline conditions. Aquat Bot 142:78–86

Hasanuzzaman M, Bhuyan M, Nahar K, Hossain M, Mahmud J, Hossen M, Masud A, Moumita FM (2018) Potassium: a vital regulator of plant responses and tolerance to abiotic stresses. Agronomy 8:31

He K, Zhao X, Chi X, Wang Y, Jia C, Zhang H, Zhou G, Hu R (2019) A novel *Miscanthus* NAC transcription factor MlNAC10 enhances drought and salinity tolerance in transgenic Arabidopsis. J Plant Physiol 233:84–93

Horie T, Hauser F, Schroeder JI (2009) HKT transporter-mediated salinity resistance mechanisms in *Arabidopsis* and monocot crop plants. Trends Plant Sci 14:660–668

Horie T, Yoshida K, Nakayama H, Yamada K, Oiki S, Shinmyo A (2001) Two types of HKT transporters with different properties of Na^+ and K^+ transport in *Oryza sativa*. Plant J 27:129–138

Hou X, Lee LYC, Xia K, Yan Y, Yu H (2010) DELLAs modulate jasmonate signaling via competitive binding to JAZs. Dev Cell 19:884–894

Isayenkov S, Isner JC, Maathuis FJM (2010) Vacuolar ion channels: roles in plant trition and signalling. FEBS Lett 584:1982–1988

Jiang C, Belfield EJ, Cao Y, Smith JAC, Harberd NP (2013) An *Arabidopsis* soil-salinity–tolerance mutation confers ethylene-mediated enhancement of sodium/potassium homeostasis. Plant Cell 25:3535–3552

Joint Research Centre, European Commission (2018) World atlas of desertification. https://wad.jrc.ec.europa.eu/soilsalinization Accessed 20 Oct 2020

Katschnig D, Bliek T, Rozema J, Schat H (2015) Constitutive high-level *SOS1* expression and absence of *HKT1;1* expression in the salt-accumulating halophyte *Salicornia dolichostachya*. Plant Sci 234:144–154

Katschnig D, Broekman R, Rozema J (2013) Salt tolerance in the halophyte *Salicornia dolichostachya* Moss: growth, morphology and physiology. Environ Exp Bot 92:32–42

Kim W-Y, Ali Z, Park HJ, Park SJ, Cha J-Y, Perez-Hormaeche J, Quintero FJ, Shin G, Kim MR, Qiang Z, Ning L, Park HC, Lee SY, Bressan RA, Pardo JM, Bohnert HJ, Yun D-J (2013) Release of SOS2 kinase from sequestration with GIGANTEA determines salt tolerance in *Arabidopsis*. Nat Commun 4:1352

Köster P, Wallrad L, Edel KH, Faisal M, Alatar AA, Kudla J (2019) The battle of two ions: Ca^{2+} signalling against Na^+ stress. Plant Biol 21:39–48

Koyro H-W (1997) Ultrastructural and physiological changes in root cells of *sorghum* plants (*Sorghum bicolor* x *S. sudanensis* cv. Sweet Sioux) induced by NaCl. J Exp Bot 48:693–706

Kramer DM, Cruz JA, Kanazawa A (2003) Balancing the central roles of the thylakoid proton gradient. Trends Plant Sci 8:27–32

Kumar A, Mann A, Lata C, Kumar N, Sharma PC (2019) Salinity-induced physiological and molecular responses of halophytes. In: Dagar J, Yadav R, Sharma P (eds) Research developments in saline agriculture. Springer, Singapore, pp 331–356

Lian C, Yao K, Duan H, Li Q, Liu C, Yin W, Xia X (2018) Exploration of ABA responsive miRNAs reveals a new hormone signaling crosstalk pathway regulating root growth of *Populus euphratica*. Int J Mol Sci 19:1481

Liu H-H, Tian X, Li Y-J, Wu C-A, Zheng C-C (2008) Microarray-based analysis of stress-regulated microRNAs in *Arabidopsis thaliana*. RNA 14:836–843

Liu W, Tai H, Li S, Gao W, Zhao M, Xie C, Li W-X (2014) bHLH122 is important for drought and osmotic stress resistance in *Arabidopsis* and in the repression of ABA catabolism. New Phytol 201:1192–1204

Liu X, Cai S, Wang G, Wang F, Dong F, Mak M, Holford P, Ji J, Salih A, Zhou M, Shabala S, Chen Z-H (2017) Halophytic NHXs confer salt tolerance by altering cytosolic and vacuolar K^+ and Na^+ in *Arabidopsis* root cell. Plant Growth Regul 82:333–351

Luan S (2009) The CBL–CIPK network in plant calcium signaling. Trends Plant Sci 14:37–42

Maathuis FJM (2014) Sodium in plants: perception, signalling, and regulation of sodium fluxes. J Exp Bot 65:849–858

Maathuis FJM, Ahmad I, Patishtan J (2014) Regulation of Na^+ fluxes in plants. Front Plant Sci 5:467

Marino D, Dunand C, Puppo A, Pauly N (2012) A burst of plant NADPH oxidases. Trends Plant Sci 17:9–15

Marques DN, dos Reis SP, de Souza CRB (2017) Plant NAC transcription factors responsive to abiotic stresses. Plant Gene 11:170–179. https://doi.org/10.1016/j.plgene.2017.06.003

Mondal TK, Ganie SA, Debnath AB (2015) Identification of novel and conserved miRNAs from extreme halophyte, *Oryza coarctata*, a wild relative of rice. PLoS One 10:e0140675

Munns R, Tester M (2008) Mechanisms of salinity tolerance. Annu Rev Plant Biol 59:651–681

Nikalje GC, Srivastava AK, Pandey GK, Suprasanna P (2018) Halophytes in biosaline agriculture: mechanism, utilization, and value addition. L Degrad Dev 29:1081–1095

Niu X, Bressan RA, Hasegawa PM, Pardo JM (1995) Ion homeostasis in NaCl stress environments. Plant Physiol 109:735–742

Pan T, Liu M, Kreslavski VD, Zharmukhamedov SK, Nie C, Yu M, Kuznetsov V, Allakhverdiev SI, Shabala S (2020) Non-stomatal limitation of photosynthesis by soil salinity. Crit Rev Environ Sci Technol 5:1–35

Peng Z, He S, Gong W, Sun J, Pan Z, Xu F, Lu Y, Du X (2014) Comprehensive analysis of differentially expressed genes and transcriptional regulation induced by salt stress in two contrasting cotton genotypes. BMC Genomics 15:760

Percey WJ, McMinn A, Bose J, Breadmore MC, Guijt RM, Shabala S (2016a) Salinity effects on chloroplast PSII performance in glycophytes and halophytes. Funct Plant Biol 43:1003

Percey WJ, Shabala L, Wu Q, Su N, Breadmore MC, Guijt RM, Bose J, Shabala S (2016b) Potassium retention in leaf mesophyll as an element of salinity tissue tolerance in halophytes. Plant Physiol Biochem 109:346–354

Proseus TE, Boyer JS (2012) Pectate chemistry links cell expansion to wall deposition in *Chara corallina*. Plant Signal Behav 7:1490–1492

Qiu Q-S, Guo Y, Quintero FJ, Pardo JM, Schumaker KS, Zhu J-K (2004) Regulation of vacuolar Na^+/H^+ exchange in *Arabidopsis thaliana* by the salt-overly-sensitive (SOS) pathway. J Biol Chem 279:207–215

Quintero FJ, Martinez-Atienza J, Villalta I, Jiang X, Kim W-Y, Ali Z, Fujii H, Mendoza I, Yun D-J, Zhu J-K, Pardo JM (2011) Activation of the plasma membrane Na/H antiporter salt-overly-sensitive 1 (SOS1) by phosphorylation of an auto-inhibitory C-terminal domain. Proc Natl Acad Sci 108:2611–2616

Quintero FJ, Ohta M, Shi H, Zhu J-K, Pardo JM (2002) Reconstitution in yeast of the *Arabidopsis* SOS signaling pathway for Na^+ homeostasis. Proc Natl Acad Sci 99:9061–9066

Redondo-Gómez S, Mateos-Naranjo E, Figueroa ME, Davy AJ (2010) Salt stimulation of growth and photosynthesis in an extreme halophyte, *Arthrocnemum macrostachyum*. Plant Biol 12:79–87

Roberts SK, Tester M (1997) Permeation of Ca^{2+} and monovalent cations through an outwardly rectifying channel in maize root stelar cells. J Exp Bot 48:839–846

Rozema J, Flowers T (2008) Crops for a salinized world. Science 322:1478–1480

Shabala S (2013) Learning from halophytes: physiological basis and strategies to improve abiotic stress tolerance in crops. Ann Bot 112:1209–1221

Shabala S, Cuin TA, Pottosin I (2007) Polyamines prevent NaCl-induced K^+ efflux from pea mesophyll by blocking non-selective cation channels. FEBS Lett 581:1993–1999

Shi H, Ishitani M, Kim C, Zhu J-K (2000) The *Arabidopsis thaliana* salt tolerance gene *SOS1* encodes a putative Na^+/H^+ antiporter. Proc Natl Acad Sci 97:6896–6901

Slama I, Abdelly C, Bouchereau A, Flowers T, Savouré A (2015) Diversity, distribution and roles of osmoprotective compounds accumulated in halophytes under abiotic stress. Ann Bot 115: 433–447

Sun J, Wang M-J, Ding M-Q, Deng S-R, Liu M-Q, Lu C-F, Zhou X-Y, Shen X, Zheng X-J, Zhang Z-K, Song J, Hu Z-M, Xu Y, Chen S-L (2010) H_2O_2 and cytosolic Ca^{2+} signals triggered by the PM H^+-coupled transport system mediate K^+/Na^+ homeostasis in NaCl-stressed *Populus euphratica* cells. Plant Cell Environ 33:943–958

Sun Y, Lindberg S, Shabala L, Morgan S, Shabala S, Jacobsen S-E (2017) A comparative analysis of cytosolic Na^+ changes under salinity between halophyte quinoa (*Chenopodium quinoa*) and glycophyte pea (*Pisum sativum*). Environ Exp Bot 141:154–160

Tao J-J, Chen H-W, Ma B, Zhang W-K, Chen S-Y, Zhang J-S (2015) The role of ethylene in plants under salinity stress. Front Plant Sci 6:5

Tester M, Davenport R (2003) Na^+ tolerance and Na^+ transport in higher plants. Ann Bot 91:503–527

Trotta A, Redondo-Gómez S, Pagliano C, Clemente MEF, Rascio N, La Rocca N, Antonacci A, Andreucci F, Barbato R (2012) Chloroplast ultrastructure and thylakoid polypeptide composition are affected by different salt concentrations in the halophytic plant *Arthrocnemum macrostachyum*. J Plant Physiol 169:111–116

Vatanparast G, Mirdehghan H, Karimi H, Vazifeshenas M (2012) Foliar application of salicylic acid, methyl jasmonate and potassium sulfate on photosynthetic characteristics and fruit quality of pomegranate. Iran Agric Res 31:23–34

Wang X, Bai J, Wang W, Zhang G, Yin S, Wang D (2020) A comparative metabolomics analysis of the halophyte *Suaeda salsa* and *Salicornia europaea*. Environ Geochem Health

Wu Y, Guo J, Cai Y, Gong X, Xiong X, Qi W, Pang Q, Wang X, Wang Y (2016) Genome-wide identification and characterization of *Eutrema salsugineum* microRNAs for salt tolerance. Physiol Plant 157:453–468

Yadav N, Shukla P, Jha A, Agarwal PK, Jha B (2012) The *SbSOS1* gene from the extreme halophyte *Salicornia brachiata* enhances Na^+ loading in xylem and confers salt tolerance in transgenic tobacco. BMC Plant Biol 12:188

Yalcinkaya T, Uzilday B, Ozgur R, Turkan I (2019) The roles of reactive carbonyl species in induction of antioxidant defence and ROS signalling in extreme halophytic model *Eutrema parvulum* and glycophytic model *Arabidopsis thaliana*. Environ Exp Bot 160:81–91

Yang Y, Guo Y (2018) Elucidating the molecular mechanisms mediating plant salt-stress responses. New Phytol 217:523–539

Yepes L, Chelbi N, Vivo J-M, Franco M, Agudelo A, Carvajal M, del Martínez-Ballesta MC (2018) Analysis of physiological traits in the response of Chenopodiaceae, Amaranthaceae, and Brassicaceae plants to salinity stress. Plant Physiol Biochem 132:145–155

Yuan F, Guo J, Shabala S, Wang B (2019a) Reproductive physiology of halophytes: current standing. Front Plant Sci 9:1954

Yuan F, Wang B (2020) Adaptation of recretohalophytes to salinity. In: Grigore MN (ed) Handbook of halophytes. Springer, Cham, pp 1–21

Yuan F, Xu Y, Leng B, Wang B (2019b) Beneficial effects of salt on halophyte growth: morphology, cells, and genes. Open Life Sci 14:191–200

Zarei M, Shabala S, Zeng F, Chen X, Zhang S, Azizi M, Rahemi M, Davarpanah S, Yu M, Shabala L (2020) Comparing kinetics of xylem ion loading and its regulation in halophytes and glycophytes. Plant Cell Physiol 61:403–415

Zeng D-E, Hou P, Xiao F, Liu Y (2015) Overexpression of *Arabidopsis XERICO* gene confers enhanced drought and salt stress tolerance in rice (*Oryza Sativa* L.). J Plant Biochem Biotechnol 24:56–64

Zhang H, Feng H, Zhang J, Ge R, Zhang L, Wang Y, Li L, Wei J, Li R (2020) Emerging crosstalk between two signaling pathways coordinates K^+ and Na^+ homeostasis in the halophyte *Hordeum brevisubulatum*. J Exp Bot 71:4345–4358

Zhang X, Zhang L, Dong F, Gao J, Galbraith DW, Song C-P (2001) Hydrogen peroxide is involved in abscisic acid-induced stomatal closure in *Vicia faba*. Plant Physiol 126:1438–1448

Zhou H, Lin H, Chen S, Becker K, Yang Y, Zhao J, Kudla J, Schumaker KS, Guo Y (2014) Inhibition of the *Arabidopsis* salt overly sensitive pathway by 14-3-3 proteins. Plant Cell 26: 1166–1182

Zhu M, Meng X, Cai J, Li G, Dong T, Li Z (2018) Basic leucine zipper transcription factor SlbZIP1 mediates salt and drought stress tolerance in tomato. BMC Plant Biol 18:83

Wastewater Pollution, Types and Treatment Methods Assisted Different Amendments. A Review

Mujahid Farid, Sidra Tul Muntaha, Muhammad Abubakar, Sheharyaar Farid, Wajiha Sarfraz, Shafaqat Ali, Zaki Ul Zaman Asam, Muhammad Zubair, and Muhammad Rizwan

Abstract Mining activities, industrial processing, domestic and agricultural use of metal and metal-containing compounds have resulted in the release of toxic metals into the atmosphere, which has become a widespread environmental problem as a result of global industrialization. Metal contamination has grave consequences for human health and the environment. Few heavy metals are toxic and lethal in trace quantities, and others are teratogenic, mutagenic and endocrine disruptors, while others cause behavioural and neurological disorders. As a result, heavy metal remediation from polluted sites could be the only viable alternative for reducing the harmful effects on ecosystem health. Reclamation methods used in the past were both costly and detrimental to the ecosystem. Phytoremediation is a modern set of technologies that utilises green plants to eliminate toxins from the atmosphere. It has been promoted as a cost-effective and noninvasive alternative to traditional engineering-based remediation methods. The use of organic chelators such as citric acid, ascorbic acid, oxalic acid etc., increased metal uptake and accumulation in plant roots, stems and leaves significantly. Thus, keeping in view the above facts, an attempt has been made in this article to review the best remediation technique, and

M. Farid (✉) · S. T. Muntaha · M. Abubakar · Z. U. Z. Asam
Department of Environmental Sciences, University of Gujrat, Hafiz Hayat Campus, Gujrat, Pakistan

S. Farid
Institute of Biomedical Sciences ABEL SALAZAR, University of Porto, Porto, Portugal

Department of Biology, Ecology and Evolution, University of Liege, Liege, Belgium

University of the Basque Country, Bilbao, Spain

W. Sarfraz
Department of Botany, Government College Women University, Sialkot, Pakistan

S. Ali · M. Rizwan
Department of Environmental Sciences and Engineering, Government College University, Faisalabad, Pakistan

M. Zubair
Department of Chemistry, University of Gujrat,Hafiz Hayat Campus, Gujrat, Pakistan

M. Hasanuzzaman et al. (eds.), *Managing Plant Production Under Changing Environment*, https://doi.org/10.1007/978-981-16-5059-8_11

the role of different amendments in increasing heavy metals uptake and accumulation in along with their ameliorating role in plant stress by supporting its normal growth and functioning.

Keywords Organic acid · Heavy metals · Accumulation · Phytoremediation

1 Introduction

Water system of horticultural land with slime is a regular practice among forming nations that has been triggered prolonged metal fixations in soil and plant that may decline crop produce and nature of harvest delivered (Singh et al. 2007). Heavy metals, such as copper, zinc, nickel, lead and chromium, are fundamental elements that limit percolation reusing to the agrarian land. They have the capability of being amassed in human tissue and bioamplification through the evolved ways of life and have the mechanism to bring change in human wellbeing and can cause different disorders (Alvarez and Steinbüchel 2002). Microorganisms are exceptionally grumpy to poisonous quality of soil and this led to new branches or research, e.g., the bioavailability of overwhelming metals, ecotoxicology and presence of basis species (Giller et al. 2009). Lower bioavailability frequently diminishes heavy metals proficiency of phytoextraction. Natural chelators can be utilised to improve this strategy by expanding the dissolvability of metal (Ehsan et al. 2014). The aptitude of phytoextraction relies on the high biomass and high bioavailability of metal for plant take-up. Bioavailability is influenced by soil factors (pH, redox potential, cation trade limit and soil type and soil surface) and by plant factors (Sheoran et al. 2011). Phytoremediation efforts have been focusing on plant uses with the main purpose of spoiling natural pollutants, mainly rhizosphere microorganisms or removing overwhelming metals from the soil and water (Freitas et al. 2013). The emission of the heavy metals from the activities like paints, application of fertilizers on the land, mine tailings, disposal of metal-contaminated waste, leaded gasoline, animal manure, pesticides, sewerage, wastewater irrigation to the agricultural lands, combustion of untreated coal and its residues and the spillage of the petrochemical compounds on the soil cause the heavy metal pollution (Adrees et al. 2015).

The harmful metals in water, air and soil are a worldwide problem due to the state in which they come in from characteristic and anthropogenic metals sources that involve the extraction and processing of minerals, particularly the mechanical process of phosphate rock and the waste and use of phosphate compost in the ground (Javied et al. 2009). The uptake of metal is dependent on plant species. It is well-known that the metals are aqueous chemistry and is very difficult to control trace amount of metal toxicity by the uptake by aquatic organisms. In such an environment where nutrients are present in a large amount, the bioavailable fraction of metals binds with these nutrients making anions. The metal uptake is also affected when there is competition between the nutrient cation and the metals for the uptake by the plant uptake sites (Göthberg et al. 2004). The quality of the soil and the degree of contamination of the soil have a remarkable impact on the number of essential metals, such as mercury, cadmium and heavy metals, which accumulate in spinach leaves (Chaney et al. 2000). The use of EDTA considerably reduced the danger caused by Cu in *B. napus*. EDTA increased the absorption of Cu, indicating that it is better to use it for the phytoextraction of Cu, since it also reduces the focus of malonaldehyde and H_2O_2 on leaves and roots (Habiba et al. 2015; Farid et al. 2015).

Phytoextraction has been used to remove metals for a long time. But because of their low bioavailability, the treatment work is limited. Work is being done to enhance the bioavailability of heavy metals in different plants and enhancing the ability to extract metals in that plants. For this purpose, different amendments are being used like chelators (Nowack et al. 2006; Ashfaq et al. 2020). For the purpose of improving phytoextraction capacity, metal uptake by root is enhanced by using solubilising agents that help in enhancing the speed of transfer of metal to upper parts of plants (Lestan et al. 2008).

2 Environmental Pollution

Environmental pollution is the unwanted and undesired changes in the composition of air, water, soil in the physical, chemical and biological aspects of the environment which affects plants, animals and living organisms. Pollution can occur from heat, noise or light. The pollution could occur from natural substances and also from artificial substances (Farid et al. 2018). Some reports showed the deterioration of crops and scented plants and the high capacity of these plants to absorb and translocate significant heavy metals in pleasing parts (Rizwan et al. 2017a). In the 1980s, there was an exciting contrast in the use of agricultural business space for various purposes for the following reasons: (1) little developmental advantage (2) the organic heavy metals spoke of a positive non-mediation disposition towards the agricultural industry with no direct blessing for farmers (CCME 2007). There are large amounts of overwhelming metals and consistent characteristic contaminations (POPs) associated with the waste (Oomen et al. 2002).

2.1 Types of Pollutants

There are three kinds of pollutants which are organic, inorganic and biological pollutants. There are also pollutants that belong to other categories, but these all receive importance because of the adverse impacts they pose on the environment. The correlation between the world's population and pollution production is directly proportional to each other as when the population increase then there is also an increase in pollution across the world. The whole situation leads to a point where pollution is a significant problem across the world (Masindi and Muedi 2018).

2.2 Soil Pollution

Overwhelming metal-spoiled soil and water pollution the earth and human affluence; the issue has required an incredibly innovative method. Overwhelming metal-spoiled soils have spread considerably over decades due to waste and sewage from nonnatural sources (Vakili and Aboutorab 2013). The amount of Pb on earth has been increased because of the prosperity concerns about Pb and regardless of how you see it in our condition (Succuro 2010). The emission of the heavy metals from the activities like paints, application of fertilizers on the land, mine tailings, disposal of metal-contaminated waste, leaded gasoline, animal manure, pesticides, sewerage, wastewater irrigation to the agricultural lands, combustion of untreated coal and its residues and the spillage of the petrochemical compounds on the soil cause the heavy metal pollution is huge (Luo et al. 2014).

2.3 Soil Pollutants

Contamination of the soil by some common mixtures is a phenomenal danger to the earth and human prosperity due to the disturbing influence of the soil structure and the restriction of green, fauna and microflora. Due to their high bioaccumulation potential (Gregoraszczuk et al. 2003), polychlorinated biphenyls (PCBs) are one of the most dangerous pollutants for enthusiasts, as this could be a normal endocrine disorder (Decastro et al. 2006). Heavy metals and metal handling can be a major source of metal contamination in nature (Navarro et al. 2008). Regular pollution in mining areas is primarily related to the physical disruption to the scene, including residue from the spilled mine, accumulated release and transmission of the destructive mine to the pipelines (Adrees et al. 2015). The extraordinary accumulation of considerable metals in the plant soil around different cities has been a major issue (Pruvot et al. 2006).

Overwhelming metals, e.g. Pb, Cr, Hg and Zn generally indicate metals and metalloids whose density is more than 5 g/cm^3 (Lei et al. 2005). Metalloids, for example, arsenic (As) heavy metals, are normally included in the significant heavy metal's characterization of the metal due to the similarities in the properties of the compounds and the common direct (Komnitsas and Modis 2006). Metal overload is hidden, consistent and irreversible (Kim et al. 2014). This type of pollution can also undermine the prosperity of animals and humans using advanced lifestyle technology (Hu et al. 2009). It has become a major problem in different parts of the world (Li 2006). There is no doubt that various assessments have shown that sources of generous metal contamination in nature generally come from these anthropogenic heavy metal sources (Wei and Yang 2010; Farid et al. 2018). In China, wastewater, sludge application and metal mineral excavation and disinfection exercises are the most common sources of metal contamination in soils (Li et al. 2006). Significant heavy metal concentrations in rice fields and nurseries exceeded the most reasonable obsessions for Chinese topsoil (Zhuang et al. 2009).

2.4 *Wastewater*

A lot of wastewater has been created in recent decades, mainly due to anthropogenic activities. The source of heavy metals mainly such as agricultural practices, urbanisation and industrialisation (Renuka et al. 2013; Farid et al. 2017). Permanent disposal of the wastewater without satisfactory treatment can lead to real contamination problems. Onė of the serious problems associated with the continued release of wastewater into water is the supposed phenomenon of eutrophication: the improvement of water resources in food supplements, mainly heavy metals such as argon and phosphorus. This phenomenon is responsible for the improvement in algal bloom, the multiplication of marine plants, the lack of oxygen and the loss of key species, which leads to the complete degradation of the biological freshwater systems (Ruiz et al. 2013). In addition to spoiling the freshwater environment, not developing these vegetation's can be a general threat to the wellbeing (Cai et al. 2013).

Modern wastewater has been the subject of extensive research by the scientific network as it poses a real threat to amphibian, air and soil conditions, particularly about the well-being of humans and animals (Bahnmüller et al. 2015). Metals, that are difficult to evacuate due to their water solubility or the proximity of deadly and stubborn mixtures (Ahmad et al. 2019). These mixtures can contain aromatics, sulphur, heavy metals and argon, and oxygen, double-linked and deeply subatomic mixtures (Liang and Guo 2010) (Table 1

2.4.1 Sources of Waste Water

There are many sources of water contamination which include industrialisation and urbanisation which are the two major sources for heavy metal's contamination. The municipal, industrial and urban runoff contains heavy metals and is added up in the water bodies becoming a source of heavy metal contamination (Masindi and Muedi 2018).

Elegant wastewater pollution is largely shown in the associated perspectives: (1) influencing drinking and groundwater assets, endangered maritime assets

Table 1 Characteristics of raw wastewater

Colour (pt.-Co)	100–250
Turbidity (NTU) pH	50–100
COD (mg/L)	10.5–11
	250–600

Source: Liu et al. (2010)

(2) endangered human well-being (3) pollution (4) influencing plant production; (5) The destruction of the characteristic prospect and probably even due to the combination of the safety of the oil burner leads to this emergency (Poulopoulos et al. 2005).

2.4.2 Textile Industry Wastewater

The developing human population has created some difficulties for the global economy, especially about the natural preservation and security of vitality (Kassim and Meng 2017). The global economy is mainly dependent on limited and inexhaustible fossil fuels (Colo et al. 2016). Textile wastewater contains a multitude of colours and synthetic additives, which can pose real dangers for ecological beneficiaries. Pakistan needs substantial wastewater treatment and recovery to meet its residential and mechanical water needs (Maqbool et al. 2016) (Table 2).

2.4.3 Tannery Industry Wastewater

Tannery industries are similarly considered to be the notable point of the source of biological contamination from their potentially destructive and unsafe liquid waste, which gives a negative image of leather tanneries (LT) in the general population. The wastewater derived from the LTs is represented by a darker veil, a hostile fragrance, a high pH value and oxygen requirement for the substance, a biochemical oxygen requirement, large-scale separated solids, chromium, sulphate, phosphate, heavy metals trade and an arrangement of incredibly toxic normal technical mixtures and overwhelming metals (Saxena et al. 2016; Afshan et al. 2015). TWW is used in the production of lands, and this preparation offers opportunities for bioaccumulation of harmful metals at dynamically higher trophic levels (Lofrano et al. 2013).

Table 2 Effects and characteristics of textile wastewater

Process	Chemical discharge	Pollutants	Health effects
Singe	Benzene PVA gum, Starch	Resin starch glucose	Carcinogenic, mutagenic, effect on the central nervous system
Bleaching	Detergents, hydrogen NAOH	Sodium silicate	Exposure to a large amount of time can affect kidney and liver
Dying	Sulphates and salts	Dyes sulphates	Eyes and respiratory issues
Printing	Nitrates, phosphates and dyes	Starch gum, PVA, colours	Harmful health hazards
Finishing	Fats, silicon	Starch finishing agent	Suppression of haematological system

Sources: Imtiazuddin (2018)

2.5 Wastewater Treatment

Advances in treatment, such as gravity distribution, centrifugation, coagulation, buoyancy, adsorption, natural strategies, filtration techniques and hot oxidation, have been used to fully meet the requirements of treatment enactment (Poyatos et al. 2010; Farid et al. 2017) but they lead to results such as profligacy. Syrups or salts that require further treatment can be expensive and dangerous. In addition, due to the proximity of significant amounts of refractory mixtures such as aromatic hydrocarbons, some types of wastewater cannot be treated directly by organic advances (Mota et al. 2009). In this way, one of the main goals of the twenty-first century is to create simple, protected, affordable, effective technologies and the most important nondestructive factor for the treatment of nature (Boczkaj et al. 2017).

2.6 Heavy Metals

Improving yields for the use of human or trained heavy metals has contaminated soils can lead to absorption and assembly of the following metals in the appetising plant, which poses a consequent threat to human and animal welfare (Monika and Sawicka-Kapusta 2004). Polycyclic aromatic hydrocarbons (PAHs) are unavoidable normal poisons. Climatic PAHs can cause breathing problems that weaken the aspiration limit and cause bronchitis (Tsapakis et al. 2003). Decomposition of soil and groundwater by overwhelming metals is a common problem that discusses situations and general prosperity (Merdoud et al. 2016). The terrible waste generation and unhealthy handling are responsible for the increasing proportion of prodigious metals in soils and groundwater (Rizwan et al. 2017b). Some breakthroughs have been proposed for the remediation of significant heavy metals, metal targets in poor condition. In the meantime, electrokinetic remediation has achieved some success even in soils with low permeability (Reddy and Cameselle 2009).

This discussion is equally relevant to: (1) the expansion of soil preparation as techniques for heavy metals for incorporating biodegradable waste from landfill disposal (Küpper and Kochian 2009; Smith 2009), as the increasing confirmation shows that the large metal pollution from depleted areas that harm residents' health and productivity in the affected areas (Kachenko and Singh 2006). Soil remediation is essentially cultivated by physically expelling soils from degraded sites for landfilling, cremation or in-situ adjustment of connections (Belluck et al. 2006).

2.6.1 Inorganic Heavy Metals

Different inorganic heavy metals alterations, for example, lime materials, phosphate blends and earth materials are utilized to immobilise overpowering metals and improve soil conditions to empower filthy soils to be replanted (Kumpiene et al.

2007). Treatment applications moved the feast of Pb from the exchangeable segment to the carbonate and oxide outline and diminished heavy metals the ruinous extraction limit of Pb. Pb solubility is regularly lower in nonacidic soils and the progressions used can reduce their versatility to a certain extent (McGrath et al. 2002).

Regular constituents, e.g. wastewater overflow, dense rejection, mulch and topmost soil enhanced by the physical thinking of the soil by expanding the containment of water retention and additionally providing plant additions in a structure with the moderate release (Park et al. 2011). A broad confederacy of characteristic mixtures in the DOC is related to the schedule of metal-soluble buildings (Wong 2003), it has been regularly observed that the addition of compost builds up the activity of the microbe in the soil as assessed by prolonged biological activities (Tejada 2009).

2.6.2 Organic Pollutants

Biodegradable substances present in the environment are organic pollutants. These pollutants are naturally found in the environment and then increased by anthropogenic activities to meet their own needs. There are some organic pollutants that are found in large concentrations as the result of human activities are PCB's, food waste, herbicides, pesticides, fertilizers, chemicals, etc. the organic pollutants have become the major problem of the environment across the world and therefore seeking lots of attention. There are some properties of the organic pollutants that make them more persistent in the environment, as they have the ability to bioaccumulate in different parts of the environment which then cause toxic effects (Masindi and Muedi 2018).

Dangerous citrus organic heavy metals products have been utilised to improve the discharge of metals from the dirt because of their quick accessibility, as they are modestly petulant (Amir et al. 2020). Juang and Shiau (2000) showed that citrus organic heavy metals products up to 80% of the metals dispense with copper from filthy wood dissipate. Yang et al. (2001) demonstrated that citrus evacuation is helpful for expelling a couple of metals, for example Pb, Zn, Cu and Mn from debased soil. The stripping of the considerable metals was cultivated by a mixture of metal solubility by complexing with citrate and transport by the improved electro-osmotic current (Gent et al. 2004). Due to its complexation with other cations, it is less unbelievable when it comes to metal removal (Evangelou et al. 2008).

3 Treatment of Heavy Metals

Phytoremediation is seen as an intelligent and friendly innovation for the treatment of soils and waters that are contaminated by essential metals/metalloids (Salt et al. 1998). Complex methods, including combined natural and composite techniques, as well as the expansion of biosurfactants or potentially EDTA, could be important for remediation of heavily contaminated soil (Viisimaa et al. 2013).

3.1 *Phytoremediation*

Phytoremediation as an ecological development option is particularly important among wetlands because it uses plants and their related microorganisms to recover hydrocarbon degraded water and is more amusing than conventional heavy metals cleaning methods (Barrutia et al. 2011; Khair et al. 2014). This process requires less labour and equipment than other forms but may damage the natural plant's growth process as it accumulates pollutants. Similarly, the site can be cleansed without tunnelling or pulling the ground or extracting groundwater, saving the unembellished slightest (EPA 2012).

3.2 *Phytoremediation Types*

Phytoremediation has several types: phytoextraction, phytodegradation, rhizofiltration, phytostabilisation and phytovolatilisation (Fayiga and Ma 2005). The advancement of phytoremediation is that helps plants to do cleaning metals of stained areas and is a hopeful development for reclaiming the clean area and conditions (Pantola and Alam 2014; Naeem et al. 2021).

3.3 *Treatment Technologies*

Phytoremediation is a development that utilises higher plants to adjust and revegetate besmirched destinations (Adreano et al. 2004; Freitas et al. 2013). Various methods and treatments are revised for the phrase phytoremediation. For instance, the technique where plants are utilised to drive normal or inorganic contaminants out of soil and water and to store them in the harvestable tissue is indicated as phytoextraction or phyto-filtration. Therefore, the procedure utilizing plants to clear contaminants through volatilisation is suggested as phytovolatilisation (Dickinson et al. 2009). Overpowering metals are immobilised in the ground, which confines your vehicle in water or waste. This advancement can improve the debasement of common toxins, for example, pesticides and hydrocarbons, through microbial developments related to plant roots that quicken the change of these contaminations into nonpoisonous structures (Pulford and Watson 2003).

3.3.1 Rhyzodegradation

This process is also called as Enhanced Rhizosphere biodegradation, plant assisted bioremediation and Phytostimulation. In this process, there is a breakdown of the organic contaminants by the soil-dwelling microbes in the soil, which is improved by the rhizosphere's occurrence. The plant roots exhaust some sugars, organic acids and alcohols as the source of carbon for the microflora and the enhancement of the growth of the microbes and their activity. The roots of plants also slacken off the soil and help in the effective transport of the water, thus increasing the microbial activity (Khair et al. 2014).

3.3.2 Phytostabilisation

Hyper accumulator plants characterised by metal expulsion can be used to improve this technology (Wong 2003). Plants with significant heavy metals, thick roots, that can absorb large amounts of contaminated soil have a much larger surface that promises soil modification and improves the microbial volatilisation of metals in the rhizosphere (Unterbrunner et al. 2007). The interactions between the hyperaccumulators establish and dominate the metals in the rhizosphere and stimulate metal transactions from the ground to the roots (Alford et al. 2010). Microorganisms that colonise the rhizosphere can drive plant improvement, redesign the availability and flexibility of metals in corrupt soils (Sessitsch et al. 2013). The development of high-throughput sequencing has disrupted environmental microbiology and made it possible to examine and request the phylogenetic disposed plan, like the structure of microbial systems, with more remarkable precision than previous techniques (Wei and Twardowska 2013).

3.3.3 Metal Immobilisation

Plants decrease the express ability and transport of toxins in the dirt, either through retention or immobilisation (Pulford and Watson 2003). Phytostabilisation can be improved by utilising soil disquieting influences, which are solid in metal immobilisation and which rapidly adjust to the examined trademark remediation of corrupted destinations (Adriano et al. 2004). The precipitation likewise controls the separating of contaminations and the rot of soil and residue. The temperature impacts both the improvement of the plant and the properties of the dirt surface, for instance, the division and course of action of the structure. As the division develops the toxic channel, the exposed, dry and uncovered soil can't distort the root of the feast (Dickinson et al. 2009).

3.3.4 Rhizoremediation

This renewal innovation, known as rhizoremediation, was characterised as "increasing decomposition rate of natural pollutants in the soil due to the improved biodegradable effect of microorganisms and heavy metals in the rhizosphere" (Shaw and Burns 2005). The roots of plants can undoubtedly provide valuable and living habitats for hydrocarbon-degrading microorganisms (Gaskin and Bentham 2010).

In the field of phytoremediation, rhizoremediation was recommended as an essential tool for the degradation of hydrocarbons in the soil (Yateem et al. 2007). In addition to observing the evacuation of pollutants, it is generally imperative during restoration forms to assess the restoration of soil well-being, which is referred to as the "boundary of the soil for the development of its capabilities" (Abhilash et al. 2016). In recent years, inorganic heavy metal's properties have been used increasingly due to their ability to influence and provide data with numerous natural components (Mijangos et al. 2006).

3.3.5 Micro Remediation

Phytoremediation is the broad term for a collection of advances that use plants and rhizosphere-related microorganisms and heavy metals to drive or change filtered contaminants from soils/wastes and used watercourses (Saxena et al. 2016). Absorption and detoxification capacity of numerous unmanageable xenobiotics, which act as "green liver" in nature (Vanek et al. 2010) An exceptionally investigated and important field within phytoremediation is the use of developed wetlands (CW) for pharmaceutical products from the wastewater of sewage treatment plants (Carvalho et al. 2014).

3.3.6 Volatilisation

The volatilisation of pollutants in the atmosphere by plant methods can be a great strategy in phytostabilising soils where high values are present. Combinations of characteristic contaminants are accessible (Ouyang 2002). Phytovolatilisation can correct metal-spoiled soils that structure unusual mixtures of hydride and methyl. Progressive efforts have focused on producing transgenic species with an expanded potential for volatilisation of Hg (Moreno et al. 2005) and Se (de Souza et al. 2002). A disadvantage of volatilisation is that there is no control over the ultimate destination of the pollutants.

3.3.7 Phyto-Extraction

Phytoextraction of metals is part of the idea of phytoremediation (Ehsan et al. 2014). It can be applied to metals such as Zn, Cd, Cu and Pb. As an innovation, the phytoextraction of metals is still in its initial disposition and is an innovation that is only economically accessible for the evacuation of Pb. For the most part, the metal Phytoextraction convention consists of the accompanying components (1) plant development in a contaminated location, (2) evacuation of biomass that is rich in harvested metals, (3) medication after harvesting and the resulting disposing biomass in harmful forms and (4) unavoidable metal recovery from metal-reinforced biomass (Blaylock and Huang 2000). The first is to completely remove contaminants from dirty stocks, and the second is to convert these toxins into harmless structures using at least one of the design advances, which are essentially elimination, distribution, extraction, electrokinesis, washing, oxidation, degradation (Wei et al. 2008) Previous author has concluded that lead stress has a very large impact in the reduction of chlorophyll, it has an impact on the disorganisation of chloroplast and reduction in the number of thylakoids and grana. It directly reduces the chlorophyll production and changes the structure of chlorophyll due to changes in key nutrients e.g., magnesium iron and cooper (Akinci et al. 2010).

4 Amendments

4.1 Chelating Agents

Chelating agents can improve the devastating removal of metals from degraded soils. The benefits of these chelators' high viability of metal extraction, the high thermodynamic safety and the dissolving power of metal buildings and the low adsorption of chelating agents in soils (Merdoud et al. 2016; Amir et al. 2020). Freitas et al. (2013) said commercial CA is very helpful for using in phytoextraction because it has very low cost and high ability to biodegrade. He also concluded that the Pb has low natural solubility and requires about 150 yrs. or more phytoextracting it without chelator will not be effective and take more time (Amir et al. 2020). In a study where they had plants with CA and without CA, they concluded that plants with CA application had higher height and dry weights than the other plant (Zaheer et al. 2015; Farid et al. 2020).

4.2 Biochar

Extensive studies on biochar have been carried out, which show that biochar, when mixed with compost, is best suited for the environment (Agegnehu et al. 2017). When biochar mixes with the soil, the pH of the soil increases, which also creates a temporary barrier to the mobility of metal ions (Cui et al. 2011). The main goal of adding biochar to the soil is to check the mobility of the metal ion (Yang et al. 2001; Latif et al. 2020).

5 Conclusions

Since toxic heavy metal pollution of soils and waters is a serious environmental issue, successful remediation methods are needed. Physical and chemical approaches for heavy metal clean up and regeneration have significant disadvantages, such as high costs, permanent changes in surface resources, degradation of native soil microflora and issues of secondary contamination. Changes in soil morphological and chemical composition have resulted from increased heavy metal concentrations. As a result, serious effects on soil fertility and the natural environment have occurred. Remediating sites polluted with dangerous heavy metals is a challenging task; additionally, unlike organic compounds, these sites do not dissolve on their own and require careful management. Using plants is being considered as an environmentally safe and cost-effective strategy for extracting heavy metals from soil and wastewater. Wide biomass and the use of chelating agents to increase bioavailability have been shown to be effective in increasing heavy metal extraction in studies. Heavy metals' bioavailability is enhanced by citric acid. Furthermore, various amendments are very effective at lowering heavy metal concentrations in plants. The type of contaminants and degree of contamination decide the remediation approaches used to remove heavy metals from the soil, so choosing the right one is crucial for successful heavy metals polluted site treatment.

References

Abhilash MR, Srikantaswamy S, Shiva Kumar D, Jagadish K, Shruthi L (2016) Phytoremediation of heavy metal industrial contaminated soil by *Spiracia oleracea* L. and *Zea mays* L. Int J Appl Sci 4:1

Adrees M, Ali S, Rizwan M, Zia-ur-Rehman M, Ibrahim M, Abbas F, Farid M, Qayyum MF, Irshad MK (2015) Mechanisms of silicon-mediated alleviation of heavy metal toxicity in plants: a review. Ecotoxicol Environ Saf 119:186–197

Adriano DC, Wenzel W, Vangronsveld J, Bolan N (2004) Role of assisted natural remediation in environmental cleanup. Geoderma 122(2):121–142

Afshan S, Ali S, Bharwana SA, Rizwan M, Farid M, Abbas F, Abbasi GH (2015) Citric acid enhances the phytoextraction of chromium, plant growth, and photosynthesis by alleviating the oxidative damages in *Brassica napus* L. Environ Sci Pollut Res 22(15):11679–11689

Agegnehu G, Srivastava AK, Bird MI (2017) The role of biochar and biochar-compost in improving soil quality and crop performance: a review. Appl Soil Ecol 119:156–170

Ahmad R, Ali S, Rizwan M, Dawood M, Farid M, Hussain A, Wijayae L, Alyemenie MN, Ahmad P (2019) Hydrogen sulfide alleviates chromium stress on cauliflower by restricting its uptake and enhancing antioxidative system. Physiol Plant

Akinci I, Akinci S, Yilmaz K (2010) Response of tomato (*Solanum lycopersicum* L.) to lead toxicity: growth, element uptake, chlorophyll and water content. Afr J Agric Res 5(6):416–423

Alford ER, Pilon-Smits EAH, Paschke MW (2010) Metallophytes-a view from the rhizosphere. Plant and Soil 337(1):33–50

Alvarez H, Steinbüchel A (2002) Triacylglycerols in prokaryotic microorganisms. Appl Microbiol Biotechnol 60(4):367–376

Amir W, Farid M, Ishaq HK, Farid S, Zubair M, Rizwan M, Raza N, Ali S (2020) Accumulation potential and tolerance response of *Typha latifolia* L. under citric acid assisted phytoextraction of lead and mercury. Chemosphere 257:127247

Ashfaq H, Abubakar M, Ghulzar H, Farid M, Yaqoob S, Komal N, Azam Z, Hamza A, Ali S, Adrees M (2020) Phytoremediation potential of oilseed crops for Lead- and nickel-contaminated soil. In: Hasanuzzaman M (ed) Plant ecophysiology and adaptation under climate change: mechanisms and perspectives II, mechanisms of adaptation and stress amelioration. Springer, Singapore, pp 801–820. https://doi.org/10.1007/978-981-15-2172-0_31

Bahnmüller S, Loi CH, Linge KL, Von Gunten UV, Canonica S (2015) Degradation rates of benzotriazoles and benzothiazoles under UV-C irradiation and the advanced oxidation process UV/H_2O_2. Water Res 74:143–154

Barrutia O, Garbisu C, Epelde L, Sampedro MC, Goicolea MA, Becerril JM (2011) Plant tolerance to diesel minimizes its impact on soil microbial characteristics during rhizoremediation of diesel-contaminated soils. Sci Total Environ 409(19):4087–4093

Belluck DA, Benjamin SL, David S (2006) Why remediate? Phytoremed Metal-Contam Soils 68:1–23

Blaylock MJ, Huang JW (2000) Phytoextraction of metals. In: Raskin I, Ensley B (eds) Phytoremediation of toxic metals. Wiley, New York, p 53

Boczkaj G, Fernandes A (2017) Wastewater treatment by means of advanced oxidation processes at basic pH conditions: a review. Chem Eng J 320:608–633

Cai T, Park SY, Li Y (2013) Nutrient recovery from wastewater streams by microalgae: status and prospects. Renew Sustain Energy Rev 19:360–369

Carvalho PN, Basto MCP, Almeida CMR, Brix H (2014) A review of plant–pharmaceutical interactions: from 483 uptake and effects in crop plants to phytoremediation in constructed wetlands. Environ Sci Pollut Res 21(20):11729–11763

CCME-Canadian Council of Ministers of the Environment (2007) Summary of a protocol for the derivation of environmental and human health soil quality guidelines. CCME, Winnipeg

Chaney RL, Li YM, Brown SL, Homer FA, Malik M, Angle JS, Baker AJM, Reeves RD, Chin M (2000) Improving metal hyperaccumulator wild plants to develop commercial phytoextraction systems: approaches and progress. In: Banuelos GS, Terry N (eds) Proceedings of the symposium on phytoremediation, fourth international conference on the biogeochemistry of trace elements. CRC Press, Boca Raton, pp 129–158

Colo MS, Guglielmino SPP, Solinas V, Salis A (2016) Consequences of microbial interactions with hydrocarbons. Oil Lipids Prod Fuels Chem 5:1–20

Cui L, Li L, Zhang A, Pan G, Bao D, Chang A (2011) Biochar amendment greatly reduces rice cd uptake in a contaminated paddy soil: a two-year field experiment. Bioresources 6(3):2605–2618

de Souza MP, Pickering IJ, Walla M, Terry N (2002) Selenium assimilation and volatilization from Selenocyanate-treated Indian mustard and musk grass. Plant Physiol 128(2):625–633

DeCastro BR, Korrick SA, Spengler JD, Soto AM (2006) Estrogenic activity of polychlorinated biphenyls present in human tissue and the environment. Environ Sci Technol 40:2819–2825

Dickinson NM, Baker AJM, Doronila A, Laidlaw S, Reeves RD (2009) Phytoremediation of inorganics: realism and synergies. Int J Phytoremed 11(2):97–114

Ehsan S, Ali S, Noureen S, Mahmood K, Farid M, Ishaque W, Shakoor MB, Rizwan M (2014) Citric acid assisted phytoremediation of cadmium by *Brassica napus* L. Ecotoxicol Environ Saf 106:164–172

EPA (2012) A citizen's guide to phytoremediation. EPA, Washington

Evangelou MWH, Ebel M, Hommes G, Schaeffer A (2008) Biodegradation: the reason for the inefficiency of small organic acids in Chelant-assisted phytoextraction. Water Air Soil Pollut 195:177–188

Farid M, Ali S, Akram NA, Rizwan M, Abbas F, Bukhari SAH, Saeed R (2017) Phyto-management of Cr-contaminated soils by sunflower hybrids: physiological and biochemical response and metal extractability under Cr stress. Environ Sci Pollut Res 24(20):16845–16859

Farid M, Ali S, Ishaque W, Shakoor MB, Niazi NK, Bibi I, Dawood M, Gill RA, Abbas F (2015) Exogenous application of EDTA enhanced phytoremediation of cadmium by *Brassica napus* L. Int J Environ Sci Technol 12(12):3981–3992

Farid M, Ali S, Rizwan M, Ali Q, Saeed R, Nasir T, Abbasi GH, Rehmani MIA, Ata-Ul-Karim ST, Bukhari SAH (2018) Phyto-management of chromium contaminated soils through sunflower under exogenously applied 5-aminolevulinic acid. Ecotoxicol Environ Saf 151(30):255–265

Farid M, Farid S, Zubair M, Rizwan M, Ishaq HK, Ali S, Ashraf U, Alhaithloul HAS, Gowayed S, Soliman MH (2020) Efficacy of Zea mays L. for the management of marble effluent contaminated soil under citric acid amendment; morpho-physiological and biochemical response. Chemosphere 240:124930

Fayiga AO, Ma LQ (2005) Using phosphate rock to immobilize metals in soil and increase arsenic uptake by hyperaccumulator *Pteris vittata*. Sci Total Environ 15(1–3):17–25

Freitas EV, Nascimento CW, Souza A, Silva FB (2013) Citric acid-assisted phytoextraction of lead: a field experiment. Chemosphere 92(2):213–217

Gaskin SE, Bentham RH (2010) Rhizoremediation of hydrocarbon contaminated soil using Australian native grasses. Sci Total Environ 408(17):3683–3688

Gent B, Bricka RM, Alshawabkeh AN, Larson SL, Fabian G, Granade S (2004) Bench- and field-scale evaluation of chromium and cadmium extraction by electrokinetics. J Hazard Mater 110(1–3):53–62

Giller KE, Witter E, McGrath SP (2009) Heavy metals and soil microbes. Soil Biol Biochem 41(10):2031–2037

Göthberg A, Greger M, Holm K, Bengtsson BE (2004) Influence of nutrient levels on uptake and effects of mercury, cadmium, and lead in water spinach. J Environ Qual 33(4):1247–1255

Gregoraszczuk EL, Grochowalski A, Chrzaszcz R, Wegiel M (2003) Congener-specific accumulation of polychlorinated biphenyls in ovarian follicular wall follows repeated exposure to PCB 126 and PCB 153. Comparison of tissue levels of PCB and biological changes. Chemosphere 50:481–488

Habiba U, Ali S, Farid M, Shakoor MB, Rizwan M, Ibrahim M, Abbasi GH, Hayat T, Ali B (2015) EDTA enhanced plant growth, antioxidant defense system, and phytoextraction of copper by *Brassica napus* L. Environ Sci Pollut Res 22(2):1534–1544

Hu RZ, Liu JM, Zhai MG (2009) Mineral resources science in China: a roadmap to 2050. Science Press, Beijing

Imtiazuddin SM (2018) Impact of textile wastewater pollution on the environment. Pakistan Textile J 68:38

Javied S, Mehmood T, Chaudhry MM, Tufail M, Irfan N (2009) Heavy metal pollution from phosphate rock used for the production of fertilizer in Pakistan. Microchem J 91(1):94–99

Juang RS, Shiau RC (2000) Metal removal from aqueous solutions using chitosan-enhanced membrane filtration. J Membr Sci 165(2):159–167

Kachenko AG, Singh B (2006) Heavy metals contamination in vegetables grown in urban and metal smelter contaminated sites in Australia. Water Air Soil Pollut 69(1):101–123

Kassim MA, Meng TK (2017) Carbon dioxide (CO_2) biofixation by microalgae and its potential for biorefinery and biofuel production. Sci Total Environ 584:1121–1119

Khair KU, Farid M, Ashraf U, Zubair M, Rizwan M, Farid S, Ishaq HK, Iftikhar U, Kim JY, Kim KW, Ahn JS, Ko I, Lee CH (2014) Investigation and risk assessment modeling of as and other heavy metals contamination around five abandoned metal mines in Korea. Environ Geochem Health 27(2):193–203

Komnitsas K, Modis K (2006) Soil risk assessment of as and Zn contamination in a coal mining region using geostatistics. Sci Total Environ 371(1–3):190–196

Kim YH, Khan AL, Kim DH, Lee SY, Kim KM, Waqas M, Lee IJ (2014) Silicon mitigates heavy metal stress by regulating P-type heavy metal ATPases, Oryza sativa low silicon genes, and endogenous phytohormones. BMC Plant Biol 14(1):1–13

Kumpiene J, Ore S, Lagerkvist A, Maurice C (2007) Stabilization of Pb- and cu-contaminated soil using coal fly ash and peat. Environ Pollut 145(1):65–373

Küpper H, Kochian LV (2009) Transcriptional regulation of metal transport genes and mineral nutrition during acclimatization to cadmium and zinc in the cd/Zn hyperaccumulator, *Thlaspi caerulescens* (Ganges population). New Phytol 185(1):114–129

Latif U, Farid M, Rizwan M, Ishaq HK, Farid S, Ali S, El-Sheikh MA, Alyemeni MN, Wijaya L (2020) Physiological and biochemical response of *Alternanthera bettzickiana* (regel) G. Nicholson under acetic acid assisted phytoextraction of lead. Plan Theory 9(9):1084

Lei M, Yue QL, Chen TB, Huang ZC, Liao X, Liu YR, Zheng GD, Chang QR (2005) Heavy metal concentrations in soils and plants around Shizuyuan mining area of Hunan province. Acta Ecol Sin 25(5):1146–1151

Lestan D, Luo CL, Li XD (2008) The use of chelating agents in the remediation of metal contaminated soils: a review. Environ Pollut 153(1):3–13

Li MS (2006) Ecological restoration of mine land with particular reference to the metalliferous mine wasteland in China: a review of research and practice. Sci Total Environ 357(1–3):38–53

Liang C, Guo YY (2010) Mass transfer and chemical oxidation of naphthalene particles with zerovalent iron activated persulfate. Environ Sci Technol 44(21):8203–8208

Liu RR, Tian Q, Yang B, Chen J (2010) Hybrid anaerobic baffled reactor for treatment of desizing wastewater. Int J Environ Sci Technol 7:111–118

Lofrano G, Meric S, Zengin GE, Orhon D (2013) Chemical and biological treatment technologies for leather tannery chemicals and wastewaters: a review. Sci Total Environ 461–462:265–228

Luo YL, GuoWS NHH, Nghiem LD, Hai FI, Zhang J, Liang S, Wang XC (2014) A review on the occurrence of micropollutants in the aquatic environment and their fate and removal during wastewater treatment. Sci Total Environ 473:619–641

Maqbool Z, Hussain S, Ahmad T, Nadeem H, Imran M, Khalid A, Abid M, Martin-Laurent F (2016) Use of RSM modeling for optimizing decolorization of simulated textile wastewater by Pseudomonas aeruginosa strain ZM130 capable of simultaneous removal of reactive dyes and hexavalent chromium. Environ Sci Pollut Res 23(11):11224–11239

Masindi V, Muedi K (2018) Environmental contamination by heavy metals. IntechOpen, Rijeka. https://doi.org/10.5772/intechopen.76082

McGrath SP, Zhao FJ, Lombi E (2002) Phytoremediation of metals, metalloids, and radionuclides. Adv Agron 75:1–56

Merdoud O, Cameselle C, Boulakradeche MO, Akretche DE (2016) Removal of heavy metals from contaminated soil by electrodialytic remediation enhanced with organic acids. Environ Sci: Processes Impacts 18(11):1440–1448

Mijangos I, Parmo RP, Albizu I, Garbisu C (2006) Effects of fertilization and tillage on soil biological parameters. Enzyme Microb Technol 40(1):100–106

Monika DP, Sawicka-Kapusta K (2004) Histopathological changes in the liver, kidneys, and testes of bank voles environmentally exposed to heavy metal emissions from the steelworks and zinc smelter in Poland. Environ Res 96(1):72–78

Moreno FN, Anderson CWN, Stewart RB, Robinson BH (2005) Mercury volatilisation and phytoextraction from base-metal mine tailings. Environ Pollut 136:341–352

Mota A, Albuquerque LF, Beltrame L, Chiavone-Filho O, Machulek A, Nascimento C (2009) Advanced oxidation processes and their application in the petroleum industry: a review. Braz J Petroleum Gas 2(3):122–142

Naeem N, Khalid N, Sarfraz W, Ejaz U, Yousaf A, Rizvi ZF, Ikram S (2021) Assessment of Lead and cadmium pollution in soil and wild plants at different functional areas of Sialkot. Bull Environ Contam Toxicol

Navarro MC, Perez-Sirvent C, Martínez-Sánchez MJ, Vidal J, Tovar PJ, Bech J (2008) Abandoned mine sites as a source of contamination by heavy metals: a case study in a semi-arid zone. J Geochem Explor 96(2–3):183–193

Nowack B, Schulin R, Robinson BH (2006) Critical assessment of chelant-enhanced metal phytoextraction. Environ Sci Technol 40(17):5225–5232

Oomen AG, Hack A, Minekus M, Zeijdner E, Cornelis C, Schoeters G, Verstraete W, Van de Wiele T, Wragg J, Rompelberg CJM, Sips AJAM, Wijnen JHV (2002) Comparison of five in vitro digestion models to study the bioaccessibility of soil contaminants. Environ Sci Technol 36(15):3326–3334

Ouyang Y (2002) Phytoremediation: modeling plant uptake and contaminant transport in the soil-plant-atmosphere continuum. J Hydrol 266:66–82

Pantola RC, Alam A (2014) Potential of Brassicaceae Burnett (mustard family; angiosperms) in phytoremediation of heavy metals. Int J Sci Res Environ Sci 2(4):120–138

Park JH, Lamb D, Paneerselvam P, Choppala G, Bolan N, Chung JW (2011) Role of organic amendments on enhanced bioremediation of heavy metal(loid) contaminated soils. J Hazard Mater 185(2–3):549–574

Poulopoulos SG, Voutsas EC, Grigoropoulou HP, Philippopoulos CJ (2005) J Hazard Mater 117(2–3):135–139

Poyatos JM, Muñio MM, Almecija MC, Torres JC, Hontoria E, Osorio F (2010) Advanced oxidation processes for wastewater treatment: state of the art. Water Air Soil Pollut 205(1): 187–204

Pruvot C, Douay F, Herve F, Waterlot C (2006) Heavy metals in soil, crops and grass as a source of human exposure in the former mining areas (6pp). J Soil Sediment 6(4):215–220

Pulford ID, Watson C (2003) Phytoremediation of heavy metal-contaminated land by trees-a review. Environ Int 29:529–540

Reddy KR, Cameselle C (2009) Overview of electrochemical remediation technologies. In: Electrochemical remediation technologies for polluted soils,sediments and groundwater. John Wiley & Sons, New York, pp 1–28

Renuka N, Sood A, Ratha SK, Prasanna R, Ahluwalia A (2013) Evaluation of microalgal consortia for treatment of primary treated sewage effluent and biomass production. J Appl Phycol 25: 1529–1537

Rizwan M, Ali S, Hussain A, Ali Q, Shakoor MB, Zia-ur-Rehman M, Farid M, Asma M (2017a) Effect of zinc-lysine on growth, yield and cadmium uptake in wheat (*Triticum aestivum* L.) and health risk assessment. Chemosphere 187:35–42

Rizwan M, Ali S, Qayyum MF, Ok YS, Adress M, Ibrahim M, Zia-ur-Reham M, Farid M, Abbas F (2017b) Effect of metal and metal oxide nanoparticles on growth and physiology of globally important food crops: a critical review. J Hazard Mater 322:2–16. https://doi.org/10.1016/j.jhazmat.2016.05.061

Ruiz F, Abad M, Bodergat AM, Carbonel P, Rodríguez-Lázaro J, González-Regalado ML, Prenda J (2013) Freshwater ostracods as environmental tracers. Int J Environ Sci Technol 10(5): 1115–1128

Salt DE, Smith RD, Raskin I (1998) Phytoremediation. Annu Rev Plant Biol 49:643–668

Saxena G, Chandra R, Bharagava RN (2016) Environmental pollution, toxicity profile and treatment approaches for tannery wastewater and its chemical pollutants. Rev Environ Contam Toxicol 240:31–69

Sessitsch A, Kuffner M, Kidd P, Vangronsveld J, Wenzel WW, Fallmann K, Puschenreiter M (2013) The role of plant-associated bacteria in the mobilization and phytoextraction of trace elements in contaminated soils. Soil Biol Biochem 60:182–194

Shaw LJ, Burns RG (2005) Rhizodeposits of *Trifolium pratense* and *Lolium perenne*: their comparative effects on 2,4-D mineralization in two contrasting soils. Soil Biol Biochem 37(5):995–1002

Sheoran V, Sheoran AS, Poonia P (2011) Role of hyperaccumulators in phytoextraction of metals from contaminated mining sites: a review. Crit Rev Environ Sci Technol 41(2):168–214

Singh G, Brar MS, Malhi SS (2007) Decontamination of chromium by farmyard manure application in spinach grown in two texturally different Cr-contaminated soils. J Plant Nutr 30:2

Smith SR (2009) A critical review of the bioavailability and impacts of heavy metals in municipal solid waste composts compared to sewage sludge. Environ Int 35(1):142–156

Succuro JS (2010) The effectiveness of using Typha latifolia (broadleaf cattail) for phytoremediation of increased levels of lead-contamination in soil (Master's thesis, Humboldt State University)

Tejada M (2009) Application of different organic wastes in a soil polluted by cadmium: effects on soil biological properties. Geoderma 153:254–268

Tsapakis M, Stephanou EG, Karakassis I (2003) Evaluation of atmospheric transport as a nonpoint source of polycyclic aromatic hydrocarbons in marine sediments of the eastern Mediterranean. Mar Chem 80(4):283–298

Unterbrunner R, Puschenreiter M, Sommer P, Wieshammer G, Tlustoš P, Zupan M, Wenzel W (2007) Heavy metal accumulation in trees growing on contaminated sites in Central Europe. Environ Pollut 148:107–114

Vakili AH, Aboutorab M (2013) The potential of *Lepidium sativum* for phytoremediation of contaminated soil with cadmium. Int J Sci Res Knowled 1(2):20–24

Vanek T, Podlipná R, Fialova Z, Petrova S, Soudek P (2010) Uptake of xenobiotics from polluted waters by plants. https://doi.org/10.1007/978-90-481-3509-7_23

Viisimaa M, Karpenko O, Novikov V, Trapido M, Goi A (2013) Influence of biosurfactant on combined chemical–biological treatment of PCB-contaminated soil. Chem Eng J 220:352–359

Wei H, Li B, Li J, Dong S, Wang E (2008) DNAzyme-based colorimetric sensing of lead (Pb2+) using unmodified gold nanoparticle probes. Nanotechnology 19(9):095501

Wei S, Twardowska I (2013) Main rhizosphere characteristics of the cd hyperaccumulator *Rorippa globosa* (Turcz.) Thell. Plant and Soil 372:669–668

Wei B, Yang L (2010) A review of heavy metal contaminations in urban soils, urban road dusts and agricultural soils from China. Microchem J 94(2):99–107

Wong MH (2003) Ecological restoration of mine degraded soils, with emphasis on metal contaminated soils. Chemosphere 50(6):775–780

Yang Y, Ratte D, Smets BF, Pignatello JJ, Grasso D (2001) Mobilization of soil organic matter by complexing agents and implications for polycyclic aromatic hydrocarbon desorption. Chemosphere 43(8):1013–1021

Yateem A, Al-Sharrah T, Bin-Haji A (2007) Investigation of microbes in the rhizosphere of selected grasses of rhizoremediation of hydrocarbon-contaminated soils. Soil Sediment Contam 16:269–280

Zaheer I, Ali S, Rizwan M, Farid M, Shakoor MB, Gill RA, Najeeb U, Iqbal N, Ahmad R (2015) Citric acid assisted phytoremediation of copper by *Brassica napus* L. Ecotoxicol Environ Saf 120:310–317

Zhuang P, McBride MB, Xia H, Li N, Li Z (2009) Health risk from heavy metals via consumption of food crops in the vicinity of Dabaoshan mine, South China. Sci Total Environ 407(5): 1551–1561

Insights into Potential Roles of Plants as Natural Radioprotectants and Amelioration of Radiations Induced Harmful Impacts on Human Health

Misbah Naz, Tariq Shah, Martin Battaglia, Mohammad Sohidul Islam, Akbar Hossain, Muhammad Aamir Iqbal, Muhammad Zahid Ihsan, Md Tahjib-Ul-Arif, Anamika Dubey, Aman Raj, Ashwani Kumar, Khalid Rehman Hakeem, Disna Ratnasekera, Muhammad Irfan, Mine Pakyürek, Arpna Kumari, Talha Javed, Rubab Shabbir, Hüseyin Arslan, Doğan Arslan, Gülen Özyazici, and Ayman E. L. Sabagh

Abstract Radiations have a significant impact on many physiological, biochemical and molecular processes in plants, animals, and humans. Several studies have revealed the beneficial and adverse effects of radiation on human health. The radiation tolerance potential of plants can be used to protect humans from different harmful radiations. However, the underlying mechanisms that enable plants as radioprotectants remain unclear. Therefore, this chapter summarizes findings related

M. Naz
Institute of Environment and Ecology, School of The Environment and Safety Engineering, Jiangsu University, Zhenjiang, Jiangsu, China

State Key Laboratory of Crop Genetics and Germplasm Enhancement, MOA Key Laboratory of Plant Nutrition and Fertilization in Low-Middle Reaches of the Yangtze River, Nanjing Agricultural University, Nanjing, China

T. Shah
INRAE, Paris, France College of Agriculture, West Virginia University, Morgantown, WV, USA

College of Tropical Crops, Hainan University, Haikou, China

M. Battaglia
Department of Animal Sciences, Cornell University, Ithaca, NY, USA

M. S. Islam
Deparment of Agronomy, Hajee Mohammad Danesh Science and Technology University, Dinajpur, Bangladesh

A. Hossain
Bangladesh Wheat and Maize Research Institute, Dinajpur, Bangladesh

M. A. Iqbal
Department of Agronomy, Faculty of Agriculture, University of Poonch Rawalakot, Rawalakot, Pakistan

M. Hasanuzzaman et al. (eds.), *Managing Plant Production Under Changing Environment*, https://doi.org/10.1007/978-981-16-5059-8_12

to the detrimental effects of electromagnetic radiation on human health and the potential role of plants in mitigating the adverse effects of radiation. There is a dire need to increase our understanding of plants' ability to reduce damages caused by radiations through their scavenging activity of free radicals, synthesis of various antioxidants, inhibiting apoptosis, and modulation of growth factors, cytokines, and redox genes. The identification and characterization of plants to tolerate radiations could provide safe, cost-effective, and sustainable radiation protection measures to human health in our surroundings.

M. Z. Ihsan
Baghdad-ul-Jadeed Campus Cholistan Institute of Desert Studies (CIDS), The Islamia University of Bahawalpur (IUB), Bahawalpur, Pakistan

M. Tahjib-Ul-Arif
Department of Biochemistry and Molecular Biology, Faculty of Agriculture, Bangladesh Agricultural University, Mymensingh, Bangladesh

A. Dubey · A. Raj · A. Kumar
Metagenomics and Secretomics Research Laboratory, Department of Botany, Dr. Harisingh Gour University (A Central University), Sagar, Madhya Pradesh, India

K. R. Hakeem
Department of Biological Sciences, Faculty of Science, King Abdulaziz University, Jeddah, Saudi Arabia

D. Ratnasekera
Department of Agricultural Biology, Faculty of Agriculture, University of Ruhuna, Matara, Sri Lanka

M. Irfan
SoMine Pakyürekil and Environmental Sciences Division, Nuclear Institute of Agriculture (NIA), Tandojam, Pakistan

M. Pakyürek
Department of Horticulture, Siirt University Faculty of Agriculture, Siirt, Turkey

A. Kumari
Department of Botanical and Environmental Sciences, Guru Nanak Dev University, Amritsar, Punjab, India

T. Javed · R. Shabbir
College of Agriculture, Fujian Agriculture and Forestry University, Fuzhou, China

H. Arslan
Department of Field Crops, Faculty of Agruculure, Harran Üniversity, Osmanbey Campus, Şanliurfa, Turkey

D. Arslan · G. Özyazici
Departmant of Field Crops, Faculty of Agriculture, Siirt University, Siirt, Turkey

A. E. L. Sabagh (✉)
Departmant of Field Crops, Faculty of Agriculture, Siirt University, Siirt, Turkey

Department of Agronomy, Faculty of Agriculture, Kafrelsheikh University, Kafrelsheikh, Egypt
e-mail: ayman.elsabagh@agr.kfs.edu.eg

Keywords Herbal plants · Human health · Indoor plants · Radiation · UV · Infrared rays

1 Introduction

In the contemporary era, mankind is surrounded by a network of natural and non-natural radiation globally which is continually increasing. According to the energy, there are two types of radiation, i.e., ionizing and nonionizing. The sensitivity to radiation has got a great deal of concern in the last decades because many electromagnetic devices were invented and introduced (Guleria et al. 2019). Ionizing radiation possesses enough energy to break down molecular bonds and transfer electrons from atoms, thereby considered as more damaging to the well-being of the organisms. Ionizing radiations have several harmful effects on human health. The rapid development of technology has greatly increased people's exposure to ionizing radiations; thus, protecting humans from the detrimental impacts of ionizing radiations is a need of hour (Jagetia 2007). Radioprotective agents or anti-radioactive substances are those that can reduce the negative effects of radiation in healthy tissues while retaining the sensitivity to radiation damage in cancerous cells (Painuli and Kumar 2016). Radiation can be defined as the energy released in the form of electromagnetic waves or particles from radioactive material (Lindell and Favaro 2013). In particular, the demand for harmful ionizing radiation is constantly increasing due to expeditious advancement in their use in radiography, nuclear science, space flights, and other modern technologies. Thereby, there is an urgent need to protect plants, animals, and humans from radiation. Exposure to ionizing radiation may cause detrimental consequences to various organs and systems viz., eyes, skin, thyroid glands, stomach, guts, lungs, and reproductive blood systems, which can cause numerous pathophysiological disorders (Fig. 1). Though all materials absorb radiations to some extent, nevertheless some are better than others to absorb specific electromagnetic radiation frequencies. For example, glass absorbs from opaque to infrared radiations (Li et al. 2014), charcoal absorbs from opaque to visible light, while high-energy radiations (like X-rays) can be absorbed by denser materials such as lead (Rebois and Ray 2012). The air, on the other hand, absorbs almost all the radiation coming from space. Future research should focus on understanding the mechanistic role of different types of radiation in our daily life to reduce their impacts on human health without using an electronic instrument, which has high effects on radiations (Durante and Cucinotta 2011). In this regard, plants are the most suitable and economical source to absorb the harmful radiation in the air and mitigate the impacts of air pollution and climate change on human health. The most practical way is the identification and selection of viable candidates along with their evaluation for radioprotection (Jagetia 2007; Ayyanar and Subash-Babu 2012). After the critical evaluation of the anti-inflammatory, antioxidant, antimicrobial, immunomodulatory, free radical scavenging, or anti-stress properties of a substance that can be considered as a potential radioprotective agent (Jagetia and Baliga 2003). With this background, the work is designed to provide the basics of radiation

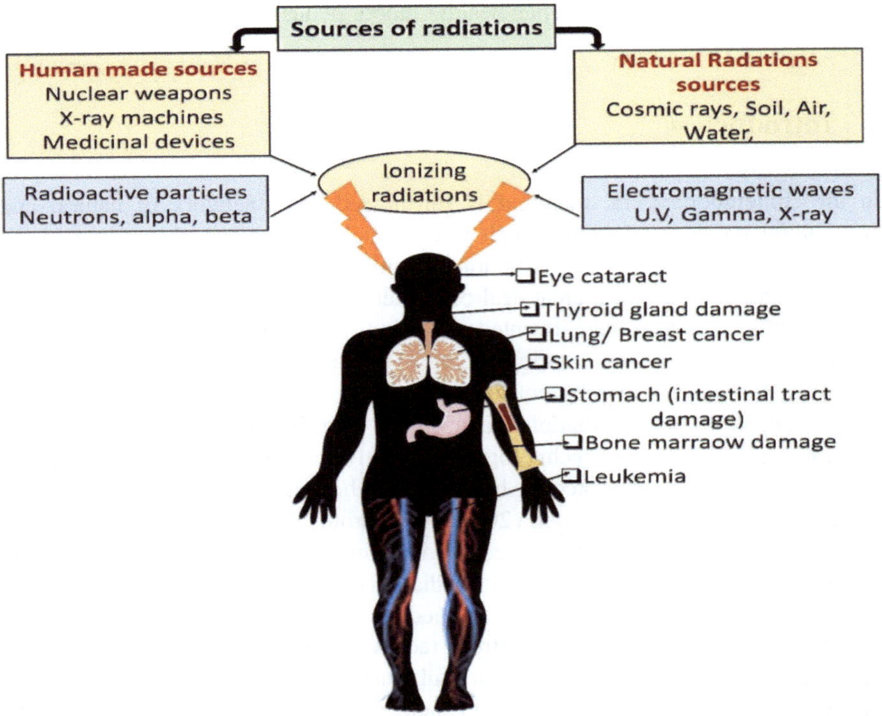

Fig. 1 Sources of ionizing radiations and their adverse impacts on the human body

exposure-mediated consequences in humans along with the radioprotection potential of various plants. Moreover, this chapter is an outcome of an extensive literature survey, so authors have tried to provide a clear image plus gaps in this particular area which is still scarce.

2 Mechanisms of Radiation Damage

Ionizing radiations damage various cells, tissues, and organs through a cascade of molecular events often triggered by free radicals known as reactive oxygen species (ROS) (Mates 2000). Exposure to radiation leads to DNA damage in terms of base damage, single-strand breaks (SSBs) or double-strand breaks (DSBs), DNA–DNA or protein crosslinks, which is ultimately responsible for altered genomic expression, cell death, protein modification, genomic instability, and senescence (Devasagayam et al. 2004). Genomic instability often leads to cancer, mutations, and childbirth defects. Among them, DSBs are considered a very lethal result of ionizing radiation.

3 Ionizing and Nonionizing Radiations

Radiation can be either ionizing (long wavelength with low frequency and thus contains lower energy) or nonionizing (short wavelength with high frequency that has higher energy) (Ng 2003; Zamanian and Hardiman 2005). Nonetheless, ionizing radiation (IR) contains ample energy for producing ions at the molecular level. Radiations are of three types: (1) Alpha (α) radiation, consisting of alpha particles that are emitted from radioactive isotopes; (2) Beta (β) radiation, which transmits a high-energy electron with a negative charge that has greater penetrating power than alpha radiations; and (3) Gamma (γ) radiation, which is electromagnetic radiation like visible light, ultraviolet (UV) light, and radio waves (Zamanian and Hardiman 2005). Ionizing radiations contain electromagnetic energy (Dartnell 2011). IR is a type of electromagnetic radiation, which is generated when atoms absorb and release energy that could be coming from sunlight, fire, iron, visible light, x-rays, γ rays, and domestic utilization of heating artifacts (Cho et al. 2009; Zamanian and Hardiman 2005). IR tends to ionize the atoms of the living organisms, damaging living tissues and causing alterations in the genome of living organisms (Desouky et al. 2015). However, living organisms possess the ability to efficiently repair this damage and undo the genetic alterations caused by IR. Although IR can kill almost all forms of life, nevertheless it is not effective against all types of viruses (Gomes et al. 2017). The γ-ray is electromagnetic radiation with very high energy that can be transmitted into human and animal bodies and causes mutation (Shaban et al. 2017). However, γ-ray can also be useful to improve the morphophysiological characteristics and productivity of several crops such as sorghum, maize, and barley (Goron and Raizada 2015). Furthermore, in crop breeding, physical mutagens (UV light, X-rays neutrons, and alpha-beta particles), and thermal neutrons, especially γ-rays, are more useful than chemical mutagens (ethyl methanesulfonate) to develop mutant lines (Beyaz and Yildiz 2017). The γ-rays can influence seed germination by inducing free radicals in seeds (Zani et al. 2017).

Exposure to UV radiation often leads to cell death due to persistent DNA damage in the longer run. Recently, a gene product known as p53 has been discovered to be involved in regulating the cell cycle (Matsumura and Ananthaswamy 2004). In case of repairable damage to cells caused by UV radiations, p53 activates the repair machinery to restore cell's functioning. However, in case of irreparable damages, the cells are subjected to apoptosis, which is the programmed death of damaged cells (Lieberman 2008). Molecular studies have further revealed that multiple genes are involved in the programmed death of damaged cells under long-term exposure to UV radiation (Nawkar et al. 2013). However, repeated and prolonged exposure to UV light leads to the synthesis of more thymine dimers in the DNA, which multiplies the risk of incorrect repairs of DNA bases, and the utter disruption or malfunctioning of cellular processes (Friedberg 2003). Subsequently, cells died in case of severe damage, while a small-scale incorrect repairing leads to cancerous cells' formation.

4 Indoor Plants as Radiation Protectors

Small gardens at homes, offices, and study rooms can be used not only for ornamental purposes but also for other benefits such as flower therapy and health repair (Hall and Dickson 2011). Plants absorb harmful radiations emitted around our surroundings from electric and electronic gadgets like smartphones, laptops, laser machines, and microwave ovens. (Schmor 2011). Specific indoor anti-radiation plants constitute a significant source to detoxify air polluted with radiation. It also helps to increase the human metabolism and immune system by a positive impact on health and reduce the frequency of stress by minimizing headaches (Portelance et al. 2001). A non-exhaustive list of plants with a strong anti-radiation activity includes cactus, spider plant, aloe vera, sunflower, snake plant, and rubber plant (Arora et al. 2005; Astuti et al. 2020), as shown in Table 1.

5 Radioprotective Ability of Botanicals Against the Ionizing Radiations

The results obtained from various in vitro and in vivo studies indicate that various plants (such as *Ocimum sanctum*, *Gingko biloba*, *Panax ginseng*, *Amaranthus paniculatus*, *Tinospora cordifoila*, *Hippophae rhamnoides*, *Centella asiatica*, *Podophyllum hexandrum*, *Emblica officinalis*, *Piper longum*, *Phyllanthus amarus*, *Mentha piperita*, *Mentha arvensis*, *Zingiber officinale*, *Syzygium cumini*, *Aegle marmelos, Carica papaya, Apium graveolens, Camellia sinensis, Curcuma longa, Caesalpiniadigyna, Aphanamixis polystachya*, and *Ageratum conyzoides*) can protect against radiation-induced DNA damage, lipid peroxidation, and lethality (Jagetia 2007). Apart from these herbal plants, there are various ornamental herbs and shrubs such as *Hedera, Piper, Ficuslyrata, Brassicasps. Heveabrasiliensis, Dracaena, Chlorophytum comosum, Ceratopteris thalictroides, Lithops pseudotruncatella,* and a few grass species as *Cypress* and *Asparagus,* and legumes like soybean (*Glycine max)* that possess the ability to absorb harmful radiations and thus can play a crucial role to combat lethal irradiations. When a molecule absorbs energy through interaction with radiation, it is considered activated. In this energy-rich state, it may experience various abnormal chemical reactions, which are usually unavailable under thermal equilibrium, directly or indirectly, thus altering molecule structure. In some cases, radiation absorption is enough to get rid of an electron, thus leading to bond breakage at the molecular level (Fig. 2).

Table 1 Brief description of the anti-radiation activity of some indoor plants

Plants	Functions	References
Cactus compressus	Absorbs computer radiation as well as helps to absorb radiations coming from other sources or nearby cell towers	Nobel and Hartsock (1983), Darling (1989)
Piper sarmentosum	Highly effective plant for absorbing ambient radiation in-home or office	Astuti et al. (2020)
Chlorophytum comosum	Cleaning out various harmful gasses such as formic acid and aldehyde, and useful for reducing radiation at home	Munyao et al. (2020), Li et al. (2019), Inbathamizh (2020)
Dracaena trifasciata	Absorbsover 100 different types of poisons. A strong anti-pollutant, and occasionally can be used as an antidote to certain forms of radiation	Boraphech and Thiravetyan (2015), Wolverton et al. (1995), Yan et al. (2012)
Dypsis lutescens	Purify the air containing toxins and reduces pollution and electronic rays	Adeel et al. (2019), Chowdhury et al. (2017)
Aloe barbadensis	It absorbs some radiations, which could include the radiations from electromagnetic fields EMFs	Kumar et al. (2009), Ahmadi (2012); Korać and Khambholja (2011), Mishra, et al. (2011), Gupta (2013), Ray et al. (2013)
Helianthus annuus	Considered the most efficient plant in the world for absorbing radiations	Hussain et al. (2017)
Ficus elastic	Protect the humans from harmful EMF radiations	Priyadarshan (2011), Cruz et al. (2014), Pan et al. (2020)
Nelumbo nucifera	Absorbs radiations from electronic devices	Fernández-Mazuecos and Glover (2017)
Dracaena trifasciata	It absorbs radiation emitted from computers	Giri et al. (2019), Kaur and Misra (2014)
Hedera helix	Ivy is one of the best absorbents of radiations available in nature. It can absorb up to 90% of benzene in the air within 24 h. It can either block or eliminate radiation	Di and Wang (1999), Hoyano (1988), Kenai et al. (2020)
Asparagus densiflorus	Highly effective to absorb electromagnetic radiations	Maeda et al. (2010), Kapoor (2017)
Brassica juncea	Absorbs electromagnetic radiations	Beggs (1986), O'Connell et al. (2004), Kovacs and Keresztes (2002)
Dypsis lutescens	Removes toxins from the air	Inbathamizh (2020)
Chamaedorea seifrizii	Actively removes toxins at a fast pace and absorbs heat from the environment	Inbathamizh (2020)

(continued)

Table 1 (continued)

Plants	Functions	References
Spathiphyllum wallisii Magnolia champaca, Hibiscusrosa-sinensis; Elephantopus scaber Sesamum indicum	Ensures thermal comfort in a residential environment	Yoo et al. (2006), Ma (2019), Kumar et al. (2013), de Silva and Tencomnao (2018); Koocheki et al. (2016), Lorenzo et al. (2018)

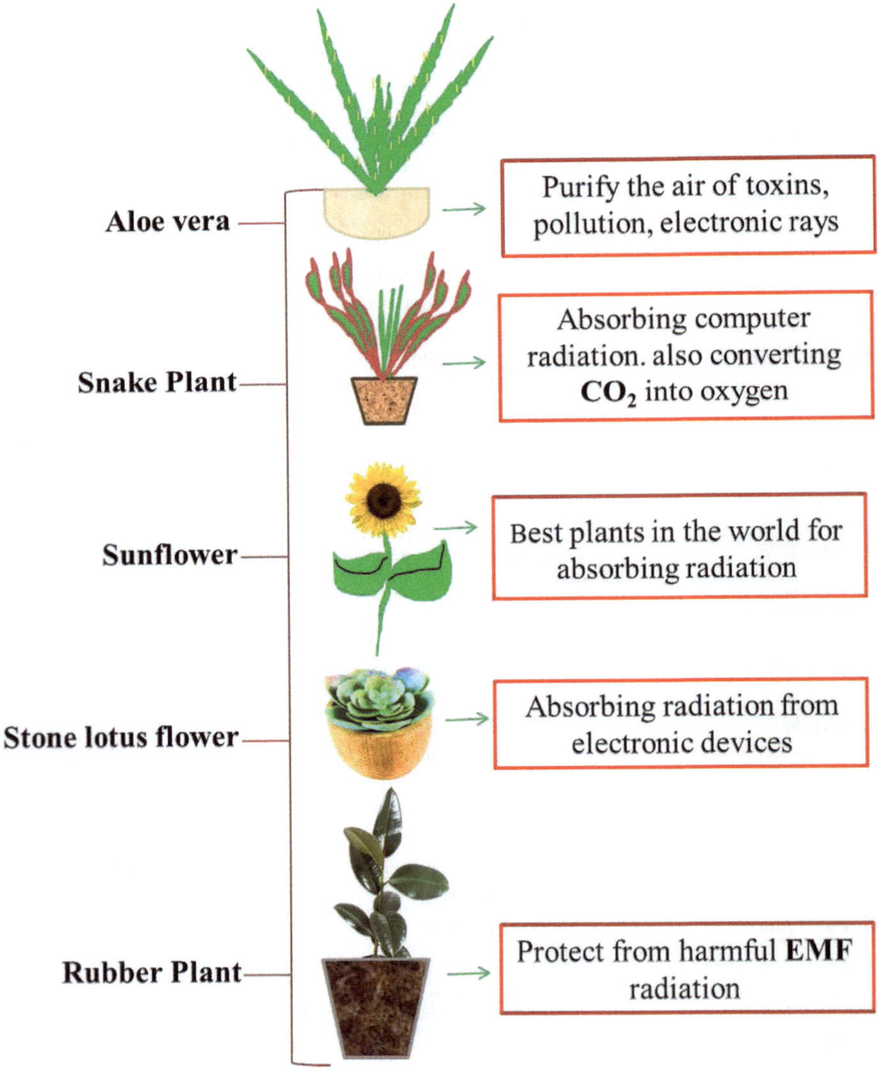

Fig. 2 Anti-radiation activities of some indoor plants

6 Importance of Radio-Protectors for Human Health

Radioprotectors entail cumulative measures to ensure human health against the harmful and damaging effects of radiations, especially ionizing radiations (Spitz and Albert 2001). These effectively protect humans from harmful radiations, including β, γ, UV, and radio waves. Ionizing radiations impart injuries that lead to malfunction of biological systems and therefore, it is pertinent to employ pharmacologically dynamic radioprotectors to offer adequate protection against harmful radiations. The upregulation of repair activity of damaged DNA can also constitute a neutralizing strategy against the damaging effects of radiations (Adhikari et al. 2007). Another neutralizing mechanism against the detrimental impact of ionizing radiations involves protein kinase (PK)-C inactivation and downregulation of factors responsible for the damages at the molecular level (Erickson III et al. 2015). Radioprotectors help humans and help plants by absorbing harmful radiation above its optimum range, as these radiations create stressful conditions in plants, leading to the accumulation of toxic compounds and the formation of ROS in plants. These harmful radiations increase the content of polyphenols in legumes by 8.5-folds (Sreekumar 2007) and drop the ATP content in cells of *Solanum lycopersicon* by 30%, affecting all ATP dependent pathways and leading to the formation of toxic compounds in plant cells. Exposure to harmful radiation like UV-A, UV-B, and EMF reduces plants' oxidative response to high-salt concentration in *Triticum aestivum* (Dawood et al. 2021). Radioprotectant plants can absorb most of these radiations, helping other plants carry out their normal metabolism and indirectly benefitting humans and other species dependent on them from toxic compounds (Ali et al. 2015).

7 Possible Mechanisms of Radioprotection Adopted by Plants

Global warming and the increasing intensity of UV radiation due to ozone depletion are drastically affecting the growth of plants. The extent of these adverse effects usually varies depending on the wavelength and duration of exposure of plants to radiations (Ballare et al. 2011). Plants tend to respond to light through their photoreceptors and subsequently exhibit photomorphogenic development. In addition to photosynthetically active radiations (PAR; 400–700 nm), plants get subjected to UV radiations comprised of UV-A (320–390 nm), UV-B (280–320 nm), and UV-C (below 280 nm). The ozone layer protects the earth from UV-C radiation. The UVR8 protein acts as a receptor for UV-B radiation in plants. A lower UV-B exposure level initiates signaling through UVR8 and induces secondary metabolite genes involved in protection against the UV-B, while its higher dosages are detrimental to plants (Nawkar et al. 2013). Research reports that exposure to UV radiations generates ROS in mitochondria and chloroplast as a byproduct of essential energy-generating processes such as photosynthesis and respiration (Nawkar et al. 2013). Plant responses upon exposure to different UV radiations are activated through the involvement of different hormones. The salicylic acid levels increased in plants when exposed to UV-B radiations. Exposure to radiation does not alter jasmonic acid levels but increases plants tissue sensitivity towards the jasmonic acid. In plants, salicylic acid, jasmonic acid, and ethylene levels are changed in response to diverse external stresses such as ozone exposure and UV-B (Soheila 2000; Liu et al. 2012). The UV exposure represses a specific set of up-down genes regulated by the UV effects (Gomes et al. 2018). The underlying mechanism behind radioprotection offered by plants is the synthesis of vital chemical compounds such as polyphenols and numerous other antioxidants that perform scavenging activity for radiation-induced free radicals (Bhat et al. 2015). The synthesis of polyphenols in response to radiation exposure helps plants to upregulate antioxidants (catalase, superoxide dismutase, glutathione peroxidase, glutathione transferase, etc.) and mRNAs thereby reducing the oxidative stress caused by the ionizing radiations.

The putative mechanisms for radioprotection involve the production of antioxidants (-SH, GHS, GST, CAT, COD, mRNA GSH, GST) (Surapaneni and Jainu 2014). Remarkably, a variety of plants and herbs possess the ability to serve as radioprotectors, which may be exploited to cope with the adverse effects of radiations. There is a dire need to increase our understanding of plants' ability to reduce damages caused by radiations through their scavenging activity of free radicals, synthesis of various antioxidants, inhibiting apoptosis, and modulation of growth factors, cytokines, and redox genes. The evaluation techniques based on fractionation can generate advanced information regarding plants' ability to radio protect human beings. Advancement in nonprotein sulphydryl groups and reduction in lipid peroxidation indicates radioprotective activity. The plants and herb may also inhibit activation of mitogen-activated protein kinase (MAPK), protein kinase C (PKC), nitric oxide (NO), cytochrome P-450, and several other genes that are responsible for inducing damage after irradiation (Jagetia 2007). The anticancer and radioprotective properties of several plant-based natural products have been explored (Fig. 3). Due to the limitation of cost and side effects, there is an urgent need to explore safe, effective, and economic radiation protection agents, especially those of plant origin (Painuli and Kumar 2016).

8 Conclusion

The usefulness of UV radiations and other forms of primers in medical, engineering, agricultural, and other applied fields, and how humans rely on electronic technology in the modern era is well established. This chapter highlighted some detrimental impacts of radiation on human health. Various indoor plants and herbs have the potential to absorb and tolerate these harmful radiations through an array of biochemical and molecular mechanisms. Most of these plants and herbs have the potential as future radioprotectors which are also important for radiation control.

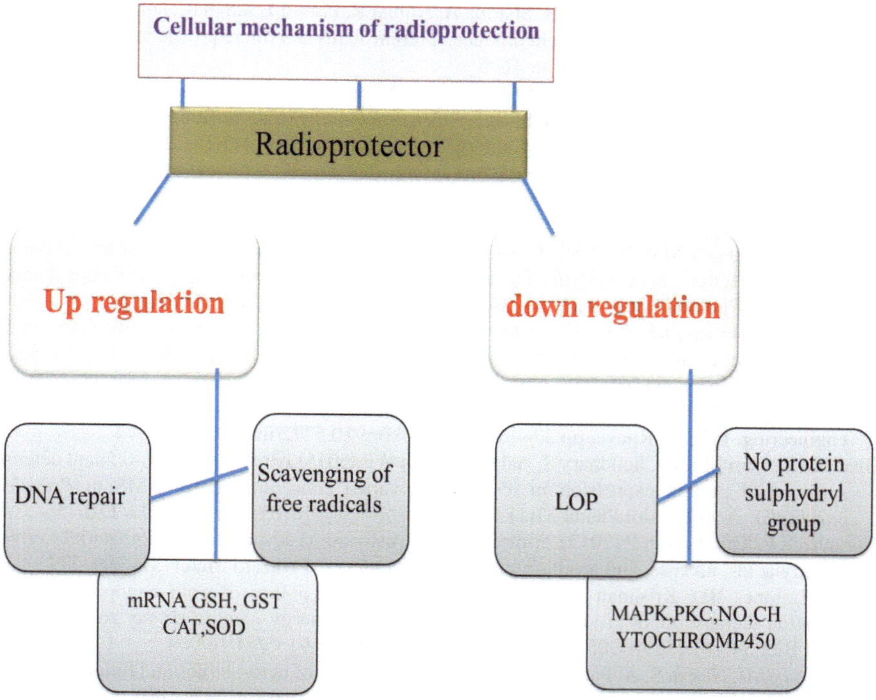

Fig. 3 Cellular mechanisms of radioprotection in plants (adapted from Painuli and Kumar 2016)

The identification, characterization, and plantation of such plants in our surroundings are crucial to protect human health from radiation.

Conflict of Interest No conflict of interest is hereby declared.

References

Adeel S, Rehman FU, Rafi S, Zia KM, Zuber M (2019) Environmentally friendly plant-based natural dyes: extraction methodology and applications. In: Plant and human health. Springer, Cham, pp 383–415. https://doi.org/10.1007/978-3-030-03344-6_17

Adhikari S, Priyadarsini KI, Mukherjee T (2007) Physico-chemical studies on the evaluation of the antioxidant activity of herbal extracts and active principles of some Indian medicinal plants. J Clin Biochem Nutr 40:174–183. https://doi.org/10.3164/jcbn.40.174

Ahmadi A (2012) Potential prevention: Aloe vera mouthwash may reduce radiation-induced oral mucositis in head and neck cancer patients. Chinese J Integ Med 18:635–640. https://doi.org/10.1007/s11655-012-1183-y

Ali H, Ghori Z, Sheikh S, Gul A (2015) Effects of gamma radiation on crop production. In: Crop production and global environmental issues. Springer, New York, pp 27–78. https://doi.org/10.1007/978-3-319-23162-4_2

Arora R, Gupta D, Chawla R, Sagar R, Sharma A, Kumar R, Prasad J, Singh S, Samanta N, Sharma RK (2005) Radioprotection by plant products: present status and future prospects. Phytother Res 19:1–22. https://doi.org/10.1002/ptr.1605

Astuti SD, Tirtana RD, Mahmud AF, Mawaddah A, Yasin M (2020) Ultraviolet (UV) activation effect on antibacterial agents of red betel (*Piper crocatum*) extract to Streptococcus mutans. J Phys Conf Ser 1445:012004. https://doi.org/10.1088/1742-6596/1445/1/012004

Ayyanar M, Subash-Babu P (2012) Syzygium cumini (L.) Skeels: a review of its phytochemical constituents and traditional uses. Asian Pac J Trop Biomed 2:240–246. https://doi.org/10.1016/s2221-1691(12)60050-1

Ballare CL, Caldwell MM, Flint SD, Robinson SA, Bornman JF (2011) Effects of solar ultraviolet radiation on terrestrial ecosystems. Patterns, mechanisms, and interactions with climate change. Photochem Photobiol Sci 10(2):226–241. https://doi.org/10.1039/c0pp90035d

Beggs CJ, Schneider-Ziebert U, Wellmann E (1986) UV-B radiation and adaptive mechanisms in plants. In stratospheric ozone reduction, solar ultraviolet radiation and plant life. Springer, Berlin, Heidelberg, pp 235–250. https://doi.org/10.1007/978-3-642-70090-3_18

Beyaz R, Yildiz M (2017) The use of gamma irradiation in plant mutation breeding. In: Plant engineering. InTech, Rijeka, pp 33–46. https://doi.org/10.5772/intechopen.69974

Bhat TM, Ansari MYK, Choudhary S, Aslam R, Bhat WF (2015) Alteration in anti-oxidant defense system and protein expression in response to varied concentrations of EMS in *Psoralea corylifolia*. Acta Physiol Planta 37(1):1707. https://doi.org/10.1007/s11738-014-1707-5

Boraphech P, Thiravetyan P (2015) Trimethylamine (fishy odor) adsorption by biomaterials: effect of fatty acids, alkanes, and aromatic compounds in waxes. J Hazard Mater 284:269–277

Cho SH, Jones BL, Krishnan S (2009) The dosimetric feasibility of gold nanoparticle-aided radiation therapy (GNRT) via brachytherapy using low-energy gamma−/x-ray sources. Phys Med Biol 54(16):4889–4905. https://doi.org/10.1088/0031-9155/54/16/004

Chowdhury MJ, Nasrin S, Al Faruque MA (2017) Significance of agro-textiles and future prospects in Bangladesh. Eur Sci J 13:21. https://doi.org/10.19044/esj.2017.v13n21p139

Cruz MD, Christensen JH, Thomsen JD, Müller R (2014) Can ornamental potted plants remove volatile organic compounds from indoor air? A review. Environ Sci Pollut Res 21(24): 13909–13928. https://doi.org/10.1007/s11356-014-3240-x

Darling MS (1989) Epidermis and hypodermis of the saguaro cactus (*Cereus giganteus*): anatomy and spectral properties. Am J Biol 76(11):1698–1706. https://doi.org/10.1002/j.1537-2197.1989.tb15155.x

Dartnell LR (2011) Ionizing radiation and life. Astrobiology 11(6):551–582. https://doi.org/10.1089/ast.2010.0528

Dawood MF, Tahjib-Ul-Arif M, Sohag AAM, Latef AAHA, Ragaey MM (2021) Mechanistic insight of allantoin in protecting tomato plants against ultraviolet c stress. Plan Theory 10(1):11. https://doi.org/10.3390/plants10010011

de Silva MB, Tencomnao T (2018) The protective effect of some Thai plants and their bioactive compounds in UV light-induced skin carcinogenesis. J Photochem Photobiol B Biol 185:80–89. https://doi.org/10.1016/j.jphotobiol.2018.04.046

Desouky O, Ding N, Zhou G (2015) Targeted and non-targeted effects of ionizing radiation. J Radiat Res Appl Sci 8(2):247–254. https://doi.org/10.1016/j.jrras.2015.03.003

Devasagayam TPA, Tilak JC, Boloor KK, Sane KS, Ghaskadbi SS, Lele RD (2004) Free radicals and antioxidants in human health: current status and future prospects. J Assoc Physicians India 52:794–804

Di HF, Wang DN (1999) Cooling effect of ivy on a wall. Exp Heat Trans 12(3):235–245. https://doi.org/10.1080/089161599269708

Durante M, Cucinotta FA (2011) Physical basis of radiation protection in space travel. Rev Mod Phys 83(4):1245–1281. https://doi.org/10.1103/revmodphys.83.1245

Erickson DJ III, Sulzberger B, Zepp RG, Austin AT (2015) Effects of stratospheric ozone depletion, solar UV radiation, and climate change on biogeochemical cycling: interactions and feedbacks. Photochem Photobiol Sci 14(1):127–148

Fernández-Mazuecos M, Glover BJ (2017) The evo-devo of plant speciation. Nat Ecol Evol 1(4): 1–9. https://doi.org/10.1038/s41559-017-0110

Friedberg EC (2003) DNA damage and repair. Nature 421:436–440. https://doi.org/10.1038/nature01408

Giri BS, Sarowgi A, Kaushik Y, Pal A, Jaiswal A, Kumari S, Singh H, Sonwani R, Thivaharan V, Singh RS (2019) Indoor potted plant based biofilter: performance evaluation and kinetics study. Indian J Exp Biol 57:11

Gomes LR, Menck CF, Leandro GS (2017) Autophagy roles in the modulation of DNA repair pathways. Int J Mol Sci 18(11):2351. https://doi.org/10.3390/ijms18112351

Gomes SI, Roca CP, Scott-Fordsmand JJ, Amorim MJ (2018) Identifying conserved UV exposure genes and mechanisms. Sci Rep 8(1):1–10. https://doi.org/10.1038/s41598-018-26865-9

Goron TL, Raizada MN (2015) Genetic diversity and genomic resources available for the small millet crops to accelerate a new green revolution. Front Plant Sci 6:157. https://doi.org/10.3389/fpls.2015.00157

Guleria R, Bhushan B, Guleria A, Bhushan A, Dulari P (2019) Harmful effects of ionizing radiation. Int J Res Appl Sci Engineer Technol 7(12):887–889

Gupta D (2013) UV absorbing properties of some plant derived extracts. Res J Chem Environ Sci 1: 34–36

Hall CR, Dickson MW (2011) Economic, environmental, and health/Well-being benefits associated with green industry products and services: a review. J Environ Hort 29(2):96–103. https://doi.org/10.24266/0738-2898-29.2.96

Hoyano A (1988) Climatological uses of plants for solar control and the effects on the thermal environment of a building. Energ Buildings 11(1–3):181–199. https://doi.org/10.1016/0378-7788(88)90035-7

Hussain F, Iqbal M, Shah SZ, Qamar MA, Bokhari TH, Abbas M, Younus M (2017) Sunflower germination and growth behavior under various gamma radiation absorbed doses. Acta Ecol Sin 37(1):48–52. https://doi.org/10.1016/j.chnaes.2016.09.009

Inbathamizh L (2020) Indoor medicinal plants: beneficial biocatalysts for air filtration and bioremediation–a review. Int J Green Pharm 14:2

Jagetia GC (2007) Radioprotective potential of plants and herbs against the effects of ionizing radiation. J Clin Biochem Nutr 40(2):74–81. https://doi.org/10.3164/jcbn.40.74

Jagetia GC, Baliga MS (2003) Evaluation of the radioprotective effect of the leaf extract of *Syzygium cumini* (Jamun) in mice exposed to a lethal dose of γ-irradiation. Food Nahrung 47(3):181–185. https://doi.org/10.1002/food.200390042

Kapoor M (2017) Managing ambient air quality using ornamental plants—an alternative approach. Univ J Plant Sci 5(1):1–9. https://doi.org/10.13189/ujps.2017.050101

Kaur A, Misra AK (2014) Impact of indoor surface materials and environment on perceived air quality. J Environ Human 1:1

Kenai MA, Libessart L, Lassue S, Defer D (2020) Impact of plants obscuration on energy balance: theoretical and numerical study. J Build Eng 29:101112. https://doi.org/10.1016/j.jobe.2019.101112

Koocheki A, Mahallati MN, Solouki H, Karbor S (2016) Evaluation of radiation absorption and use efficiency in substitution intercropping of sesame (*Sesamum indicum* L.) and mung bean (*Vigna radiata* L.). Adv Plants Agril Res 3(5):001–009

Korać RR, Khambholja KM (2011) Potential of herbs in skin protection from ultraviolet radiation. Pharmacog Rev 5(10):164. https://doi.org/10.4103/0973-7847.91114

Kovacs E, Keresztes A (2002) Effect of gamma and UV-B/C radiation on plant cells. Micron 33(2): 199–210. https://doi.org/10.1016/s0968-4328(01)00012-9

Kumar MS, Datta PK, Gupta SD (2009) In vitro evaluation of UV opacity potential of *Aloe vera* L. gel from different germplasms. J Nat Med 63(2):195–199

Kumar SR, Arumugam T, Anandakumar C, Balakrishnan S, Rajavel D (2013) Use of plant species in controlling environmental pollution. Bull Env Pharmacol Life Sci 2(2):52–63

Li J, Niu L, Zheng Z, Yan F (2014) Photosensitive graphene transistors. Adv Mater 26(31): 5239–5273. https://doi.org/10.1002/adma.201400349

Li J, Zhong J, Zhan T, Liu Q, Yan L, Lu M (2019) Indoor formaldehyde removal by three species of *Chlorophytum comosum* under the long-term dynamic fumigation system. Environ Sci Pollut Res 26(36):36857–36868. https://doi.org/10.1007/s11356-019-06701-x

Lieberman HB (2008) DNA damage repair and response proteins as targets for cancer therapy. Curr Med Chem 15(4):360–367. https://doi.org/10.2174/092986708783497328

Lindell IV, Favaro A (2013) Electromagnetic media with no dispersion equation. In: 2013 international symposium on electromagnetic theory. IEEE, New York, pp 188–190

Liu X, Chi H, Yue M, Zhang X, Li W, Jia E (2012) The regulation of exogenous jasmonic acid on UV-B stress tolerance in wheat. J Plant Growth Regul 31(3):436–447. https://doi.org/10.1007/s00344-011-9253-5

Lorenzo GA, Mascarini L, Gonzalez MN, Lalor E (2018) Sand mulching and its relationship with soil temperature and light environment in the cultivation of *Lilium longiflorum* cut flower. Sci Hortic 240:453–459. https://doi.org/10.1016/j.scienta.2018.06.025

Ma S (2019) Flower forcing in banana shrub (Michelia skinneriana Dunn.) and bougainvillea (bougainvillea wild.). Doctoral dissertation, Mississippi State University, Mississippi

Maeda T, Honda K, Sonoda T, Motoki S, Inoue K, Suzuki T, Oosawa K, Suzuki M (2010) Light condition influences rutin and polyphenol contents in asparagus spears in the mother-fern culture system during the summer–autumn harvest. J Japanese Soc Hort Sci 79(2):161–167. https://doi.org/10.2503/jjshs1.79.161

Mates JM (2000) Effects of antioxidant enzymes in the molecular control of reactive oxygen species toxicology. Toxicology 153(1-3):83–104. https://doi.org/10.1016/s0300-483x(00)00306-1

Matsumura Y, Ananthaswamy HN (2004) Toxic effects of ultraviolet radiation on the skin. Toxicol Appl Pharmacol 195(3):298–308. https://doi.org/10.1016/j.taap.2003.08.019

Mishra AK, Mishra A, Chattopadhyay P (2011) Herbal cosmeceuticals for photoprotection from ultraviolet B radiation: a review. Trop J Pharma Res 10:3. https://doi.org/10.4314/tjpr.v10i3.7

Munyao JN, Dong X, Yang JX, Mbandi EM, Wanga VO, Oulo MA, Saina JK, Musili PM, Hu GW (2020) Complete chloroplast genomes of *Chlorophytum comosum* and *Chlorophytum gallabatense*: genome structures, comparative and phylogenetic analysis. Plan Theory 9(3): 296. https://doi.org/10.3390/plants9030296

Nawkar GM, Maibam P, Park JH, Sahi VP, Lee SY, Kang CH (2013) UV-induced cell death in plants. Int J Mol Sci 14(1):1608–1628. https://doi.org/10.3390/ijms14011608

Ng KH (2003) Non-ionizing radiations-sources, biological effects, emissions and exposures. In: Proceedings of the international conference on non-ionizing radiation at UNITEN, pp 1–16

Nobel PS, Hartsock TL (1983) Relationships between photosynthetically active radiation, nocturnal acid accumulation, and CO_2 uptake for a crassulacean acid metabolism plant, *Opuntia ficus-indica*. Plant Physiol 71(1):71–75

O'Connell MG, O'leary GJ, Whitfield DM, Connor DJ (2004) Interception of photosynthetically active radiation and radiation-use efficiency of wheat, field pea and mustard in a semi-arid environment. Field Crop Res 85(2–3):111–124. https://doi.org/10.1016/s0378-4290(03)00156-4

Painuli S, Kumar N (2016) Prospects in the development of natural radioprotective therapeutics with anti-cancer properties from the plants of Uttarakhand region of India. J Ayurveda Integr Med 7(1):62–68. https://doi.org/10.1016/j.jaim.2015.09.001

Pan L, Wei S, Lai PY, Chu L (2020) Effect of plant traits and substrate moisture on the thermal performance of different plant species in vertical greenery systems. Build Environ 175:106815. https://doi.org/10.1016/j.buildenv.2020.106815

Portelance L, Chao KC, Grigsby W, Bennet H, Low D (2001) Intensity-modulated radiation therapy (IMRT) reduces small bowel, rectum, and bladder doses in patients with cervical cancer receiving pelvic and para-aortic irradiation. Int J Radiat Oncol Biol Phys 51(1):261–266. https://doi.org/10.1016/s0360-3016(01)01664-9

Priyadarshan PM (2011) Biology of Hevea rubber. CABI, Wallingford, pp 1–6

Ray A, Gupta SD, Ghosh S (2013) Evaluation of anti-oxidative activity and UV absorption potential of the extracts of *Aloe vera* L. gel from different growth periods of plants. Ind Crop Prod 49:712–719. https://doi.org/10.1016/j.indcrop.2013.06.008

Rebois RV, Ray K (2012) Ionizing radiation and radioactive materials in health and disease. In: Veterinary toxicology: basic and clinical principles. Elsevier, London, p 391

Schmor P (2011) Review of cyclotrons for the production of radioactive isotopes for medical and industrial applications. In: Reviews of accelerator science and technology: volume 4: accelerator applications in industry and the environment. World Scientific Publishing Company, Singapore, pp 103–116. https://doi.org/10.1142/9789814383998_0005

Shaban NZ, Zahran AMA, El-Rashidy FH, Kodous ASA (2017) Protective role of hesperidin against γ-radiation-induced oxidative stress and apoptosis in rat testis. J Biol Res-Thessaloniki 24(1):5. https://doi.org/10.1186/s40709-017-0059-x

Soheila AH (2000) Plant responses to ultraviolet-B (UV-B: 280–320 nm) stress: what are the key regulators? Plant Growth Regul 32(1):27–39

Spitz H, Albert RE (2001) Ionizing radiation. In: Patty's toxicology. Wiley, New York, pp 1–22. https://doi.org/10.1002/0471435139.tox023.pub2

Sreekumar PK (2007) Identification of radio-protective activity in the extract of Indian green mussel, Pernaviridis. Doctoral dissertation, Goa University, India

Surapaneni KM, Jainu M (2014) Comparative effect of pioglitazone, quercetin and hydroxy citric acid on the status of lipid peroxidation and antioxidants in experimental non-alcoholic steatohepatitis. J Physiol Pharmacol 65(1):67–74

Wolverton BC, Wolverton JD (1995) Indoor humidifier and air purifier. U.S. Patent No. 5,433,923. U.S. Patent and Trademark Office, Washington, DC

Yan S, Hua H, Yang C, Zhang Y (2012) Effect of $LaCl_3$ on resistance and absorptive capacity to formaldehyde of indoor ornamental plants under formaldehyde stress. Agricl. Sci Technol 13(12):2607

Yoo MH, Kwon YJ, Son KC, Kays SJ (2006) Efficacy of indoor plants for the removal of single and mixed volatile organic pollutants and physiological effects of the volatiles on the plants. J Am Soc Hort Sci 131(4):452–458. https://doi.org/10.21273/jashs.131.4.452

Zamanian A, Hardiman CJHFE (2005) Electromagnetic radiation and human health: a review of sources and effects. High Freq Electr 4(3):16–26

Zani D, Dondi D, Araújo S, Mondoni A, Balestrazzi A (2017) Impact of γ-rays on seed germination/short-term storage in four native alpine species: correlation with free radical and antioxidant profiles. Radi Phys Chem 131:86–94. https://doi.org/10.1016/j.radphyschem.2016.11.001

Development of Rangeland Conservation and Sustainable Management Practices Under Changing Climate

Mujahid Farid, Sameera Iqbal, Dua Nouman Rana, Hina Mushtaq, Wajiha Sarfraz, Maria Islam, Nighat Raza, Muhammad Ramzan, Mubeen Umer, and Zahid Saqib

Abstract Rangeland plays an important role for the agroecosystem. They preserve the biodiversity and some of the plant species that are used for the medicines. Worldwide rangelands absorb 30% of the globally carbon. But due to increase in the population and change in the climate is causing the destruction of the rangelands. When population increases it requires more food and other resources. A good management system helps to reduce soil erosion. Due to the current climate scenario, the government of Pakistan focuses on the conservation of the rangelands and requires proper management plans policies. Rangeland policies and implementations are necessary for the sustainable management. As we cannot cope with the climatic variations but we can make better plans to protect the rangelands and their habitats. Degradation of the rangelands not only affects the direct users, but it also affects the environmental services. This article deals with the significance and conservation of the rangelands. The article also deals with sustainable management plans under the changing climate conditions and the effects of climate change on the rangelands are also described. Besides this, other factors which are affecting the productivity of the rangelands are also discussed in the article. But the conservation of the rangelands, making of a sustainable management plans, and implementation strategies is a difficult task for a developing country with a low economic status.

Keywords Rangeland · Biodiversity · Climate change · Sustainable

---

M. Farid (✉) · S. Iqbal · D. N. Rana · H. Mushtaq · M. Islam · M. Ramzan · M. Umer · Z. Saqib
Department of Environmental Sciences, University of Gujrat, Hafiz Hayat Campus, Gujrat, Pakistan

W. Sarfraz
Department of Botany, Government College Women University, Sialkot, Pakistan

N. Raza
Department of Food Science and Technology, Muhammad Nawaz Shareef University of Agriculture, Multan, Pakistan

M. Hasanuzzaman et al. (eds.), *Managing Plant Production Under Changing Environment*, https://doi.org/10.1007/978-981-16-5059-8_13

1 Introduction

One of the significant kinds of land is rangeland which is reasonable for foraging purposes. It covers 40% of the world's land. Except Antarctica, it is found on all the continents. It is the main source of food for animals and also provides shelter to them. Rangeland covers natural grasslands, savannas, wetlands, and certain shrub and chaparral plant networks. Plant species are also used for manufacturing purposes, pharmacological purposes, and charcoal. Rangelands are not the same as pasture land. At least ten million km^2 of the earth is secured by rangeland going from desert to sloping areas. They contain unsatisfactory conditions for development purposes (Boone et al. 2011). During twentieth century traditional system of rangeland management was demised, as new challenges occurred. Sustainable rangeland management is affected by the susceptibility of the rangeland users. The vulnerability of rangeland users affects Sustainable Rangeland Management (SRM) (Bedunah and Angerer 2012; Gharibvand et al. 2015). Rangeland supports nearly 200 million livestock with approximately 960 million livestock and is disclosing their facilities and remunerations (World Bank 2006). Climate change will also affect the productivity of the amphitheater. Various processes are applied to plant development, such as active adaptive management (Walters and Holing 1990). It is uncertainty-based management. It detects the effect and corrects it before a serious condition develops a Cross-monitoring of adaptive management.

Long-term monitoring is not an easy task. Theater also plays an important role in livestock production. Cattle represent an economic investment in agricultural resources. The amphitheater is also used for cultural recognition and performs significant spiritual, cultural. and social utilities (Vetter et al. 2020; Bayer et al. 2004; Ainslie 2005; Salomon 2011). Management of natural resources has been done through three paradigms (Weddell 2002). For maintenance of sustainable production, the utilitarian method is used which is a preservation-based method (Pinchot 1947). The protectionist paradigm use for the protection of nature while the third paradigm use for the management of the ecosystem (Leopold 1949). These provide ecological management and reduce soil erosion. Active adaptive management is a process used to deal with uncertainties (Marshall 2015). It is a scientific-based management process (Walters and Holling 1990). Rangeland management can take place by preserving the proper stocking rate, proper circulation of animals, and use of proper kind of grazing animals. Farmers managed the uniform grazing conditions by increasing the desired grasses (Fuhlendorf et al. 2009).

2 Types of Rangelands

Rangelands are mostly enclosed by species of perennial grasses, grass-like plants, forbs, and shrubs. There are five basic types of rangelands worldwide, viz. natural pasture, desert shrubland, savannah woodland, forest, and tundra. No shrubs or trees growing on prairies. Desert shrubs are the most extensive and varied. Savannah is a variation between wild prairie and woods and consists of incongruous, underdeveloped shrubs and mixed shrubs between trees. Trees grow taller than savanna. The tundra is a landless area in the Arctic or in the mountains.

3 Importance

The most important source of fodder is rangeland and then diversity gives safe homes. They are an important source of protein and food. Rangeland additionally fills in waterfalls, which receive rainfall, which flows into rivers and tributaries for a long time. Provide habitats for wildlife and soil life. Rangeland provides value for conservation. Healthy cover of natural vegetation provides real and commercial cover for soil and water management (Morgan et al. 2007). Woodlands are forced to land. Many ancient foods, medicines, and valuable compounds are produced from natural flora, including sugar, nuts, seeds, turpentine, rubber, quinine, digitalis, gums, and toxins for the control of pests and semi-sites. The global livestock industry depends on natural grazing lands (Lal 2004). Domestic animals get 75% of their feed from natural pastures. The world's agricultural ecosystems (rangeland, crops, and pastures), soil biota, and extraterrestrial biosphere have not been successfully maintained and are important carbon dioxide reservoirs that carry renewable energy sources through carbon and moderate climate change (Fatima et al. 2020). Rangelands can be huge carbon sinks because they are one of the most widely distributed scenarios on the planet. Snow covers about 30% of the world's surface and up to 30% of terrestrial carbon reserves (Schuman et al. 2002). Rangelands can extract 198 million tons of CO_2 from the environment each year.

4 Factors Affecting the Productivity of Rangeland

4.1 Climate Change

Climate change also affects the properties of rangeland (Polley et al. 2013). The rainfall pattern of the area also changes due to climatic conditions (Christensen et al. 2007). Climatic factors like temperature, humidity, precipitation, light power, and elevation control the nourishment of plants. Plants are also dependent upon the dirt for their mineral additions; climatic variables impact respiration, absorption, photosynthesis, and breakdown to the point that the mineral of plants has the capacity to adjust by climatic factors instead of the fact that developed on a similar soil (Ezaz et al. 2020). During some of the recent decades, normal temperatures have extended with less chilly days, increasingly hot days, and expanded rainfall over a great part of the world. Unexpected incidents such as dry spells, heat waves, and exceptional precipitation occasions are predicted to turn out to be progressively normal (Smith et al. 2000). CO_2 concentrations also increase. By increasing the concentration of CO_2, the process of photosynthesis effect and transpiration rate decrease and enhance the efficiency of plants (Smith et al. 2014; Ziska et al. 2005). Climatic conditions impact the rangelands which are varying by region (Briske et al. 2015).

4.2 Temperature

Temperature is the estimation of the warmth energy that is accessible from sun-based radiation, it is an important factor for the plants in which organic action and growth occur inside just a thin scope of temperatures, between 32°F (00°C) and 122°F (50°C) (Barbour et al. 1987). Extreme temperatures limit the reactions in light and denatured the protein structure. Photosynthesis takes place at 32°F and low temperature limits the organic responses and water also becomes unreachable and available energy is insufficient (Bakht et al. 2020).

4.3 Water (Precipitation)

Water is a necessary part of living organisms and it is naturally important on the grounds that it is an important power in molding climatic patterns and biochemically significant in light of the fact that it is a fundamental part in physical processes (Brown 1995). The active component of plant cells is water, which is usually more than 80% of the new weight of herbaceous plants. Water is essential for the infinite maintenance of plant tissues and for plant growth (Young 2006).

4.4 Water Deficiency

The climatic situations in the Northern Great Plains cause conditions when plants experience water pressure. Rain insufficiency periods in which 75% or less of the drawn-out mean rainfall is received are called droughts. Times of dry spell circumstances can occur for the whole year or a total developing season. Lack of water that occurs for a month can affect the plant's development. Water scarcity situations during the months of May, June, and July are not common. August, September, and October experience water scarcity that is for the greater part of the time and are not reliable for positive water relations. The relationship of water during this last segment of the developing season may limit plant development and herbage biomass aggregation (Manske 2000). Continuous late seasons of water insufficiency may limit bush and tree development more than grass development.

4.5 *Water Stress*

Temperature and precipitation perform their action collectively to stimulate the physical and environmental status of range plants. The stability among rainfall and probable evapotranspiration adopts a plant's natural condition. Precipitation-evapotranspiration levels work together and influence the rates of the carbon and nitrogen cycles (Yousaf et al. 2021). Dissipation rates are dependent on temperature: as normal temperature reduces; evaporation rate also reduces. The mixed-grass prairie area has a more prominent evapotranspiration demand than precipitation. The tallgrass prairie area has more prominent precipitation than evapotranspiration demand. Water deficiency exists during the periods when the precipitation is lower than evapotranspiration demand. In the situation of water lack circumstances, the amount of water loss from transpiration exceeds the rate of water assimilation by the roots and therefore plants experience water pressure (Maalik et al. 2020). Water pressure changes from a little reduction in water potential to the deadly furthest reaches of dryness. In spite of the fact that range plants have components that support to reduce the destruction from water pressure (Young 2006). The yearly variability in temperature, precipitation, and period of water lack, which thus influence the degrees of water pressure (Brown 1995).

4.6 *Plant Water Stress*

Plants facing water pressure conditions respond at several inhibitory levels in relationship to the seriousness of the water lack. The beginning periods of water pressure moderate shoot and leaf development. Leaves give signs of withering, collapsing, and staining. As water pressure builds, catalyst movement decreases and the development of important mixes eases back or stops. The stomata start to close, and the rate of transpiration and the rate of photosynthesis become decline. Translocation is substantially reduced as water pressure rises. At the point when the water pressure gets severe, it affects the plant's growth. Plant death happens when the meristems become dried out past the limits required to keep up cell bloat and biochemical action (Brown 1995).

4.7 *Drought*

Drought creates a complex web of effects on the rangeland. The effects of drought are usually classified as direct or indirect. Decreased crop, rangeland, and forest productivity. Drought can lead to problems in forests due to pests and diseases and reduce plant growth. Forest and range fires increase significantly during an extended drought, affecting humans, wildlife and putting them at high risk. Drought has also economic, environmental, and social impacts.

4.8 Light

Light is an ultimate source of energy and the most important environmental aspect manipulating the plant's development. Variations in quality, force, and span of light stimulus plant growth. Light force varies particularly with season and time of day as a result of changes in the point of rate of the sun's beams and the separation of light goes through the air (Lal 2011). Light power likewise alters with the amount of humidity and cloudy spread on the grounds that air dampness ingests and scatters light beams. Most of the range plants require full daylight or elevated levels of daylight for best development; shading can diminish or restrain the development of range plants (Young 2006).

4.9 Wild Life

Numbers of rangelands are cleared by natural life. Wildlife also performs a vital role in rangeland management. Traditional systems to improve rangeland settings and wildlife natural surroundings have pulled in increasing attention as society has changed. With respect to the rangeland and wildlife, game farming is certainly a significant activity. This includes buffalo creation, which is turning out to be progressively better known in the United States and Canada. In South Africa 2001, various untamed life farms have been set up since the 1960s since they offer higher salaries than domesticated animals cultivating. One purpose behind the extension of game farming, aside from economic consideration, is the way that local animals are better adjusted to nearby conditions (searching, water, atmosphere), especially in territories with extraordinary atmosphere conditions (Bakht et al. 2020).

5 Impact of Grazing on Rangeland Ecosystem

Feeding animals have some immediate and backhand effects that can cure or damage. Ranger affects Ranger by expelling vegetation, depositing and depositing soil by foot movement, and preserving minerals and nutrients in the form of animal urine, feces, or carcasses. Rangeland grazing by domestic animals, sometimes in combination with others, alters the structure and structure of the Rangelands. Well-maintained fodder contributes to attractive plants, improves the environment for natural life, reduces weed infestation, reduces land for recovery, reduces the accumulation of mulch, increases the natural problem of the soil, and reduces the fuel pile exiting from the control flame. Fodder systems for high fodder and poor maintenance can expel plants that need it, reduce water intrusion into the dust, increase soil erosion, reduce water quality, along with weed infiltration, and make the plant network structure a less attractive location. In this method, the effect of fodder depends on when and how it occurs.

6 Effects of Climate Change on Rangelands

Environmental change has been viewed as an essential worldwide natural danger that will restrain the appropriation and wealth of the plant populations around the world (Pearson and Dawson 2003). Exploration suggested that by 2100, impacts identified with environmental change might be the essential fundamental factor of the decrease in worldwide biodiversity and biological system administrations (Metzger et al. 2006). Due to this misfortune in biodiversity might be credited to the failure of local plants to endure variations in the atmosphere designs, for example, helpless flexibility of certain species to occasional variances in temperature and the precipitation (Belgacem and Louhaichi 2013)."

6.1 IPCC and Climate Change

The situation of Climate Change generated by the Intergovernmental Panel on Climate Change (IPCC) distinguished a consistent increment in the worldwide mean surface temperature about 2 °C–6 °C above the pre-mechanical levels by 2100 (IPCC 2012). In light of this expansion, environmental variation situations anticipate more prominent variances in worldwide means that soil surface temperatures and the transient precipitation designs that might actuate progressively extreme flooding delayed dry season designs, and affecting biological system administrations at worldwide scales (IPCC 2012). Lately, the natural outcomes of continuous environmental change designs have been discussed (Belgacem and Louhaichi 2013). Rangelands remain profoundly vulnerable due to the effects of climate change because of constrained water accessibility and higher air and soil temperatures. Effects on the rangeland vegetation contain diminished development rates, lesser photosynthetic rates, disabled mineral assimilation, low tissue recovery, and expanded convergences of optional metabolites, for example, ginsenosides and polyphenols (Jochum et al. 2007). Therefore, diminished vegetation spread and lower plant biodiversity increment the weakness of rangelands to anticipated variances in environmental change (Belgacem and Louhaichi 2013; Hudson et al. 2014). Because of this, it is regularly emphasized by shallow soils with low supplement content that cutoff points plant development and spread and expands disintegration potential (Hudson et al. 2014). Environmental change and variability are a significant worry for touching frameworks around the world. Predicted increases in the recurrence and seriousness of extreme atmosphere occasions (e.g., heat pressure, dry season, and flooding) just as dehydrated conditions in part of the world, particularly in dry and semi bone-dry districts (Herrero et al. 2016; Kitoh and Endo 2016), stand relied upon to have noteworthy adverse outcomes on crowd populations because of diminishes in the feed and water amount and quality, declined regenerative execution, heat pressure, and expanded ailment rate and mortality (Rojas-Downing et al. 2017; Thornton et al. 2009). These decreases in creature rates compromise nearby vocations, particularly in areas that are reliant on domesticated animals as a wellspring of food or salary."

6.2 Precipitation Fluctuations

For the rest of the century, despite the positive and negative patterns, inter- and intra-yuk annual rainfall fluctuations generally spread over plain areas around the world (Sloat et al. 2018). By all accounts, the uncertainty of the Aces appears to be increasing year by year, with the largest of those districts being important for achieving animal feed and neighborhood food markets in the Sahel, Somalia, Kenya, Zimbabwe, and Australia (Sloat et al. 2018). Furthermore, variations in climate can balance the effects of change on average factors (IPCC 2012). Despite the economic, social, and environmental risks that are associated with such rainfall fluctuations, the effects of fluctuations on climate change and short- and long-term congestion factors are well understood (Thornton et al. 2009).

7 Uncertainties Regarding Climate Change and Rangeland Strategies

Rangeland frameworks are powerless against climatic variations. The elements of rangeland vegetation, and as a result animal's creation, remain profoundly delicate to the atmosphere which mean atmosphere patterns, yet besides and critically atmosphere inconstancy. High inter-annual (year-to-year) atmosphere changeability makes huge uncertainties in reach gracefully, and in this manner speaks to a test for the group the executives (Sayre et al. 2013; Marshall 2015). Increments in the intra-yearly (inside year) atmosphere fluctuation may likewise influence animal creation, even though reviews setting up connections between atmosphere irregularity and domesticated animal elements and profitability are scant and frequently constrained to the examination of a dry spell and flood occasions. Studies concentrated on vegetation have anyway discovered that adjustments in the occasional atmosphere, examples can have both positive or adverse effects on over the ground biomass and scavenge quality, contingent upon the idea of the change and the agro-ecological setting (Craine et al. 2012; Peng et al. 2013; Zeppel et al. 2014). Rangelands are likewise compromised by the environmental change-driven woody plant infringement. Notwithstanding changing rangeland's biological systems benefits, these elements sway on ruminant creation frameworks since woody rummage is more earnestly to truly accessible for cows, sheep and fewer tasteful, edible, and nutritious than herbaceous plants (Guan et al. 2014; Prevéy and Seastedt 2014)."

Rangeland people group defenselessness depends on atmosphere impacts on biological systems forms as well as on the capacity of these networks to change because of or adapt to stressors, for example, versatile limit (Gallopín 2006; Marshall 2015). A specific quality of the rangeland frameworks contrasted and the other food creation frameworks is that they are for the most part situated in remote territories with hardly any individuals, who will, in general, have constrained versatile limit (Thomas and Twyman 2005; Godber and Wall 2014; Marshall 2015). That is why rangelands are typically not enough reasonable for other food creation types. Rangeland's sensitivity to weather patterns has been recognized through modeling studies around the world. Rainfall variability for a highly variable year compared to low variable areas supports a lower livestock storage rate (Sloat et al. 2018). Over the past era, interannual rainfall variability in global grasslands has generally increased. The variability from year-to-year normalization was found to be adversely correlated with the differential vegetative index, which is a proxy suitable for vegetation development (Sloat et al. 2018). Conversely, there is no difference among the herbs, shrubs, and trees.

7.1 *Community Vulnerability*

The vulnerability of the rangeland communities depends not on the effects of the climatic conditions but also effect on the abilities of these communities to adapt or respond to stress or adaptability (Peng et al. 2013; Marshall 2015). The distinctive feature of the Rangeland systems compared to the other food production systems is that they are frequently in remote areas with limited adaptive capacity (Thomas and Twyman 2005). Rang spheres are generally not appropriate for any other food product type. The above-mentioned features create these systems and associations that are specifically dependent on weather hazards (Reid et al. 2014). Merging Rangeland flora analysis with the information on Rangeland socio-economic frameworks can help to gain insight into climate impacts on such systems (Godber and Wall 2014; Marshall 2015).

7.2 *G-Range Model*

To better understand how much the global theater is threatened by climatic change, the Global Range Land model runs on G-Range climate change conditions.

Vegetation production modeled from the G-range then interacts with the spatially clear universal livestock, economic, and demographic datasets, providing a greater understanding of the susceptibility of the Rangelands to climatic change, which provides ecosystem processes and is dependent on both socio-economic climate effects.

8 Management Practices in Northern and Southwest Great Plains

The Great Plains ranges from a semiarid climate in the west to a humid, humid climate in the east and a cool temperature in the north to a warm temperature in the south. Cool-C3 grass in northern latitudes, most hot and drought-resistant shrubs in the south from mid-C4 grasses to mid-latitudes (Terri and Stowe 1976; Joyce et al. 2001). The growing period of the northern Great Plains is 110 d. Approximately 80% of the land in the Great Plains is used for agriculture, with more than half being rainforests and pastures (Ojima et al. 2002). The average temperatures in the region were "low" cold, with more "hot" days and increased rainfall in the region. Annual rainfall is expected to increase in the northern Great Plains, but severe events such as drought and severe rainfall are common. Temperature is expected to continue to rise, leading to an increase in northern areas (Karl et al. 2009). Southwest landscapes dominate Chihuahua, Sonoran, and Mojave deserts. The desert vegetation of the area leads to the highlands of the Colorado Plateau and leads to more wood and tree vegetation and meeting areas throughout the area. Livestock and allied agriculture have been important features of the region since the eighteenth century (Guido 2009). Rainfall changes and their associated consequences for southwestern society are more important and uncertain than the northern Great Plains. Water is already a limited resource, and every change in rainfall patterns is seemed to have a strong impact on plant production and the community composition.

8.1 Management Practices in the Northern Great Plains

The land managers in the northern Great Plains are expected to minimize the effects of climate change because its effects are minimal. With the continuous increase of atmospheric CO_2, precipitation in the north and long growing periods indicate an increase in fodder production. Although the number of cattle in the Rain lands is likely to increase, as a result of the ranking, for grazing the amount of available land will not change. In the short term, more cattle on earth will have a greater environmental response. As productivity increases, the ratio of the rate of the return on investment on the farm and the total return on the farm increases. As fodder is thought to occur in large quantities, its value decreases, while the value of other products remains uncertain. Entertainment is expected to increase over time (Bowker et al. 1999) and the value generated by entertainment will increase. At the same time, there are progressive and adverse effects on the ecosystem. The density of roads and human structures is expected to increase and the bare ground (chance of destruction) to decrease. In this scenario, the other effects on the ecosystem are very small. Given these changes, it is reasonable to expect a steady or small increase in theater improvement practices as a higher return on investment. Increased rainfall may stabilize or reduce investment in renovating theaters, making current restoration efforts more effective and denying the need for further intervention. As the demand for recreational opportunities increases, more investment can be expected in recreational facilities and infrastructure. Consequently, in the conditions of the North Great Plains, there might be little incentive to alter economic policies to help the wrestling sector. Further educational and technical assistance is needed to increase public participation in land use laws and policies and to address these emerging issues.

8.2 *Management Practices in Southwest Great Plains*

The southwest amphitheater is usually limited by rainfall. Annual rainfall is bimodal, every winter, and spring-early thawing and monsoons occur in the month of July and August (Swetnam and Betancourt 1998). Winter rains are essential to recharge soil moisture; however, summer rains mainly limit rangeland-producing capacity and provide fodder for fodder. Livestock herd adjustment provides the primary Rangeland management tool in the Southwest. Stacking rates are based on the current productivity and the residual biomass from the previous year's use (Paulsen and Ares 1961). Almost all livestock need to be removed from the affected amphitheater during a severe drought. Shrub invasion of desert grasslands is partly promoted by rainfall (Swetnam and Betancourt 1998), and by prolonged increasing CO_2 (Morgan et al. 2007) and temperature (Shaw et al. 2000) in some areas. Shrubs can reduce fodder production well and separate shrubs quickly, which is an early stage. State and transition models can be used as a tool to better understand and respond to landscapes for climate change. Fodder quality is a factor influencing Rangeland management in all areas. In the southwest, fodder quality is associated with precipitation (Cable and Shumway 1966). Land managers can also take advantage of forest quality during the pacification period (Vavra and Raleigh 1976) and before milking. In winter it is usually increased due to mild weather. As time goes on and the weather increases, the winter cold becomes more possible (Smith et al. 2014).

9 Conservation Strategy

Water resources, soil formation, and plant communities have been depleted due to overexploitation of rangelands in the Pacific Northwest (Charnely et al. 2018). Therefore, preserving high-quality and high-performance theater throughout the region is an important first step in supporting productive livestock operations.

9.1 Improving Landscape Connectivity

The theater supports climate change adaptation by using the Mosaic of Habitat to "support the landscape and facilitate habitats, animals, and other organisms in the landscape" to reduce fragmentation of the landscape. Landscape connectivity improves the resilience of rangelands to climate change by maintaining or improving ecosystem diversity for vegetation and water resources (Galvin et al. 2008). Landscape connectivity supports migratory species (plants, animals, and insects), pollinators, and landscape buffers needed to support variable habitats in the natural ecosystems that promote biodiversity and environmental stress (Janowiak et al. 2016). Tools help to improve landscape connectivity include creating natural habitat corridors for migratory plants, animals, and insect species, as well as developing and maintaining effective partnerships with landscape planning that add value to landscape connectivity.

9.2 Mixed Rangeland and Cropland

Integrating livestock into established cropland operations will increase access to additional livestock feed, reduce feed costs, eliminate manure concentrated areas and improve overall agricultural efficiency. These practices can be improved by planting and supporting the development of drought and heat-resistant species. In addition, mixed cropland and amphitheater activities can improve amphitheater elasticity to higher levels of CO_2.

9.3 Soil Health

Range soils are essential for a working amphitheater, can reduce production costs, and provide essential ecosystem services (Provenza 2008). Soils support biological life and diversity and play key roles in carbon sequestration. Climate change, directly and indirectly, increases nutrient loss to soil health, reduces water retention, and limits filtration during extreme heat, drought, and heavy rainfall (Farid et al. 2018).

9.4 Native Grasses

The diversity of natural plants in the rangelands reduces the risk of catastrophic events (wildfires, disease, and pests) and improves the sustainability of livestock production. In addition, permanent non-native grass also reduces the risk of wildfire. Local meadows and shrubs are disappearing from the western United States due to invasive species, altered fire systems, and the onset of growth (Shock et al. 2015). Increasing concentrations of atmospheric CO_2 and temperature reduce the quality, productivity, and species composition of native grasses in the sagebrush steppe (Augustine et al. 2018).

10 Restoration Strategy

The restoration of the rangeland ecosystem, in particular, is important in creating resilience in barren lands. Effective planning, monitoring, and evaluation of resilience tools and methods are essential (SERI 2004).

10.1 Upland Restoration

Restoration of tall plant, shrub, and tree species improves the health of the rangeland ecosystem. Many organizations specialize in land reclamation, especially with tribes. Organizations such as Trees, Water and People, and the Red Cloud Renewable Energy Center support the restoration of forests on tribal lands with native tree species. These applications help to provide shade for cattle, increase erosion control, improve water quality, increase carbon, improve habitat, and increase biodiversity in the rangeland.

10.2 Fire

Rangeland operators face many challenges in relation to fires: resource shortages; Adequate local and national policies; Institutional barriers; the legacy of poor fire management; and climate change is at an increased risk of fire. Not only does the fire directly endanger livestock, property, and human life, but it can significantly impact an agricultural or community economy by losing access to land (Stasiewicz et al.

2018). The Rangeland Fire Protection Associations (RFPA) have been developed to synthesize cooperation between community and state and federal land managers, empowering private citizens to help them adapt to changing fire arrangements on public lands.

10.3 Invasive Species

The technical and physical resources of the terrain, which may affect the management objectives of the invasive species, should be available to land managers. The dominance of reindeer, mainly comprised of native plants or permanently introduced species to change the composition of invasive species, is time-consuming, costly, and can yield variable results.

11 Rangeland Management to Conserve Pattern and Process

The conservation of natural resources is described as evolving through three consecutive models (Callicott 1990; Weddell 2002). Autitarian paradigm relies on the conservation to sustain long-lasting (sustainable) production, with the goal of providing more to maximum people (Pinchot 1947). Gifford Pinchot is deliberated popular for this perspective, which relies on the conservation to sustain financial strength. Inspired by the spirituality of conservation, it is resulting from the ideas of Ralph Waldo Emerson, Henry David Thoreau, and John Muir, the conservation method aims to protect nature by isolating or burning land, national parks, and forest areas from humans. The terms utilitarianism and protectionism are often seen as dialectical approaches. Among them, the third example emphasizes on the ecosystem management, process conservation, and processes (feeding, reentry, water cycling, nutrient cycling, etc.) with the goal of ultimately sustaining a complete suite of biodiversity (Leopold 1949). We claim that the preservation of paradigm and process paradigm is a rational substitute to a useful example for the theatrical profession. However, the conservation-oriented model is not unique or does not fully match the historical evidence of the profession (Krausman 1996). Supervision of the amphitheater has recently focused on amphitheater health, which has led to the conservation management based on reducing bare ground and soil stabilization (Pellant et al. 2005). By erosion, Rangeland is the greatest threat to biodiversity conservation. Conservation should consider variegated and large-scale specimens on areas experiencing severe fracture and/or racial invasion (Fuhlendorf et al. 2002). All types of ecosystems around the world have interactions with arthropods, birds, mammals, amphibians, and reptiles, clearly supporting the notion that diversity is the source of biodiversity and therefore the basis for the conservation of hierarchical lands and other ecosystems (Fuhlendorf et al. 2006).

11.1 Grazing Intensity

The intensity of forage (the ratio of the underground primary production to forage animals) is yet considered to be the most significant principle of forage management (Milchunas and Lauenroth 1993; Holechek et al. 2004). Although feed intensity and storage rate are not synonymous, among them; the two are frequently discussed together because the perceptions are very high. Various experimental findings have shown that optimal animal gain per unit area can be achieved by very large storage, optimal gain over light storage for animals, and 25–30% of the optimal medium for fodder harvesting (i.e., medium) used by pets) (Hart et al. 1988; Heitschmidt and Taylor 1991; Torell et al. 1991). Achieving moderate variable deployment due to high variable international climate patterns is the goal of challenging equilibrium ecosystems. In utility management, "it is affordable" storage (i.e., moderate use) maintains major fodder species, reduces soil loss, and enhances economic returns. Therefore, conservation of fodder intensity is contraindicated because full storage rates must be at suitable rates to sustain biodiversity. This inconsistency can be solved in the preservation and process patterns by considering grazing as a disruptive procedure that can be used in complex landscapes (Fuhlendorf and Engle 2001; Archibald et al. 2005).

11.2 Distribution of Grazing in Space and Time

The management objective of many forage systems, known as the "Management Middle" (Fuhlendorf et al. 2006), stimulates the equal use of highly productive fodder species while sustaining the effective use of these species and is used here as well (Stoddart et al. 1975; Bailey, 2004; Fuhlendorf et al. 2009). Concentration on an equal use of space and times of emergence from the hierarchical management development at a time when water and other attractive animals are a primary concern on the ranch. In order to preserve large landscapes, especially areas around specific water and mineral areas are often left open for moderate grazing (Vallentine, 2001). While still needed in some cases (e.g., Riparian areas), this vision has been established into a standard that does not remain a historical masterpiece to meet the full set of conservation goals. Not all elements of the Rangeland ecosystem have a "reasonable" storage rate, which spread over equally to the distribution of fodder in space and time. When animals are permitted to graze in a large landscape at a moderate storage rate, their scattering over space and time is very variable and water, top endophytic characteristics, vegetation structure, composition, and past disturbances (Heitschmidt and Taylor 1991). The utilitarian example of equal distribution of fodder over time and space is largely impossible to sustain or increase biodiversity and productivity in the Rangeland.

11.3 Grazing Distribution Factors

There are various factors that are described under as;

1. Animal-like and square feeding habits, especially those related to length and slope stability.
2. Water development site.
3. Appointment of salt and mineral.
4. Lawn compatibility.
5. Vegetation type, location of shade.
6. Combination of range sites and range condition classes.
7. Fence design, lawn shape, grazing system.
8. Storage density and current winds.

12 Principles for Conservation of Pattern and Process on Rangeland Ecosystems

We suggest the resulting principles of amphitheater preservation of models and processes. We know that these principles are completely non-existent, and are not anticipated to completely substitute all traditional principles of hierarchical (fodder) management. Instead, we hope to introduce these principles to create a new conservation model.

1. Preservation of the large continuous tracks of rangelands is important to preserve patterns and processes involved so that disturbance processes can interrelate with complex scenarios and build multilevel mosaics.
2. Fodder strength (i.e., storage rate) is the prime factor of the fodder effect on a single range.
3. In one scenario, constant distribution of fodder over time and space cannot be achieved. Fodder distribution management should aim at diversity in the landscape like a variable mosaic.
4. Transfer mosaics are required to maintain the structure and function of the ecosystem and to achieve many goals. A single state, state, phase, or subsequent phase management can maintain maximum and livestock production, but may not promote biodiversity or several uses.
5. Conservation of the rangelands should eventually consider all species of animals and plants. The individual species and groups can be used as analytical indicators of management response, but plants and animals should not be considered as

large-scale species or target management objectives throughout the whole landscape.

6. Ecosystem structure and disturbance regimes such as fodder are important for climate and soil to function. If we want to preserve biodiversity, we must look at them as interactive processes.

13 Grazing Management Strategies that Influence Livestock Distribution

13.1 Rotational Grazing

Animals are raised on two or more pastures in a rotating pasture, with each pasture grazing one or more times during the growing season. Applying rotating fodder does not improve livestock distribution; however, rotational fodder can be used to influence fodder distribution, as it affects the distance to water, lawn size, homogeneity of fodder, and storage density. Preventing rotating forage in continuous grazing pastures does not affect the first four factors. Rotational forage implementation can be designed to influence these factors when existing pastures are subdivided into new pastures. Storage density (i.e., the number of animals per unit area at any given time) increases when rotating fodder is replaced by continuous fodder.

13.2 Stoking Density

Rotating fodder naturally increases storage density because cattle do not spread over large pastures or many small pastures, but are integrated at some point in some subset of the entire grazing area. Increasing storage density leads to better fodder distribution and crop efficiency due to competition for limited fodder (Reinkensmeyer et al. 2007).

13.3 Flash Grazing

The use of high storage density is not only associated with the rotating feed. Flash or crowd feed is a short-term, one-time grazing event at high storage density that improves the distribution of forage animals within the forage, which is adversely affected by poor forage distribution. Convenient areas are mostly used, but low areas should also be grazed due to high storage density. Cattle should be attracted to the area when they appear to be littered, and unused mature waste should be removed from unused areas (Holechek et al. 2004).

14 Fire as Rangeland Ecosystem Process

Utility management as a plant management tool mainly uses fire to control unwanted plants (Scifres and Hamilton 1993). Agni clearly maintains vegetarian dominance in most meadows, but although the theater often has vegetarian dominance with B-return intervals, ways to restore diversity and increase biodiversity can be used to change fodder patterns (Anderson et al. 2006; Fuhlendorf et al. 2006). Most Rangeland meadows and other species respond to the intensity of grazing with a return, season, and intensity of grazing (Fuhlendorf et al. 2006; Reinkensmeyer et al. 2007). Some landscapes may have limited effects on the restoration of overgrown rhizomes (e.g., the closed-canopy juniper forest uses its sensitivity to move to a new state (Brom-occupied Great Basant shrubs), but once this degraded landscape is interactive and restorative specimens should be conservation-oriented. In the meantime, placing them in a relatively stable state of endangered communities will prevent them from preserving only endemic species. Therefore, research and management should always be limited to focusing on the management of historical plant communities without considering spatial and temporal patterns of disruption procedures.

15 Recommendations

1. Fodder trees should be planted in maximum numbers and they should also be preserved.
2. Grazing activities should be banned in the spring season when vegetation is recovering.
3. Leave the areas which are totally damaged and focus on the rangeland which is getting damaged.
4. Rangeland management system requires long-term monitoring, which is a difficult process, so relevant information should be gathered for the management decisions.
5. For better management practices the correct data should be provided to the land managers.
6. Policies should be made according to the climatic patterns of the particular rangeland.
7. Conservation of the rangelands and a better management plan can help to mitigate the negative consequences.
8. Conservation departments should aware the people about the condition of the rangelands.
9. As some of the rangelands in Baluchistan might convert into the arid, so land managers should make proper planning.

16 Conclusions

The use of the rangelands without any management plan will cause negative consequences and will surely cause rangeland degradation. The rangelands of Pakistan are on a decreasing site due to a poor management system. There are many factors by which the rangelands are getting affected and the climate change is one of them. Many other factors are arising due to climate change like droughts, change in rainfall patterns and a rise in the temperature which are affecting the diversity of the rangelands. The other main factor is the overgrazing by which the vegetation cover is getting removed. A good way to improve the quality or the degraded rangelands is to stop the grazing activities. The negative consequences of the degradation of the rangelands are not getting monitored, documented, or discussed with the society.

References

Ainslie A (2005) Farming cattle, cultivating relationships: cattle ownership and cultural politics in Peddle District, Eastern Cape. Social Dyn 31:129–156

Anderson RH, Fuhlendorf SD, Engle DM (2006) Soil N availability in tall grass prairie under the fire-grazing interaction. Rangel Ecol Manage 59:625–631

Archibald S, Bond WJ, Stock WD, Fairbanks DHK (2005) Shaping the landscape: fire-grazer interactions in an African savanna. Ecol Appl 15:96–109

Augustine D, Blumenthal D, Springer T, LeCain D, Gunter S, Derner J (2018) Elevated CO_2 induces substantial and persistent declines in forage quality irrespective of warming in mixed grass prairie. Ecol Appl 28(3):721–735

Bailey DW (2004) Management strategies for optimal grazing distribution and use of arid rangelands. J Anim Sci 82:147–153

Bakht S, Safdar K, Khair KU, Fatima A, Fayyaz A, Ali SM, Munir H, Farid M (2020) Response of major food crops under drought stress; physiological and biochemical response. In: Agronomic crops - volume 3: stress responses and tolerance. Springer, Cham, pp 94–116

Barbour MG, Burk JH, Pitts WD (1987) Method of sampling the plant community. In: Terrestrial plant ecology. Benjamin/Cummings Publishing, Menlo Park, CA

Bayer W, Alcock R, Dladla F, Gilles P, Masondo M, Mkhize P, Mtshali E, Ntombela L (2004) A study of indigenous livestock management in rural KwaZulu–Natal, South Africa. Unpublished report. Mdukatshani: Mdukatshani Rural Development Project

Bedunah DJ, Angerer JP (2012) Rangeland degradation, poverty, and conflict: howcan rangeland scientists contribute to effective responses and solutions? Rangeland Ecol Manag 65 (6):606–612

Belgacem AO, Louhaichi M (2013) The vulnerability of native rangeland plant species to global climate change in the West Asia and north African regions. Clim Change 119(2):451–463

Boone RB, Galvin KA, BurnSilver SB, Thornton PK, Ojima DS, Jawson JR (2011) Using coupled simulation models to link pastoral decision making and ecosystem services. Ecol Soc 16:6

Bowker JM, English DBK, Cordell HK (1999) Projections of outdoor recreation participation to 2050. In: Cordell HK (ed) Outdoor recreation in American life: a national assessment of demand and supply trends. Sagamore Publishing, Champaign, pp 323–350

Briske DD, Joyce LA, Polley HW, Brown JR, Wolter K, Morgan JA, McCarl BA, Bailey DW (2015) Climate- change adaptation on rangelands: linking regional exposure with diverse adaptive capacity. Front Ecol Environ 13:249–256

Brown RW (1995) The water relations of range plants: adaptations to water deficits. In: Bedunah DJ, Sosebee RE (eds) Wildland plants: physiological ecology and developmental morphology. Society for Range Management, Denver, pp 291–413

Cable DR, Shumway RP (1966) Crude protein in rumen contents and in forage. J Range Manage 19:124–128

Callicott JB (1990) Whither conservation ethics? Conserv Biol 4(1):15–20

Charnely S, Gosnell H, Wendel KL, Rowland MM, Wisdom MJ (2018) Cattle grazing and fish recovery on US federal lands: can social-ecological systems science help? Front Ecol 16 (S1):11–22

Christensen JH, Hewitson B, Busuioc A, Chen A, Gao X, Held I, Jones R, Kolli RK, Kwon WT, Laprise R, Magaña Rueda V (2007) Regional climate projections. In: Solomon S, Qin D, Manning M, Marquis M, Averyt K, Tignor MMB, Miller J, LeRoy H, Chan Z (eds) Climate change 2007: the physical science basis. Contribution of Working Group I to the Fourth Assessment Report of the Intergovernmental Panel on Climate Change. Cambridge University Press, Cambridge, pp 847–940

Craine JM, Nippert JB, Elmore AJ, Skibbe AM, Hutchinson SL, Brunsell NA (2012) Timing of climate variability and grassland productivity. Proc Natl Acad Sci U S A 109(9):3401–3405

Ezaz Z, Azhar R, Rana A, Ashraf S, Farid M, Mansha A, Naqvi SAR, Zahoor FA, Rasool N (2020) Current trends of phytoremediation in wetlands: mechanisms and applications. In: Plant eco-physiology and adaptation under climate change: mechanisms and perspectives II, mechanisms of adaptation and stress amelioration. Springer, Singapore, pp 747–765

Farid M, Ali S, Zubair M, Saeed R, Rizwan M, Sallah-Ud-Din R, Azam A, Ashraf R, Ashraf W (2018) Glutamic acid assisted phyto-management of silver contaminated soils through sunflower; physiological and biochemical response. Environ Sci Pollut Res 25(25):25390–25400

Fatima A, Farid M, Alharby HF, Bamagoos AA, Rizwan M, Ali S (2020) Efficacy of fenugreek plant for ascorbic acid assisted phytoextraction of copper (cu); a detailed study of cu induced morpho-physiological and biochemical alterations. Chemosphere 251:126424

Fuhlendorf SD, Engle DM (2001) Restoring heterogeneity on rangelands: ecosystem management based on evolutionary grazing patterns. Bioscience 51:625–632

Fuhlendorf SD, Engle DM, Kerby J, Hamilton R (2009) Pyric herbivory: rewilding landscapes through the recoupling of fire and grazing. Conserv Biol 23:588–598

Fuhlendorf SD, Harrell WC, Engle DM, Hamilton RG, Davis CA (2006) Should heterogeneity be the basis for conservation? Grassland bird response to fire and grazing. Ecol Appl 16 (5):1706–1716

Fuhlendorf SD, Woodward AJW, Leslie DM, Shackford JS (2002) Multi-scale effects of habitat loss and fragmentation on lesser prairie-chicken populations of the US southern Great Plains. Landsc Ecol 17:617–628

Gallopín GC (2006) Linkages between vulnerability, resilience, and adaptive capacity. Glob Environ Chang 16(3):293–303

Galvin KA, Reid RS, Behnke RH, Hobbs NT (2008) Fragmentation in semi-arid and arid landscapes: consequences for human and natural systems. Springer, Dordrecht

Gharibvand KH, Azadi H, Witlox F (2015) Exploring appropriate livelihood alternatives for sustainable rangeland management. Rangel J 37(4):345–356

Godber OF, Wall R (2014) Livestock and food security: vulnerability to population growth and climate change globe. Chan Biol 20(10):3092–3102

Guan K, Good SP, Caylor KK, Sato H, Wood EF, Li H (2014) Continental-scale impacts of intra-seasonal rainfall variability on simulated ecosystem responses in Africa. Biogeosciences 11:6939–6954

Guido Z (2009) Cattle and climate: ranching in the arid southwest. Southwest Clim Outlook 8:3–5

Hart RH, Samuel JJ, Test PS, Smith MA (1988) Cattle, vegetation, and economic responses to grazing systems and grazing pressure. J Range Manage 41:282–286

Heitschmidt RK, Taylor CA (1991) Livestock production. In: Heitschmidt RK, Stuth JW (eds) Grazing management: an ecological perspective. Timber Press, Portland, pp 162–177

Herrero M, Addison J, Bedelian C, Carabine E, Havlík P, Henderson B, Thornton PK (2016) Climate change and pastoralism: impacts, consequences, and adaptation. Rev Sci Tech 35:417–433

Holechek J, Pieper RD, Herbel CH (2004) Range management: principles and practices, 5th edn. Prentice Hall, Upper Saddle River, p 587

Hudson LN, Newbold T, Contu S, Hill SL, Lysenko I, De Palma A, Phillips HR, Senior RA, Bennett DJ, Booth H, Choimes A (2014) The PREDICTS database: a global database of how local terrestrial biodiversity responds to human impacts. Ecol Evol 4(24):4701–4735

Intergovernmental Panel on Climate Change (IPCC) (2012) Managing the risks of extreme events and disasters to advance climate change adaptation. Cambridge University Press, Cambridge. https://doi.org/10.1017/CBO9781139177245

Janowiak M, Dostie D, Wilson M, Kucera M, Howard Skinner R, Hatfield J, Hollinger D, Swanston C (2016) Adaptation resources for agriculture: responding to climate variability and change in the midwest and northeast. Technical Bulletin 1944. U.S. Department of Agriculture, Washington, DC

Jochum GM, Mudge KW, Thomas RB (2007) Elevated temperatures increase leaf senescence and root secondary metabolite concentrations in the understory herb Panax quinquefolius (Araliaceae). American J Bot 94(5):819–826

Joyce L, Aber J, McNulty S, Dale V, Hansen A, Irland L, Neilson R, Skog K (2001) Potential consequences of climate variability and change for the forests of the United States. In: National Assessment Synthesis Team (ed) Climate change impacts on the United States. Cambridge University Press, Cambridge, pp 489–524

Karl TR, Melillo JM, Peterson TC (2009) Global climate change impacts in the United States: a state of knowledge report from the U.S. global change research program. Cambridge University Press, New York

Kitoh A, Endo H (2016) Changes in precipitation extremes projected by a 20-km mesh global atmospheric model. Weather Clim Extrem 11:41–52

Krausman PR (1996) In: Krausman PR (ed) Rangeland wildlife. Society for Range Management, Denver, pp 245–279

Lal R (2004) Soil carbon sequestration to mitigate climate change. Geoderma 123:1–22

Lal R (2011) Sequestering carbon in soils of agro-ecosystems. Food Policy 36:33–39

Leopold AS (1949) A sand county almanac and sketches here and there. Oxford University Press, New York, p 295

Maalik U, Farid M, Zubair M, Ali S, Rizwan M, Shafqat M, Ishaq HK (2020) Rice production, augmentation, escalation and yield under water stress. In: Agronomic crops - volume 3: stress responses and tolerance. Springer, Singapore, pp 117–128

Manske LL (2000) Environmental factors to consider during planning of management for range plants in the Dickinson, North Dakota, region 1892–1999. NDSU Dickinson Research Extension Center. Range research report DREC 00-1018c. Dickinson, ND, pp 36

Marshall N (2015) Adaptive capacity on the northern Australian rangelands. Rangel J 37 (6):617–622

Metzger MJ, Rounsevell MDA, Acosta Michlik L, Leemans R, Schröter D (2006) The vulnerability of ecosystem services to land-use change. Agric Ecosyst Environ 114:69–85

Milchunas GG, Lauenroth WK (1993) Quantitative effects of grazing on vegetation and soils over a global range of environments. Ecol Monogr 63:327–366

Morgan JA, Milchunas DG, LeCain DR, West MS, Mosier A (2007) Carbon dioxide enrichment alters plant community structure and accelerates shrub growth in the shortgrass steppe. Proc Natl Acad Sci U S A 104:14724–14729

Ojima DS, Lackett JM (2002) Preparing for a changing climate: the potential consequences of climate variability and change—central Great Plains. Report for the global change research program. Colorado State University, Fort Collins

Paulsen HA Jr, Ares FN (1961) Trends in carrying capacity and vegetation on an arid southwestern range. J Range Manage 14:78–83

Pearson RG, Dawson TP (2003) Predicting the impacts of climate change on the distribution of species: are bioclimate envelope models useful? Glob Ecol Biogeogr 12(5):361–371

Pellant M, Shaver P, Pyke DA, Herrick JE (2005) Interpreting indicators of rangeland health. Version 4. US Department of the Interior–Bureau of Land Management National Science and Technology Center, Denver

Peng S, Piao S, Shen Z, Ciais P, Sun Z, Chen S, Bacour C, Peylin P, Chen A (2013) Precipitation amount, seasonality and frequency regulate carbon cycling of a semi-arid grassland ecosystem in Inner Mongolia, China: a modeling analysis. Agric For Meteorol 178-179:46–55

Pinchot G (1947) Breaking new ground. Harcourt, Brace and Co, New York, p 522

Polley HW, Briske DD, Morgan JA, Wolter K, Bailey D, Brown JR (2013) Climate change and north American rangelands: trends, projections, and implications. Rangel Ecol Manage 66:493–511

Prevéy JS, Seastedt TR (2014) Seasonality of precipitation interact with exotic species to alter composition and phenology of a semi-arid grassland. J Ecol 102:1549–1561

Provenza FD (2008) What does it mean to be locally adapted and who cares anyway? Am Soci Anim Sci 86:271–284

Reid RS, Fernández-Giménez ME, Galvin KA (2014) Dynamics and resilience of range lands and pastoral peoples around the globe. Annu Rev Env Resour 39:217–242

Reinkensmeyer DP, Miller RF, Anthony RG, Marr VE (2007) Avian community structure along a mountain big sagebrush successional gradient. J Wild Manag 71:1057–1066

Rojas-Downing MM, Nejadhashemi AP, Harrigan T, Woznicki SA (2017) Climate change and livestock: impacts, adaptation, and mitigation. Clim Risk Manag 16:145–163

Salomon ML (2011) Keeping cattle in a changing rural landscape: communal rangeland management in Okhombe, KwaZulu-Natal, South Africa. PhD thesis, University of KwaZulu-Natal, Pietermaritzburg, South Africa

Sayre NF, McAllister R, Bestelmeyer BT, Moritz M, Turner MD (2013) Earth stewardship of rangelands: coping with ecological, economic, and political marginality. Front Ecol Environ 11:348–354

Schuman GE, Janzen HH, Herrick JE (2002) Soil carbon dynamics and potential carbon sequestration by rangelands. Environ Pollut 116:391–396

Scifres CJ, Hamilton WT (1993) Prescribed burning for brush land management: the South Texas example. Texas A&M University Press, College Station, p 246

Shaw MR, Loik ME, Harte J (2000) Gas exchange and water relations of two Rocky Mountain shrub species exposed to a climate change manipulation. Plant Ecol 146:197–206

Shock CC, Feibert EBG, Shaw N, Shock M, Saunders LD (2015) Irrigation to enhance native seed production for Great Basin restoration. Nat Areas J 35(1):74–82

Sloat LL, Gerber JS, Samberg LH, Smith WK, Herrero M, Ferreira LG, Godde CM, West PC (2018) Increasing importance of precipitation variability on global livestock grazing lands. Nat Clim Chan 8:214–218. https://doi.org/10.1038/s41558-018-0081-5

Smith SD, Charlet TN, Zitzer SF, Abella SR, Vanier CH, Huxman TE (2014) Long- term response of a Mojave Desert winter annual plant community to a whole-ecosystem atmospheric CO_2 manipulation (FACE). Glob Chang Biol 20:879–892

Smith SD, Huxman TE, Zitzer SF, Charlet TN, Housman DC, Coleman JS, Fenstermaker LK, Seemann JR, Nowak RS (2000) Elevated CO_2 increases productivity and invasive species success in an arid ecosystem. Nature 408:79–82

Society for Ecological Restoration International Science & Policy Working Group (SERI) (2004) The SER international primer on ecological restoration. Society for Ecological Restoration International, Tucson

Stasiewicz AM, Paveglio TB (2018) Wildfire management across rangeland ownerships: factors influencing rangeland fire protection association establishment and functioning. Rangel Ecol Manage 71(6):727–736
Stoddart LA, Smith AD, Box TW (1975) Range management, 3rd edn. Springer, New York, p 532
Swetnam TW, Betancourt JL (1998) Mesoscale disturbance and ecological response to decadal climatic variability in the American southwest. J Climate 11:3128–3147
Terri JA, Stowe LG (1976) Climatic patterns and the distribution of C4 grasses in North America. Oecologia 23(1):1–12
Thomas DSG, Twyman C (2005) Equity and justice in climate change adaptation amongst natural resource-dependent societies. Glob Environ Chang 15(2):115–124
Thornton PK, van de Steeg J, Notenbaert A, Herrero M (2009) The impacts of climate change on livestock and livestock systems in developing countries: a review of what we know and what we need to know. Agr Syst 101:113–127
Torell LA, Lyon KS, Godfrey EB (1991) Long-run versus short-run planning horizons and the rangeland stocking rate decision. American J Agric Econom 73:795–807
Vallentine JF (2001) Grazing management, 2nd edn. Academic Press, San Diego
Vavra M, Raleigh RJ (1976) Coordinating beef cattle management with the range forage resource. J Range Manage 29:449–452
Vetter S, Goodall VL, Alcock R (2020) Effect of drought on communal livestock farmers in KwaZulu-Natal, South Africa. Afr J Range Forage Sci 37(1):93–106
Walters CJ, Holling CS (1990) Large- scale management experiments and learning by doing. Ecology 71(6):2060–2068
Weddell BJ (2002) Conserving living natural resources: in the context of a changing world. Cambridge University Press, Cambridge, p 426
World Bank (2006) Agriculture investment sourcebook. Module 5: investment in sustainable natural resource. Management for Agriculture. www.worldbank.org/agsourcebook
Young TP (2006) Declining rural populations and the future of biodiversity: missing the forest for the trees? J Int Wildl Law Policy 9:319–334
Yousaf A, Khalid N, Aqeel M, Nomal A, Naeem N, Sarfraz W, Ejaz U, Qaiser Z, Khalid A (2021) Nitrogen dynamics in wetland systems and its impact on biodiversity. Nitrogen 2:196–217
Zeppel MJB, Wilks JV, Lewis JD (2014) Impacts of extreme precipitation and seasonal changes in precipitation on plants. Biogeosciences 11:3083–3093
Ziska LH, Reeves JB, Blank B (2005) The impact of recent increases in atmospheric CO2 on biomass production and vegetative retention of cheatgrass (*Bromus tectorum*): implications for fire disturbance. Glob Chang Biol 11:325–332

Biological Nitrogen Fixation: An Analysis of Intoxicating Tribulations from Pesticides for Sustainable Legume Production

Zahoor Ahmad, Rana Muhammad Sabir Tariq, Musarrat Ramzan, Muhammad Adnan Bukhari, Ali Raza, Muhammad Aamir Iqbal, Ram Swaroop Meena, Mohammad Sohidul Islam, Oksana Sytar, Ntsomboh-Ntsefong Godswill, Allah Wasaya, Kulvir Singh, Akbar Hossain, Muhammad Ali Raza, Mirza Hasanuzzaman, Sipan Soysal, Murat Erman, Fatih Cig, Mustafa Ceritoğlu, Semih Açıkbaş, Özge Uçar, Aynur Bilmez Özçinar, Rukiye Kılıç, and Ayman EL Sabagh

Abstract A score of pedo-environmental factors serves as limiting elements for the biological nitrogen fixation (BNF) process in root nodules of leguminous plants. Since the advent of the green revolution, pesticides have been considered indispensable for keeping crop pests below the economic threshold level to ensure sustainable production of field crops for the rapidly increasing world population. However, pesticide application has also been associated with adverse effects on plant growth and development besides causing a detrimental reduction in microbial community dynamics. Rhizobium strains that are host-specific are no exception to this threat and

Z. Ahmad
University of Central Punjab, Bahawalpur Campus, Bahawalpur, Pakistan

R. M. S. Tariq
Department of Agriculture & Agribusiness Management, University of Karachi, Karachi, Pakistan

M. Ramzan
Department of Botany, The Islamia University of Bahawalpur, Bahawalpur, Pakistan

M. A. Bukhari
Department of Agronomy, University College of Agriculture and Environmental Sciences, The Islamia University of Bahawalpur, Bahawalpur, Pakistan

A. Raza
Oil Crops Research Institute, Chinese Academy of Agricultural Sciences (CAAS), Wuhan, Hubei, China

M. A. Iqbal
Department of Agronomy, Faculty of Agriculture, University of Poonch Rawalakot, Rawalakot, Pakistan

R. S. Meena
Department of Agronomy, Institute of Agricultural Sciences, Banaras Hindu University, Varanasi, Uttar Pradesh, India

M. Hasanuzzaman et al. (eds.), *Managing Plant Production Under Changing Environment*, https://doi.org/10.1007/978-981-16-5059-8_14

are negatively influenced by different pesticides, especially fungicides, which seriously affect the functioning of the nitrogen (N) fixation process. Pesticides containing different synthetic chemicals affect symbiotic nitrogen fixation (SNF), and consequently, the amount of N fixed. This leads to reliance on crop plants primarily on N available in soil solution. Ultimately, reduced soil fertility leads to deteriorate crop productivity and quality of the produce. The objective of this review has been to synthesize, explore and critically analyze the effects of pesticide applications and their physiological impacts on BNF in legumes for sustainable crop production to strengthen food security for the increasing world population. Our review elucidates that indiscriminate use of agrochemicals could result in an undesirable environment for the healthy survival of symbiotic and symbiotic organisms leading to a corresponding reduction in exopolysaccharide (EPS) synthesis leading to poor atmospheric N fixation and thus affecting the whole agroecosystems. Therefore, by giving due consideration to the harmful effects of pesticides, farmers' awareness about the safe usage of agrochemicals might be among the top priorities to conserve the environment besides harmonically preserving living organisms.

M. S. Islam
Department of Agronomy, Hajee Mohammad Danesh Science and Technology University, Dinajpur, Bangladesh

O. Sytar
Department of Plant Biology, Kiev National University of Taras Shevchenko, Kiev, Ukraine

N.-N. Godswill
Department of Plant Biology, Faculty of Science, University of Yaounde, Yaounde, Cameroon

A. Wasaya
College of Agriculture, Bahauddin Zakariya University, Bahadur Sub-Campus, Layyah, Pakistan

K. Singh
Punjab Agricultural University, Faridkot, Punjab, India

A. Hossain
Bangladesh Wheat and Maize Research Institute, Dinajpur, Bangladesh

M. A. Raza
College of Agronomy, Sichuan Agricultural University, Chengdu, China

M. Hasanuzzaman
Department of Agronomy, Faculty of Agriculture, Sher-e-Bangla Agricultural University, Sher-e-Bangla Nagar, Dhaka, Bangladesh

S. Soysal
Kurtalan Vocational High School, Siirt University, Siirt, Turkey

M. Erman · F. Cig · M. Ceritoğlu · S. Açıkbaş · Ö. Uçar · A. B. Özçinar · R. Kılıç
Department of Field Crops, Faculty of Agriculture, Siirt University, Siirt, Turkey

A. E. Sabagh (✉)
Department of Field Crops, Faculty of Agriculture, Siirt University, Siirt, Turkey

Department of Agronomy, Faculty of Agriculture, Kafrelsheikh University, Kafrelsheikh, Egypt
e-mail: ayman.elsabagh@agr.kfs.edu.eg

Keywords Legume · Productivity · Biological nitrogen fixation · Pesticides

Abbreviations

ATP	Adenosine triphosphate
BNF	Biological nitrogen fixation
CT	Conventional tillage
EPS	Exopolysaccharide
N	Nitrogen
NT	Non-tillage
OM	Organic matter
ROS	Reactive oxygen species
SNF	Symbiotic nitrogen fixation

1 Introduction

Pesticides have been deemed indispensable for obtaining the economic yield of field crops since the inception of the green revolution leading to the production of sufficient food for exponential human population growth over the globe. The importance of biological nitrogen fixation (BNF) by legumes to improve crop production and the quality of legumes needs to be looked upon from a broader perspective. The BNF is an environmentally friendly process actively used for sustainable agriculture (Stagnari et al. 2017; Vanlauwe et al. 2019). Due to the costly chemical fertilizers in developing countries need to be focused on integrated nutrient management through legumes involve in the cropping system. BNF by legumes has the potential to increase the main benefits in terms of reducing or dispensing the need for mineral N without loss of total output (Salim and Raza 2020). Some legumes also can solubilize otherwise unavailable forms of phosphate by excreting organic acids from their roots, in addition to improving soil fertility. Legumes also help to restore soil organic matter and reduce pests and diseases when used in rotation with nonleguminous crops (Das and Ghosh 2012). Various studies suggest that the rhizobial growth attributes, as well as the exopolysaccharides (EPSs) production, are severely affected by the various soil and foliar-applied pesticides. It is a well-established production symbiotic relationship; it can be distorted by the harmful effect of pesticides (Ahemad and Khan 2013; Sultana et al. 2019). The chemicals keep the plants safe by managing various insect pests besides preventing diseases from being transmitted by the pests. In some cases, they cause enormous harm to the soil microflora (Gupta and Diwan 2017).

Due to indiscriminate pesticide applications in agroecosystems, a considerable quantity of their residues usually gets accumulated in the pedosphere. Thus, pesticide contamination of farmlands had been a significant concern at the global level with specific reference to countries having high consumption. Nevertheless, the environmental fate of pesticides on crops has been described by Ahemad and Khan (2013).

Although grain production has tremendously increased over the past 20 years, because of the better cultivars, the widespread use of pesticides, synthetic N fertilizers, and irrigation, etc. which have been an integral part of the green revolution strategy. This increase in agricultural output seems to be unsustainable because of declining crop yields and the environmental impacts of modern agricultural practices (Fox et al. 2007; Meena and Lal 2018; Raza et al. 2019). It is a matter of concern that pesticide residues can potentially impart severe deleterious effects on crop quality, human beings, and the environment (Hajslova and Zrostlikova 2003; Tang et al. 2018). The fact remains that pesticides play a crucial role in keeping numerous plant diseases and other insects below the injury threshold level. Apart from field crops, the production of fruits and vegetables could be reduced because of the pest attack in the absence of pesticides application (Prodhan et al. 2018). As per an estimate, about 20 to 40% of crop yields might be reduced globally due to plant pests and diseases (FAO 2016). To overcome such situations, farming communities have been using pesticides intensively as the most convenient and economical way to control insect pests and diseases. A significant chunk of masses consuming inorganically grown agricultural produce have reported numerous health problems and complexities. Also, a variety of factors such as lack of appropriate awareness about pesticide dosages, the recommended time of application and stage of the crop, etc. have made the matter from bad to worse (Prodhan et al. 2018). The objective of this review is to synthesize, explore and critically analyze the effects of pesticide application and their physiological impacts on BNF in legumes for sustainable crop production to enhance food security for the ever-growing world population.

2 Area and Production of Legumes Across the Globe

As we compared grain legumes in the past, their acreage, production, and productivity were low due to the poor potential yield of the erstwhile varieties, besides serious biotic and abiotic stresses. However, now the production of grain legumes has increased globally. The area under cultivation of some of the major legumes is ~220 Mha, with a productivity of 1.7 MT ha^{-1} (FAOSTAT 2016). Among the various grain legumes, dry beans, which include all types of *Phaseolus* beans, have the highest area under cultivation after soybean, i.e., 27 Mha (Islam and Adjesiwor 2017). Akibonde and Maredia (2011) observed that in Pakistan, Myanmar, and India, black gram (*Vigna mungo*), moong beans (*Vigna radiata*), and dew bean (*Vigna aconitifolia*) have a higher position, whereas, in China, only mung beans (*Vigna radiata*) are the key dry bean followed by common beans (*Phaseolus vulgaris*). However, in other parts of the world such as Africa, Latin America, and the Caribbean, common beans have a significant role in the production. Another grain legume crop is chickpea (*Cicer arietinum*), which is grown in over 57 countries across the world under different pedo-climatic conditions (Li et al. 2015). The South and Southeast Asian countries have been dominated by chickpea production, which has increased tremendously in the recent past. In South Asia, India persistently remains one of the largest chickpea producers accounting for 68% (7.9 Mha) of the chickpea area worldwide (Merga and Haji 2019).

Besides, Pakistan, along with Turkey, are also key producers of chickpea (*Cicer arietinum*) who contribute over 1.1 M T annually (Nedumaran et al. 2015). Chickpea yield on an average remains 1 T ha^{-1}, while the corresponding figure for South and SouthEast Asia remains 0.812 T ha^{-1} (Nedumaran et al. 2015). Cowpea is another leguminous crop grown across the globe, especially in the semiarid and arid regions. It is a drought-tolerant crop and well adapted to dry farming regions where most of the legumes fail to thrive well (Akibonde and Maredia 2011). The area under cultivation for cowpea crop is about 10 Mha with major growing continents like Asia, Africa, and Latin America. Nigeria is a major producer of cowpea (*Vigna unguiculata*) covering 3 Mha area under cultivation with 2.6 M T production in the year 2008–2010. Pigeonpea (*Cajanus cajan*) plays a significant role in grain legume crops. It is cultivating under rainfed conditions of Asia, Eastern and Southern Africa, and the United States. The production of pigeon pea is concentrated in Asia, about 4.0 Mha recorded during the year 2008–2010 and with a yield of 3.3 MT.

3 Importance of Legumes

Legumes crops are significant not only due to their nutritional value but also due to their role in the economy (Nedumaran et al. 2015). Legumes crops are rich in carbohydrates, proteins, vitamins, minerals, fibers, and essential amino acids. In addition, legumes play an important function in nitrogen fixing, which finally enhances the fertility of soil during the process of symbiotic (Hossain et al. 2020). In addition to several benefits to soil, grain legumes assist in climate change mitigation, which has become a severe threat to the sustainability of modern agriculture (Kumar et al. 2018). It can enhance the availability of nutrient, soil structure, decrease the incidence of disease and mycorrhizal colonization promotion (Hossain et al. 2016; Das et al. 2018). Legumes can lower greenhouse gas emissions (nitrous oxide and carbon dioxide) in comparison to N-fertilization-based farming along with assisting in carbon sequestration in soils (Alpmann et al. 2013; Angus et al. 2015; Jeuffroy et al. 2013). Legumes release around 5–7 times lesser greenhouse gases per unit area compared with other crops (Jensen et al. 2012). Grain legumes can increase the soil organic carbon (SOC) and N concentration in cropping systems (Lemke et al. 2007; Garrigues et al. 2012; La-Favre and Focht 1983). Legumes have been reported to be equally competitive to cereals as far as utilization of agricultural and environmental resources are concerned and thus have the potential for inclusion in modern cropping systems to decrease N-fertilization requirements (PlazaBonilla et al. 2017).

Legumes hold numerous unprecedented features that have the potential to impart sustainability to modern agricultural systems by conserving nonrenewable resources and restoring them to remain functional for an extended period (Meena et al. 2020). The conservation system ensures minimum soil disturbance and provides soil cover through effective rotation. Some grain legume species also have tap root systems, which assist in solubilizing soil nutrients through the action of root exudates leading to a significant increase in the uptake along with their recycling in soil horizons (Hobbs et al. 2007).

4 Role of Legumes in Human Nutrition

Grain legumes are excellent sources of energy and vitamins (Singh et al. 2017) though the protein quality is not at par to meat, owing to their lesser amounts of essential amino acids such as methionine. However, this deficiency can be easily rectified through the mixing of grain legumes with cereals in the diet. Soybeans and peanuts are rich sources of lysine and vegetable oils used for numerous purposes such as cooking oil, margarine, mayonnaise, and salad dressings (Li et al. 2016; Salinas Valdés et al. 2015). Legumes are a cheap and prominent source of various nutrients that could substitute animal proteins (Stagnari et al. 2017). Although animal proteins contain a higher concentration of saturated fats, fats found in legumes are mostly of the unsaturated class. Legumes offer many essential minerals along with dietary fiber, as well as many phytochemicals that impart positive effects on human health. Soybean came to the center stage owing to its phytoestrogen contents, which are also called isoflavones (Messina 1999). The term bean is now applied in a general way to many other related legume plants such as faba-bean (*Vicia faba*). In the English language, some beans are also being referred to as some other plants of the nonlegume family, e.g., coffee beans, castor beans, and cocoa beans (which resemble bean seeds), and vanilla beans (which resemble the pods). The dried seeds of legumes contain higher protein contents along with carbohydrates (Bender 2006). Grain legumes, which are roughly spherical, flat, and round, are therefore invariably classified as nuts, peas, and beans. Lentils differ from this pattern and are multicolored (green, yellow, orange, black, and/or brown). It is pertinent to mention that the seed colors of lentils have nothing to do with the nutritional value (Messina 1999).

5 Importance of Nitrogen on the Growth and Development of Plants

Nitrogen is one of the most required macronutrients in the plant's biological systems. It plays an important role in plant biology and the shortage of nitrogen results in chlorosis and a decrease in over plant growth causing the reduction in production (Molla et al. 2019; EL Sabagh et al. 2020). It is an important factor of chlorophyll, enzymes, and proteins (Mondal et al. 2021; Perveen et al. 2021). Naturally, N exists in inert gaseous form, making 78% of the atmosphere. Although N continues to remain one of the most deficient nutrients in all ecosystems (Batut et al. 2011; Glick 2012), it can, however, be utilized by certain microbes, which are referred to as N-fixing microorganisms. These microbes transform the inert N to ammonia (NH_3), which can then be absorbed and subsequently assimilated by plants into various biochemical compounds. This transformation through N fixation requires energy for breaking the triple-bonded N and converting it to plant usable form. It is worth mentioning that over 12% of N gets fixed in the same way (Galloway et al. 2008). Alternatively, N fixation may also occur through metabolic activities in prokaryotes, which are called diazotrophs. BNF has been reported to be the sole source of biologically available N, which is produced in terrestrial ecosystems amounting to over 140 MMT annually (Galloway et al. 2008). For N fixation, two types of bacteria (free-living and symbiotic) are present in the soil around the roots of plants or present in association with plant roots. Both types of bacteria influence plant roots, though they are physiologically diverse (Reed et al. 2011). N fixation by free-living bacteria is also restricted due to the oxygen sensitivity of the nitrogenase enzyme in addition to the presence of antagonistic microbes that compete for food (Ryu et al. 2020).

Nitrogen is a major building block of various complex compounds such as enzymes, proteins, etc. act as a catalyst for the chemical reactions, transports the electrons, and also initiates the photosynthesis process in plant bodies. Various plant physiological functions are carried out in the presence of N. Nitrogen is also responsible for green chlorophyll pigment in plants, improves leaves growth, stem elongation, as well as, overall vegetative growth. The rapid increase in plant growth, fruiting quality, root density, increment in protein concentration of fodder crops could be the result of N application. Its application improves the availability of other nutrients such as potassium and phosphorus (Bloom 2015).

6 Role of Legumes in Biological Nitrogen Fixation

Grain legumes are an indispensable constituent of human diets. They are also recognized as a good source of feed for animals and also referred to as the poor man's meat (Iqbal et al. 2019; Umeta et al. 2005). The mature dried seeds are used as food, feed, or processed into multiple edible products (Nedumaran et al. 2015). The plant of legume crops has the advantage of fixing atmospheric N with the help of soil microorganisms for their own needs, and hence they increase soil fertility by reducing the input cost of fertilizer during crop farming. This is why they are frequently sown as intercrops with cereals. There are different species of legumes, having around 20,000 types of plants, and most of these species have hundreds of cultivars. It is thought that more than 40,000 different bean cultivars are stored in the gene bank alone (Nedumaran et al. 2015). This makes it one of the largest plant families in the Kingdom *Plantae*. The legume crops include all forms of peas and beans from the *Leguminosae* or *Fabaceae* family. The *Leguminosae* plant's family includes grain legumes and is also known as pulses. They produce dried seeds and are primarily planted for their edible purpose. The word pulse comes from the Latin word, *puls*, which means the seeds. The pod seeds of legume plants usually split into two halves. Legumes have a variety of shapes, sizes, and colors (Stagnari et al. 2017). However, agricultural research has been continuously making efforts on increasing cultivated area, yields, and production in these locations besides lowering the crop losses in legumes to achieve food security. Owing to their diverse role in farming systems and nutritional safety, the research on grain legumes could exert critical effects on nutritional safety and soil fertility (Nedumaran et al. 2015).

7 The Beneficial Roles of Bacteria in BNF

N-fixing bacteria can transform the atmospheric N into fixed N that is readily available to the plant for utilization in critical metabolic processes. These microbes play a crucial role in the BNF process, thus regulating the N cycle in nature. N-fixing bacteria are generally categorized into two types: The free-living bacteria that do not establish a symbiotic relationship with plants, also known as nonleguminous bacteria. They include species like *Cyanobacteria, Nostoc* and *Anabaena,* and the genera include *Azotobacter*, *Clostridium,* and *Beijerinckia*. The second type of N-fixing bacteria consists of the symbiotic (leguminous) species. They are further categorized according to their host plant families, such as (a) *Rhizobium which are established on* various members of the *Fabaceae* family; (b) *Frankia* which generally harbor dicotyledonous species mainly including actinorhizal plants; and (c) *Azospirillum* species which infect the roots of cereal grasses.

The symbiotic N-fixing species have been thoroughly investigated due to their exceptional potential for supplying N through the BNF process carried out at the root zone of the host plants. These bacteria penetrate and multiply inside the root hairs and initiate nodule formation, which are tumor-like organs. These nodules later serve as the BNF factories. Bacteria inside the nodules convert free N into ammonia that is utilized by host plants for healthy growth. This process has been commercialized, and legume seeds are artificially treated with commercial strains of *Rhizobium* species with profound effects on microbially deficient soils (Kumar et al. 2018; Encyclopaedia Britannica 2019).

Mutualistic N fixation is an important phenomenon for fixing atmospheric N sustainably, and major enzyme nitrogenase is responsible for fixation and conversion of atmospheric N into ammonical form (Oldroyd et al. 2011; Udvardi and Poole 2013). Plants of leguminous species have special organs in their roots called nodules, where enzymes are present. This fixation occurs due to the mutualistic host plant relationship between *Mesorhizobium*, *Bradyrhizobium,* and Rhizobia. Such symbiotic relation between the host plant and Rhizobacteria gives benefit to both participants as the host plant gives the energy required for the growth of bacteria, while in response, bacteria fix atmospheric N into a plant-available form. This symbiotic relationship is economically beneficial because it might reduce the fertilizer input cost (Van Hameren et al. 2013).

8 Pesticides and Their Impact on Sustainable Crop Production

Pesticides can be classified based on targeted agents. For example, herbicides are used to control herbs/weed plants, insecticides for keeping insects below the threshold level, fungicides for the control of different types of seed-borne and soil-borne fungi, rodenticides for killing the rodent (animals), molluscicides against animals of the Mollusca group, and nematicides for controlling a variety of nematode worms. Based on chemical formulation, pesticides can be broadly categorized as organochlorines, carbamates, organophosphates, pyrethroids, and neonicotinoids (Shakir et al. 2018). The classification of pesticides based on chemical structures is presented in Foo and Hameed (2010)). The impact of different chemicals on N-fixation in a variety of legumes has been documented in Table 1.

According to Silver and Riley (2001), different routes by which pesticides contaminate the environment are (a) Air/wind drift by which pesticides are carried away from the target area; (b) Volatilization, a process through which pesticides get evaporated from the soil and plant foliage and are re-deposited on nontarget vegetation or soil; (c) Leaching, the movement of pesticides through soil due to water movement which gets in-filtered into soil's deeper horizons; and (d) Runoff through which pesticides get washed off from the soil or plants by rainfall and continue to be deposited into nearby water bodies.

9 Factors Affecting Pesticide Toxicity in Plants' Rhizosphere

Pesticide applications affect the populations and activities of soil microorganisms. These effects vary by the concentration, active ingredients, toxicity, and mode of use of the pesticides. Generally, soil fumigants and fungicides offer a more drastic challenge to the soil microbial communities as compared to herbicides. Long-term experiments and overpopulation dynamics have indicated that pesticides do not

Table 1 Impact of various agrochemicals on the rate of N-fixation in legume plants

Type of chemical	Chemical used in the study	Studied plant	Impact on N fixation	References
Fungicide	• Carbendazim • Thiram	Red clover (*Trifolium pratense* L.)	44–63% (−)	Niewiadomska (2004)
Herbicide	• Imazetapir			
Insecticide	• methyl parathion Dichlorodiphenyltrichloroethane (DDT)	Alfalfa (*Medicago sativa*) var. Iroquois	~50% (−)	Fox et al. (2007)
Pesticide	• pentachlorophenol (PCP) • Chrysin			
Fungicide	• bisphenol			
Herbicide	• Flumetsulam • Imazethapyr • Fluazifop-P • Butroxydim • Haloxyfop-R • Diflufenican • Trifluralin	Field pea (*Pisum sativum* L.)	34–60% (−)	Drew et al. (2008)
Insecticide	• Fipronil • Pyriproxyfen	Chickpea (*Cicer arietinum* L.)	15–28% (−)	Ahemad and Khan (2009)
Fungicide	• Benomyl • Captan • Carbendazin • Carboxin • Difenoconazole • Thiabendazole • Thiram • Tolylfluanid	Soybean (*Glycine max*)	15% (−)	Campo et al. (2009)
Insecticide	• Fipronil • Pyriproxyfen	Pea (*Pisum sativum* L.)	20–27% (−)	Ahemad and Khan (2010)
Fungicide	• Captan	Chickpea (*Cicer arietinum* L.)	N rate 1.58% (+), Grain yield 9.6% (+)	Sharma (2012)
Insecticide	• Chlorpyrifos • Endosulfan			
Insecticide	• Thiamethoxam (THIA)	Soybean (*Glycine max*)	Seed germination and N-fixation (14–30%) (+)	Zhou et al. (2013)
Pesticide	• Chlorpyrifos • Cypermethrin	Mung bean (*Vigna radiata*)	29–64% (−)	Singh et al. (2015)

(−): decreased, (+): increased

cause critical declines in microbial populations. However, their growth and physiological activities are retarded to some extent. A flow chart showing the sequence of events from pesticide entry into a plant to the death of the plant (Ahemad et al. 2009).

10 Pesticide Concentrations in Soil

Previously, it has been documented that a higher rate of pesticides causes more pronounced losses to the soil microbes (Muturi et al. 2017). However, the recommended dose of the pesticides for field applications, when tested in the laboratory experiments, was found to be satisfactory (Damalas and Eleftherohorinos 2011). Limitations in the field application of pesticide experiments could be their uneven distribution in soil. However, generally, it is either lower or safe. Thus, a better understanding of the impact of pesticides needs more augmentation of the concentration of the pesticide in the tested soils and microbial population dynamics in advance. The distribution of pesticides in the soil is also a factor that determines the fate of soil microbes. The amount of the active ingredients per square meter of the soil, the chemistry of the pesticide, the rate, and methods of application determine the fate of the microbes inside the soil (Meena et al. 2020a). Earlier studies have indicated that preplant soil application of 1–2 lb. acre^{-1} of pesticide distributes it over the soil surface about 1.5–3 inches with an average concentration of 1–4 ppm. However, postemergence pesticide applications initially become more concentrated at the soil surface. Later, it is gradually removed by rainfall or irrigation water (Milan et al. 2019). Seed treatment with fungicides and granular pesticides is less distributed compared to the drench or preemergence pesticides. Moreover, they are also subjected to biodegradation. These factors help to compensate for the pesticide toxicity in the soil. Adsorption or soil binding also reduces the direct impact of pesticides over the rhizosphere.

11 Pesticides and Soil Microbial Activity

The response of soil microorganisms can be classified as tolerant (resistance) or sensitive (susceptible). Pesticides adversely affect the vulnerable microbial species by interfering with physiological and biochemical processes. For example, the herbicide dinoseb disrupts the electron transport chain (ETC) process that ultimately deprives the cells of ATP (Roach and Krieger-Liszkay 2014). Dinoseb has also been reported to invade or interfere with soil algae (Marin-Morales et al. 2013; Mitran et al. 2018).

Many nonsystemic fungicides act in a nonselective manner. Since fungicides have been designed to target fungi, hence there are ample chances that it will affect the nontarget species as well. However, systemic fungicides are more selective in this case. Metalaxyl and CGA 29212 [methyl-N- (2-chloroacetyl)-N-(2,6-xylxyl)-

DL-alinate) have been found inhabiting the ribonucleic acid (RNA) synthesis in soil communities. Around 25–30% of soil fungi and actinomycetes are reduced due to benomyl application. Fungal species are known to be more affected by pesticide applications as compared to the bacteria, though, in some cases, bacterial populations could be promoted (Agnihotri 1971). Contact herbicides are less inhibitory to the soil microbes. Hence they become toxic under specific conditions; for instance, when they are spilled or excessively applied, they even become injurious to the host plants (Machado-Neto 2015).

Glyphosate, amitrole, and the imidazolinone and sulfonylurea herbicides act by inhibiting the synthesis of essential amino acids in target plants and soil microbes (Vencill et al. 2012; Shaner et al. 2018). Dinitroaniline herbicides and carbamates disrupt cell division by inhibiting the formation of microtubules in plants and some fungi (Tadeo et al. 2008; Chatterji et al. 2011). Similarly, the antifungal activity of alachlor, metolachlor, and propachlor has also been reported by Munoz et al. (2011), Elsayed et al. (2014).

12 Effect of Pesticides on Soil Nitrogen Availability

Management of soil N is crucial for crop production. Nevertheless, the impacts of herbicides, fungicides, insecticides, and nematicides on N transformations in the soil have been extensively studied by several researchers (Srinivasulu et al. 2012; Ju et al. 2017; Madakka et al. 2017). It was concluded that most of these pesticides could not interfere with N availability and nitrification unless applied more than the recommended doses. However, soil fumigants, including methyl bromide and/or chloropicrin, have a significant impact on N mineralization and nitrification (Pingmei et al. 2012; Li et al. 2017). Even after a long period, N mineralization was found to be recovered, but nitrification remained depressed, indicating the loss of N-fixing species in the soil. Dithiocarbamates, if applied repeatedly or applied in overdoses, may inhibit N mineralization, thus reducing the ammonium and nitrate levels in soil (Černohlávková et al. 2009; Walia et al. 2014). Corke and Thompson (1970) reported a 2–4 days lag in the nitrification process after a low dose application of 3,4-dichloroaniline. The above-mentioned chemicals may pose toxicity to the soil microbes at concentrations as little as 10 ppm. Higher rates than this may inhibit soil respiration and glucose metabolism. This phenomenon indicates the nonselective behavior of pesticides over beneficial soil fungi, N-fixing bacteria, and actinomycetes.

13 Effects of Pesticides on Nitrogen-Fixing Bacteria

13.1 Growth

It is estimated that legume crops get their 30–70% N by root nodulating symbiotic bacteria (Ham et al. 1975). Conventional pesticides for legume crops include seed and foliar fungicides and a few herbicides. These chemicals potentially retard the growth of N-fixing *Rhizobium* and *Brady Rhizobium* bacterial species, ultimately affecting the plant health. Seed priming with fungicides also poses an alarming situation for the leguminous species of N-fixing bacteria. High concentrations (≥100 ppm) of captan, carboxanilidoamino-thiazole, oxycarboxin, carboxin, fuberidazole, folpet, thiabendazole, and propiconazole have been reported to inhibit the growth of *R. trifolii* (Heinonen-Tanski et al. 1982). In a few instances, folpet, captan, and propiconazole were found inhibitory even at ≥10 ppm.

13.2 Root Nodulation

The effects of fungicide seed treatments on nodulation are variable. Truman, sodium diethyldithiocarbamate, and carbaryl have been reported to inhibit the nodulation by *Rhizobium sp*. However, it was dependent upon the concentration of the applied fungicide (Cevheri et al. 2011; Kisiel and Kępczyńska 2016). In some studies, carboxin, captan, and PCNB have reduced nodule formation due to their seed-priming effects (Dubey 2006; Kalia and Gosal 2011). It has been reported that dichlone caused reduced N-fixation due to less nodule formation (Eberbach 2018). However, certain cases have been reported of little or no toxicity of some fungicides on the nodulation process in leguminous crops. In a previous study, thiram and captan slightly reduced the nodule formation in soybean tap roots, but this inhibition was not observed in chloranil, or ethyl mercury chloride treated plants (Aamil et al. 2004; Narayanasamy 2013; Eberbach 2018). Similarly, it is noted that seed treatment of captan and thiram had no significant role in nodulation by *B. japonicum*. Apart from other N-fixing bacteria, *Rhizobium* strains have been found more sensitive to the fungicide's applications during the seed treatment process. Even the recommended doses of PCNB, carboxin, thiram, and captan can inhibit the nodulation process and survival of these species within few hours of treatment (Cevheri et al. 2011; Kisiel and Kępczyńska 2016). Therefore, researchers suggested the wise use of fungicides for *Rhizobium sp*. in legume crops. The use of fungicide-resistant strains of *Rhizobium* has already been recommended for optimized yield in a few legumes (Deaker et al. 2004; Duc et al. 2015; Shahid et al. 2020). Herbicides have also been known to inhibit the nodulation process of *Rhizobium* and *Bradyrhizobium* in legumes (Zawoznik and Tomaro 2005; Brechenmacher et al. 2008; Eberbach 2018). Though, it is believed that herbicide-induced retardation in the nodulation process is indirect. They are correlated to the plant response to the herbicides, or it might be that herbicide treatment of plants indirectly affects the nodulating species systemically. It is also suggested that herbicide injuries cause an overall reduction in photosynthesis, which ultimately minimizes the nodulation process (de María et al. 2005; Zobiole et al. 2012).

14 Strategies to Maximize Nitrogen Fixation Associated with Pesticide Application

Sustainable agriculture seeks to improve N fixation and nutrient-use efficiency besides reducing the environmental costs associated with pesticide application. One of the strategies to optimize BNF is the improvement of pasture productivity by liming and fertilization, which favors the growth of legume and ryegrass to the detriment of the native grasses (Meena et al. 2015). The increase in the proportion of ryegrass and clover may not only improve the quality of the available forage but, in the case of the clover, stimulates N contribution from BNF, thereby reducing the need for N fertilizer (Campillo et al. 2005).

Another strategy is based on utilizing alternate N sources, including BF. N fixation through nonsymbiotic means by employing soilborne free-living bacteria may assist in fulfilling a part of the N requirement. This approach is especially suitable for low-input farming systems globally. The isotope-based direct methods detected agronomically significant quantities of N_2 fixation both in perennial grass systems and annual crops as well (Mitran et al. 2018; Fatima et al. 2019). The novel molecular technologies may support the development of new plant-diazotrophic combinations for specific environments and more sustainable exploitation of N-fixing bacteria as an inoculant for agriculture. A well-structured diversity for the nifH-harboring diazotroph community was shown in all plant parts of summer-growing perennial grasses such as *Panicum coloratum* L. cv. *Bambatsi* (Bambatsi panic), *Chloris gayana* Kunth cv. *Katambora* (Rhodes grass) and *Digitaria eriantha* Steud. cv. *Premier* (Premier digit grass). Such is an example of a functional endophytic community (Gupta et al. 2019). Compared to the commensal bacteria, new diazotroph mutants with enhanced capabilities to excrete ammonium are being successfully used to promote the growth of cereal plants (maize, rice, wheat, and sorghum) (Rosenblueth et al. 2018). There is a current perspective of introducing nitrogenase genes into significant nonleguminous crop plants as a more realistic shorter-term strategy to better synchronize plant-microbe interactions for enhanced N_2 fixation, especially when the N needs of the plant are greatest (Ropera and Gupta 2016).

A comprehensive field trial has been performed involving the technique of seed inoculum by utilizing N-fixing bacterium isolated from sugarcane intercellular juice, which is not only non-nodulating but also non-rhizobial. *Gluconacetobacter diazotrophicus* indicates that it can significantly improve yields of wheat, maize, oilseed rape, and grasses, in both the presence and absence of synthetic fertilizers.

Such results are based on an increased rate of photosynthesis with a possible combination of intracellular SNF with additional plant growth factors (Meena et al. 2018; Dent and Cocking 2017).

The other management practices that influence BNF in agricultural production systems are intercropping systems. Here we will discuss some pesticides that improve the ability of N-fixing bacteria and are useful for BNF. Appropriate agronomic management holds the potential to reduce the drastic effects of pesticides on soil microbes and the BNF process (Babalola 2010). The first and foremost thing that needs to be kept into consideration is the threshold level of different insect pests. The application of pesticides should only be recommended when economic losses are feared to outstrip environmental concerns. An agronomic management plan which integrates different approaches instead of using only pesticides can offer a biologically and economically viable option to reduce the heavy use of pesticides. It is worth mentioning that fungicides are especially harmful to rhizobacteria and thus need to be used with due care (Peiffer et al. 2013; Meena et al. 2018). Besides, to strive for a reduction in the amount of pesticide use, it is also direly needed to consider the residual life of various pesticides. After a thorough examination, only those pesticides should be applied, which have a minimum residual life, especially in the soil. Besides, pesticide residues have been reported to reduce the harvest yields along with increasing the need for more and more synthetic fertilizers application, thereby leading to the significant raising of costs for farmers. The higher cost of production is not only disadvantage as heavy uses of pesticides contribute significantly to environmental pollution as well. The observations may also explain a trend in the past 40 years toward stagnant crop yields despite record-high use of pesticides and synthetic fertilizers worldwide. The connection between different types of pesticides and N-fixation by legumes has indicated that this pristine and natural interaction between bacterial strains involved in BNF and host plants are being jeopardized by various types of pesticides that are being put into the soil, while agronomic management has the potential to reverse this trend. Different insecticides belonging to the groups of chlorinated hydrocarbon and organophosphate have been thoroughly studied to find out their effect on N-fixing bacteria under both conditions (in isolated states as well as with the host plant). A study has revealed that insecticides severely damage rhizobacteria, though different insecticides vary in lethality, and thus only those insecticides which minimally inflict the BNF process should be used (Babalola et al. 2007).

A study has shown that agronomic management practices have the potential to reduce the drastic impact of fungicides on the functioning of root nodules. It was concluded that different cultivars of forage legumes having specified genomic setup could significantly affect the rhizobial population in the rhizosphere, which may lead to higher N-fixation. The release of a single exogenous glucosinolate from legumes improved the microbial community on transgenic *Arabidopsis* roots in which fungal and α-proteobacteria were predominantly affected, as revealed by denaturing gradient gel electrophoresis (DGGE). Another study reported that the ABC transporter mutant of *Arabidopsis*, abcg30, has the potential to produce root exudates containing higher concentrations of phenolic compounds and low amounts of sugars, which give a boost to a unique rhizosphere microbiome (Rincon-Florez et al. 2013).

15 Impacts of Agronomic Practices on the Physiology of Legume Nitrogen Fixation

Usually, legumes are affected by the abundance of *Rhizobia* in the soil. Their interaction depends on the soil properties (Farhangi-Abriz and Torabian 2018), and land preparation before cultivation (Sidiras et al. 1999). Rowland et al. (2015) observed that nodules in *Arachis hypogeal* L. were smaller in CT, compared to conservation tillage. Tillage helps to reduce soil temperature as high temperature could accelerate the senescence of soybean nodules (Bordeleau and Prevost 1994). Weisz and Sinclair (1994) elucidated that nitrogenase activity is significantly influenced by soil temperature. Aranjuelo et al. (2007) reported that depression of root hair formation and reduced adherence of bacteria to root hair occurs as secondary inhibitions. Matus et al. (1997) studied the influence of tillage on N-fixation and found that it was higher by 10 and 30% for lentil and pea, respectively, when they were grown using zero tillage compared to conventional tillage. According to Matus et al. (1997), lentil fix 12% more N than less diversified crop rotations.

When barley was grown in rotation with vetch, N uptake was increased by 61% (Papastylianou 2004). Li et al. (2018) in an 8 years rotation study compared legume monoculture systems to crop rotation with annual legumes such as chickpea (*Cicer arientinum* L.), pea (*Pisum sativum* L.), and lentil (*Lens culinaris* L.). However, legume monoculture improved soil N accumulation by 38% average as compared to diversified systems. Organic mulching on the soil surface mainly influences environmental factors (water content, temperature, nutrient status, aeration, etc.), the growth, and functions of legumes (Lipiec and Hatano 2003). The mulching with straw indirectly affects nodulation and N_2 fixation by changing soil physical, chemical, and biological environments (Siczek and Lipiec 2011).

A study has been conducted using soybean to evaluate soil compaction, and mulching effects of nodulation and nitrogenase activity, and protein content results revealed that nitrogenase activity decreased with increasing soil compaction. Whatever the compaction level, mulching increased the nitrogenase activity in Rhizobia. Soil compaction and mulching enhanced the large nodules greater than0.41 cm) and dry weight of individual nodules (Siczek and Lipiec 2011). In the field conditions, nodule number and dry mass reduced with the increase of fertilizer N in soybean. Nodulation and N fixation are adversely affected by the addition of NO_3^- or NH_4^+ fertilizers (Zahran 1999). Addition of 10 kg N ha^{-1} depressed acetylene reduction in feather moss (Rousk et al. 2014). Inorganic N addition tends to stimulate nitrogenase activity in cowpea at the vegetative stage (Hassan 2017). High nitrogenase activity was recorded with 25 kg N ha^{-1} and more than 100 kg N ha^{-1} prolongedly reduced nitrogenase activity. In another study, the optimum rate of N for mung bean was

40 kg N ha^{-1} (Dudeja 2014). Bradirhizobium-soybean symbiosis increases N concentration and number of root nodules when calcium and magnesium silicate were applied as fertilizer (Steiner et al. 2018).

16 Conclusion

This chapter portrays some pedo-environmental factors that affect BNF activity, taking place in root nodules of the leguminous species under pesticide application. It appraises the correlation between the harmful effects of such factors on plant growth and development. These factors can cause a severe decrease in microbial population and activity. Synthetic pesticides have an adverse effect on SNF and the amount of N fixed through BNF in different legume species. It is suggested to understand the interactive effects of soil N status concerning preemergence and postemergence pesticide application in soil. This will help in weed and pest control sustainably. Moreover, it is suggested to identify microorganisms tolerant to soil pesticide residues in addition to abiotic stresses which can establish symbiotic relationships with leguminous crop species. This will boost crop productivity and secure food security.

Conflict of Interest No conflict of interest is hereby declared.

References

Aamil M, Zaidi A, Saghir Khan M (2004) Fungicidal impact on chickpea: Mesorhizobium Symbiosis. J Environ Sci Health 39:779–790. https://doi.org/10.1081/lesb-200030867

Agnihotri VP (1971) Persistence of captan and its effects on microflora, respiration, and nitrification of a forest nursery soil. Can J Microbiol 17:377–383

Ahemad M, Khan MS (2009) Effect of insecticide-tolerant and plant growth-promoting Mesorhizobium on the performance of chickpea grown in insecticide stressed alluvial soils. J Crop Sci Biotechnol 12:217–226

Ahemad M, Khan MS (2010) Comparative toxicity of selected insecticides to pea plants and growth promotion in response to insecticide-tolerant and plant growth promoting *rhizobium leguminosarum*. Crop Prot 29:325–329

Ahemad M, Khan MS (2013) Pesticides as antagonists of rhizobia and the legume-rhizobium symbiosis: a paradigmatic and mechanistic outlook. Biochem Mol Biol 1:63–75

Ahemad M, Khan MS, Zaidi A, Wani PA (2009) Remediation of herbicides contaminated soil using microbes. In: Khan MS, Zaidi A, Musarrat J (eds) Microbes in sustainable agriculture. Nova Sci Pub, New York

Akibonde S, Maredia M (2011) Global and regional trends in production, trade and consumption of food legume crops. Department of Agricultural, Food and Resource Economics, Michigan State University, Michigan, p 83

Alpmann D, Braun J, Schafer BC (2013) Analyse einer Befragung unter erfolgreichen Körnerleguminosen anbauern im konventionellen Landbau. Erste Ergebnisse aus dem

Forschungsprojekt LeguAN. In: DLG W (ed) Im Fokus. Heimische Kornerleguminosen vom Anbau bis zur Nutzung, Berlin

Angus JF, Kirkegaard JA, Hunt JR, Ryan MH, Ohlander L, Peoples MB (2015) Break crops and rotations for wheat. Crop Pasture Sci 66:523–552

Aranjuelo I, Irigoyen JJ, Sánchez-Díaz M (2007) Effect of elevated temperature and water availability on CO_2 exchange and nitrogen fixation of nodulated alfalfa plants. Environ Exp Bot 59: 99–108

Babalola OO (2010) Beneficial bacteria of agricultural importance. Biotechnol Lett 32:1559–1570

Babalola OO, Sanni AI, Odhiambo GD, Torto B (2007) Plant growth-promoting rhizobacteria do not pose any deleterious effect on cowpea and detectable amounts of ethylene are produced. World J Microbiol Biotechnol 23:747–752

Batut J, Mergaert P, Masson-Boivin C (2011) Peptide signalling in the rhizobium-legume symbiosis. Curr Opin Microbiol 14:181–187

Bender DA (2006) Benders' dictionary of nutrition and food technology, 8th edn. Woodhead Publishing, New York, p 552

Bloom AJ (2015) The increasing importance of distinguishing among plant nitrogen sources. Curr Opin Plant Biol 25:10–16

Bordeleau LM, Prevost D (1994) Nodulation and nitrogen fixation in extreme environments. Plant and Soil 161:115–215

Brechenmacher L, Kim MY, Benitez M, Li M, Joshi T, Calla B, Lee SH (2008) Transcription profiling of soybean nodulation by *Bradyrhizobium japonicum*. Mol Plant-Microbe Int 21:631–645

Campillo R, Urquiaga S, Undurraga P, Pino I, Boddey RM (2005) Strategies to optimise biological nitrogen fixation in legume/grass pastures in the southern region of Chile. Plant and Soil 273: 57–67

Campo RJ, Araujo RS, Hungria M (2009) Nitrogen fixation with the soybean crop in Brazil: compatibility between seed treatment with fungicides and bradyrhizobial inoculants. Symbiosis 48:154–163

Černohlávková J, Jarkovský J, Hofman J (2009) Effects of fungicides mancozeb and dinocap on carbon and nitrogen mineralization in soils. Ecotoxicol Environ Saf 72:80–85

Cevheri C, Küçük Ç, Çetin E (2011) Fungicide, antibiotic, heavy metal resistance and salt tolerance of root nodule isolates from *Vicia palaestina*. Afr J Biotechnol 10:2423–2429

Chatterji BP, Jindal B, Srivastava S, Panda D (2011) Microtubules as antifungal and antiparasitic drug targets. Expert Opin Ther Pat 21:167–186

Corke CT, Thompson FR (1970) Effects of some phenylamide herbicides and their degradation products on soil nitrification. Can J Microbiol 16:567–571

Damalas CA, Eleftherohorinos IG (2011) Pesticide exposure, safety issues, and risk assessment indicators. Int J Environ Res Public Health 8(5):1402–1419

Das A, Ghosh PK (2012) Role of legumes in sustainable agriculture and food security: an Indian perspective. Outlook Agric 41(4):279–284

Das S, Ali MM, Rahman MH, Khan MR, Hossain A, Ayman ES, Barutcular C (2018) Soil test based with additional extra nutrients increased the fertility and productivity of wheat-mung bean-aman rice cropping pattern in the high Ganges River floodplain of Bangladesh. Bulg J Agric Sci 24:992–1003

de María N, de Felipe MR, Fernández-Pascual M (2005) Alterations induced by glyphosate on lupin photosynthetic apparatus and nodule ultrastructure and some oxygen diffusion related proteins. Plant Physiol Biochem 43:985–996

Deaker R, Roughley RJ, Kennedy IR (2004) Legume seed inoculation technology-a review. Soil Biol Biochem 36:1275–1288

Dent D, Cocking E (2017) Establishing symbiotic nitrogen fixation in cereals and other non-legume crops: the greener nitrogen revolution. Agric Food Secur 5:6–7. https://doi.org/10.1186/s40066-016-0084-2

Drew EA, Gupta VVSR, Roget DK (2008) Herbicide use, productivity, and nitrogen fixation in field pea (*Pisum sativum*). Aust J Agr Res 58:1204–1214

Dubey SC (2006) Integrating bioagents with plant extract, oil cake and fungicide in various modes of application for the better management of web blight of urdbean. Arch Phytopathol Plant Prot 39:341–351

Duc G, Agrama H, Bao S, Berger J, Bourion V, De Ron AM, Tullu A (2015) Breeding annual grain legumes for sustainable agriculture: new methods to approach complex traits and target new cultivar ideotypes. Crit Rev Plant Sci 34:381–411

Dudeja SS (2014) Fertilizer N and nitrogen fixation in legume-rhizobium symbiosis. Ann Biol 9:2

Eberbach PL (2018) The effect of herbicides and fungicides on legume-rhizobium symbiosis In: Altman J Pesticide interactions in crop production: beneficial and deleterious effects. CRC Press, New York, 183-212

EL Sabagh A, Hossain A, Islam MS, Fahad S, Ratnasekera D, Meena RS, Wasaya A, Yasir TA, Ikram M, Mubeen M, Fatima M, Nasim W, Çığ A, Çığ F, Erman M, Hasanuzzaman M (2020) Nitrogen fixation of legumes under the family Fabaceae: adverse effect of abiotic stresses and mitigation strategies. In: Hasanuzzaman M, Araújo S, Gill S (eds) The plant family Fabaceae. Springer, Singapore, pp 75–111

Elsayed OF, Maillard E, Vuilleumier S, Imfeld G (2014) Bacterial communities in batch and continuous-flow wetlands treating the herbicide S-metolachlor. Sci Total Environ 499:327–335

Encyclopaedia Britannica (2019) Nitrogen-fixing bacteria. https://www.britannica.com/science/nitrogen-fixing-bacteria

FAO (Food and Agriculture Organization) (2016) Food and agriculture organization. FAO, Rome

FAOFAOSTAT (2016) FAOSTAT. http://faostat3.fao.org/download/Q/QC/E. Accessed 6 Jul 2016

Farhangi-Abriz S, Torabian S (2018) Biochar improved nodulation and nitrogen metabolism of soybean under salt stress. Symbiosis 74(3):215–223

Fatima P, Mishra A, Hari Om Bholenath S, Kumar P (2019) Free living nitrogen fixation and their response to agricultural crops. In: Kaushik BD, Deepak K, Shamim MD (eds) Biofertilizers and biopesticides in sustainable agriculture. Apple Academic Press, Palm Bay, p 460

Foo KY, Hameed BH (2010) Detoxification of pesticide waste via activated carbon adsorption process. J Hazard Mater 175:1–11

Fox JE, Gulledge J, Engelhaupt E, Burow ME, McLachlan JA (2007) Pesticides reduce symbiotic efficiency of nitrogen-fixing rhizobia and host plants. Proc Natl Acad Sci USA 104:10282–10287

Galloway JN, Townsend AR, Erisman JW, Bekunda M, Cai Z, Freney JR, Sutton MA (2008) Transformation of the nitrogen cycle: recent trends, questions, and potential solutions. Science 320(5878):889–892

Garrigues E, Corson MS, Walter C, Angers DA, van der Werf H (2012) Soil quality indicators in LCA: method presentation with a case study. In: Corson MS, van der Werf HMG (eds) Proceedings of the 8th international conference on life cycle assessment in the Agri-food sector, 1–4 October 2012. INRA, Saint Malo, pp 163–168

Glick BR (2012) Plant growth-promoting bacteria: mechanisms and applications. Forensic Sci 2012:963401. https://doi.org/10.6064/2012/963401

Gupta P, Diwan B (2017) Bacterial exopolysaccharide mediated heavy metal removal: a review on biosynthesis, mechanism and remediation strategies. Biotechnol Rep 13:58–71

Gupta V, Zhang B, Penton CR, Yu Jand Tiedje JM (2019) Diazotroph diversity and nitrogen fixation in summer active perennial grasses in a Mediterranean region agricultural soil. Front Mol Biosci 6:115. https://doi.org/10.3389/fmolb.2019.00115

Hajslova J, Zrostlikova J (2003) Matrix effects in ultra-trace analysis of pesticide residues in food and biotic matrices. J Chrom A 1000:181–197

Ham GE, Liener IE, Evans SD, Frazier RD, Nelson WW (1975) Yield and composition of soybean seed as affected by N and S fertilization. Agron J 67:293–297

Hassan MA (2017) Application of biofertilization and biological control for cowpea production. Ann Agric Sci Moshtohor 55:271–286

Heinonen-Tanski H, Oros G, Kesckes M (1982) The effect of soil pesticides on the growth of red clover rhizobia. Acta Agric Stand 32:283–288

Hobbs PR, Ken S, Raj G (2007) The role of conservation agriculture in sustainable agriculture. Philosophical transactions of the Royal Society B. Biolo Sci 1491:543–555. https://doi.org/10.1007/978-981-13-0253-4_1

Hossain A, EL Sabagh A, Erman M, Fahad S, Islam T, Bhatt R, Hasanuzzaman M (2020) Nutrient Management for Improving Abiotic Stress Tolerance in legumes of the family Fabaceae. In: Hasanuzzaman M, Araújo S, Gill S (eds) The plant family Fabaceae. Springer, Singapore, pp 393–415

Hossain MS, Hossain A, Sarkar MAR, Jahiruddin M, Teixeira da Silva JA, Hossain MI (2016) Productivity and soil fertility of the rice–wheat system in the high Ganges River floodplain of Bangladesh is influenced by the inclusion of legumes and manure. Agric Ecosyst Environ 218: 40–52

Iqbal MA, Hamid A, HussainI SMH, Ahmad T, Khaliq A, Ahmad Z (2019) Competitive indices in cereal and legume mixtures in a south Asian environment. Agron J 111(1):242–249

Islam MA, Adjesiwor AT (2017) Nitrogen fixation and transfer in agricultural production systems. In: Amanullah FS (ed) Nitrogen in Agriculture-Updates. Intech Open, Rijeka. https://doi.org/10.5772/intechopen.71766

Jensen ES, Peoples MB, Boddey RM, Gresshouse PM, Hauggaard-Nielsen H, Alves BJ, Morrison MJ (2012) Legumes for mitigation of climate change and the provision of feedstock for biofuels and biorefneries-a review. Agron Sustain Dev 32:329–364

Jeuffroy MH, Baranger E, Carrouée B, de Chezelles E, Gosme M, Hénault C, Cellier P (2013) Nitrous oxide emissions from crop rotations including wheat, oilseed rape and dry peas. Biogeosciences 10(3):1787–1797

Ju C, Xu J, Wu X, Dong F, Liu X, Tian C, Zheng Y (2017) Effects of hexaconazole application on soil microbe's community and nitrogen transformations in paddy soils. Sci Total Environ 609: 655–663

Kalia A, Gosal SK (2011) Effect of pesticide application on soil microorganisms. Arch Agron Soil Sci 57:569–596

Kisiel A, Kępczyńska E (2016) Medicago truncatula Gaertn. As a model for understanding the mechanism of growth promotion by bacteria from rhizosphere and nodules of alfalfa. Planta 243:1169–1189

Kumar S, Meena RS, Lal R, Yadav GS, Mitran T, Meena BL, Dotaniya ML, EL Sabagh A (2018) Role of legumes in soil carbon sequestration. In: Meena RS et al (eds) Legumes for soil health and sustainable management. Springer, Singapore, pp 109–138. https://doi.org/10.1007/978-981-13-0253-4_4

La-Favre JS, Focht DD (1983) Conservation in soil of H_2 liberated from N_2 fixation by H up-nodules. Appl Environ Microbiol 46:304–311

Lemke RL, Zhong Z, Campbell CA, Zentner RP (2007) Can pulse crops play a role in mitigating greenhouse gases from north American agriculture? Agron J 99:1719–1725

Li H, Rodda M, Gnanasambandam A (2015) Breeding for biotic stress resistance in chickpea: progress and prospects. Euphytica 204:257. https://doi.org/10.1007/s10681-015-1462-8

Li J, Deng M, Wang Y, Chen W (2016) Production and characteristics of biosurfactant produced by bacillus pseudomycoides BS6 utilizing soybean oil waste. Int Biodeter Biodegr 112:72–79

Li J, Huang B, Wang Q, Li Y, Fang W, Yan D, Cao A (2017) Effect of fumigation with chloropicrin on soil bacterial communities and genes encoding key enzymes involved in nitrogen cycling. Environ Pollut 227:534–542

Li J, Liu K, Zhang J, Huang L, Coulter JA, Woodburn T, Li L, Gan Y (2018) Soil-plant indices help explain legume response to crop rotation in a semiarid environment experiment. Front Plant Sci 9:1–13. https://doi.org/10.3389/fpls.2018.01488

Lipiec J, Hatano R (2003) Quantification of compaction effect on soil physical properties and crop growth. Geoderma 116:107–136. https://doi.org/10.1016/S0016-7061(03)00097-1

Machado-Neto JG (2015) Safety measures for handlers/workers against herbicide intoxication risk, herbicides, physiology of action, and safety, andrew price, jessica kelton and lina sarunaite. Intech Open, Rijeka. https://doi.org/10.5772/61464

Madakka M, Jayaraju N, Rangaswamy V (2017) Changes in the metabolic activities of two agricultural soils as influenced by the pesticides and insecticides combination. Appl Soil Ecol 120:169–178

Marin-Morales MA, de Campos B, Ventura C, Miyuki HM (2013) In: Price AJ, Kelton JA (eds) Toxicity of herbicides: impact on aquatic and soil biota and human health, herbicides-current research and case studies in use. Intech Open, Rijeka. https://doi.org/10.5772/55851

Matus A, Derksen DA, Walley FL, Loeppky HA, van Kessel C (1997) The influence of tillage and crop rotation on nitrogen fixation in lentil and pea. Can J Plant Sci 77(2):197–200

Meena R, Yadav R, Reager M (2015) Temperature use efficiency and yield of groundnut varieties in response to sowing dates and fertility levels in western dry zone of India. Ame Exp Agric 7:170–177. https://doi.org/10.9734/ajea/2015/13689

Meena RS, Kumar S, Datta R, Lal R, Vijayakumar V, Britnicky M, Sharma MP, Singh GS, Jahariya MK, Jangir CK, Pathan SI, Dokulilova PV, Marfo TD (2020) Impact of agrochemicals on soil microbiota and management: a review. Landarzt 9:34. https://doi.org/10.3390/land9020034

Meena RS, Kumar V, Yadav GS, Mitran T (2018) Response and interaction of *Bradyrhizobium japonicum* and *arbuscular mycorrhizal* fungi in the soybean rhizosphere: a review. Plant Growth Regul 84:207–223

Meena RS, Lal R (2018) Legumes and sustainable use of soils. In: Meena RS, Das A, Yadav GS, Lal R (eds) Legumes for soil health and sustainable management. Springer, New York. https://doi.org/10.1007/978-981-13-0253-4_1

Meena RS, Lal R, Yadav GS (2020a) Long-term impacts of top soil depth and amendments on soil physical and hydrological properties of an Alfisol in Central Ohio, USA. Geoderma 363: 1141164. https://doi.org/10.1016/j.geoderma.2019.114164

Merga B, Haji J (2019) Economic importance of chickpea: production, value, and world trade. Cog Food Agri 5(1):1615718. https://doi.org/10.1080/23311932.2019.1615718

Messina MJ (1999) Legumes and soybeans: overview of their nutritional profiles and health effects. Am J Clin Nutr 70(3 Suppl):439S–450S

Milan M, Ferrero A, Fogliatto S, Piano S, Negre M, Vidotto F (2019) Oxadiazon dissipation in water and topsoil in flooded and dry-seeded rice fields. Agron 9(9):557. https://doi.org/10.3390/agronomy9090557

Mitran T, Meena RS, Lal R, Layek J, Kumar S, Datta R (2018) Role of soil phosphorus on legume production. In: Meena et al (eds) Legumes for soil health and sustainable management. Springer, New York. https://doi.org/10.1007/978-981-13-0253-4_15

Molla MSH, Nakasathien S, Ali MA, Khan A, Alam MR, Hossain A, Farooq M, El Sabagh A (2019) Influence of nitrogen application on dry biomass allocation and translocation in two maize varieties under short pre-anthesis and prolonged bracketing flowering periods of drought. Arch Agron Soil Sci 65(07):928–944

Mondal M, Skalicky M, Garai S, Hossain A, Sarkar S, Banerjee H, Kundu R, Brestic M, Barutcular C, Erman M, EL Sabagh A, Laing AM (2021) Supplementing nitrogen in combination with rhizobium inoculation and soil mulch in Peanut (Arachis hypogaea L.) production system: part II. Effect on phenology, growth, yield attributes, pod quality, profitability and nitrogen use efficiency. Agronomy 10(10):1513. https://doi.org/10.3390/agronomy10101513

Munoz A, Koskinen WC, Cox L, Sadowsky MJ (2011) Biodegradation and mineralization of metolachlor and alachlor by Candida xestobii. J Agric Food Chem 59:619–627

Muturi EJ, Donthu RK, Fields CJ, Moise IK, Kim CH (2017) Effect of pesticides on microbial communities in container aquatic habitats. Sci Rep 7:44565. https://doi.org/10.1038/srep44565

Narayanasamy P (2013) Biological disease management systems for agricultural crops. In: Biological management of diseases of crops. Springer, Dordrecht, pp 189–235

Nedumaran S, Abinaya P, Jyosthnaa P, Shraavya B, Rao P, Bantilan C (2015) Grain legumes production, consumption and trade trends in developing countries. In: ICRISAT research program, markets, institutions and policies. International Crops Research Institute for the Semi-Arid-Tropics, Patancheru

Niewiadomska A (2004) Effect of Carbendazim, Imazetapir and Thiram on Nitrogenase activity, the number of microorganisms in soil and yield of red clover (*Trifolium pratense* L.). Polish J Environ Stud 13:403–410

Oldroyd GE, Murray JD, Poole PS, Downie JA (2011) The rules of engagement in the legume-rhizobial symbiosis. Annu Rev Genet 45:119–144

Papastylianou I (2004) Effect of rotation system and N fertiliser on barley and vetch grown in various crop combinations and cycle lengths. Agric Sci 142:8–41

Peiffer JA, Spor A, Koren O, Jin Z, Tringe SG, Dangl JL, Buckler ES, Ley RE (2013) Diversity and heritability of the maize rhizosphere microbiome under field conditions. Proc Natl Acad Sci U S A 110:6548–6553

Perveen S, Ahmad S, Skalicky M, Hussain I, Habibur-Rahman M, Ghaffar A, Shafqat Bashir M, Batool M, Hassan MM, Brestic M, Fahad S, EL Sabagh A (2021) Assessing the potential of polymer coated urea and Sulphur fertilization on growth, physiology, yield, oil contents and nitrogen use efficiency of sunflower crop under arid environment. Agronomy 11(2):269. https://doi.org/10.3390/agronomy11020269

Pingmei Y, Teng Z, Yuan L, Dongdong Y, Liangang M, Qiuxia W, Aocheng C (2012) Effects of methyl bromide fumigation on community structure of denitrifying bacteria with nitrous oxide reductase gene (nosZ) in soil. Afr J Microbiol Res 6:2095–2100

PlazaBonilla D, Jean-Marie N, Didier R, Eric J (2017) Innovative cropping systems to reduce N inputs and maintain wheat yields by inserting grain legumes and cover crops in southwestern France. Eur J Agron 82:331–341

Prodhan MD, Akon MW, Alam SN (2018) Determination of pre-harvest interval for quinalphos, malathion, diazinon and cypermethrin in major vegetables. J Environ Anal Toxicol 553:2161–0525

Raza A, Razzaq A, Mehmood SS, Zou X, Zhang X, Lv Y, Xu J (2019) Impact of climate change on crops adaptation and strategies to tackle its outcome: a review. Plan Theory 8(2):34. https://doi.org/10.3390/plants8020034

Reed SC, Cleveland CC, Townsend AR (2011) Functional ecology of free-living nitrogen fixation: a contemporary perspective. Annu Rev Ecol Evol Syst 42:489–512. https://doi.org/10.1146/annurev-ecolsys-102710-145034

Rincon-Florez VA, Carvalhais LC, Schenk PM (2013) Culture-independent molecular tools for soil and rhizosphere microbiology. Diversity 5:581–612

Roach T, Krieger-Liszkay A (2014) Regulation of photosynthetic electron transport and photoinhibition. Curr Protein Pept Sci 15(4):351–362. https://doi.org/10.2174/1389203715666140327105143

Ropera MM, Gupta VVSR (2016) Enhancing non-symbiotic N2 fixation in agriculture. Open Agri J 10:7–27

Rosenblueth M, Ormeño-Orrillo E, López-López A, Rogel MA, Reyes-Hernández BJ, Martínez-Romero JC, Reddy PM, Martínez-Romero E (2018) Nitrogen fixation in cereals. Front Microbiol 9:1794. https://doi.org/10.3389/fmicb.2018.01794

Rousk K, Jones DL, DeLuca TH (2014) Exposure to nitrogen does not eliminate N_2 fixation in the feather moss *Pleurozium schreberi* (Brid.) mitt. Plant and Soil 374:513–521

Rowland DL, Smith C, Cook A, MasonA SA, Bennett J (2015) Visualization of peanut nodules and seasonal nodulation pattern in different tillage systems using a minirhizotron system. Peanut Sci 42:1–10

Ryu MH, Zhang J, Toth T, Khokhani D, Geddes BA, Mus F, Voigt CA (2020) Control of nitrogen fixation in bacteria that associate with cereals. Nat Microbiol 5:314–330

Salim N, Raza A (2020) Nutrient use efficiency (NUE) for sustainable wheat production: a review. J Plant Nutr 43:297–315

Salinas Valdés A, De la Rosa MJ, Serna Saldívar SO, Chuck Hernández C (2015) Yield and textural characteristics of panela cheeses produced with dairy vegetable protein (soybean or peanut) blends supplemented with transglutaminase. J Food Sci 80:S2950–S2956

Shahid M, Khan MS, Zaidi A (2020) Fungicide toxicity to legumes and its microbial remediation: a current perspective. In: Pesticides in crop production: physiological and biochemical action. John Wiley & Sons, Hoboken, pp 15–33

Shakir SK, Irfan S, Akhtar B, ur Rehman S, Daud MK, Taimur N, Azizullah A (2018) Pesticide-induced oxidative stress and antioxidant responses in tomato (*Solanum lycopersicum*) seedlings. Ecotoxicology 27(7):919–935

Shaner DL, Bascomb NF, Smith W (2018) Imidazolinone-resistant crops: selection, characterization, and management. In: Herbicide-resistant crops. CRC Press, Boca Raton, pp 143–158

Sharma P (2012) Influence of pesticide-treated seeds on survival of *Mesorhizobium* sp. Cicer, symbiotic efficiency and yield in chickpea. Plant Prot Sci 48:37–43

Siczek A, Lipiec J (2011) Soybean nodulation and nitrogen fixation in response to soil compaction and surface straw mulching. Soil Tillage Res 114(1):50–56. https://doi.org/10.1016/j.still.2011.04.001

Sidiras N, Avgoulas C, Bilalis D, Tsougrianis N (1999) Effects of tillage and fertilization on biomass, roots, N-accumulation and nodule bacteria of vetch (*Vicia sativa* cv. Alexander). Agron Crop Sci 182:209–2016

Silver J, Riley B (2001) Environmental impact of pesticides commonly used on urban landscapes. Northwest Coali Altern Pesti 3:8–16

Singh B, Singh JP, Shevkani K, Singh N, Kaur A (2017) Bioactive constituents in pulses and their health benefits. J Food Sci Technol 54:858–870. https://doi.org/10.1007/s13197-016-2391-9

Singh S, Gupta R, Kumari M, Sharma S (2015) Nontarget effects of chemical pesticides and biological pesticide on rhizospheric microbial community structure and function in *Vigna radiata*. Environ Sci Pollut Res 22:11290–11300

Srinivasulu M, Mohiddin GJ, Subramanyam K, Rangaswamy V (2012) Effect of insecticides alone and in combination with fungicides on nitrification and phosphatase activity in two groundnut (*Arachis hypogeae* L.) soils. Environ Geochem Health 34:365–374

Stagnari F, Maggio A, Galieni A (2017) Multiple benefits of legumes for agriculture sustainability: an overview. Chem Biol Technol Agric 4:1–13. https://doi.org/10.1186/s40538-016-0085-1

Steiner F, Mario ZA, Bush A, Diego Muniz da Silva S (2018) Silicate fertilization potentiates the nodule formation and symbiotic nitrogen fixation in soybean. Agron Sustain Dev 48:212–221

Sultana T, Anowara B, Humaira A (2019) Effect of pesticides on exopolysaccharide (EPS) production, antibiotic sensitivity and phosphate solubilization by Rhizobial isolates from *Sesbania bispinosa* in Bangladesh. Afr J Agric Res 14:1845–1854. https://doi.org/10.5897/AJAR2019.14304

Tadeo JL, Sánchez-Brunete C, González L (2008) Pesticides: classification and properties. In: Analysis of pesticides in food and environmental samples. CRC Press, Boca Raton, pp 13–46

Tang W, Wang D, Wang J, Wu Z, Li L, Huang M, Yan D (2018) Pyrethroid pesticide residues in the global environment: an overview. Chemosphere 191:990–1007

Udvardi M, Poole PS (2013) Transport and metabolism in legume-rhizobia symbioses. Annu Rev Plant Biol 64:781–805

Umeta M, West CE, Fufa H (2005) Content of zinc, iron, calcium and their absorption inhibitors in foods commonly consumed in Ethiopia. J Food Compos Anal 1:803–817

Van Hameren B, Hayashi S, Gresshoff PM, Ferguson BJ (2013) Advances in the identification of novel factors required in soybean nodulation, a process critical to sustainable agriculture and food security. J Plant Biol Soil Health 1(1):1–6

Vanlauwe B, Hungria M, Kanampiu F, Giller KE (2019) The role of legumes in the sustainable intensification of African smallholder agriculture: lessons learnt and challenges for the future. Agric Ecosyst Environ 284:106583

Vencill WK, Nichols RL, Webster TM, Soteres JK, Mallory-Smith C, Burgos NR, McClelland MR (2012) Herbicide resistance: toward an understanding of resistance development and the impact of herbicide-resistant crops. Weed Sci 60:2–30

Walia A, Mehta P, Guleria S, Chauhan A, Shirkot CK (2014) Impact of fungicide Mancozeb at different application rates on soil microbial populations, soil biological processes, and enzyme activities in soil. Scientific World J 2014:1–9. https://doi.org/10.1155/2014/702909

Weisz PR, Sinclair TR (1994) Soybean nodule gas permeability, nitrogen fixation and diurnal cycles in soil temperature. Plant Soil 161:115–125

Zahran HH (1999) Rhizobium -legume Symbiosis and nitrogen fixation under severe conditions and in an arid climate. Microbiol Mol Biol Rev 63(4):968–989

Zawoznik MS, Tomaro ML (2005) Effect of chlorimuron ethyl on *Bradyrhizobium japonicum* and its symbiosis with soybean. Pest Manag Sci 61:1003–1008

Zhou GC, Wang Y, Zhai S, Ge F, Liu ZH, Dai YJ, Yuan S, Hou JY (2013) Biodegradation of the neonicotinoid insecticide thiamethoxam by the nitrogen-fixing and plant-growth-promoting rhizobacterium *Ensifer adhaerens* strain TMX-23. Appl Microbiol Biotechnol 97:4065–4074

Zobiole L, Kremer RJ, de Oliveira Jr RS, Constantin J (2012) Glyphosate effects on photosynthesis, nutrient accumulation, and nodulation in glyphosate-resistant soybean. J Plant Nutr Soil Sci 175:319–330

Sustainable Urban Forestry, Merits, Demerits, and Mitigation of Climate Change at Global Scale

Mujahid Farid, Asma Qadri, Umer Hayat, Muhammad Bilal Shakoor, Mohsin Abbas, Aiman Masood, Kinza Mehak, Afifa Khizar, and Nighat Raza

Abstract The worldwide environmental changes and increasing levels of carbon dioxide are some of the key natural concerns of the date. In the present article, the evaluation of urban woods organization was utilized to measure the capability of urban biological system as carbon sink and moderating environmental changes through carbon digestion, and the capability of the framework to upgrade biodiversity preservation. Recently settled regions appeared to have lower carbon stockpiling potential, while territories set up before have most elevated carbon stockpiling potential. Around 36 diverse tree species developing/planted as urban timberland were recognized, ruled by *Senna siamea*, *Azadirachta indica*, *Polyalthia longifolia*, *Leucaena leucocephala*, *Pithecellobium* and *Mangifera indica*. Aside from being common enhancement the tree species as carbon dioxide (CO_2) sink through photosynthesis and territories of ex-situ preservation through plants, urban ranger service can store huge amount of carbon anyhow biodiversity preservation particularly where they spread broad regions like parks, nurseries, and roads oversaw over extensive stretches, similar to the case in urban environments. Improved administration of urban forests will probably improve the potential for carbon storing by earthly vegetation as a method for alleviating CO_2 outflows and environmental change just as biodiversity protection. It is appeared here that such a system may contain the board practices that think about the net impact of both diminishing weakness and outflows. In the biophysical domain, these practices may incorporate, however, are not constrained to planting more trees, planting better adjusted species to the future atmosphere, lessening tree worry, among others.

M. Farid (✉) · A. Qadri · U. Hayat · M. Abbas · A. Masood · K. Mehak · A. Khizar
Department of Environmental Sciences, University of Gujrat, Hafiz Hayat Campus, Gujrat, Pakistan

M. B. Shakoor
College of Earth and Environmental Sciences, University of Punjab, Lahore, Punjab, Pakistan

N. Raza
Department of Food Science and Technology, Muhammad Nawaz Shareef University of Agriculture, Multan, Pakistan

M. Hasanuzzaman et al. (eds.), *Managing Plant Production Under Changing Environment*, https://doi.org/10.1007/978-981-16-5059-8_15

Keywords Biological system · Carbon dioxide · Biodiversity · Urban forest · Carbon sink

1 Introduction

The green plant life is a brilliant structure and power supply of forest system and persist organisms from bacteria to fungi, arthropods, and vertebrates, are the similarly essential infrastructure and conduits of vitamins and power (Adrees et al. 2015; Owusu and Asumadu-Sarkodie 2016). There has been a multiplied emphasis on sustainable towns. One thing of a sustainable metropolis is the inclusion of trees as part of the more urban environment (Afshan et al. 2015; Davoudi et al. 2009). Scientists have made important development at some point in the previous few decades in studying the role of city forests in each mitigating urbanization's impact and providing a diversity of environmental services (Fatima et al. 2020a, b; Millar et al. 2007). However logical comprehension of key components administering environment capacities over numerous scales is deficient. There are significant tradeoffs across a scale and between capacities. There's also incredible variability across metropolitan regions and biophysical regions (Maalik et al. 2020; Locatelli et al. 2015).

National Academy of Science conducted a workshop on February 25–26, 2013, In his initial comments, Gary Allen (Executive Director of the Center for Chesapeake Communities and Chair of the workshop arranging panel), noticed that expansion in populace can bring about the greater part of the accessible land that being used for constructing and "hard" framework advancement. Due to the development of urban areas, trees and the natural environment are regularly lost, and with them important biological administrations, just as all the advantages originating from those administrations. Almost a number of stakeholders are paying attention to such issues that have proven helpful in understanding our logical and specialized abilities. This incorporates various government organization programs (Zari 2010; Young and McPherson 2013).

It likewise incorporates a huge cluster of government and nearby ranger service and supervision of land, alongside private establishments, nonlegislative associations, and academic investigators. Huge numbers of the human-environment interaction arising from urbanization are not yet surely known (Ahmad et al. 2019). The main task for what's near come is to create techniques for "manageable administration" of urban biological systems that can carry a fine and healthy tree shelter and sound, protected, different conditions for the individuals existing in urban areas (Zomer et al. 2008).

The 2010 evaluation detailed that about 81% of Americans presently live in urban areas, up from 79% only 10 years sooner. Over this equivalent time period, urban populaces developed by over 12.1%, speed up the national development normal of simply 9.7%. Unmistakably we are turning into a progressively urbanized country. By lessen noise and giving spots to reproduce, urban woodlands enhance social union, incentive community renewal, and increase the value of our networks (Ashfaq et al. 2020; Leung et al. 2011).

2 What Is Forestry?

Forestry is the act of overseeing woodland lands for different utilizations, including business, horticultural, and public. Forestry is exercised in plantations and characteristic stands. The study of forestry has components that have a place in physical, social, political, and administrative science. Forestry is a significant economic segment in various industrial countries. For example, 33% of the woodland spread on the land in Germany, wood is the significant inexhaustible asset, and forestry carried the abundance of a million occupations and about €181 billion of huge worth to the German economy consistently (Cameron et al. 2012).

3 Urban Forestry

Worldwide climate warming has become an undeniable truth. The worldwide normal temperature has kept on increasing at an uncommon rate over the most recent four decades, and the worldwide mean temperature is probably going to increment by 1.1–6.0 °C in the following century, particularly in regions where urbanization is increasing. Dangerous atmospheric warming straightforwardly gives rise to the expansion in air temperature and surface temperature, which will influence the survival, development, advancement, and multiplication of urban vegetation. Because of the distinctive development stages and formative conditions, the adjustment techniques of seedlings to the environment are not completely relevant to adult trees (Fig. 1). In this manner, it is important to examine the reaction of urban greening plants to warming (Ezaz et al. 2020; Depietri and McPhearson 2017).

Urban forestry is commonly characterized as the science, art, and innovation of leading trees and woods assets in all over urban network biological systems for the aesthetic, physiological, and sociological advantages trees give to the community (Amir et al. 2020). Urban ranger service has a more extended past in North America, depend on the customs of shade tree the management. Also, urban ranger service has gotten increasingly systematized in North America. Urban ranger service in Europe has manufactured strongly on a century-long custom of 'town ranger service' (Heller and Zavaleta 2009).

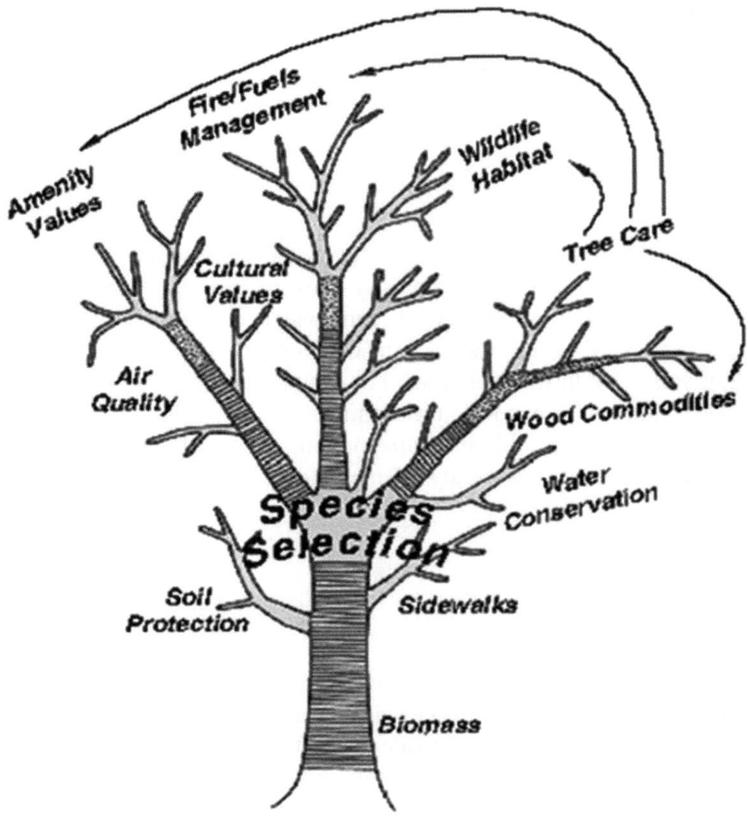

Fig. 1 The elements of sustainability in urban forestry (Thompson et al. 1994)

4 What Is "Sustainable Urban Forest?"

The Sustainable Urban Forest incorporates everything expected to guarantee that the whole forest framework accomplishes and keeps up a healthy expanse and structure adequate to give the ideal advantages, or environment administrations, after some time. This can incorporate such crossing areas as waste depletion and reusing, management of stormwater, vitality use, air and water quality, wildlife natural habitat, general well-being, monetary feasibility, social value, etc. (Blum and Janaki 2017).

At the nearby metropolitan level, obviously, the supporting monetary assets themselves are frequently in extraordinary need of a sustainable supply. Generally, when financial plans are stressed or contracting, numerous districts have seen satisfactory consideration of the urban forests, beyond fundamental tree risk management, as a luxury, they can't manage. However, that image is changing with an expanded awareness of the wide-going advantages of trees, particularly when a dollar value is put on those advantages (Bakht et al. 2020); Perceiving and esteeming the substantial administrations gave by the urban forest, plainly there is a lot to be obtained by a healthy forest—and genuine, quantifiable, and unavoidable expenses if the venture isn't made to look after it. Creation and the administration of urban forests to accomplish manageability is the drawn-out objective of urban foresters just as chose authorities, business pioneers, and residents (Raman et al. 2012).

5 Green Infrastructure

"A deliberately organizes and managed framework of native lands, running terrains and other wide areas that preserve environmental values and tasks and supply correlated advantages to human beings" (Niemelä et al. 2010).

Green foundation uses vegetation to give a portion of the elements of traditional grey infrastructure, for example, water treatment plants and management framework of stormwater, however, with extra natural, economic, and social advantages. Comprehensively characterized, each tree and the whole urban woods can be viewed as a green framework. The green framework can be intended to address a wide scope of concerns, including water amount and quality, air contamination, vitality preservation, GHG emissions, and general health. In numerous pieces of the nation, the green framework is frequently intended to address urban stormwater issues—thus the more explicit term "green stormwater foundation" (Price et al. 2015).

6 Elements of Sustainability in Urban Forestry

There are the following elements of urban forestry (Swart et al. 2003):

1. Selection of species and diversity
2. Inventory and planning of landscape
3. Tree cover and wood consumption
4. Public relation and maintenance

6.1 Selection of Species and Diversity

The significance of appropriate species of tree determination is likely the utmost effective component of sustainable development.

The following picture represents the relationships among different species determination and urban forestry standards and exercises. For instance, those species which are not chosen deliberately with the planting place and natural impacts as a primary concern frequently bring about intrusion with cities (e.g., walkways, electrical cables, lighting, and traffic signs), dangerous fire scenes, cost of high tree conservation, air quality, and short life issues (Khalid et al. 2020; Swart et al. 2003).

These harmful impacts can be stayed away from through genuine choice and decent variety. Joining these connections in the species determination procedure can extraordinarily upgrade sustainability. Obviously, the whole framework, that is the urban forest biological system, rises above its parts when working appropriately. From a considerably more extensive point of view, appropriate species choice must liken to an expanded mix of animal types. This implies making arrangements for a given site the "best" must incorporate environment level organizing to keep away from animal husbandry impact on the land. An ongoing study reveals that the 90% of trees planted in California, in around three-fourths of reacting urban areas, comprised of only five or less species. Metropolitan foresters are working with organizers and with different city administrations in a different turn of events and modification of the current city foundation to suit urban forestry's service's needs (Swart et al. 2003).

6.2 *Inventory and Planning of Landscape*

Another component of sustainable urban ranger service incorporates stock and arrangement. An ongoing report showed that just 38% of the city foresters distinguished with any exactness what number of road trees was in their city. Steenberg et al. (2019) stated that seeing in what way the urban woodland is formed and allocated (e.g., the age, appropriation of urban trees, and condition) will give significant data on following

- Trees support sequences.
- The possessions on further city administrations (e.g., electrical cables and street lights).
- The woodland construction and organization for pathogen "spread" models.
- Strategic arranging and planning.
- The urban woodlands for open collaborations the initial phase in the arranging of the forest is to comprehend the position of the current woods. Ordinary data documented on urban forest inventories are species, area, age, size, ailment, and support history. Extra helpful things are live wood, accessibility and nature of planting site, nearness of above wires, and explanation behind trimming. One of the significant objectives in forestry service is to deal with forests so yields of every woodland esteem are supported over the long haul to make a "managed" forest. The expression "guideline" is used in deciding and the consistency and manageability of the forest's yields.

The standards utilized in planning urban woods to accomplish supportable yields and qualities are genuinely easy to see yet challenging to execute (Steenberg et al. 2019). The standards are as follows.

- class age control appropriation via greatest (pivot) age by classes
- class age control and species dispersion topographically
- regulates the executives 'applies to keep up woods well-being and development.

In wildland settings, this is a complex function because of the various configuration of the woodland and various use value goals forced on the scenery. In wildland, the executives; GIS (Geographic Information Systems) has been an important apparatus in actualizing besides assessing attempts to construction and deal with a maintainable forest environment. The genuine capability of a geographic information system relies upon its capacity to assemble numerous strata of asset information "overlaps" so the client can understand the influences of strategy and the board techniques before execution.

6.3 Tree Cover and Wood Consumption

Appropriate service, management, elimination, and urban forest planting, containing related bushes, are significant achievements in making progress toward sustainability. Urban forestry can't endure such expenses or a "cost-picture." Over the utilization of reasonable species choices and usage of urban forests deposits, trees activities can be economically sustainable and, now and again, gainful. The situation is urging to take into consideration that dumping wood leftover has deteriorated as of late (Steenberg et al. 2019).

The utilization of pruning as covering, firewood, and different biomass has developed in modern years. A portion of the explanations behind this is

- Expanding landfill expenses (over $20 per cubic yard or about $45 per ton, and rising).
- Community realization of waste problems (the government objective is to decrease strong leftovers half by 2000).
- Land-fill banning (accessible land-fill space is exceptionally obliged in certain city areas).

Usage of garbage removal as a strong wood crude material increases the value of the urban woodland (and income to the urban) when the quality of wood, species, and volume is adequate (Hiloidhari et al. 2019).

6.4 Public Relations and Maintenance

Sustainable development can't be accomplished dependent happening on the initial three components. It must be acknowledged when a strong base of help is accessible from people in general and chose authorities. A large amount can be consumed on expanding simulations to plan sustainable urban forests. In any case, it is a cost without advantage except if equivalent energy is disbursed on building political help in the networks (Joosten et al. 2012).

6.4.1 Internal Organization and Funding

Another significant methodology is to teach the town forestry workers, comprising the field team, on the long-term estimation of high feature, on-opening up to the public relations. There is not a viable alternative for growing great development of public with communities (Joosten et al. 2012).

6.4.2 External Organization and Funding

Various techniques for constructing a political help depend on the general population must be very effective. A genuinely new methodology in building open contribution and backing is to the network of city forestry arranging and exercises over the association of homeowner (HOAs) and other comparative locality elements. Meanwhile, a large portion of the city woodland is under isolated consideration, efforts to include isolated residents and gatherings will make the best influence in cultivating circumstances (Joosten et al. 2012).

7 Advantages of Urban Forestry

7.1 Livelihood

Urban forests conditions give tastefully satisfying environmental factors, expanded happiness regarding regular day-to-day existence, and a more prominent feeling of association among individuals and the common habitat. Trees are among the most significant highlights that contribute to the stylish nature of private avenues and network parks. Urban forests make significant commitments to the essentialness and character of a city, neighborhood, or on the other hand region. Besides, the demonstration of planting and thinking about trees, when attempted by inhabitants, yields significant social benefits and a more grounded feeling of network. Also, strengthening to improve neighborhood conditions in downtowns has been credited to inclusion in urban ranger service endeavors (Joosten et al. 2012).

7.2 Possible Expansion

Diminished pressure and better physical well-being for urban inhabitants have been related to the nearness of urban trees and backwoods. Scenes with trees furthermore, other vegetation has created progressively loosened up the mental states in people than scenes without these characteristic highlights. Emergency clinic patients with window perspectives on trees recuperated altogether quicker and with fewer complexities than similar patients without such perspectives. It results in the expansion of the areas covering urban forests (Joosten et al. 2012).

7.3 *Source of Material*

Urban forests are the source of providing useful material like food, wood, and medicines, etc.

- Forest foods include vegetables, honey, fruits, spices, and nuts. Timberland nourishments are sound, as they have numerous significant nutrients. Sometimes individuals experience difficulty in getting food or cash, for instance, if a flood ruins their yields. In the event that woodlands close by have numerous normal nourishments, individuals don't need to go hungry.
- Forests are the source of fuelwood as wood is the source of energy for heating and cooking purposes. Forests also provide fodder for grazing animals and cattle.
- Medicinal plants are important for fulfilling the health-based needs of people. Forest trees contain a wide variety of bioactive compounds which have the potential to use as antioxidants and anticancer medications. Some trees which are used to make medicines include tulsi, peppermint, neem, lavender, and cinnamon (Fatima et al. 2020a, b; Joosten et al. 2012).

7.4 *Job Opportunities*

Urban forests assets contribute to the monetary value of a neighborhood, city, or development. By improving the quality of the environment, trees add to expanded property estimations, deals by organizations, and work. Urban forests also provide jobs to flourish local economy, limbs, and woods from trees are useful in making many wooden products or fuel that can be used by residents, providing many jobs in this process (Joosten et al. 2012).

7.5 *Contributes to Industrialization*

Trees diminish vitality requirements for warming or cooling structures by shading structures in the midyear, lessening summer air temperatures by blocking winds. Trees likewise can build warming needs, be that as it may, by concealing structures in the winter whenever planted in inappropriate areas near structures. The vitality impacts of trees differ with territorial atmosphere and their area around the structure (Espeland and Kettenring 2018).

Urban trees decrease carbon dioxide a significant nursery gas, by legitimately expelling it from the climate and putting away the carbon in the trees as biomass. By decreasing building vitality use, trees can likewise decrease the discharge of carbon dioxide from power plants. Tree-upkeep exercises frequently require the utilization of petroleum products that discharge carbon dioxide, nonetheless, and inappropriately found trees around structures can expand vitality requests and resulting outflows of carbon dioxide. When there is a lower risk of environmental pollution, it will expand the activities like industrialization (Olsen 2007).

8 Disadvantages

8.1 Destroy Habitat of Animals and Birds

Land advancement fundamentally modifies the urban scene influencing plant and wildlife populations, and timberland biodiversity and well-being. Advancement can lead to quick decreases in tree populations freeing from timberland stands, can change species structure, e.g., tree planting after advancement can expand tree populations, e.g., tree planting in the past cleared regions, and can adjust the urban condition, e.g., increment or abatement in air temperatures (Olsen 2007).

People are presently answerable for the causing of changes in the conditions that hurt wildlife. We introduce exotic species into biological habitats; these activities expel resources and domains from plants and animals. Human action regularly changes or wrecks the natural surroundings that wildlife needs to cope with (Olsen 2007).

8.2 Flood and Fires

Uncontrolled forest fires can make huge harm to trees and timberlands and significantly change the urban scenery, particularly in urban zones adjoining wildlands. Large population development and urban extensions in California, for instance, have prompted a significant increment in fire starts in wildland urban interface zones. What's more, the blending of trees with made structures in these zones fundamentally confuses and limits the alternatives accessible for fire concealment exercises and management practices for vegetation used to decrease fire hazards. Urban forests have been damaged by excessive floods, wind, snow, and ice storms (Pires 2019).

8.3 Build-Up of GHG

Air and water pollutants can affect tree fitness in urban regions if pollutant concentrations attain damaging degrees. Forests had been proven to be laid low with air pollutants mainly from local deposition of nitrogen, sulfur, ozone, and hydrogen (Farid et al. 2020a, b). Ozone has been reported to decrease tree development, reduce protection from bark creepy crawly and increment defenselessness to dry season. Air pollutants also can beautify tree boom via accelerated levels of carbon dioxide or by providing essential plant nutrients, which include sulfur and nitrogen (Duguma et al. 2014).

8.4 Affect the Discovery of Herbal Medicines

Factors like environmental pollution, excessive floods, and storms damage the quality of plants or even causing the death of several plants resulting in difficulty in the discovery of herbal plants.

8.5 Affect Water Cycle

Urban forests have an important role in maintaining hydrological processes by slowing down the precipitation flow to the ground. They are beneficial in reducing the volume and rate of storm runoff and damage from flooding. During massive rain occasions, the percentage of rainfall interception can drop to a totally small percent as the maximum of the rain reaches the floor. But excessive flooding, storms, wildfires, and developmental activities may result in disturbing groundwater tables hence affect the water cycle (Alberti 2010; Duguma et al. 2014).

8.6 Environmental and Community Harms

It is expected that climate change is to provide changed precipitation patterns, hotter air temperatures, and greater precipitation events and extreme temperatures. These weather changes can cause changes in city forest composition and feature the potential to exacerbate other urban forest threats like pests and invasive species (Farid et al. 2017a, b). Climate change has the ability to adjust urban forests, now not handiest through species changes, however also thru direct results from storms and floods that can kill huge quantities of the woodland in relatively brief time periods. Urban wooded area managers will need to understand and adapt to capability species shifts and changes to the surroundings to supply sustainable and healthful urban forests underneath future climatic situations (Duguma et al. 2014).

Because a group of people at once manage a maximum of the city forest, the decisions of the managers directly have an effect on city forest health and composition. Improper choices associated with the selection of species, locations of trees, and protection can result in disputes with the population of the city and infrastructure, harm to many trees, and terrible health of trees that can cause premature mortality of trees. Actions taken by many urban landowners can threaten urban forests, however, they can also help to flourish city wooded area health and sustainability if the right tree management is conducted (Farid et al. 2019; Dale et al. 2020).

9 Urban Forest Management Plan

Trees on roads and on the other openly claimed places oversaw by open works organizations give a large number of tasteful and natural advantages to residents, organizations, and guests the same. Past shade and magnificence, trees additionally have down-to-earth benefits and genuine money-related worth those urban areas once in a while are uninformed of—your urban timberland offers important open types of assistance and could be the cost of over a million dollars. In contrast to other open foundation parts, appropriately planted and kept up trees increment in esteem after some time (Latif et al. 2020; Dale et al. 2020). An urban wood, the executive's plan, in light of ongoing tree stock information and examination of accessible staff, hardware, and spending assets, is a basic device for ensuring this important asset. Urban woodland the executives plan is an activity plan; it gives open works organizations point by point data, suggestions, and assets expected to viably and proactively oversee open trees (Kenney et al. 2011; Dale et al. 2020).

The motivation behind having an urban wood the executives plan is to guarantee that a network will appreciate the advantages of trees through appropriate arboricultural methods and the board rehearses. The goal of the game plan is to state what is relied upon to manage the urban woods and to portray activities and organizations required to execute these obligations. In the event that an administration plan depends on examination from an exact tree stock and created with contribution from open works staff, agroforestry specialists, and the residents, at that point the open work's organization liable for the urban woodland will acknowledge numerous advantages (Farid et al. 2015, 2017c; Keenan 2015).

9.1 Increased Public Safety

Every open work organization realizes that a huge piece of their essential crucial to guarantee well-being and oversee hazards identified with an open foundation. A tree stock and the executive's plan will give arrangements of trees requiring need expulsion and pruning that a supervisor can do inside the constraints of financial plan and time. The stock can be utilized in this way to screen trees for dangers on a nonstop premise. By actualizing suggestions made in the administration plan, storm harm dangers will likewise decay (Blum and Janaki 2017).

9.2 *Increased Efficiency*

When a stock has recognized the work to be done and an administration plan has endorsed a support program, a supervisor can execute that work in a significantly more effective way than previously. By booking all work in an offered zone to be done simultaneously (instead of by responding to single demands) the investment funds in movement and arrangement time are considerable, with chronicled models appearing around a 50% decrease in cost—particularly when an arrangement of rotational work, as well as safeguard, upkeep is received. There is likewise expanded effectiveness in the workplace made by utilizing an electronic stock to find and control records and select and plan work. The proficient reaction to resident demands and questions likewise improves client assistance (Blum and Janaki 2017).

9.3 *Justify Budget*

Urban woodland the board plan gives the information and investigation expected to decide explicit degrees of financing for tree support and tree planting anticipated over a multiyear time span. With precise information, a supervisor can build up, organize, and legitimize yearly spending demands. The undertakings and related expenses are obviously illuminated in the arrangement and can be bolstered by point-by-point records. Numerous open works supervisors have discovered that they have a lot more noteworthy accomplishments with spending demands that depend on the investigation of excellent information. Likewise, a decent stock gives a strong premise to concede applications (Duinker et al. 2013).

9.4 *Components of Management Plan*

The segments and varieties of urban timberland the board plans are many, contingent upon the formative phase of the urban ranger service program inside an open works office. For the most part, these components are incorporated or tended to in the arrangement (Blum and Janaki 2017).

1. Tree stock information and examination
2. Tree stock and planning information the executives programming

3. Tree chance decrease/crisis storm reaction plan
4. A tree board or warning committee improvement
5. Advertising and training
6. Urban woodland cost/advantage investigation

The parts and varieties of urban woods the executive's plans are many, contingent upon the formative phase of the urban ranger service program inside an open works organization. For the most part, these components are incorporated or tended to in the arrangement:

9.4.1 Tree Inventories

Open tree inventories are a factually dependable overview of freely possessed and oversaw trees, used to decide the area and the specific or evaluated estimations of amount, quality, well-being, and patterns of the urban woods, just as a portrayal of other urban woodland characteristics, for example, potential planting destinations, utilities present, and hardscape highlights.

Information normally gathered during a stock incorporates:

- Location
- Species
- Diameter
- Condition
- Maintenance need
- Sidewalk and other hardscape harm
- Insect and malady issues
- Potential planting destinations

Checklists are commonly finished via prepared certified experienced stock arborists. The tree trait and area information are commonly gathered utilizing handheld PCs, geographic data frameworks (GIS) information, or potentially geographic situating frameworks (GPS) hardware.

Sorts of checklists are expending on the size of locale and assets, there are various kinds of inventories that can be cultivated to furnish you with precise bookkeeping of open trees.

9.4.2 Statistical Sample Inventories

A factually stable, arbitrary example of urban timberland is a practical method of getting a general image of the condition of the trees. For the most part, acquiring information from between 3% and 6% of road miles or potentially open property land will create results that are exact to within 10% of what a total stock would deliver.

9.4.3 Partial Inventories

Fractional inventories gather tree information on 100% of the option to proceed miles or sections of land, yet just in explicit territories of a network. At the point when spending plans are constrained, this methodology can be successful and moderate. The openworks office chooses which characterized regions of the city or province are stocked: specific wards, neighborhoods, locale, memorable territories, and so forth. Utilizing halfway inventories permits the organization to spread the stock procedure over some stretch of time contingent upon accessible assets and assets.

9.4.4 Using and Managing the Inventory Data

Utilizing financially accessible tree the board GIS-based resource the executives programming programs, basic database programs or PC spreadsheet programs, open works organizations can utilize the stock information to make work reports, plan tree upkeep and planting assignments, track costs, and proficiently react to resident solicitations.

Overseeing and refreshing stock information and work requests can involve a noteworthy venture of time and cash, so open works directors need to deliberately consider who will play out this undertaking, and what yields are wanted, and afterward, select a framework that is perfect with the current organization abilities and methods. At the point when the correct tree stock information the executive's framework is chosen, open works supervisors can utilize the information for long-go, proactive wanting to guarantee the proceeded with magnificence, essentialness, well-being, and endurance of every single open tree.

9.4.5 Inventory Data Analysis

A noteworthy part of an urban woodland board plan is an expert investigation of the tree stock information. By and large, the factual investigation is performed bringing about various tables and diagrams delineating the tree populace's attributes. At that point, in light of that investigation and the outcomes, upkeep and planting needs are created, and by and large administration proposals are made for a multi-year time frame. Following is a depiction of the stock information investigation part of an administration plan.

Characteristics of Population

The open urban woods are a complex, between the related arrangement of site conditions, trees, and other foundation parts. By knowing this dynamic framework is significant for legitimate dynamic with respect to proper tree care works on, planting choices, and urban woodland the executives. The open tree populace attributes segment of an administration plan gives understanding into the current organization and state of a stocked tree populace (Khair et al. 2020).

The qualities of the urban woods incorporate conditions, species, sizes, and other related tree and site factors. By distinguishing the species, size, and state of trees in the urban woods, much is uncovered about the woodland's structure, relative age, and well-being. It is significant for open works directors to know the sorts of trees just as the quantity of trees present. Species synthesis information is fundamental since tree species fluctuate extensively in the future and upkeep needs. The kinds of trees present in a network extraordinarily influence tree upkeep exercises and financial plans. Correspondingly, tree distance, size, and class information help to characterize the general age and size appropriation of the all-out tree populace.

By breaking down and utilizing this data, open works and urban woodland supervisors can estimate patterns, envision support needs, a financial plan for tree-related consumptions, and build up a reason for long-go arranging. Knowing urban timberland populace qualities encourages dynamic, which at that point permits appropriate and ideal move to be made for danger decrease on the open rights-of-way, preventive upkeep to lessen storm harm, and making arrangements for required tree planting activities. This guarantees a steady and various tree populace for what's to come.

Planting Program

Urban woodland the executive's plans address planting needs likewise and can utilize stock information to create and control open tree planting programs. Tree species choice and planting area assignments are noteworthy segments of an urban ranger service program. Choices of what type of tree to plant and where to plant it is basic because of the drawn out effect of those choices.

The tree stock uncovers the quantity of empty planting locales, the size and sorts of these areas, the current species conveyance, and other appropriate information. The urban timberland the executive's plan takes a gander at this information to build up a general planting technique and address numerous issues identified with new care and tree planting. The arrangement recognizes the territories with the best requirement for development, suggests species fitting for the accessible planting spaces, talks about explicit upkeep plans for recently building up trees, and gives specialized data about legitimate tree planting procedures.

Insect Disease, Threat, and Control

American urban communities and provinces have managed creepy crawly and sickness dangers to open woodlands for in excess of 100 years. Verifiably, numerous networks have endured noteworthy tree misfortune and harm from such dangers as the chestnut curse, Dutch elm ailment, and the rover moth. The twenty-first century and the new overall economy convey new threats to our urban forest areas, for instance, the Emerald Ash Borer, Asian Long-Horned Beetle, and Sudden Oak Death.

Through cautious examination of neighborhood conditions and species structure, arrangements in the administration plan can be incorporated to endeavor to alleviate the disturbance to its urban woodland brought about by the current or possible creepy crawly and illness pervasions. Adopting a proactive strategy to these sorts of dangers empowers the open works office to address open and private needs in a proficient and powerful way (Rizwan et al. 2017a, b).

American urban areas and regions have managed creepy crawly and infection dangers to open woods for in excess of 100 years. Truly, numerous networks have endured huge tree misfortune and harm from such dangers as the chestnut curse, Dutch elm ailment, and the rover moth. The twenty-first century and the new worldwide economy carry new dangers to our urban woodlands, for example, the Emerald Ash Borer, Asian Long-Horned Beetle, and Sudden Oak Death.

Through cautious investigation of neighborhood conditions and species pieces, arrangements in the administration plan can be incorporated to endeavor to moderate the disturbance to its urban backwoods brought about by the current or possible bug and illness pervasions. Adopting a proactive strategy to these sorts of dangers empowers the open works office to address open and private needs in a productive and viable way (Kolehmainan et al. 2003).

9.4.6 Tree Inventory and Mapping Data

Automated office and resource inventories, area data, and work request frameworks are basic apparatuses utilized by open works directors consistently. Overseeing tree stock data isn't that unique in relation to dealing with some other open foundation part and there is an assortment of modernized frameworks and programming projects to help in this errand.

On the most fundamental level, tree stock information can be entered and kept up in any straightforward spreadsheet or database programming program. These projects are economical, simple to utilize, and typically as of now exist on most office PCs. Basic information arranging and questioning can rapidly give data on urban woods conditions and assignments.

All the more normally, tree stock information and planned area data are best kept up and oversaw utilizing monetarily accessible programming programs explicitly intended for urban woodland executives. These projects are tweaked for the open work's organization to encourage refreshing and altering, and are able to do in a flash giving helpful data and creating reports, for example,

- Work chronicles and expenses for each tree
- Citizen service and data demands
- Work orders

- Available planting destinations
- Maps
- Tree valuation

As an administration device, a mechanized tree stock and information the executives programming program advance productive designation of work teams and hardware; speeds up reactions to support demands; recognizes dangers; encourages exact cost investigation; gives information for conveying people in general, chosen authorities, and different divisions; can give data expected to allow tracks and applications; extends succeeding work programs and the requirement financial plans.

The administration plan will for the most part survey the requirements, capacities, and duties of the open works office and make a suitable suggestion of what individual or a mix of programming projects and information the executives' frameworks is directly for the organization.

9.4.7 Tree Risk Reduction Plan

The urban woods the board plan can and ought to have areas committed to urban timberland chance decrease and a crisis reaction and recuperation plan that gives data about general tree hazard decrease and offers headings to the open work's organization during an extraordinary tempest crisis.

When building up a crisis the board plan, managing genuine open well-being and medical problems is a conspicuous segment, however, remembering trees and woody flotsam and jetsam for relief endeavors must not be ignored. At the point when disastrous catastrophes, for example, cyclones, ice tempests, tropical storms, and serious straight-line winds strike a metropolitan community, a large number of cubic yards of flotsam and jetsam are delivered. Vegetation can represent roughly 30% of this garbage volume.

An exhaustive urban wood the executive's program incredibly decreases storm perils through legitimate planting, preventive upkeep, and efficient hazard decrease. In any case, when catastrophes happen, a crisis plan as an addendum to this arrangement can give strong information, realities, and conventions to guarantee administration progression and ideal recuperation and reclamation. The general target is to make a crisis readiness program that subtleties improved approaches and systems, expanding the effectiveness and efficiency of crisis storm reaction tasks.

Hazard decrease plans can likewise deliver dangers to open well-being, and open works operational obligations and the issues that are non-storm crises, for example,

- Clearing leaves and woody garbage from canals and tempest channels
- Sidewalk, road, and building clearance guidelines
- Line-of-sight conflicts for street and well-being signage

- Blockage of road lights and traffic lights
- Conflicts with overhead and underground utilities

Both the crisis reaction plan and hazard decrease plans ought to be made as a synergistic exertion between every single key office and partner in the network. With the open work's division as the lead, data and contribution from police and fire, parks, buying, city or district organization, controlling service organizations, nearby and state crisis the board offices, and temporary workers ought to be acquired and thought about when building up these plans (Depietri and McPhearson 2017).

9.4.8 Urban Forest Cost/Benefit Analysis

The open trees developing in any network are significant metropolitan assets. They give substantial and elusive advantages to differing administrations, for example, contamination control, vitality decrease, stormwater the executives, property estimations, untamed life environment, instruction, and feel. Already, the administrations and advantages trees gave in the urban and rural settings were viewed as unquantifiable. Be that as it may, by utilizing broad logical investigations and viable examination, these advantages would now be able to be unhesitatingly determined utilizing models contained in I-Tree programming and current tree stock data.

The I-Tree set up of free programming devices was as of late discharged by the U.S. Woodland Service and can be utilized to evaluate and oversee network timberlands. With these apparatuses, open works and urban woodland directors can precisely measure the advantages of urban woodlands and comprehend and balance the expenses of dealing with urban timberland.

An urban wood the board plan that incorporates such a money-saving advantage investigation will help the open works supervisor:

- By getting up the economic evaluations of street trees utilizing yearly spending plan and use the information to survey the administration program.
- Justify funding and perform vital getting ready for the urban forest area.
- Obtain more public support for the value of trees to financial turn of events, natural well-being, and personal satisfaction issues in the network.
- Determine the annual amount of pollution removed by the urban timberland, the percent of air quality improvement, the measure of carbon sequestered, the measure of vitality utilization decreases, and evaluated increments in property estimations and feel.

This sort of cost/advantage examination may give open works administrators the avocation for more consideration and financing for urban ranger service arranging, plan, the executives, and upkeep. The science behind these models and kind of investigation is sound and has been distributed in peer-checked on diaries. The test presently is to apply the science to upgrade the personal satisfaction in our networks by improving the condition and degree of the urban timberland.

10 Mitigation of Climate Change at Global Scale

Atmosphere alleviation alludes to the decrease of GHG outflows. As talked about, urban backwoods diminish GHG discharges by catching carbon from the air and decreasing vitality use. In any case, carbon stockpiling by urban trees is definitely not a critical commitment to diminish worldwide, national, or even nearby outflows. In any case, it's anything but an inconsequential commitment. In urban backwoods, the executives can expand carbon catch by expanding the urban shelter spread. Greater and more youthful trees catch more carbon, and the urban woods could be improved to follow such a development and age structure. In addition, carbon catch can be expanded by species determination. The improvement of a carbon-animal type's choice lattice is pivotal for this methodology (Bakht et al. 2020). The game plan of trees according to structures could likewise be improved to add to vitality productivity. At long last, urban woodland support likewise discharges emanations, and decrease around there would include handling the innovative, social, and financial components included. Atmosphere moderation contemplations have been all around described previously (Abdollahi et al. 2000). Truth be told, some urban woodland the board plans created in North America allude to such administration rehearses (Seattle Gov 2007). In any case, the operationalization of such reaction in the executives is hard to address without a coupled alleviation adjustment approach, as the effects of environmental change may make light of a considerable lot of the GHG outflow decrease impacts (Farid et al. 2018a, b).

10.1 Adaptation

Adjustment is the modification of a framework accordingly or in expectation to changing ecological conditions that rely upon the framework's weakness, level of effect, level of hazard, and versatile limit. There are two features of an adjustment reaction in an urban timberland setting: altering the urban woodland to change and utilizing urban backwoods to assist urban areas with adjusting to change. With respect to the principal feature, the adjustment in a normal environment ought not to mean come back to a past characteristic state. With the environmental change, the thought of characteristic state is tested and atmosphere baselines for timberland the executives can never again be considered as perfect (Spittlehouse and Stewart 2003). Adjustment to environmental change in urban woods suggests taking administration choices considering an anticipated atmosphere situation and acclimating to vulnerability. Concerning the subsequent aspect, an urban backwoods adjustment reaction would be inadequate without full thought of the more extensive urban woodland network, mainly, the individuals and framework inside (Wilby and Perry 2006; Johnston 2004).

An atmosphere the board methodology can be centered around diminishing the framework's atmosphere weakness and expanding its atmosphere versatile limit

(Adger et al. 2007) as indicated by the contemplations. The two reactions lessen the level of atmosphere sway and the degree of atmosphere chance. One of the most significant components in a versatile procedure is the choice of atmosphere strong species. Building up an animal type determination framework, based on a wide scope of standards, is at the center of this component. While the rules of scanning for southern seeds in northern nations might be reasonable (Yang 2009), additionally dry spell and ice versatility advise the authority in choosing sufficient urban tree species (Roloff et al. 2009). More measures may come into thought at a specific second for animal types determination. For instance, after storm Juan hit Nova Scotia, Canada, in September 2003, over 70% of the developed trees of Point Pleasant Park in the city of Halifax were lost. While thinking about the natural recuperation of the recreation center, the supervisors considered future atmosphere situations for the territory. In light of a couple of rules, American basswood (*Tilia americana* L.) and Butternut (*Juglans cineria* L.) were considered appropriate to be planted in the recreation center. Helping the movement of these species to Nova Scotia through planting in looming decades would appear to be a sensible versatile reaction to environmental change.

Atmosphere adjustment may incorporate different practices, for example, improving species blend. Guidelines for species blend as of now exist to handle nuisances and infections and specify that nobody species ought to speak to over 15% of all species in the urban woodland (Miller and Miller 1991). The game plan of urban woods corresponding to the framework could likewise be improved. Atmosphere demonstrating contemplates have indicated that the vicinity of vegetation to the city's foundation helps in directing the urban microclimate a vital component in improving the impacts of an expansion in temperature in the urban domain (Gill et al. 2007; Cummings et al. 2007). Keeping biological hallways to build the natural availability of the urban woodlands could likewise decide nine flexibilities (McKinney 2002). Different components of urban backwoods structure, for example, age structure have not been all around examined, just as factor stages of the notable administration procedures above. Future exploration is greatly required around there. The social element of urban woodland adjustment includes decreasing weakness in zones of establishments, proprietorship, value, training, which is identified with versatile limit, support, among others. Tending to such issues may include the modification of establishments, meeting, incorporation, and strengthening, and raising the degree of significance of the urban woods inside policy management, and network administration. In any case, how these components might be indicated involves banter. Urban timberland adjustment additionally implies that administration must react to the urban backwoods esteems that individuals in the urban woodland network would prefer to support considering the environmental change. Studies or urban woodland esteems that consider an environmental change reaction involve future exploration. At last, the financial component of urban woods includes decreasing weakness in regions of valuation of urban timberland capacities and advantages, institutional spending plan, among different issues identified with common asset financial matters (Sajjad et al. 2020). Tending to such issues may include the change of institutional financial plan, the improvement of innovation and development, cost-effectiveness, among others. In any case, how these components might be determined involves future examination and contextual investigations.

11 Conclusions and Recommendations

Urban timberlands can be valuable both in moderating environmental change and in helping urban communities adjust to higher temperatures and different effects of environmental change. Urban trees diminish the measure of ozone harming substances noticeable all around by sequestering carbon dioxide and by lessening the measure of vitality expected to warm and cool structures. These jobs can be evaluated at the size of individual trees or whole urban communities (McPherson et al. 2005) and furthermore for states (McPherson and Simpson 2003).

The general impact of urban trees is to cool the neighborhood condition during summer. Trees give conceal that shields some daylight from arriving at the surface beneath their overhangs. At the point when trees conceal structures, this capture can lessen summer interest for cooling, which in numerous urban communities is controlled by ozone-depleting substance discharging petroleum products, for example, gaseous petrol or coal. Shade around cooling units can decrease vitality use by mostly precooling air before it enters the structure. Trees additionally cool the air through evapotranspiration (i.e., the vanishing of water through their leaves). Decreased outflows from concealing and evapotranspiration can be considerable, particularly in urban communities with long cooling seasons and many cooled structures drawing their power from coal-fueled plants. For cooling intentions, it's imperative to plant trees close to structures and deliberately find them so they conceal the structure, particularly from the setting sun in summer. Planting trees where they do little to give conceal, for example, on the north side of structures, can be counterproductive in light of the fact that they can keep breezes from overcoming (Depietri and McPhearson 2017).

Then again, planting to help shield the structure from winning breezes in winter can help decrease warming expenses in certain urban areas. Winter conceal, in any case, can expand the utilization of non-renewable energy sources for warming, possibly astounding investment funds from wind security. Deciduous trees close to structures permitted more winter warming than evergreens, however, they despite everything hindered an expected 30% of approaching daylight in the wake of shedding their leaves. This thinks about blocking 85% of approaching daylight in summer. Trees can likewise help relieve a dangerous atmospheric deviation by sequestering the ozone-depleting substance carbon dioxide. As trees photosynthesize, they gather carbon dioxide from the air and convert it into carbon-based items, for example, sugar and cellulose. Carbon dioxide enters through the stomata on their leaves, joins with water in a substance response fueled by daylight, and changes these into sugars, discharging oxygen all the while. A great part of the carbon-based materials become fixed as wood, albeit some are breathed back as carbon dioxide or are utilized to make leaves that are in the long run shed by the tree (Blum and Janaki 2017).

The decay of leaves and branch trimmings must be considered while evaluating a tree's carbon sequestration esteem. When trees bite the dust or are chopped down, they start to decay, restoring a portion of the put away carbon to the environment. The pace of deterioration varies enormously dependent on species and the destiny of the wood. Wood that is chipped and applied as mulch decays generally rapidly, while wood rescued for use in wood items, for example, furniture, can endure flawless for a long time or more before step-by-step deteriorating. Furthermore, the support of urban trees can deliver ozone-depleting substance emanations using gas and diesel energized by vehicle armadas, and by gas-fueled gear, for example, cutting tools, shredders, stump removers, and leaf blowers. Commonly, carbon dioxide discharged because of tree planting, support, and other related exercises add up to around 2–5% of yearly decreases got through carbon sequestration and diminished force plant emanations. Urban woodlands can assist urban areas with adjusting to rising temperatures and other environmental change impacts, remembering an expansion for heatwaves.

References

Abdollahi M, Dehpour A, Kazemian P (2000) Alteration by cadmium of rat submandibular gland secretory function and the role of the L-arginine/nitric oxide pathway. Pharmacol Res 42(6):591–597

Adger WN, Agrawal S, Mirza MMW, Conde C, O'brien KL, Pulhin J, Takahashi K (2007) Assessment of adaptation practices, options, constraints and capacity. In: Climate change 2007: impacts, adaptation and vulnerability. Contribution of Working Group II to the Fourth Assessment Report of the Intergovernmental Panel on Climate Change. Cambridge University Press, pp 719–743

Adrees M, Ali S, Rizwan M, Zia-ur-Rehman M, Ibrahim M, Abbas F, Farid M, Qayyum MF, Irshad MK (2015) Mechanisms of silicon-mediated alleviation of heavy metal toxicity in plants: a review. Ecotoxicol Environ Saf 119:186–197

Afshan S, Ali S, Bharwana SA, Rizwan M, Farid M, Abbas F, Abbasi GH (2015) Citric acid enhances the phytoextraction of chromium, plant growth, and photosynthesis by alleviating the oxidative damages in *Brassica napus* L. Environ Sci Pollut Res 22(15):11679–11689

Ahmad R, Ali S, Rizwan M, Dawood M, Farid M, Hussain A, Wijayae L, Alyemenie MN, Ahmad P (2019) Hydrogen sulfide alleviates chromium stress on cauliflower by restricting its uptake and enhancing antioxidative system. Physiol Planta 168:289–300

Alberti M (2010) Maintaining ecological integrity and sustaining ecosystem function in urban areas. Curr Opin Environ Sustain 2(3):178–184

Amir W, Farid M, Ishaq HK, Farid S, Zubair M, Rizwan M, Raza N, Ali S (2020) Accumulation potential and tolerance response of *Typha latifolia* L under citric acid assisted phytoextraction of lead and mercury. Chemosphere 257:127247

Ashfaq H, Abubakar M, Ghulzar H, Farid M, Yaqoob S, Komal N, Azam Z, Hamza A, Ali S, Adrees M (2020) Phytoremediation potential of oilseed crops for lead- and nickel-contaminated soil. In: Plant ecophysiology and adaptation under climate change: mechanisms and perspectives II, Mechanisms of adaptation and stress amelioration. Springer, Singapore, pp 801–820

Bakht S, Safdar K, Khair KU, Fatima A, Fayyaz A, Ali SM, Munir H, Farid M (2020) Response of major food crops under drought stress; physiological and biochemical response. In: Agronomic crops - volume 3: stress responses and tolerance. Springer, Singapore

Blum J (2017) Urban forests: ecosystem services and management. CRC Press, New York

Boone CG, Buckley GL, Grove JM, Sister C (2009) Parks and people: an environmental justice inquiry in Baltimore, Maryland. Ann Am Assoc Geog 99:4

Cameron RW, Blanuša T, Taylor JE, Salisbury A, Halstead AJ, Henricot B, Thompson K (2012) The domestic garden–its contribution to urban green infrastructure. Urban For Urban Green 11(2):129–137

Cumming AB, Nowak DJ, Twardus DB, Hoehn R, Mielke M, Rideout R (2007) Urban forests of Wisconsin 2002: pilot monitoring project 2002. State and private forestry report NA-FR-05-07. U.S. Department of Agriculture, Forest Service, Northeastern Area, Newton Square, PA, p. 33

Dale A, Robinson J, King L, Burch S, Newell R, Shaw A, Jost F (2020) Meeting the climate change challenge: local government climate action in British Columbia, Canada. Clim Policy 20(7): 866–880

Davoudi S, Crawford J, Mehmood A (Eds.) (2009) Planning for climate change: strategies for mitigation and adaptation for spatial planners. Earthscan

Depietri Y, McPhearson T (2017) Integrating the grey, green, and blue in cities: nature-based solutions for climate change adaptation and risk reduction. In: Nature-based solutions to climate change adaptation in urban areas. Springer, Cham, pp 91–109

Duguma LA, Minang PA, van Noordwijk M (2014) Climate change mitigation and adaptation in the land use sector: from complementarity to synergy. Environ Manag 54(3):420–432

Duinker PN, Burbidge EL, Boardley SR, Greig LA (2013) Scientific dimensions of cumulative effects assessment: toward improvements in guidance for practice. Environ Rev 21(1):40–52

Espeland EK, Kettenring KM (2018) Strategic plant choices can alleviate climate change impacts: a review. J Environ Manag 222:316–324

Ezaz Z, Azhar R, Rana A, Ashraf S, Farid M, Mansha A, Naqvi SAR, Zahoor FA, Rasool N (2020) Current trends of phytoremediation in wetlands: mechanisms and applications. In: Plant eco-physiology and adaptation under climate change: mechanisms and perspectives II, mechanisms of adaptation and stress amelioration. Springer, Singapore

Farid M, Ali S, Akram NA, Rizwan M, Abbas F, Bukhari SAH, Saeed R (2017c) Phyto-management of Cr-contaminated soils by sunflower hybrids: physiological and biochemical response and metal extractability under Cr stress. Environ Sci Pollut Res 24(20):16845–16859

Farid M, Ali S, Ishaque W, Shakoor MB, Niazi NK, Bibi I, Dawood M, Gill RA, Abbas F (2015) Exogenous application of EDTA enhanced phytoremediation of cadmium by Brassica napus L. Int J Environ Sci Technol 12(12):3981–3992

Farid M, Ali S, Rizwan M, Ali Q, Abbas F, Bukhari SAH, Saeed R, Wu L (2017a) Citric acid assisted phytoextraction of chromium by sunflower; morpho-physiological and biochemical alterations in plants. Ecotoxicol Environ Saf 145:90–102

Farid M, Ali S, Rizwan M, Ali Q, Saeed R, Nasir T, Abbasi GH, Rehmani MIA, Ata-Ul-Karim ST, Bukhari SAH (2018b) Phyto-management of chromium contaminated soils through sunflower under exogenously applied 5-aminolevulinic acid. Ecotoxicol Environ Saf 151(30):255–265

Farid M, Ali S, Rizwan M, Saeed R, Tauqeer HM, Sallah-Ud-Din R, Azam A, Raza N (2017b) Microwave irradiation and citric acid assisted seed germination and phytoextraction of Nickel (Ni) by Brassica napus L.; morpho-physiological and biochemical alterations under Ni stress. Environ Sci Pollut Res 24(25):2150–2164

Farid M, Ali S, Rizwan M, Yasmeen T, Arif MS, Riaz M, Saqib M, Zia Ur Rehman M, Ayub MA (2020a) Combined effects of citric acid and 5-aminolevulinic acid in mitigating chromium toxicity in sunflower (Helianthus annuus l.) grown in cr spiked soil. Pak J Agri Sci 57:477

Farid M, Ali S, Saeed R, Rizwan M, Ali B, Azam A, Ashraf Hussain A, Ahmad I (2019) Combined application of citric acid and 5-aminolevulinic acid improved biomass, photosynthesis and gas exchange attributes of sunflower (Helianthus annuus L.) grown on chromium contaminated soil. Int J Phytorem 21:1–8

Farid M, Ali S, Zubair M, Saeed R, Rizwan M, Sallah-Ud-Din R, Azam A, Ashraf R, Ashraf W (2018a) Glutamic acid assisted phyto-management of silver contaminated soils through sunflower; physiological and biochemical response. Environ Sci Pollut Res 25(25):25390–25400

Farid M, Farid S, Zubair M, Rizwan M, Ishaq HK, Ali S, Ashraf U, Alhaithloul HAS, Gowayed S, Soliman MH (2020b) Efficacy of Zea mays L. for the management of marble effluent contaminated soil under citric acid amendment; morpho-physiological and biochemical response. Chemosphere 240:124930

Fatima A, Farid M, Alharby HF, Bamagoos AA, Rizwan M, Ali S (2020a) Efficacy of fenugreek plant for ascorbic acid assisted phytoextraction of copper (Cu); a detailed study of Cu induced morpho-physiological and biochemical alterations. Chemosphere 251:126424

Fatima A, Farid M, Safdar K, Fayyaz A, Ali SM, Adnan S, Nawaz M, Munir H, Raza N, Zubair Z (2020b) Loss of agro-biodiversity and productivity due to climate change in continent Asia: a review. In: Plant ecophysiology and adaptation under climate change: mechanisms and perspectives, general consequences and plant responses. Springer, Singapore

Gill SE, Handley JF, Ennos AR, Pauleit S (2007) Adapting cities for climate change: the role of the green infrastructure. Built Environ 33(1):115–133

Heller NE, Zavaleta ES (2009) Biodiversity management in the face of climate change: a review of 22 years of recommendations. Biol Conserv 142(1):14–32

Hiloidhari M, Baruah DC, Kumari M, Kumari S, Thakur IS (2019) Prospect and potential of biomass power to mitigate climate change: a case study in India. J Clean Prod 220:931–944

Johnston M (2004) Impacts and adaptation for climate change in urban forests. Paper presented at the 6th Canadian urban forest conference. p. 15

Joosten H, Tapio-Biström ML, Tol S (2012) Peatlands: guidance for climate change mitigation through conservation, rehabilitation and sustainable use. Food and Agriculture Organization of the United Nations, Rome

Keenan RJ (2015) Climate change impacts and adaptation in forest management: a review. Ann Forest Sci 72(2):145–167

Kenney WA, van Wassenaer PJE, Satel AL (2011) Criteria and indicators for strategic urban forest planning and management. Arboricul Urban For 37(3):108–117

Khair KU, Farid M, Ashraf U, Zubair M, Rizwan M, Farid S, Ishaq HK, Iftikhar U, Ali S (2020) Citric acid enhanced phytoextraction of nickel (Ni) and alleviate *Mentha piperita* (L.) from Ni induced physiological and biochemical damages. Environ Sci Pollut Res 27:270110–227022

Khalid A, Farid M, Zubair M, Rizwan M, Iftikhar U, Ishaq HK, Farid S, Latif U, Hina K, Ali S (2020) Efficacy of Alternanthera bettzickiana to remediate copper and cobalt contaminated soil physiological and biochemical alterations. Int J Environ Res 14:243–255

Kolehmainen J, Black GC, Saarinen A, Chandler K, Clayton-Smith J, Träskelin AL, Lehesjoki AE (2003) Cohen syndrome is caused by mutations in a novel gene, COH1, encoding a transmembrane protein with a presumed role in vesicle-mediated sorting and intracellular protein transport. Am J Hum Genet 72(6):1359–1369

Latif U, Farid M, Rizwan M, Ishaq HK, Farid S, Ali S, El-Sheikh MA, Alyemeni MN, Wijaya L (2020) Physiological and biochemical response of Alternanthera bettzickiana (Regel) G. Nicholson under acetic acid assisted phytoextraction of lead. Plants 9(9):1084

Leung DY, Tsui JK, Chen F, Yip WK, Vrijmoed LL, Liu CH (2011) Effects of urban vegetation on urban air quality. Landsc Res 36(2):173–188

Locatelli B, Pavageau C, Pramova E, Di Gregorio M (2015) Integrating climate change mitigation and adaptation in agriculture and forestry: opportunities and trade-offs. WIREs Clim Chan 6(6): 585–598

Maalik U, Farid M, Zubair M, Ali S, Rizwan M, Shafqat M, Ishaq HK (2020) Rice production, augmentation, escalation and yield under water stress. In: Agronomic crops—volume 3: stress responses and tolerance. Springer, Singapore

McKinney ML (2002) Urbanization, biodiversity, and conservation: the impacts of urbanization on native species are poorly studied, but educating a highly urbanized human population about these impacts can greatly improve species conservation in all ecosystems. BioScience 52(10): 883–890

McPherson EG, Simpson JR (2003) Potential energy savings in buildings by an urban tree planting programme in California. Urban For Urban Green 2(2):73–86

McPherson G, Simpson JR, Peper PJ, Maco SE, Xiao Q (2005) Municipal forest benefits and costs in five US cities. J Forest 103(8):411–416

Millar CI, Stephenson NL, Stephens SL (2007) Climate change and forests of the future: managing in the face of uncertainty. Ecol Appl 17(8):2145–2151

Miller RH, Miller RW (1991) Planting survival of selected street tree taxa. J Arboricul 8:13–23

Niemelä J, Saarela SR, Söderman T, Kopperoinen L, Yli-Pelkonen V, Väre S, Kotze DJ (2010) Using the ecosystem services approach for better planning and conservation of urban green spaces: a Finland case study. Biodivers Conserv 19(11):3225–3243

Nowak DJ (2000) The interactions between urban forests and global climate change. In: Abdollahi KK, Ning ZH, Appeaning A (eds) Global climate change and the urban forest. Gulf Coast Climate Change Assessment Council (GCRCC) and Franklin Press, Baton Rouge, pp 31–44

Olsen KH (2007) The clean development mechanism's contribution to sustainable development: a review of the literature. Clim Chan 84(1):59–73

Owusu PA, Asumadu-Sarkodie S (2016) A review of renewable energy sources, sustainability issues and climate change mitigation. Cogent Eng 3(1):1167990

Pires JCM (2019) Negative emissions technologies: a complementary solution for climate change mitigation. Sci Total Environ 672:502–514

Price A, Jones EC, Jefferson F (2015) Vertical greenery systems as a strategy in urban heat island mitigation. Water Air Soil Pollut 226(8):1–11

Raman VVS, Iniyan S, Goic R (2012) A review of climate change, mitigation and adaptation. Renew Sust Energ Rev 16(1):878–897

Rizwan M, Ali S, Hussain A, Ali Q, Shakoor MB, Zia-ur-Rehman M, Farid M, Asma M (2017a) Effect of zinc-lysine on growth, yield and cadmium uptake in wheat (Triticum aestivum L.) and health risk assessment. Chemosphere 187:35–42

Rizwan M, Ali S, Qayyum MF, Ok YS, Adress M, Ibrahim M, Zia-ur-Reham M, Farid M, Abbas F (2017b) Effect of metal and metal oxide nanoparticles on growth and physiology of globally important food crops: a critical review. J Hazard Mater 322:2–16. https://doi.org/10.1016/j.jhazmat.2016.05.061

Roloff A, Korn S, Gillner S (2009) The climate-species-matrix to select tree species for urban habitats considering climate change. Urban Forest Urban Green 8(4):295–308

Sajjad A, Jabeen F, Farid M, Fatima Q, Akbar A, Ali Q, Hussain I, Iftikhar U, Farid S, Ishaq HK (2020) Biochar: a sustainable product for remediation of contaminated soils. current trends of phytoremediation in wetlands: mechanisms and applications. In: Plant ecophysiology and adaptation under climate change: mechanisms and perspectives II, mechanisms of adaptation and stress amelioration. Springer, Singapore

Seattle Government Superintendent's Report (2007) Seattle board of park commissioners meeting minutes, August 9, 2007. Department of Parks and Recreation

Steenberg JW, Duinker PN, Nitoslawski SA (2019) Ecosystem-based management revisited: updating the concepts for urban forests. Landsc Urban Plan 186:24–35

Swart R, Robinson J, Cohen S (2003) Climate change and sustainable development: expanding the options. Clim Policy 3:19–40

Thompson R, Pillsbury NH, Hanna RJ (1994) The elements of sustainability in urban forestry. The Department

Wilby RL, Perry GL (2006) Climate change, biodiversity and the urban environment: a critical review based on London, UK. Prog Phys Geogr 30(1):73–98

Yang J (2009) Assessing the impact of climate change on urban tree species selection: a case study in Philadelphia. J Forest 107(7):364–372

Young RF, McPherson EG (2013) Governing metropolitan green infrastructure in the United States. Landsc Urban Plan 109(1):67–75

Zari MP (2010) Biomimetic design for climate change adaptation and mitigation. Archit Sci Rev 53(2):172–183

Zomer RJ, Trabucco A, Bossio DA, Verchot LV (2008) Climate change mitigation: a spatial analysis of global land suitability for clean development mechanism afforestation and reforestation. Agric Ecosyst Environ 126(1–2):7–80

Temperate Forage Legumes Production, Weeds Dynamics, and Soil C:N Economy Under Organic Wastes

Muhammad Aamir Iqbal, Asif Iqbal, Seyithan Seydoşoğlu, Nizamettin Turan, Mehmet Arif Ozyazici, and Ayman EL Sabagh

Abstract Environmental pollution caused by leaching and volatilization of mineral fertilizers from agricultural fields especially in temperate climates has aggravated under changing climate. Utilizing organic wastes from poultry sheds and dairy farms for crops production can be a way forward towards their eco-friendly disposal and curbing environmental hazards. This review synthesizes and analyzes the use of organic manures for forage legumes production and their impact on feed value, soil fertility, microbial biomass, weeds infestation, and economic turnouts in temperate regions. The feasibility of adopting organic wastes as a conventional source of plant nutrients or continuity of their use as a fringe farming activity has also been assessed. The increment in the forage yield of legumes such as white clover, lucerne, red clover, sub-clover, birdsfoot trefoil, etc. under chemical fertilizers widened the yield gap between organic and mineral manures management systems. However, organic materials improved the nutritional quality of temperate legumes owing to increased activity of proteinase enzyme, better absorption, and utilization of nitrogen and phosphorous, production of effective carboxyl and hydro-carboxyl, β-glucosidase activity along with the presence of growth-promoting hormones. Organic fertilization effectively enhanced soil carbon sequestration and microbial biomass, while labile carbon was increased by mineral fertilizers. However, organic manures

M. A. Iqbal (✉)
Department of Agronomy, Faculty of Agriculture, University of Poonch Rawalakot, Rawalakot, Pakistan

A. Iqbal
Department of Agronomy, Faculty of Agriculture, University of Agriculture Faisalabad, Faisalabad, Pakistan

S. Seydoşoğlu · N. Turan · M. A. Ozyazici
Department of Field Crops, Faculty of Agriculture, Siirt University, Sıirt, Turkey

A. EL Sabagh (✉)
Department of Field Crops, Faculty of Agriculture, Siirt University, Sıirt, Turkey

Department of Agronomy, Faculty of Agriculture, University of Kafrelsheikh, Kafr El Sheikh, Egypt
e-mail: ayman.elsabagh@agr.kfs.edu.eg

M. Hasanuzzaman et al. (eds.), *Managing Plant Production Under Changing Environment*, https://doi.org/10.1007/978-981-16-5059-8_16

recorded higher dry matter of weeds than traditional farming systems which reduced forage yield by 35–50%. Availability at farm, regional and global levels along with nutritional composition, biological viability, and economic competitiveness of organic wastes continue to remain challenges in their wide-scale adoption as plant nutrients source.

Keywords Cattle slurry · Poultry litter · Farmyard manure · Soil organic carbon · Temperate agriculture

1 Introduction

Two vital attributes of temperate forage legumes are biological nitrogen fixation and superior nutritive value compared to temperate grasses which are being extensively exploited in modern intensive farming systems. However, accumulation of nitrates in underground and surface water owing to nitrate leaching (Misselbrook et al. 2013; Arbacauskas et al. 2018; Are et al. 2018; Žydelis et al. 2019), phosphates deposition in terrestrial, and aquatic ecosystems (Merbach and Schulz 2013; Andersen et al. 2014) and invasion of new insect-pests (Zavattaro et al. 2017) are some of the horrendous challenges posed by chemical fertilizers under intensified farming systems of temperate regions (Delin and Engström 2010; Aziz et al. 2013; Kumar et al. 2013; Khaliq and Abbasi 2015; Iqbal et al. 2017; Mondal et al. 2020). Globally, utilization of organic wastes for crops production has been deemed as an effective, eco-friendly, and pro-environment strategy for their disposal (Beaudoin et al. 2005; Schröder et al. 2013; Anglade et al. 2015). But despite extensive research, supportive policies, favorable public opinion, and media coverage, organic wastes from poultry and dairy industries continue to remain unutilized (Foissy et al. 2013; Ahmad et al. 2014; Abdou et al. 2016; Lenka et al. 2016; Keenan et al. 2014; Rehman et al. 2017; Yogesh et al. 2017) especially in temperate regions (Bečka et al. 2004; Valkama et al. 2013). The small share of organically produced feed with a higher price tag constitutes a limiting factor from the consumer's point of view (Iqbal et al. 2019a, b; Sohail et al. 2021), while lower and variable yields, little demand, and hurdles in switching to organic wastes utilization are the challenges from the perspectives of producers (Islam et al. 2020; Iqbal et al. 2018a, b; Yang et al. 2015; Zhang et al. 2015).

Compared to chemical fertilizers, organic manures are cheaper and their availability at the farm and regional levels reduce the risk of short supply (Choudhary et al. 2013; Bhogal et al. 2016; Iqbal et al. 2016; Sinha et al. 2019). Organic materials (poultry litter, buffalos and cows farmyard slurry, press mud, molasses and bagasse, crop residues, leaf litter, sewage sludge, humic acid, biochar, slaughtering houses wastes, sugarcane trash, rice husk, sawdust, vermicompost, biogas slurry, granular humus, coir pith, seaweed extract, moringa leaf extract, micro-herbal mixtures, etc.) used as biodynamic compost offer one of the most economical, biologically viable, farmers friendly, and pro-environment option for their disposal (Benoit et al. 2014; Ruibo et al. 2015).

Organic manures provide nutrients especially nitrogen and phosphorous over a longer period of time (Engström et al. 2014; Jasim and Mhanna 2014). Increased supply of nitrogen triggered the photosynthetic process and ultimately biomass was increased. Readily available phosphorous present in organic wastes helped to improve root growth and contributed to energy transfer during metabolism within plants (Abusuwar and Daur 2014; Staugaitis et al. 2016). Micro-nutrients presence in poultry litter and bovine's farmyard slurry increased the absorption and utilization efficacy of macro-nutrients along with contributing to the buildup of soil organic carbon. Increased roots growth, improved physicochemical properties of soil along enhanced soil microorganism activity are some of the other benefits rendered by well-composted organic wastes (Arjumand et al. 2013). In addition, organic manures were effective in increasing the permeability of cell membranes which promoted nutrients transmission to the site of development and cell growth which reduced the number of aborted reproductive parts in many legumes. Organic wastes rich in various amino-acid, vitamins, and other minerals served as co-factor for producing a variety of hormones which increased the growth and nutritional quality of crops (Anju and Vijayalakshmi 2013; Saravanan et al. 2013). Organic wastes from poultry and dairy industries applied solely or in amalgamation with reduced doses of synthetic fertilizers recorded significantly lesser nitrate leaching and nitrous oxide emissions compared to solo chemical fertilizers (Benoit et al. 2014).

The use of organic wastes for vegetables and cash crops production has gained ground across the globe but little attention has been paid to organic production of leguminous forages for dairy animals under temperate climatic conditions. This systematic review synthesizes and appraises the advances on the feasibility of using organic wastes for obtaining the comparable forage legumes yields, nutritional quality, and economic returns as that of chemical fertilizers. The ultimate aim was to assess the sustainability to forage legumes production systems by utilizing organic wastes and finally leading to the prediction of transition of organic manuring from infringing farming activity to a full-fledge utilization for temperate leguminous forages.

2 Forage Legumes Productivity Under Organic Manures Management Systems

Organic manures are criticized for not yielding comparable forage yields as that of chemical fertilizers under temperate agro-ecological conditions. Farmyard manure (FYM) applied as top dressing at the rate of 170 kg N ha^{-1} enhanced dry matter of red clover at first harvesting. The subsequent cuttings were not influenced by top-dressed FYM, while N fixation was also reduced. Soil nitrogen in higher concentration reduced the N fixation process that halted increment in the dry matter at second and third cuttings of French bean (Sarma et al. 2014; Emilie et al. 2016). However, these findings remained inclusive as the combined use of two or more organic materials was not assessed. Integrated fertilization management involving organic wastes and reduced doses of chemical fertilizers yielded higher biomass than solo organic manures (Dixit et al. 2017). Soybean and groundnut sown in intercropping with cassava under integrated fertility management (chemical fertilizers + composted organic manures) yielded 27% higher biomass than organic manures. In addition, soybean was comparatively more responsive to organic manures in terms of biomass production (Aziz et al. 2013). The organic systems which mainly relied on N fixed by legumes witnessed a decrease of 50% in forage biomass than the conventional manure systems. Nitrogen was the major limiting factor in decreasing the productivity of organically maintained legumes as there was a nonsignificant difference of N leaching in both nutrient management systems (Adeoye et al. 2011; Elnesairy et al. 2016). Similarly, forage yield and soil fertility remained superior in traditional cropping systems in comparison to organic farming systems under cold temperate conditions (Bell et al. 2012). Vermicompost (2 t ha^{-1}) and farmyard manure (2.5 t ha^{-1}) recorded forage yield similar to synthetic fertilizers (elemental N:P @ 20:40). However, it was suggested that pre-sowing soil fertility status may interfere in the judicious assessment of organic manures performance for forage legumes production (Joshi et al. 2016).

Low inherited nutrient levels and rapidly depleting soil fertility continues to remain a constant challenge to temperate forage legumes production (Benke et al. 2017). This has led to a nutrient capital buildup strategy through the inclusion of organic materials for legumes production (Abdou et al. 2016). Sole application of pig manure (PM) applied at the rate of 8 t ha^{-1} was effective in increasing soil N, P, and K along with Ca and Mg, but forage biomass did not remain at par to inorganically fertilized legumes. However, PM (8 t ha^{-1}) applied in conjunction with chemical NPK (60:40:20 kg ha^{-1}) yielded comparable forage yield as that of solo inorganic fertilizers (Omotoso 2020). Similarly, FYM (75 m^3 ha^{-1}) was found effective in increasing the bulk density, field capacity, and water availability along with providing nutrients slowly and gradually over a longer period of time which increased forage yield by 38%. It was also concluded that the efficacy and performance of FYM were quadrupled with the application of potassium (96 kg ha^{-1}) (Abdel Salam and Salem 2012). Similarly, 80 kg ha^{-1} of P_2O_5 application in combination with organic spray and vermicompost increased legume forage yield by 29% (Senthilkumar and Sivagurunathan 2012; Kumar et al. 2013).

Biodynamic composts from poultry sheds and cattle farmyards have the potential to supply macro as well as micronutrients over a longer period of time which results in extensive vegetative growth. Cow dung applied in conjunction with cow urine, milk, curd, and vegetable oil effectively enhanced the number of leaves and fresh weight per plant, while root and shoot lengths were increased by 23%. Various species of effective microorganisms acted as growth triggering agents, while chemical fertilizers reduced their population in the short and long term (Nira and Hamaguchi 2012; Saritha et al. 2013). Similarly, legumes wet and dry weights along with the number of leaves per plant were significantly improved by composted municipal solid waste (10 t ha^{-1}) applied in amalgamation to cow manure (3 t ha^{-1}) and chicken manure (2 t ha^{-1}) (Fatahi et al. 2014). However, there is a dire need to

Table 1 Forage biomass production of legumes as influenced by different organic manures applied solely or in amalgamation with reduced doses of synthetic fertilizers under varying agro-environmental conditions

Forage legumes	Type of organic manures	Herbage yield (%)	References
Red clover (*Trifolium pratense*)	Farm yard manure (6 t ha^{-1})	+23	Hatch et al. (2014)
Stylo (*Stylosanthes guianensis*)	Farm yard manure(10 t ha^{-1}) + P (50 kg ha^{-1}) + K (50 kg ha^{-1})	+16	Ahmed et al. (2012)
Cowpea (*Vigna Unguiculata* L. Walp)	Pig manure (6 t ha^{-1}) + NPK (60 kg ha^{-1})	+29	Olusegun (2014)
Alfalfa/Lucerene (*Medicago sativa*)	Seed pelleted with composted poultry manure + *Rhizobium melilotti*	+7	Abusuwar and Ihsanullah (2013)
Berseem clover/Egyptian clover (*Trifolium alexandrinum*)	Poultry litter (8 t ha^{-1})	+11	Badry et al. (2016)
Alfalfa/Lucerene (*Medicago sativa*)	Poultry manure (2.1 t ha^{-1}) + rock phosphate (398 kg ha^{-1}) + rhizobium (2 kg ha^{-1}) + Phosphobacteria (2 kg ha^{-1})	+13	Bama (2016)
Berseem clover/Egyptian clover (*Trifolium alexandrinum*)	Farm yard manure (10 t ha^{-1}) + Sulphur (30 kg ha^{-1}) + boron (4 kg ha^{-1}) + molybdenum (1 kg ha^{-1})	+16	Pal (2015)
Soybean (*Glycine max*)	Farm yard manure (10 t ha^{-1}) + lime (2 t ha^{-1}) + phosphorous (60 kg ha^{-1})	+19	Verde et al. (2013)
Soybean (*Glycine max*)	Poultry manure (5 t ha^{-1}) + single super phosphate (40 kg ha^{-1}) + mycorrhizal inoculation	+14	Adigun and Babalola (2017)
Cluster bean (*Cyamopsis tetragonoloba*)	Farm yard manure (37.5 mg per pot) + zinc (10 mg per pot)	+29	Singh et al. (2015)
Moth bean (*Vigna aconitifolia*)	Farmyard manure (5 t ha^{-1}) + phosphorous (60 kg ha^{-1})	+6	Asif et al. (2017)
Broad bean (*Vicia faba*)	Manure (50% of nitrogen requirement) + urea (40 kg ha^{-1})	+19	Rafaat et al. (2015)
Field bean (*Dolichos lablab*)	Farm yard manure (10 t ha^{-1})	12	Nagabhushanam and Raja (2016)

identify and optimize the dose of local organic materials that are rich in nutrients for producing comparable forage yield of legumes as that of chemical fertilizers under varied agroclimatic conditions. Furthermore, in-depth research is needed for improving the efficacy of organic wastes through improved composting techniques along with application timing keeping in view the cool temperature which reduces the decomposition rate of organic wastes (Table 1).

3 Nutritional Quality of Forage Legumes Under Short- and Long-Term Application of Organic Manures

Forage quality is of the utmost importance for enhancing milk and meat production (Akdeniz et al. 2019; Iqbal et al. 2017, 2019b; Bakken et al. 2014). Organic manures improved quality traits especially protein content of different temperate leguminous forages compared to chemical fertilizers (Gierus et al. 2012; Pholsen et al. 2014; Mohammadi 2015). The combined application of cattle manure (20 m^3 ha^{-1}) and seaweed extract resulted in improved crude protein content (93%), while chicken manure (CM) applied in amalgamation with humic acid yielded significantly higher biomass (62%) of broad bean and red clover. Sufficient nitrogen provided by CM over a longer period of time improved agro-qualitative traits, while optimum phosphorous increased root growth which led to higher absorption and utilization of moisture and micro-nutrients (Jasim and Mhanna 2014). Humic acid produced by the decomposition of organic materials increased effective hydroxyl and carboxyl which supported cell division. Furthermore, organic manures enhanced the activity of various enzymes such as proteinase which converted considerably higher nitrogen into protein (Anju and Vijayalakshmi 2013).

Organic manures increased nitrogen concentration in plants tissues which improved crude protein and total carbohydrates (Gierus et al. 2012; Bakken et al. 2014). Poultry litter and farmyard manure (applied at the rate of 8 tons ha^{-1} in four equal splits) significantly improved protein and total carbohydrates of berseem clover and annual ryegrass (Pholsen et al. 2014; Salama 2015). Similarly, red clover and alfalfa forage yield and quality were improved with poultry litter (PL) (9 Mg ha^{-1}) owing to the significant increase in particulate and total soil carbon. FYM applied in combination with rhizobium, phosphate solubilizing bacteria, and *Azospirillium* enhanced nutritional quality of forage due to better absorption and utilization efficiency of nitrogen and phosphorous. Significantly higher nitrogen fixation was attributed to improved agro-qualitative traits of forage legumes (Pal 2015). Similarly, organically grown clover gave a higher concentration of crude protein (17%) and in-vitro digestibility (3%) along with a 7% lower concentration of neutral detergent fiber (Steinshamn et al. 2016). Similarly, cattle urine applied in conjunction with wastes from palm oil mill produced green forage of puero (*Pueraria javanica*) with better quality (higher crude protein, total ash, digestibility, etc.), while fiber was decreased in comparison to untreated cattle urine. Along with N and K content, the presence of growth-promoting hormones (auxins and indole acetic acid) in cattle urine was attributed to improve nutritional quality. Growth promoting hormones also triggered root bending, cleavage, and lengthening which resulted in enhanced BNF, and ultimately protein synthesis was increased. It was suggested that the nitrogen content of puero was increased by 24% under organic wastes which favor its inclusion in animal feedstuffs (Mudhita et al. 2016). Thus, improved quality of leguminous forages favors the utilization of organic wastes even if forage productivity has not been at par to solo chemical fertilizers. These encouraging findings warrant the use of organic wastes for quality forage production under temperate climate provided optimum dose optimization of organic manures of animal and plant origin could be done for temperate forage legumes.

4 Weed Biomass in Forage Legumes Under Organic Manures

Weeds pose a serious threat to temperate forage legumes (especially lucerne and red clover) at early growth stages under organic manures (Bullied et al. 2006; Koehler-Cole et al. 2017; Schipanski et al. 2017). Weeds are the most important forage yield-limiting factor (Maikštėnienė et al. 2009) only after to N deficiency in temperate regions (Arlauskienė and Maikštėnienė 2004). Organically fertilized fields recorded 2–3 Mg ha^{-1}of weeds dry matter which reduced forage legumes yield by 35–50%. However, legumes such as alfalfa and red clover grown as relay crops with wheat and manured organically effectively reduced weed dry matter by 36–59%. In addition, reduction in aerial biomass of weeds was found to be linearly correlated to legumes biomass as cover crops offered lesser space to weeds (Camille et al. 2013).

The extensive growth of forage legumes triggered by higher soil N availability under organic wastes can give higher canopy cover which may reduce weed's access to light. Legume-grass mixtures under organic manures recorded much higher canopy cover which reduced the availability, interception, and absorption of photo-synthetically active radiation (PAR) by weeds which led to reduced density and dry matter biomass of both narrow and broadleaf weeds. Furthermore, low productivity soils witnessed higher weeds suppression under organic manures compared to high productivity soils. In addition, spreading type of legumes resulted in much lower weeds biomass than the erect type of forge legumes in organically managed fields (El-Karamany et al. 2012). Thus, spreading types of temperate legumes are recommended to be evaluated for determining the comparative efficacy of organic wastes because considerable lower weed infestations may be expected owing to lesser available space.

5 Soil Carbon and Nutrients Economy Under Organically Grown Forage Legumes

Soil fertility depicts the potential (inherited or managed) of soil to supply essential plant nutrients (macro and micro) in optimum concentration (Dind et al. 2012; Wiesmeier et al. 2015). Although, chemical fertilizers increased crops yield still soil was degraded under exhaustive temperate grasses (Henneron et al. 2014). Organic farming was instrumental in increasing soil carbon sequestration by 4.5 times than inorganic fertilizers. Organic farming resulted in accumulating soil carbon when restorative crops like lucerne, red clover, etc. were added in crop rotation (Andreas et al. 2012). In addition, there existed an inverse relationship between soil N and that fixed by temperate legumes. Green manuring of red clover and mulching in addition to the top dressing of FYM reduced N fixation (60 kg ha^{-1}) than solo FYM (170 kg of N ha^{-1}). Alfalfa under organic wastes increased soil acidity which drastically reduced phosphorous and potassium availability. Furthermore, it was also revealed that even long-term organic manuring was not effective in keeping nitrogen and phosphorous at an optimum level, however, potassium remained marginally deficient (Cavalli et al. 2014; Inwood et al. 2015).

Similarly, organically managed alfalfa caused severe depletion of phosphorous, while soil organic carbon (SOC) was reduced (30 t ha^{-1}) in comparison to traditional manure management systems. It was suggested that C inputs must be maintained for organically grown alfalfa, otherwise, SOC could deplete sharply (Benoit et al. 2014). In addition, binary application of press mud (PM) and FYM significantly increased Walkley-Black carbon and particulate organic carbon which multiplied total organic carbon, while labile carbon and microbial biomass carbon were increased by chemical NPK. It was concluded that different fractions of carbon were interrelated and linearly correlated with mineral N and pigeon pea yield (Nintu et al. 2013; Mcroberts et al. 2016). Soybean extracted significantly higher phosphorous (P) from poultry manure amended soil which was attributed to higher P released from manures (Dadson et al. 2014).

Similarly, composted dairy manure (23 Mg ha^{-1}) improved soil organic carbon (13–24%) and total soil N (10%) up to 5–10 cm soil depth, however, it was suggested that the quality of the compost determines soil N and C fractions (rapidly, slowly and resistant to mineralization) (Mikha et al. 2017). Though organically managed legumes were not effective in reducing N leaching, the SOC depletion rate was decreased owing to lower soil pH. However, it was concluded that organic wastes application was an effective choice for boosting soil C sequestration under cold temperate conditions. The agronomic efficacy of N (AEN) (10 kg of yield kg^{-1} of N) and use efficiency of P (UEP) (7%) under organic manuring was much lower than conventional fertilization (AEN = 30 kg of yield kg^{-1} of N and UEP = 36%) (Kendall et al. 2015). Thus, utilization of organic wastes may assist to preserve and restore soil fertility in the long run which can lead to its sustainable use despite intensive cultivation under cooler temperatures.

6 Soil Microbial Community and Biological Nitrogen Fixation in Organically Managed Forage Legumes

Soil fertility management has a distinct impact on soil microbial community in terms of microbial biomass and community structure under temperate conditions (Bradley et al. 2006; Sánchez et al. 2016). Cereal-legumes crop rotation strategy under long-term organic manuring was effective in increasing the relative abundance of fungi, bacteria, and protozoa by 36 times in comparison to reduced tillage and conventional soil fertility management systems. Furthermore, the combined effect of forage legumes, soil amendment, microbial substrate, tillage system, and crop rotation was more pronounced for micro-biotic characteristics, soil organic carbon, N accumulation, and forage yield. Green manuring and composted manures applied at appropriate rates were also effective in reducing the drastic impact of heavy tillage on soil microbial community under alfalfa crop (Ghimire et al. 2014).

The decomposition of organic materials is driven by the microbial community but soil microbial species also get affected by the type, nature, and decomposition status of organic materials (Patricia et al. 2016). The soil under organically grown faba bean recorded the highest soil microbial biomass, β-glucosidase activity, and carbon substrate utilization along with bacterial physiological diversity in comparison to cereal crops. It was also suggested that β-glucosidase activity was a more sensitive and reliable indicator of soil health compared to soil microbial biomass or soil organic carbon and thus it should be used to assess the performance of organic manures (Newton and Yoong 2016). Similarly, biological nitrogen fixation (BNF) by forage legumes was significantly higher under organically managed pastures in the wake of low phosphorous content. Better arbuscular mycorrhizal fungi root colonization under organic manures enhanced nutrient uptake by legume plants (Schneider et al. 2016). Similarly, the BNF process in spring forage peas was increased by 21% under Humustim organic fertilizer pre-sowing seed treatment applied at the rate of 1.5 L/t (Vasileva et al. 2017). However, experimental evidences are scarce regarding the quantification of temperate soil microbial population and their impact on BNF under organically managed annual and perennial forage legumes.

7 Economic Returns of Organically Grown Forage Legumes

There are very few studies on the economic viability of organic wastes for temperate forage legumes production necessitating in-depth researches. Organically managed cassava-common bean intercropping recorded considerably higher economic turnouts (US$ 400–700 ha^{-1}) and marginal rate of return (1.6–2.7) than NPK fertilizers while replacing common bean with groundnut reduced net profit by US$ 200–300. When soybean was intercropped with cassava under composted organic manures, profitability declined owing to the reduction in cassava yield and a comparatively longer maturity period of soybean (Arjumand et al. 2013; Yogesh et al. 2017). The future of organic wastes utilization was reported to be directly dependent on making it economically competitive to traditional farming systems (Kingery et al. 2013; Sacco et al. 2015). The competitiveness of organic feed could be increased by enhancing productivity, market demand, and reducing the cost of production. In addition, forage production systems are believed to remove more nutrients from the soil than other crop alternates. Growing forage crops on wastes of the livestock sector may strengthen the economic standing of dairy farmers and forage growers along with limiting the negative impacts of ruminant's production on the environment.

8 Impediments and Potential Remedies

Availability of organic manures at farm, regional and global levels may serve as a decisive factor in advocating, adopting, and recommending their integrated use with synthetic fertilizers or as a solo source of plant nutrients under temperate conditions. The availability of quality organic materials at critical growth stages of crops under varied cropping systems continues to remain one of the biggest challenges for researchers and promoters of organic manuring. Future research needs to broaden its sphere for including diversified sources of raw materials to be used as organic manures under temperate agro-ecological conditions (Sharma et al. 2014). By-products of local agriculture-associated industries (sugar, oil, etc.) and other sources like municipal wastes, crop residues, tree leaves, and litter must be tested for their nutritional composition and boosting their quality with advanced composting techniques. Only local raw materials utilized as organic manures can boost the organic production of temperate forages and food legumes along with solving the environmental problems caused by these wastes.

Nitrogen is regarded as the most important yield-limiting factor for forage legumes (Rhohit et al. 2013) as it is required for vegetative growth as per varietal potential while N deficiency resulted in an early switching to the reproductive stage at the cost of vegetative growth (Anderson 2015). Different organic manures failed to provide N in optimum quantities and ultimately forage yield was reduced to a great extent (Rabbi et al. 2014). In addition, soil N buildup under organic manuring in short term has also not been validated. In the short run, organic manures applied in conjunction with one-third of recommended inorganic fertilizers yielded comparable forage yields (Abdou et al. 2016). Legumes inclusion in crop rotation may also be beneficial in preserving and restoring soil N stockpiles. Extensive research for developing advanced composting techniques and breeding more responsive cultivars of forage legumes may assist in placing organic materials as a conventional source of plant nutrients in times to come.

Excessive use of chemical fertilizers under intensive farming systems has degraded soil over the period of time. Despite cognizance of the fact that soil contained definite reserves of nutrients which got depleted owing to ill-planned farming practices, erosion, and suboptimal input, no serious efforts have been concentrated in developing environmental-friendly strategies for restoring nutrients capital. The success of organic manures has mainly been allied to nitrogen provision (Dixit et al. 2017), but legumes need much higher phosphorous (to be used in numerous physiological processes including photosynthesis, starch and glucose utilization, energy storage and transfer, etc.) than cereals, thus organic manures must be developed as a complete plant nutrition package for their wide-scale adoption.

Biological viability is one of the key indicators for determining the potential and future role of organic wastes for temperate legumes production. Considerably lower yield and very high variability in the productivity of forage legumes under organic materials pose a serious hurdle in their adoption (Hatch et al. 2014). In addition, more area under organically managed forage legumes is needed to feed the world's ruminants which are destined to reduce natural and semi-natural ecosystems across the globe. This challenge may be met with extensive research for boosting crops yield under organic manures of different origins (Steinshamn et al. 2016). Lastly, dose optimization along with the time of application of composted materials may decrease the variability in crops yield.

Economic competitiveness is another hurdle in switching and adopting organic manures owing to lower yields (Kumar et al. 2013). The use and future of organic materials as the sole plant nutrients source lies in making organic manuring competitive to chemical fertilizers. This challenge can be addressed by reducing the cost of production under organic plant nutrient management systems and raising the market value of feed at the same time (Nintu et al. 2013). The increase in demand for feed produced with organic materials can be achieved with effective media awareness campaigns with respect to environmental and health benefits.

9 Conclusions

Overemphasis on the use of chemical fertilizers has seriously deteriorated agricultural lands along with causing an un-precedent complexity of environmental pollution, while crops productivity has become stagnant after a gradual slowdown in yield increment. Organic manures are crucial for imparting sustainability to current yield and profit-oriented farming systems in temperate regions by maintaining input-output nutrient cycles along with balancing the ecosystems of temperate regions. It is high time to suitably remold farming strategies and appropriately modify agro-technologies for integrating organic resources in temperate forage legumes production systems. The utilization of livestock wastes for producing forages is a way forward to eco-system restoration, biodiversity preservation, and environmental improvement. The future of organic manures largely depends on extensive research for determining local organic resources and optimizing their doses for forage production in the form of a technology package to be adopted at the farm, regional, and global levels. Awareness campaigns and demand generation of organically grown forage legumes should follow a complete and easy to practice technology package suitable for temperate legume growers. Organically produced forage legumes with high premiums are bound to broaden the scope of organic manures to be used as conventional plant nutrients in times to come, while associated risks of lower productivity by organic materials need to be assessed to bring it at par to mineral fertilizers in conventional farming systems. Lastly, extensive use of organic materials has also associated risks related to their losses to water bodies and air along with considerably lower use efficiency under temperate climate which needs in-depth studies to promote their utilization for temperate legumes production.

Conflict of Interest No conflict of interest is hereby declared.

References

Abdel Salam MA, Salem HM (2012) Interaction between potassium and organic manure application on growth of cowpea (*Vigna unguiculata* L.) and soil properties in newly reclaimed sandy soil. J World Agric Sci 8:141–149

Abdou GN, Ewusi Mensah M, Nouri FM, Tetteh EY, Safo, Abaidoo RC (2016) Nutrient release patterns of compost and its implication on crop yield under Sahelian conditions of Niger. Nutr Cycl Agroecosyst 105:117–128

Abusuwar A, Daur I (2014) Effect of cow and poultry manures on yield, quality and seed production of two alfalfa cultivars under saline arid environment. J Food Agric Environ 12(2):747–751

Abusuwar AO, Ihsanullah D (2013) Effect of seed pelletting with organic manures and rhizobia on the performance of two alfalfa cultivars grown in saline environment. Leg Res 38(4):513–518

Adeoye PA, Adebayo SE, Musa JJ (2011) Growth and yield response of cowpea (*Vigna unguiculata* L.) to poultry and cattle manures as amendments on sandy loam soil plot. J Agric For 6:218–221

Adigun MO, Babalola OA (2017) Influence of bradyrhizobium and mycorrhiza on growth, yield and phosphorus use efficiency on soybean under manure application. J Adv Biol 3(3):1–11

Ahmad M, Zahir ZA, Jamil M, Nazli F, Latif M, Akhtar MF (2014) Integrated use of plant growth promoting rhizobacteria, biogas slurry and chemical nitrogen for sustainable production of maize under salt–affected conditions. J Pak Bot 46:375–382

Ahmed SA, Halim RA, Ramlan MF (2012) Evaluation of the use of farmyard manure on a Guinea grass (*Panicum maximum*) – Stylo (*Stylosanthes guianensis*) mixed pasture. J Pertanika Trop Agric Sci 35(1):55–65

Akdeniz H, Hosaflioglu I, Koç A, Hossain A, Iislam M, Iqbal M, Imtiaz H, Gharib H, EL Sabagh A (2019) Evaluation of herbage yield and nutritive value of eight forage crop species. App Ecol Environ Res 17:5571–5581

Andersen HE, Blicher-Mathiesen G, Bechmann M, Povilaitis A, Iital A, Lagzdins A, Kyllmar (2014) Mitigating diffuse nitrogen losses in the Nordic–Baltic countries. Agric Ecosyst Environ 195:53–60

Anderson RL (2015) Suppressing weed growth after wheat harvest with under seeded red clover in organic farming. Renew Agric Food Syst 31:185–190

Andreas G, Adrian M, Matthias H, Colin S, Andreas F, Nina B, Paul M, Matthias S, Pete S, Nadia S, El-Hage UN (2012) Enhanced top soil carbon stocks under organic farming. Proc Natl Acad Sci U S A 109:18226–18231

Anglade A, Gilles B, Josette G (2015) Relationships for estimating N2 fixation in legumes: incidence for N balance of legume–based cropping systems in Europe. Ecosphere 6(3):1–24

Anju S, Vijayalakshmi A (2013) Residual effects of integrated nutrient management with farmyard manure, coir pith and press mud compost on cluster bean. J Int Sci Nat 4:405–407

Arbacauskas J, Masevicienė A, Zickienė L, Staugaitis G (2018) Mineral nitrogen in soils of Lithuania's agricultural land: comparison of oven-dried and field-moist samples. Zem Agric 105(2):99–104

Are M, Tanel K, Are S, Alar A, Endla R (2018) The interaction of soil aggregate stability with other soil properties as influenced by manure and nitrogen fertilization. Zem Agric 105(3):195–202

Arjumand BSS, Ananth NB, Puttaiah ET (2013) Effectiveness of farmyard manure poultry manure and fertilizer–NPK on the growth parameters of French bean (*Phaseolus vulgaris* L.). J Cur Res 1:31–35

Arlauskienė A, Maikštėnienė S (2004) the effect of preceding crops and organic fertilizers on the occurrence of short-lived weeds in different agrosystems. Zem Agric 88(4):102–116

Asif I, Amanulla A, Mazhar I, Ikramullah I (2017) Integrated use of phosphorus and organic matter improve fodder yield of Moth bean (*Vigna aconitifolia* Jacq.) under irrigated and dryland conditions of Pakistan. J Agri 4(1):10–15

Aziz M, Ali A, Aezum T, Chesti AT, Peer MH (2013) Effect of nutrient management on lysine and linoleic acid content of soybean [*Glycine max* (L.) Merill]. Leg Res 36:158–161

Badry HH, Heba SAS, Ahmed R (2016) Nutritional status and indigenous mycorrhizal infection of berseem clover and barley fertilized with poultry litter and compost in an organic farming system. J Alex Sci Exchang 37(4):738–746

Bakken AK, Vaga M, Hetta M, Randby AT, Steinhsamn H (2014) Feed value of restrictedly and extensively fermented organic grass–clover silages from spring and summer growth. Grass Sci 19:603–605

Bama SK (2016) Effect of different nutrient sources on fodder yield, quality and soil fertility status of lucerne grown soil. For Res 41(4):222–227

Beaudoin N, Saad JK, Van Laethem C, Machet JM, Maucorps J, Mary B (2005) Nitrate leaching in intensive agriculture in northern France: effect of farming practices, soils and crop rotations. Agric Ecosyst Environ 111:292–310

Bečka D, Vašák J, Kroutil P, Štranc P (2004) autumn growth and development of different winter oilseed rape variety types at three input levels. Plant Soil Environ 50(4):168–174

Bell LW, Sparling B, Tenutab M, Entz MH (2012) Soil profile carbon and nutrient stocks under long–term conventional and organic crop and alfalfa–crop rotations and re–established grassland. Agric Ecosyst Environ 158:156–163

Benke AP, Rieps AM, Wollmann I (2017) Fertilizer value and nitrogen transfer efficiencies with clover–grass ley biomass-based fertilizers. Nutr Cycl Agroecosyst 107:395–401

Benoit M, Josette G, Juliette A, Gilles B (2014) Nitrate leaching from organic and conventional arable crop farms in the Seine Basin (France). Nutr Cycl Agroecosyst 100:285–299

Bhogal A, Williams JR, Nicholson FA, Chadwick DR, Chambers KH, Chambers BJ (2016) Mineralization of organic nitrogen from farm manure applications. Soil Use Manage 32:32–43

Bradley K, Drijber RA, Knops J (2006) Increased N availability in grassland soils modifies their microbial communities and decreases the abundance of arbuscular mycorrhizal fungi. Soil Biol Biochem 38(7):1583–1595

Bullied WJ, Van Acker RC, Marginet AM, Kenkel NC (2006) Agronomic and environmental factors influence weed composition and canola competitiveness in southern Manitoba. J Can Plant Sci 86(2):591–599

Camille A, Marie-Helene J, Florian C, Christophe D (2013) Relay–intercropped forage legumes help to control weeds in organic grain production. J Eur Agron 49:158–167

Cavalli D, Bechini L, Gallina PM (2014) measuring and modeling soil carbon respiration following repeated dairy slurry application. Soil Sci Soc Am 78:1414–1425

Choudhary VK, Kumar SP, Bhagawati R (2013) Influence of organic nutrient sources on growth, seed yield and economics of cowpea under mid hills of Arunachal Pradesh. J Food Leg 26(3&4):51–54

Dadson RB, Javaid I, Hashem FM, Joshi J (2014) Potential of fodder soybean genotypes for phosphorus removal in poultry manure–enriched soils. J Plant Nutr 38(1):108–115

Delin S, Engström L (2010) Timing of organic fertilizer application to synchronise nitrogen supply with crop demand. Acta Agric Scand Sec B Soil Plant Sci 60:78–88

Dind X, Han X, Liang Y, Qiao Y, Li L, Li N (2012) Changes in soil organic carbon pools after 10 years of continue manuring combined with chemical fertilizer in a Mollisol in China. Soil Tillage Res 122:36–41

Dixit AK, Sunil K, Rai AK, Palsaniya DR, Mukesh C (2017) Nutrient management in fodder sorghum + cowpea–chickpea cropping system for higher system productivity and nutrient use. Range Manag Agrofor 38:82–88

El-Karamany MF, Elwea TA, Bakry BA (2012) Effect of mixture rates on forage mixture of Egyptian clover (*Trifolium alexandrinum*) with triticale (X *triticosecale wittmack*) under newly reclaimed sandy soil. J Aus Basic Appl Res 6:40–44

Elnesairy NNB, Abubaker J, Mahmod H, Mukhtar N (2016) The impact of bradyrhizobium, farmyard manure and inorganic nitrogen on growth and yield of guar. World J Agric Res 4: 56–63

Emilie M, Denis AA, Martin C, Jean L, Denis P, Philippe R, Gabriel L, Marie-Line L, Léon-Étienne P (2016) Greater accumulation of soil organic carbon after liquid dairy manure application under cereal–forage rotation than cereal monoculture. Agric Ecosyst Environ 233:171–178

Engström L, Stenberg M, Wallenhammar A, Ståhl P, Gruvaeus I (2014) Organic winter oilseed rape response to N fertilisation and preceding agroecosystem. Field Crop Res 167:94–101

Fatahi E, Hamid RM, Mohammad MA (2014) Effect of organic fertilizer on wet weight, dry weight and number of leaves in cowpea. J Novel Appl Sci 3(4):440–443

Foissy D, Jean-Francois V, Christophe D (2013) Managing nutrient in organic farming system reliance on livestock production for nutrient management of arable farmland. Org Agric 3(3): 183–199

Ghimire R, Norton JB, Stahl PD, Norton U (2014) Soil microbial substrate properties and microbial community responses under irrigated organic and reduced–tillage crop and forage production systems. PLoS One 9:e103901

Gierus M, Kleen J, Loges R, Taube F (2012) Forage legume species determine the nutritional quality of binary mixtures with perennial ryegrass in the first production year. Anim Feed Sci Technol 172:150–161

Hatch D, Joynes A, Roderick S, Shepherd M, Goodlass G (2014) Effects of cutting, mulching and applications of farmyard manure on the supply of nitrogen from a red clover/grass sward. Org Farm 4:15–24

Henneron L, Bernard L, Hedde M (2014) Fourteen years of evidence for positive effects of conservation agriculture and organic farming on soil life. Agron Sustain Dev 35:1–13

Inwood SEE, Gary EB, David MB (2015) Forage performance and soil quality in forage systems under organic management in the southeastern United States. J Agron 107:1641–1652

Iqbal MA, Asif I, Muhammad A, Javaid A (2016) Comparative study on temporal and spatial complementarity and profitability of forage sorghum–soybean intercropping systems. Custos Agronegocio 12:2–18

Iqbal MA, Bethune BJ, Iqbal A, Abbas RN, Aslam Z, Khan HZ, Ahmad B (2017) Agro–botanical response of forage sorghum–soybean intercropping systems under atypical spatio–temporal pattern. J Pak Bot 49:987–994

Iqbal MA, Hamid A, Ahmad T, Hussain I, Ali S, Ali A, Ahmad Z (2019a) Forage sorghum-legumes intercropping effect on growth, yields, nutritional quality and economic returns. Bragantia 78(1):82–95

Iqbal MA, Hamid A, Hussain I, Siddiqui MH, Ahmad T, Khaliq A, Ahmad Z (2019) Competitive indices in cereal and legume mixtures in a South Asian environment. Agron J 111(1):242–249

Iqbal MA, Hamid A, Muzammil H, Siddiqui A, Hussain I, Ahmad T, Ishaq S, Ali A (2019b) A meta-analysis of the impact of foliar feeding of micronutrients on productivity and revenue generation of forage crops. Planta Daninha 37:e019189237

Iqbal MA, Iqbal A, Abbas AN (2018a) Spatio-temporal reconciliation to lessen losses in yield and quality of forage soybean (*Glycine max* L.) in soybean-sorghum intercropping systems. Bragantia 77(2):283–291

Iqbal MA, Muzammil H, Siddiqui A, Afzal S, Ahmad Z, Maqsood Q, Khan RD (2018b) Forage productivity of cowpea [*Vigna unguiculata* (L.) Walp] cultivars improves by optimization of spatial arrangements. Rev Mex De Cien Pecuar 9(2):203–219

Islam MS, Hossain A, Timsina J, Saif H, Sarker MMR, Khan ASMMR, Barutçular C (2020) Feasibility and financial viability study of an intensive mustard–Mungbean–transplanted Aus rice–transplanted Aman rice cropping system in a non-saline coastal ecosystem of Bangladesh. Philip Agric Sci 103(1):73–83

Jasim AH, Mhanna QL (2014) Effect of some organic fertilizers treatments on dry seed yield of broad bean (*Vicia faba* L). J Agron 7:218–222

Joshi D, Gediya KM, Patel JS, Birari MM, Shivangini G (2016) Effect of organic manures on growth and yield of summer cowpea [*Vigna unguiculata* (L.) Walp] under middle Gujarat conditions. Agric Sci Digest 36(2):134–137

Keenan CM, Quirine MK, David P, Tran TH, Nguyen HQ, Nguyen XB, Charles FN, Debbie JRC (2014) Impact of forage fertilization with urea and composted cattle manure on soil fertility in sandy soils of south–Central Vietnam. J Int Agron. https://doi.org/10.1155/2016/4709024

Kendall RAJ, Long DA, Collins HP, Pierce FJ, Chatterjee A, Smith JL, Young SL (2015) Soil carbon dynamics of transition to Pacific northwest cellulosic ethanol feedstock production. Soil Sci Soc Am 79:272–281

Khaliq A, Abbasi MA (2015) Soybean response to single or mixed soil amendments in Kashmir, Pakistan. Agron J 107:887–895

Kingery WL, Wood CW, Delaney DP, Williams JC, Mullins GL, Van Santen E (2013) Implications of long–term land application of poultry litter on tall fescue pastures. J Prod Agric 6:390–395

Koehler-Cole K, James RB, Charles AF, Charles AS, Erin EB, Stephen PB (2017) Clover green manure productivity and weed suppression in an organic grain rotation. Renew Agric Food Sys 32(5):474–483

Kumar A, Singhal SK, Singh V, Kumar N, Sharma VK (2013) Impact of rock–phosphate enriched pressmud and biogass slurry on yield, phosphorus nutrition and utilization by soybean (*Glycine max*) in a typic hapluwstept. Leg Res 36:79–83

Lenka S, Narendra KL, Amar BS, Singh B, Jyothi R (2016) Global warming potential and greenhouse gas emission under different soil nutrient management practices in soybean–wheat system of central India. Environ Sci Pollut Res 24:4603–4612

Maikštėnienė S, Arlauskienė A, Velykis A, Satkus A (2009) Enchancement of competitive ability of cereals towards weeds by means of crop rotations. Zem Agric 96(2):23–34

Mcroberts KC, Quirine MK, David P, Tran TH, Nguyen HQ, Nguyen XB, Charles FN, Debbie JRC (2016) Impact of forage fertilization with urea and composted cattle manure on soil fertility in sandy soils of south–Central Vietnam. Int J Agron 26:1–14

Merbach I, Schulz E (2013) Long-term fertilization effects on crop yields, soil fertility and sustainability in the static fertilization experiment Bad Lauchstädt under climatic conditions 2001–2010. Arch Agron Soil Sci 59(8):1041–1057

Mikha MM, Dwi PW, Tunsisa TH, Joe EB, Jessica GD (2017) Influence of composted dairy manure and perennial forage on soil carbon and nitrogen fractions during transition into organic management. Agri 7:2–20

Misselbrook T, Agustin DP, David C (2013) Opportunities for reducing environmental emissions from forage–based dairy farms. Agric Food Sci 22(1):23–35

Mohammadi K (2015) Grain oil and fatty acids composition of soybean affected by nano–iron chelate, chemical fertilizers and farmyard manure. Arch Agron Soil Sci 61:1593–1600

Mondal M, Skalicky M, Garai S, Hossain A, Sarkar S, Banerjee H, Kundu R, Brestic M, Barutcular C, Erman M, EL Sabagh A, Laing AM (2020) Supplementing nitrogen in combination with rhizobium inoculation and soil mulch in peanut (*Arachis hypogaea L.*) production system: part ii. Effect on phenology, growth, yield attributes pod quality, profitability and nitrogen use efficiency. Agronomy 10(10):1513

Mudhita IK, Nafiatul U, Subur PSB, Endang B, Cuk TN, Kustono IGSB, Jeffrie W (2016) Effect of Bali cattle urine on legume cover crop puero (*Pueraria javanica*) productivity on an east Borneo oil palm plantation. Pak J Nutr 15:406–411

Nagabhushanam U, Raja V (2016) Productivity and economics of maize (*Zea mays* L.) – field bean (*Dolichos lablab* (L.) Roxb.) intercropping system as influenced by organic manure, supplemental irrigation and different fertility management approaches. Environ Eco 34(2):787–791

Newton ZL, Yoong KS (2016) Soil microbial properties during decomposition of pulse crop and legume green manure residues in three consecutive subsequent crops. Can J Soil Sci 96:413–426

Nintu M, Brahma SD, Mahesh CM, Dhyan S, Siba PD, Rakesh KT, Brij MS (2013) Effect of induced defoliation in pigeon pea, farmyard manure and sulphitation pressmud on soil organic carbon fractions, mineral nitrogen and crop yields in a pigeonpea–wheat cropping system. Field Crop Res 154:178–187

Nira R, Hamaguchi H (2012) Nitrogen accumulation in soybean [*Glycine max* (L.) Merr.] Is increased by manure compost application in drained paddy fields as a result of increased soil nitrogen mineralization. Soil Sci Plant Nutr 58:764–771

Olusegun OS (2014) Influence of NPK 15–15–15 fertilizer and pig manure on nutrient dynamics and production of cowpea, *Vigna unguiculata* L. Walp Am J Agric For 2(6):267–273

Omotoso O, Olatunji O, Obameso O, Odufoye A, Fajemisin A, Alokan A (2020) Nutritional value and acceptability of some selected forages in the derived Savanna Zone of Nigeria as ruminant feed. J Range Sci 10(4):426–433

Pal MS (2015) Effect of macro and micro–nutrients on herbage yield and quality of berseem (*Trifolium alexandrinum* L.) in tarai region of indo–Ganagetic plains. Indian J Plant Soil 2:77–80

Patricia IA, Laura Y, Amy TA (2016) Do soil organisms affect aboveground litter decomposition in the semiarid Patagonian steppe Argentina. Ecohealth 168:221–230

Pholsen S, Rodchum P, Sommart K, Ta-Un M, DEB H (2014) Dry matter yield and quality of forages derived from three grass species with and without legumes using organic production methods. J Khon Kaen Agric 42(1):65–80

Rabbi SMF, Wilson BR, Lockwood PV, Daniel H, Young IM (2014) Soil organic carbon mineralization rates in aggregates under contrasting land uses. Geoderma 216:10–18

Rafaat JG, Runak AH, Maki MA (2015) Effect of fertilizer application on vegetative growth characters of broad bean (*Vicia faba* L.). J Int Plant Anim Environ Sci 6(1):103–108

Rehman KR, Abdul C, Minminv Z, Longyu X, Xiaopeng AS, Abdul W, Hui L, Wu Y, Ziniu JZ (2017) Conversion of mixtures of dairy manure and soybean curd residue by black soldier fly larvae (*Hermetia illucens* L.). J Clean Prod 154:366–373

Rhohit K, Choudhary SK, Jitendra G, Singh P (2013) Effect of fertilizer and bio–fertilizers on growth, yield and economics of cowpea. Ann Plant Soil Res 15(2):177–178

Ruibo S, Xisheng G, Daozhong W, Haiyan C (2015) Effects of long–term application of chemical and organic fertilizers on the abundance of microbial communities involved in the nitrogen cycle. Appl Soil Ecol 95:171–178

Sacco D, Moretti B, Monaco S, Grignani C (2015) Six–year transition from conventional to organic farming, effects on crop production and soil quality. J Eur Agron 69:10–20

Salama HAS (2015) Interactive effect of forage mixing rates and organic fertilizers on the yield and nutritive value of berseem clover (*Trifolium alexandrinum* L.) and annual ryegrass (*Lolium multiflorum* lam.). Agric Sci 6:415–425

Sánchez CD, Tein B, Eremeev V, Luik A, Kauer K, Reintam E, Kahu G (2016) Winter cover crop effects on soil structural stability and microbiological activity in organic farming. Biol Agric Hort 32(3):170–181

Saravanan P, Sathish KS, Ignesh A (2013) Effect of organic manures and chemical fertilizers on the yield and macronutrient concentrations of green gram. Int J Pharm Sci Inven 2(1):18–20

Saritha M, Vijayakumari B, Hiranmai YR, Kandari LS (2013) Influence of selected organic manures on the seed germination and seedling growth of cluster bean (*Cyamopsis tetragonoloba* (L.) Taub). J Sci Technol Arts Res 2:16–21

Sarma I, Phukon M, Borgohain R, Goswami J, Neog M (2014) Response of French bean (*Phaseolus vulgaris* L.) to organic manure, vermicompost and biofertilizers on growth parameters and yield. J Asian Hort 9(2):386–389
Schipanski ME, Mary EB, Ebony GM, Jayson H, Denise MF, Jason PK, David AM, Richard GS (2017) Balancing multiple objectives in organic feed and forage cropping systems. Agric Ecosyst Environ 239:219–227
Schneider KD, Derek HL, Else KB, Paul RV (2016) Vegetative composition, arbuscular mycorrhizal fungi root colonization, and biological nitrogen fixation distinguish organic and conventional perennial forage systems. J Agron 109(4):1697–1706
Schröder JJ, Visser W, Assinck FB, Velthof GL (2013) Effects of short-term nitrogen supply from livestock manures and cover crops on silage maize production and nitrate leaching. Soil Use Manage 29:151–160
Senthilkumar PK, Sivagurunathan P (2012) Comparative effect on bacterial biofertilizers on growth and yield of green gram (*Phaseolus radiate* L.) and cow pea (*Vigna sinensis* Edhl.). J Int Cur Microbiol Appl Sci 1:34–39
Sharma S, Jat NL, Puniya MM, Shivran AC, Choudhary S (2014) Fertility levels and bio–fertilizers on nutrient concentration, uptake and quality of groundnut. Ann Agric Res New Ser 35(1): 71–74
Singh K, Arun PS, Ajay KM, Manoj P, Parveen K, Ashwani SA, Vijayata S (2015) Growth dynamics yield and nutrient uptake of fodder guar in relation to FYM and zinc fertilization. Eco Environ Conser 21(1):151–154
Sinha AK, Ghosh A, Dhar T, Bhattacharya PM, Mitra B, Rakesh S et al (2019) Trends in key soil parameters under conservation agriculture-based sustainable intensification farming practices in the eastern ganga Alluvial Plains. Soil Res 57(8):883–893
Sohail S, Ansar M, Skalicky M, Wasaya A, Soufan W, Ahmad Yasir T, El-Shehawi AM, Brestic M, Sohidul Islam M, Ali Raza M, EL Sabagh A (2021) Influence of tillage systems and cereals–legume mixture on fodder yield, quality and net returns under Rainfed conditions. Sustainability 13(4):2172
Staugaitis G, Narutytė I, Arbačauskas J, Vaišvila Z, Rainys K, Mažeika R, Masevičienė A, Žičkienė L, Šumskis D (2016) The influence of composts on yield and chemical elements of winter wheat and spring barley. Zem Agric 103(4):355–362
Steinshamn H, Steffen AA, Randi BF, Tor L, Torfinn T, Anne KB (2016) Yield and herbage quality from organic grass clover leys–a meta–analysis of Norwegian field trials. Org Agric 6:307–322
Valkama E, Salo T, Esala M, Turtola E (2013) Nitrogen balances and yields of spring cereals as affected by nitrogen fertilization in northern conditions: a meta-analysis. Agric Ecosyst Environ 164:1–13
Vasileva V, Todor K, Anna I (2017) Dry mass yield and amount of fixed nitrogen in some forage legume crops after treatment with organic fertilizer humustim. J Bul Agric Sci 23(5):816–819
Verde BS, Benjamin OD, Jayne NM (2013) the effects of manure, lime and P fertilizer on N uptake and yields of soybean (*Glycine max* (L.) Merrill) in the central highlands of Kenya. J Environ Sci Eng 2(2):111–116
Wiesmeier M, Hübner R, and Kögel-Knabner I (2015) Stagnic crop yields: an overlooked risk for the carbon balance of agricultural soils. Sci Total Environ 536:1045–1051
Yang J, Wei G, Shunrong R (2015) Long–term effects of combined application of chemical nitrogen with organic materials on crop yields, soil organic carbon and total nitrogen in fluvo–aquic soil. Soil Tillage Res 151:67–74

Yogesh P, Varma LR, Verma P, Joshi HN, More SG, Dabhi JS (2017) Influences of integrated use of organic and inorganic sources of nutrients on growth, flowering and yield of garden pea (*Pisum sativum* L.) cv. Bonneville. Leg Res 40(1):117–124

Zavattaro L, Bechini L, Grignani C, Van Evert FK, Mallast J, Spiegel H, Sandén T, Pecio A, Giráldez JV, Guzmán G, Vanderlinden K, D'Hose T, Ruysschaert G, Berge FM (2017) Agronomic effects of bovine manure: a review of long-term European field experiments. J Eur Agron 90:127–138

Zhang L, Til F, Jirko H, Christa H, Reiner D (2015) Comparison of energy consumption and economic performance of organic and conventional soybean production–a case study from Jilin Province China. J Integr Agric 14:1561–1572

Žydelis R, Sigitas L, Jonas V, Virmantas P (2019) Effect of organic and mineral fertilisers on maize nitrogen nutrition indicators and grain yield. Zem Agric 106(1):15–20

Atmospheric Chemistry of Aerosols and Their Role in Global Climate Change

Mujahid Farid, Tayyaba Younas, Saira Zaka, Muhammad Ahmed, Kiran Javed, Atif Hussain, Zaki Ul Zaman Asam, Sheharyaar Farid, Shafaqat Ali, and Muhammad Zubair

Abstract This chapter provides a brief discussion about the aerosol particles with their atmospheric composition, their different emission sources either natural (biogenic, volcanos, sea salt, desert dust) or anthropogenic (biomass burning, fossil fuel burning). Further, it included the different pathways through which aerosol components entered the atmosphere in the aerosol phase. The aerosols interact with clouds, radiations, and other atmospheric components in different ways which as a result will affect the climatic patterns, i.e., precipitation, temperature, etc. The atmospheric chemistry of aerosols is of great importance, mainly sulfate aerosols regarding their ability to affect the climate both in a positive and negative way. The physicochemical properties of aerosols and their lifetime in the atmosphere and the different ways in which they will affect climate are important. They affect the climate indirectly through their impact on clouds and have direct effects on climate through scattering and solar radiation's absorption into space. Keeping in view the effects of aerosols on climate and their cooling effect on the atmosphere, different ways are discussed which will help in mitigating climate change.

Keywords Aerosol · Anthropogenic · Radiations · Climate change · Absorption

---

M. Farid (✉) · T. Younas · S. Zaka · M. Ahmed · K. Javed · A. Hussain · Z. U. Z. Asam
Department of Environmental Sciences, University of Gujrat, Hafiz Hayat Campus, Gujrat, Pakistan

S. Farid
Institute of Biomedical Sciences ABEL SALAZAR, University of Porto, Porto, Portugal

Department of Biology, Ecology, and Evolution, University of Liege, Liege, Belgium

University of the Basque Country, Bilbao, Spain

S. Ali
Department of Environmental Sciences and Engineering, Government College University, Faisalabad, Pakistan

Department of Biological Sciences and Technology, China Medical University, Taichung, Taiwan

M. Zubair
Department of Chemistry, University of Gujrat, Hafiz Hayat Campus, Gujrat, Pakistan

M. Hasanuzzaman et al. (eds.), *Managing Plant Production Under Changing Environment*, https://doi.org/10.1007/978-981-16-5059-8_17

1 Introduction

The Atmosphere of the earth is evolving because of natural activities along with chemical complexity, anthropogenic and biological activities, have chemical links to Oceans, solid parts of the earth, and biota. Man-made events and activities like land-use change or industrial activities result in modification of the chemical composition of spheres (tropo-strato spheres) along with future impacts on the climate of the earth and living things. Such as the formation of ozone hole over Antarctica since the late 1970s, greenhouse gases along with earth trend, change in the amount of ozone present in troposphere, acidic deposition all are the result of burning of hydrocarbon substances, nitrogen, and Sulphur oxides in industrial states. Aerosols also contribute to changing of climatic conditions on earth (Hochella et al. 2019). Prediction is to be made for the chemical and physical properties of aerosols impacts on forcing climatic changes which may be physical or chemical properties. Atmospheric measurement is not enough to identify the significance of aerosols in climate. It is a great challenge in characterizing its occurrence and nature present in the atmosphere with models of impacts and uncertainties in the prediction of climate. Because it has a lifetime less than greenhouse gases spread in the atmosphere in wide range with different particle sizes, concentration along with composition have variability in spatial and temporal distribution (Huebert et al. 2003).

Climate change is the rise in average temperatures of the earth's surface, mainly due to the burning of fossil fuels. It is the most significant issue of our time and we are in crisis. From changing weather conditions that threaten food production to rising sea levels that increase the risk of catastrophic flooding, the effects of climate change are global and unprecedented. Without significant measures, adapting to these effects will be more difficult and costly in the future (Höök and Tang 2013).

The chemistry of the atmosphere is the branch of atmospheric research that focuses on chemical processes in the earth's atmosphere. Research in this area is essential for a better understanding of climate change, air quality, and the interactions between atmosphere and biosphere. Therefore, this area of study is at the interface of chemistry with physics and biology and includes processes that run on a temporal and spatial scale, less than a few seconds and mm compared to those that occur on the world's scale. The field is developing rapidly due to advances in a basic understanding of chemical processes in the atmosphere (Andreae and Crutzen 1997).

The accumulation of liquid or solid particle's suspension in the atmosphere is called an aerosol. Aerosols are defined more precisely separately from all other hydrometers, i.e., particles from clouds, ice crystals, and raindrops. They are always present in the atmosphere in extremely variable concentrations due to the very high heterogeneity of their sources and their relatively short residence time in the atmosphere. They radiate directly because of their diffusion and absorption of sun and infrared radiation in the atmosphere. They also modify the processes of formation of hot, icy, and mixed phases and increase the concentration of droplets and the concentration of ice particles (Boucher et al. 2013). In hot clouds, aerosols decrease the efficiency of precipitation and therefore caused indirect radiation related to these

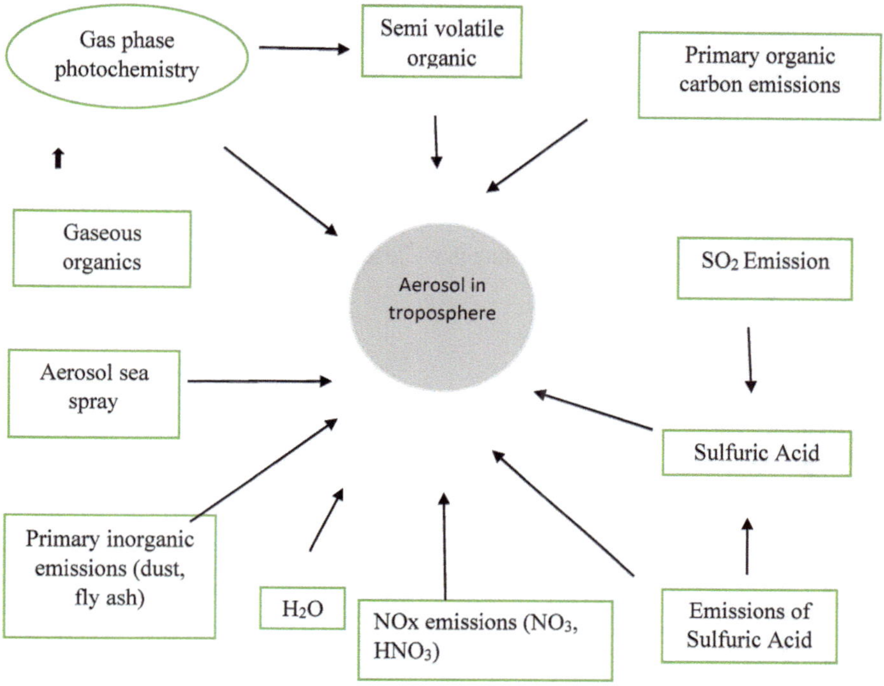

Fig. 1 Different components and ways of incorporation of tropospheric aerosols into the aerosol phase

changes in cloud properties. They have likely contributed significantly to total emissions. One of the important characteristics of aerosols is that they have a short lifespan in the atmosphere and therefore cannot simply be seen as a long-term compensation for the effects of greenhouse gases on global warming (Boucher 2015). The aerosols of the troposphere have a remarkably diverse composition that reflects the multitude of particle sources on the earth's surface. Figure 1 shows the tropospheric particles and their chemical components with their incorporating pathways in the aerosol phase. These chemical components of tropospheric aerosols include inorganic materials such as sulfate, ammonium, nitrate, sodium, chloride, trace elements, building blocks, carbonaceous materials, and water (Seinfel 2014).

2 Aerosol's Sources

2.1 Sea Salt

Sea salt is prepared by evaporating saltwater. They are also called marine aerosols by evaporating seawater, the sea salt is generated and hence this is a natural source of

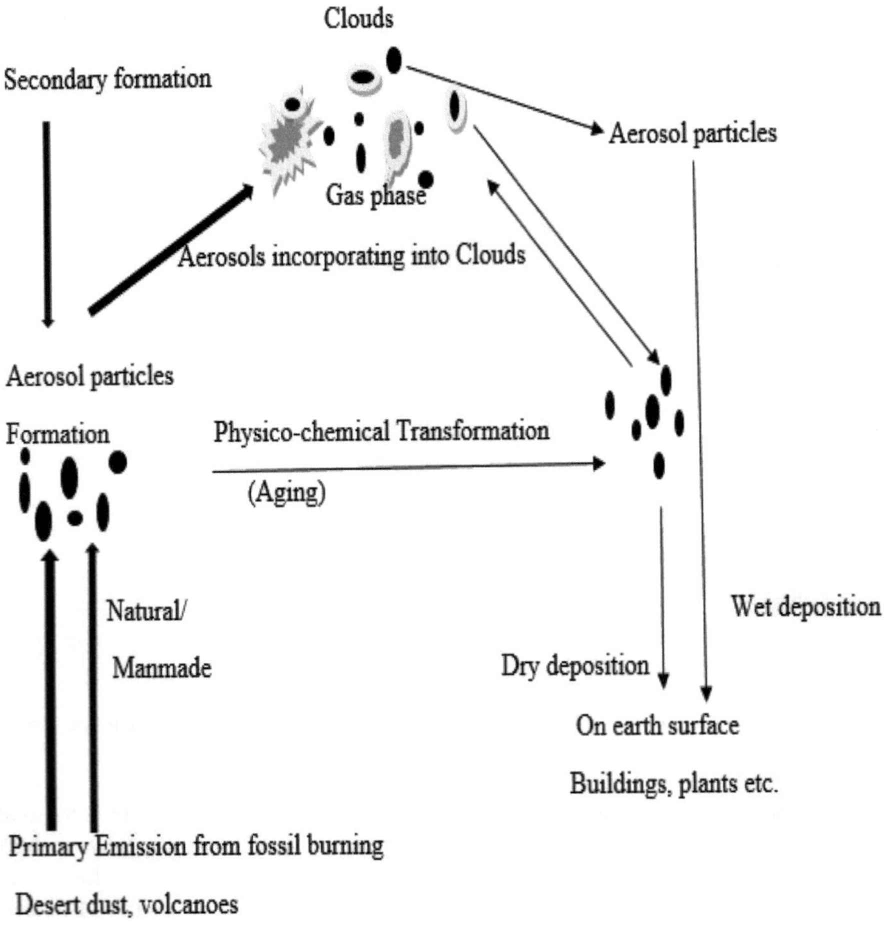

Fig. 2 Aerosol's atmospheric cycling and impacts on the climate system

sodium. Sea salt extracts from a natural source and also contains other minerals, such as potassium, magnesium, and calcium. By weight, there is 40% sodium in sea salt. The wind erosion at the surface of the sea launches fine particles of salty marine water into the air (Shepherd 2012). A small amount of the water dissipates, with the goal that the amount of salt in the molecule rises. This offers ascend to the ocean salt particles that are pretty much hydrated by the encompassing moistness. These particles are commonly called sea salt aerosols, but they may contain organic content and impurities. Therefore, it is increasingly suitable to be called as sea spray aerosols. Seaspray aerosols spread sizes that run from 100 nm to a few micrometers. The biggest particles settle down fast at the surface of the sea and are hence less significant for climate (Boucher 2015) (Fig. 2).

2.2 Dust of Deserts

Dust storms occur as often as possible over deserts and dry soils, where soil particles attach to the surface indefinitely. The grains of sand thrown into the air fall back to the ground after a few hours, but the small particles remain in suspension significantly for more than 7 days and can release a large number of kilometers in the wind. Estimates of particle lengths vary regularly from 100 nm to several micrometers. Larger particles can also rise, but they settle quickly. Desert sprays are also called mineral powders or mineral sprays. Desert dust currents are particularly dependent on the environment and weather conditions (Boucher 2015).

2.3 Volcanic Aerosols

Volcanoes can release fragments of crushed rock and minerals, commonly called volcanic ash, during explosive eruptions. These debris have sizes that typically range from a micrometer to a few millimeters. Volcanic debris can be sent through sections of 200 or 300 to 2,000 or 3,000 km, but as particles of the order of microns, they usually collapse rapidly. Its influence on the atmosphere is already limited. Volcanoes also emit sulfur-rich gases (such as SO_2 and H_2S), which oxidize in the climate to form aerosols of submicron sulfate (Skeie et al. 2011). If these sulfur-containing gases are produced in the troposphere, the residence time of the aerosol will be short, probably half a month. If the radiation is incredible enough to infuse sulfur gases into the stratosphere, volcanic aerosols have a longer life of several months to more than a year, depending on the injection site and altitude (Boucher 2015).

2.4 Biogenic Aerosols

Biogenic agents are those that result from the activities of living things. Many forms of biogenic aerosols fill the air and are spread across the planet by the wind. Pollen from trees and other plants, mild and bacterial spores, and airborne viruses fall into this category. This includes plant and insect remains pollen, spores, viruses, and bacteria. If these particles once blocked, can move from one area to another at different distances depending on their size. Seawater can also contain organic material, some of which is shifted to aerosols in the sea during the emission process. This primary organic substance is preferably present in the particles with a diameter of less than 200 nm and its amount depends on the biological activity of the seawater (Bond et al. 2013). The other main sources of aerosol precursors are terrestrial and marine ecosystems. Some phytoplankton varieties produce dimethyl sulfide (DMS), which is a gaseous compound that oxidizes in the atmosphere to form sulfur-containing aerosols. Algae and plants emit volatile organic compounds that oxidize and precipitate in the atmosphere and give the atmospheric aerosol organic matter. These sprays are called secondary biogenic sprays. Their size is generally on the order of a few tenths of a micrometer (Boucher 2015).

2.5 Biomass Burning Aerosols

Biomass is defined as any biological (vegetation, dead wood, animal waste, peat) material in the living world that can possibly be burned with the exception of fossil fuels (petroleum coal, and gas) that are formed on geological timescales. Biomass burning produces primary aerosols resulting from the incomplete combustion of organic substances. Aerosols that burn biomass include organic carbon, which binds to hydrogen and oxygen atoms, and black carbon, which has a very high carbon content. These aerosols are generally of submicronic sizes and are clearly visible in chimneys (Stocker et al. 2013). The sources of aerosols that burn biomass are both anthropogenic and natural. Biomass burning also releases gaseous compounds which are aerosol precursors, such as VOCs and SO_2, (Boucher 2015).

2.6 Aerosols from Fossil Fuel Combustion

Organic carbon, black carbon as well as sulfur dioxide are formed, when petroleum and coal products are burned, which are then converted into sulfated aerosols. These particles are mainly submicronic. Fossil fuels are the main source of power generation in industries. The need for development will increase fossil fuel burnings and as a result, there will be more submicronic particles in the atmosphere (Boucher 2015).

2.7 Nitrates

Nitrate aerosols are chemically formed in the atmosphere and their main precursors are nitric acid and ammonia. Precursors of nitric acid, ammonia, and nitric oxide have a large number of natural and anthropogenic sources. The main sources of ammonia are the feces from the synthetic fertilizers, identified wild and domestic animals, the oceans, the biomass combustion, useful plants, people and households, soils, industrial processes, and flammable minerals (Bouwman et al. 1997). Typical sources of nitrogen oxide are the combustion of fossil fuels, soils, the combustion of biomass, and lightning. With rapidly increasing nitrogen emissions, enough nitrate aerosols can be generated to compensate for the expected reduction in sulfate increase by 2100. Nitrate and ammonium aerosols offer additional particle areas for the diffusion of nitrogen. Solar radiation from incident particles so interferes with the production of photochemical oxidizing agents by modifying frequency photolysis. The formation of nitrate aerosol is caused through nitrogen radical's reactions, such as N_2O_5, NO_3, and HNO_3 heterogeneous reactions on aerosol surfaces (Bauer et al. 2007).

2.8 *Sulfates*

In the atmosphere, gaseous precursors produce sulfate aerosols by chemical reactions. The two most important sulfate precursors are dimethyl sulfide (DMS) from biogenic sources, especially marine plankton, and sulfur dioxide (SO_2) from anthropogenic sources and volcanoes and. The chemical way to convert precursors to sulfate is important as it changes the radiation's impacts. Mostly SO_2 is converted to ignited droplets, sulfate, or the gas phase, which are then evaporated. The reactions that produce sulfated aerosols can be divided into clear skies and cloudy approaches. Strategies with a clear sky react sulfur dioxide and dimethyl sulfide (DMS) in the presence of water vapor through a complex series of steps to provide sulfuric acid gas (H_2SO_4). The connection forms a micrometer fragment of the debris. This is done by condensation on existing waste or by reaction with water vapor or other sulfuric acid molecules. This conversion is called gas to particle conversion (Eliseev et al. 2009). Sulfuric acid then reacts with small portions of ammonia to form various hydrated forms of ammonium sulfate [$(NH_4)_2SO_4$]. Sulfate spray is also found in the clouds. This path begins when sulfur dioxide dissolves in existing cloud droplets. There it can be oxidized with the low concentrations of aqueous hydrogen peroxide (H_2O_2) that result from the integration of hydroxyl molecules (Schleussner, et al. 2016). The oxidative reaction then bureaucratizes sulfuric acid and its ammonium salts as a reaction. In droplets, acid sulfate is in a highly hydrated form in which the water molecules are bound to the sulfate. Evaporation removes a certain amount of moisture. Because the sulfates are suspended in water, the vaporization product is an extremely concentrated sulfate solution. The end result is a submicron aerosol droplet that is chemically no different from the aerosol obtained by converting the fuel into particles (Bauer et al. 2004).

3 Interactions of Aerosols with Climate System

Aerosols get interacted with electromagnetic radiation that always moves through the atmosphere. For aerosols to have an impact on the air conditioning system, they have to influence the energy of incoming short waves, the energy of long waves, or the internal energy flow. Solar radiation (short waves) which interacts with aerosols is deflected in all directions but is anisotropic and the process is called scattering. In the absorption process, some aerosols also absorb sunlight and convert electromagnetic electricity into heat. Aerosols also absorb and scatter terrestrial radiation (long

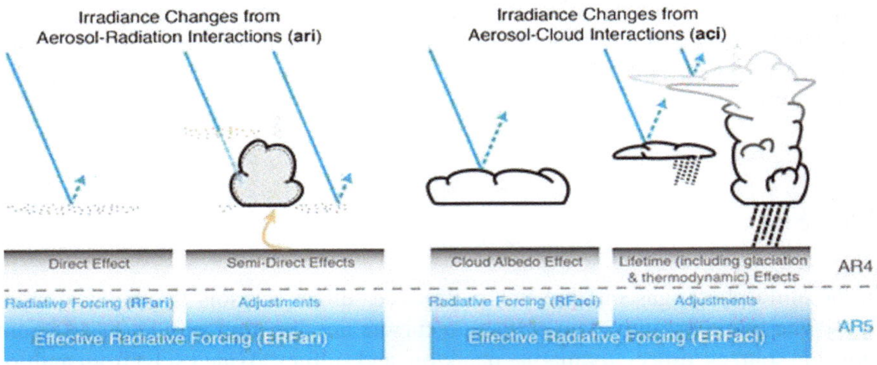

Fig. 3 Interactions of aerosol with climate (Samset 2016)

waves) emitted from the atmosphere and the earth's surface, and can emit such radiation. Aerosols partially control the concentration and size of water clouds from droplets, by affecting the microphysical properties of water clouds, in particular their role as cloud condensation nuclei on which water vapor can condense (Fuzzi et al. 2006). On the other hand, clouds also affect the populations of aerosols. Clouds of rain help in removing the aerosols from the atmosphere. A distinction is made between below-cloud scavenging, in which aerosols are captured by precipitating raindrops and in-cloud scavenging, in which aerosols get into water drops before precipitation. Although scavenging by falling snowflakes and ice crystals is much less effective, the same process applies to ice clouds. As a result, due to wet deposition clouds are an important sink for aerosols. Finally, evaporation of non-precipitating clouds releases residual cloud droplets into the atmosphere, which can be converted into aerosols and used for cloud condensation nuclei. Due to the oxidation of SO_2 in cloud droplets, the other aerosol material which is in between clouds can incorporate into cloud droplets, and cloud droplets act to form larger droplets. The process is important for convective clouds that can activate small aerosols and emit larger aerosols that can serve in other clouds as condensation nuclei (Boucher 2015) (Fig. 3).

4 Effects of Aerosols on Climatic Systems

The components of organic aerosols (OAs) make up a large, sometimes even dominant proportion of airborne particles. They affect the physicochemical properties of aerosol particles and therefore affect the atmosphere and climate by interacting with clouds, water vapor, reactive gases, precipitation, and radiation. In addition, the earth's surface affect the buildings and plants through wet and dry deposition (Gelencser and Varga 2005).

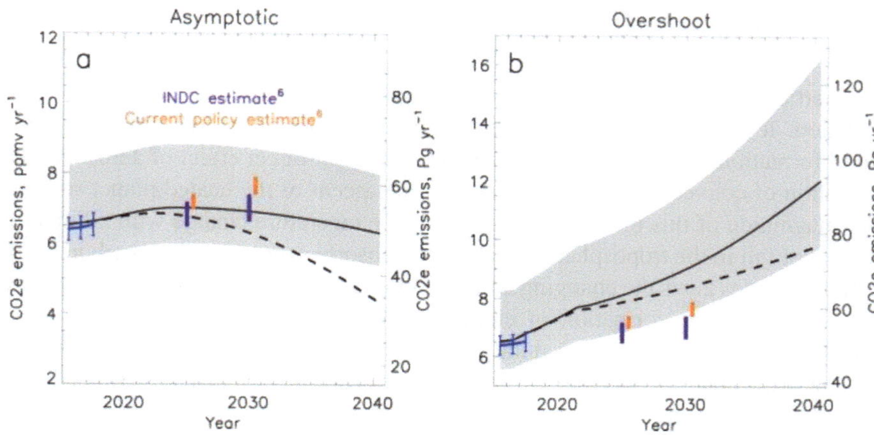

Fig. 4 Limits of CO_2 emissions for high and low aerosol scenarios solid lines for high and dashed lines for low emissions and arrange lines for current (Larson and Portmann 2019)

Atmospheric aerosols have a vital role in the climate system. They may influence the global radiation budget by directly diffusing and absorbing incoming solar radiation and outgoing infrared radiation and semi-directly by changing the structure of the atmospheric temperature and the rate of evaporation cloud droplets reflectivity by increasing the cross-sectional area of the total dispersion and weakening or increasing the precipitation regime (Fig. 4).

4.1 *Direct Effects*

Aerosols scatter and absorb short-wave radiations into space, which leads to a decrease in radiations of the sun on the surface of the earth, and causes a loss of energy and cooling of the climate system. Solar radiation's absorption is achieved by heating within the aerosol layer and also by a reduction of incoming solar radiation at the surface of the earth. Such effects occur in clear sky conditions preferentially, but not unique, and known as direct effects of aerosols. Although surface solar radiation is exposed to external and internal influences, there is solid proof that aerosols of human origin play a significant role in decadal variations in surface solar radiation (Hofmann et al. 2006).

4.2 Semi Direct Effects

The vertical temperature profile gets changed when the aerosols absorb the solar radiation. It influences the atmospheric stability, relative humidity, And thus the cloud formation. And this effect is known as the semi-direct effect of aerosols. The formation of convective clouds begins with the ascent of the heated planet surface. The magnitude of this increase depends on the temperature change with altitude or the rate of fall in the troposphere. Applying an absorbent spray such as carbon black at high altitudes heats the changing air and cools the surface, thus reducing the decay rate. Therefore, the formation of clouds can be reduced under the aerosol layer (Isaksen et al. 2009). However, at the level of the aerosol, the rise in temperature can prevent the formation of a cloud. In addition, the absorption of aerosols embedded in droplets in an existing cloud can contribute to their evaporation, which is referred to as cloud combustion. Although this is an active area of research the effects of semi-direct spraying are currently not well understood (Khain 2009).

4.3 Indirect Effects

Aerosols control the optical and microphysical properties of clouds as they serve as condensation nuclei in clouds from liquid water. Increasing the concentration of aerosols leads to an increase in the concentration of cloud condensation nuclei and in general to an increase in the concentration of cloud drops. Less absorbed solar energy and climate cooling. This effect is known as an indirect aerosol effect (Klimont et al. 2013).

4.4 Effects on Temperature

The effect of aerosols on the radiation balance leads to changes in the temperature of the earth's surface. Some researches show that the air temperature is trending above the earth's surface between 1979 and 2012 is approximately 0.26–0.27 K year^{-1}. The changes in temperature are mostly due to greenhouse gases, but the effects of aerosols cannot be overlooked. Global warming caused by greenhouse gases is partially offset by the cooling effects related with the increase in aerosols. The anthropogenic effects linked with the fast growth of short-term emissions of sulfur played a significant role in the slow warming detected between 1998 and 2008. The temperature reactions to the radioactive effects of aerosols are not evenly distributed in space and time.

4.5 *Effects on Precipitation*

Generally, increasing aerosol concentrations would decrease average precipitation. Because whether interactions with aerosols, as discussed above, indicate that aerosols can interfere with atmospheric stability and microphysics of cloud, which can lead to precipitation changes. As a result of global warming, they also contribute to the expected changes in rainfall as they can warm or cool the climate. In addition, it has been suggested that the changes in the tropical rainforest and fluctuations of the Pacific decade could be associated with recent changes in aerosol exposure from Europe and the United States to Southeast Asia (Allen et al. 2014). Change in aerosol concentration affects precipitation on two-time scales; a series of rapid effects associated with a rapid response of clouds and atmospheric stability, and another one that involves long-term changes in the temperature of the surface, and hence evaporation is connected. New model studies show that an isolated increase in black carbon initially reduces precipitation due to reduced convection. Precipitation will increase over time as the absorption of British Columbia warms the surface. Overall, however, precipitation fluctuation due to British Columbia is considered negative. On the other hand, for sulfate, there is a small change in the preliminary precipitation, but a large decrease over time as the jet cools the surface (Marcolli et al. 2004).

5 Mitigation for Climate Change

Aerosols can be used for mitigating climate change and may also increase the climate changing effect. For mitigating climate change different methods are provided here which are as follows.

5.1 *Reducing Aerosol Emissions*

Since aerosol emissions have a significant impact on the climate and the short lifespan of the atmosphere, scientists and politicians are advocating reducing emissions as strategies to curb climate change. Several studies recently found that the climatic effect that can be obtained by reducing aerosol emissions is comparatively small as compared to the long-term effects of greenhouse gas emissions reduction. Aerosols are rarely isolated, but released from the same sources. For example, black carbon and sulfur dioxide are released in a coal fire, which means that reducing these fires removes both the heating and cooling sources in the atmosphere (Ningombam et al. 2019). There is no uncertainty that reducing emissions will have health and climatic benefits. The potential to mitigate the effects of climate change is currently a debatable topic among researchers and politicians (Mircea et al. 2005).

5.2 Climate Engineering

One of the first widespread proposals for a climate engineering program was the deliberate release of sulfates and other aerosols. The idea is that the attenuation of the incoming sunlight can compensate for the additional energy that is stored due to the increased greenhouse effect. We can say simulation of the climatic effects of volcanoes using air balloons or sprays in the stratosphere will be performed (Myhre et al. 2013). Increasing the flow of sea mist droplets in areas where stratified clouds form is another aerosol proposal. Here the indirect effect of the fog will be used to whiten the clouds and reflect the sunlight. A third option, based on aerosols, is to place clouds at high altitudes, which increases the rate of evaporation. This reduces the warming of its surface as a result of the re-emission of long-wave radiation (Olivier and Peters 2018).

5.3 Issues Regarding Reduced Aerosols Emissions

When looking for positive effects on climate and air quality by reducing the pollutant emissions, two main problems arise. The first is that certain air pollutants such as sulfate and nitrate aerosols cool the atmosphere; by reducing them, warming is accelerated. It means that the inclusive impact of air pollutants on global warming and cooling must be assessed if a generally beneficial effect is sought. The IPCC regards them as short-term climate variables (SLCF), regardless of whether the increase in radiation or the energy imbalance that is affected by a change in its concentration is positive or negative. The IPCC 2 report showed that average radiation emissions from anthropogenic aerosols are negative overall, which means that reducing their emissions will increase global warming. In order to reduce air pollution and acid rain, the industrialized states mainly reduce sulfur dioxide (SO_2) emissions, which is a preliminary introduction to the sulfur mist (Paulot et al. 2018). If human air pollution is reduced as expected, the emissions reductions required to reach the target temperature of 2 °C will be much greater. In a clean air scenario (weak aerosol) containing 65% less than the anthropogenic effects of aerosols in 2100 (Fig. 5).

5.4 Uncertainty in Future Climate Mitigation Efforts

Future aerosol emissions play a crucial role in determining the permissible GHG emissions. However, the spatial and temporal heterogeneity of future aerosol emissions adds great uncertainty to the estimation of the increase in radiation. The temporary variability of volcanic eruptions will create variability of the meteorological signal from effective radiative forcing (ERF) and aerosols, but will not affect the

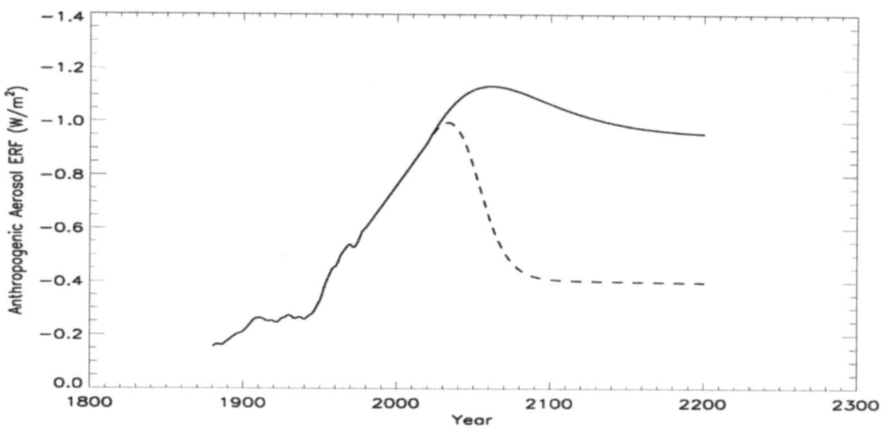

Fig. 5 Forcing scenarios of anthropogenic aerosols, solid line representing high aerosol scenario and dashed line representing low aerosol scenario (Larson and Portmann 2019)

cumulative GHG emissions, which can reach the target temperature of 2 °C. Regional changes in the type and amount of emissions of manmade aerosols can change the manmade aerosol ERF.

Directly from the amount of aerosol emitted and indirectly from changes in efficiency related to the location and type of aerosol emitted. Recent observations suggest different Regional trends in aerosol optical depth and radiation force. For these reasons, we consider a wide range of future aerosol ERFs to calculate our GHG emissions (Seinfeld and Pandis 2006). Global greenhouse gas emissions and atmospheric concentrations continue to increase. Moderate reductions in non-INDC (Intended Nationally Determined Contributions) greenhouse gas emissions may prevent uniform limiting warming to 2 °C in the near future, but aerosol emissions add great uncertainty to the timing and rate of GHG reductions. Purification of anthropogenic aerosol emissions, which is a likely scenario to reduce greenhouse gas emissions, will require much larger reductions in greenhouse gas emissions in the short term and slightly larger reductions in the long term for limiting the global warming to 2 °C above the pre-industrial stage. Regardless of the aerosol scenario, greenhouse gas emissions must be significantly reduced by 2100, and high aerosol emissions are simply time-saving (Larson and Portmann 2019).

6 Conclusion

Aerosols are suspended particles in the atmosphere, which interact with atmospheric chemistry. They are released from different anthropogenic and natural resources, i.e., volcanic, desert, biogenic, biomass burning, sulfates, and nitrates. Their emissions to the atmosphere affect the climatic patterns either directly or indirectly. As aerosols have a cooling effect in the atmosphere, their reduced emissions could enhance the global warming effect. So, mitigation for climate change is done keeping this point in view, as if we reduce the aerosols emissions to protect the atmosphere from aerosols and their precursors this may increase the warming effect.

Acknowledgments The authors are thankful to the University of Gujrat, Gujrat, Pakistan, and the Higher Education Commission (HEC) of Pakistan for providing technical and library support. The present study is part of HEC funded NRPU funded project number 2017/HEC/R&D/NRPU/8996.

References

Allen RJ, Norris JR, Kovilakam M (2014) Influence of anthropogenic aerosols and the Pacific decadal oscillation on tropical belt width. Nat Geosci 7(4):270–274

Andreae MO, Crutzen PJ (1997) Atmospheric aerosols: biogeochemical sources and role in atmospheric chemistry. Science 276(5315):1052–1058

Bauer SE, Balkanski Y, Schulz M, Hauglustaine DA, Dentener F (2004) Global modelling of heterogeneous chemistry on mineral aerosol surfaces: the influence on tropospheric ozone chemistry and comparison to observations. J Geophys Res 109:D02304. https://doi.org/10.1029/2003JD003868

Bauer SE, Mishchenko M, Lacis A, Zhang S, Perlwitz J, Metzger S (2007) Do sulfate and nitrate coatings on mineral dust have important effects on radiative properties and climate modeling? J Geophys Rev 112:D06307. https://doi.org/10.1029/2005JD006977

Bond TC, Doherty SJ, Fahey DW, Forster PM, Berntsen T, DeAngelo BJ, Flanner MG, Ghan S, Kärcher B, Koch D, Kinne S (2013) Bounding the role of black carbon in the climate system: a scientific assessment. J Geophys Res Atmos 118(11):5380–5552

Boucher O (2015) Atmospheric aerosols. In: atmospheric aerosols. Springer, Dordrecht, pp 9–24

Boucher O, Randall D, Artaxo P, Bretherton C, Feingold G, Forster P, Wyant M (2013) Clouds and aerosols. The physical science basis. Contribution of working group I to the fifth assessment report of the intergovernmental panel on climate change. Cambridge University Press, Cambridge, U.K

Bouwman AF, Lee DS, Asman WAH, Dentener FJ, Hoek KWVD, Olivier J (1997) A global high-resolution emission inventory for ammonia. Global Biogeochem 11:561–587

Eliseev AV, Mokhov I, Karpenko AA (2009) Global warming mitigation by means of controlled aerosol emissions into the stratosphere: global and regional peculiarities of temperature response as estimated in IAP RAS CM simulations. Atmos Ocean Opt 22(4):388–395

Fuzzi S, Andreae MO, Huebert BJ, KulmalaM BTC, Boy M, Kerminen VM (2006) Critical assessment of the current state of scientific knowledge, terminology, and research needs concerning the role of organic aerosols in the atmosphere, climate, and global change. Atmos Chem Phys 6(7):2017–2038

Gelencser A, Varga Z (2005) Evaluation of the atmospheric significance of multiphase reactions in atmospheric secondary organic aerosol formation. Atmos Chem Phys 5:2823–2831

Hochella MF, Mogk DW, Ranville J, Allen IC, Luther GW, Marr LC, McGrail BP, Murayama M, Qafoku NP, Rosso KM, Sahai N, Schroeder PA, Vikesland P, Westerhoff P, Yang Y (2019) Natural, incidental, and engineered nanomaterials and their impacts on the earth system. Science 363:6434

Hofmann DJ, Butler JH, Dlugokencky EJ, Elkins JW, Masarie K, Montzka SA, Tans P (2006) The role of carbon dioxide in climate forcing from 1979 to 2004: introduction of the annual greenhouse gas index. Tellus B Chem Phys Meteorol 58:614–619

Höök M, Tang X (2013) Depletion of fossil fuels and anthropogenic climate change—a review. Energy Policy 52:797–809

Huebert BJ, BatesT RPB, Shi G, Kim YJ, Kawamura K, Carmichael G, Nakajima T (2003) An overview of ACE-Asia: strategies for quantifying the relationships between Asian aerosols and their climatic impacts. J Geophys Res Atmos 108:D23

Isaksen ISA, Granier C, Myhre G, Berntsen TK, Dalsøren SB, Gauss M, Wuebbles D (2009) Atmospheric composition change: climate-chemistry interactions. Atmos Environ 43(33): 5138–5192

Khain A (2009) Notes on state-of-the-art investigations of aerosol effects on precipitation: a critical review. Environ Res Lett 4(1):015004

Klimont Z, Amann M, Rao S, Dentener F (2013) Exploring synergies of aerosol and climate mitigation strategies. Goldschmidt 2013:25–30

Larson EJL, Portmann RW (2019) Anthropogenic aerosol drives uncertainty in future climate mitigation efforts. Sci Rep 9(1):1–8

Marcolli C, Luo B, Peter T (2004) Mixing of the organic aerosol fractions: liquids as the thermodynamically stable phases. J Phys Chem A 108:2216–2224

Mircea M, Facchini MC, Decesari S, Cavalli F, Emblico L, Fuzzi S, Vestin A, Rissler J, Swietlicki E, Frank G, Andreae MO, Maenhaut W, Rudich Y, Artaxo P (2005) Importance of the organic aerosol fraction for modeling aerosol hygroscopic growth and activation: a case study in the Amazon Basin. Atmos Chem Phys 5:3111–3126

Myhre G, Myhre C, Samset BH, Storelvmo T (2013) Aerosols and their relation to global climate and climate sensitivity. Nat Knowl Proj 4:5

Ningombam SS, Larson EJL, Dumka UC, Estellés V, Campanelli M, Steve C (2019) Long-term (1995–2018) aerosol optical depth derived using ground based AERONET and SKYNET measurements from aerosol aged-background sites. Atmos Pollut Res 10(2):608–620

Olivier JGJ, Peters JAW (2018) Trends in global CO2 and total greenhouse gas emissions: 2018 report. PBL Netherlands Environmental Assessment Agency

Paulot F, Paynter D, Ginoux P, Naik V, Horowitz LW (2018) Changes in the aerosol direct radiative forcing from 2001 to 2015: observational constraints and regional mechanisms. Atmospheric Chem Phys 18:13265–13281

Samset BH (2016) Aerosols and climate. In: Oxford research encyclopedia of climate science

Schleussner C-F et al (2016) Science and policy characteristics of the Paris agreement temperature goal. Nat Climat Change 6:827–835

Seinfeld JH (2014) Tropospheric chemistry and composition-aerosols/particles

Seinfeld JH, Pandis SN (2006) Atmospheric chemistry and physics. In: Air pollution to climate change, 2nd edn. Wiley, New York

Shepherd JG (2012) Geoengineering the climate: an overview and update. Philos Trans Royal Soc 370(1974):4166–4175

Skeie RB, Berntsen TK, Myhre G, Tanaka K, Kvalevåg M, Hoyle CR (2011) Anthropogenic radiative forcing time series from pre-industrial times until 2010. Atmos Chem Phys 11(22): 11827–11857

Stocker TF, Qin D, Plattner GK, Tignor M, Allen SK, Boschung J, Midgley PM (eds) (2013) Climate change 2013: the physical science basis. Contribution of working group I to the fifth assessment report of the intergovernmental panel on climate change. Cambridge University Press, Cambridge.

Increase in Food Scarcity, Agricultural Challenges, and Their Management: Pakistan Perspectives

Mujahid Farid, Ahmad Fayyaz, Ehsan Ahmed, Memona Arooj, Shafaqat Ali, Wajiha Sarfraz, Mohsin Abbas, Alia Idrees, Madiha Shahzadi, and Zulakha Rasheed

Abstract Advances in technologies have contributed to more food production, but it is not sufficient for the current growing population. An increase in the production of growth that ultimately leads to an increase in economic growth puts a huge cost into the natural environment. Half of the forests of the earth are now depleted, and we have only 1% of the freshwater that can deplete a lot with time. Biodiversity has been eroded with time; fossil fuels that can be used for multiple purposes can contribute millions of greenhouse gases. The population is expected to be 9.9 billion in 2050, which can cause stress on the food supply. The demand for basic facilities like food will increase up to 60%, and urbanization, climate change, and degradation of soil can increase and lead to less availability of the land. It is one of the concerning issues in the world, especially in developing countries. The present study was focused to elucidate the global food scarcity and agricultural challenges faced by the current population along with the management strategies in practice to enhance food security.

Keywords Population · Biodiversity · Agriculture · Scarcity · Management strategies

---

M. Farid (✉) · A. Fayyaz · E. Ahmed · M. Arooj · M. Abbas · A. Idrees · M. Shahzadi · Z. Rasheed
Department of Environmental Sciences, University of Gujrat, Hafiz Hayat Campus, Gujrat, Pakistan

S. Ali
Department of Environmental Sciences and Engineering, Government College University, Faisalabad, Pakistan

Department of Biological Sciences and Technology, China Medical University, Taichung, Taiwan

W. Sarfraz
Department of Botany, Government College Women University, Sialkot, Pakistan

M. Hasanuzzaman et al. (eds.), *Managing Plant Production Under Changing Environment*, https://doi.org/10.1007/978-981-16-5059-8_18

1 Introduction

Survival is impossible without food, but as the number of people increases with time there is a question that is there enough food for people to eat. The world population is about 7.5 billion; more than 850 million people can face hunger and more than two billion face deficiency of food according to the report of Food and Agriculture Organization. People who face food scarcity mostly live in the roadless rural areas of the poor developing countries. Undernutrition is the condition where the person cannot receive enough calories according to the need of their body, and mostly children face the problem of undernutrition. Children under the age of 5 mostly suffer from lack of calories, and according to World Health Organization, 182 million children suffer from undernutrition, more than half of the children's death in developing countries is related to the scarcity of food (Manap et al. 2019). Global food security could be in jeopardy due to climate change and pressure on natural resources. Billions of people still face pervasive poverty, gross inequalities, joblessness, environmental degradation, disease, and deprivation that may also contribute to the scarcity of food. Advances in technologies (GMO, Green Revolution) have contributed to more food production, but it is not enough for current growing population. Increase in the production of growth that ultimately leads to increase in economic growth puts a huge cost into the natural environment. Half of the forest of earth is now depleted, and we have only 1% of the freshwater that can be depleted with time. Biodiversity has been eroded with time, and fossil fuels used for multiple purposes (also for production of food) contribute millions of greenhouse gases (Bratspies 2012).

All that negative impacts are accelerating with the passage of time and ultimately link with agriculture. Agriculture is the main problem of the negative impacts on the environment (although other factors also contribute like industrialization) because according to FAO, 60% of the world population depends on agriculture. Deforestation is mainly due to agriculture (farming purpose) and contributes to greenhouse gases, destruction of habitat, and loss of biodiversity. This can increase the rate of natural disasters fivefold since 1970. This can disturb the sustainability of the food and can cause the scarcity of food. Climate change is the major factor that can affect agriculture and forestry, and it is very difficult to predict the effect of climate change because impacts can change at different levels (Fatima et al. 2020). But a minor change in the climate can affect the productivity of the crops to a great extent. Currently, only 12 crops account for 75% of all human calories. If a pest or pathogen arises, or a vulnerability to changing climates, the entire crop becomes vulnerable because of the lack of genetic variability within it. We have seen this happen before. In the 1950s, the global banana supply was largely based on a single variety called the Gros Michel. The Panama Virus wiped it out because there was a lack of variability within the crop. The population is expected to be 9.9 billion in 2050, which can cause our supply of food into the under stress or threat. The demand for basic facilities like food will increase up to 60%, and urbanization, climate change, and degradation of soil can increase and leads to less availability of the land (Ashfaq et al. 2020). The pressure of food scarcity is high in developing countries than developed countries. But agriculture sector can be affected by natural disasters that are caused by anthropogenic activities like climate change that can disturb the monsoon pattern and affect the mango crop in Pakistan that can cause loss of millions of dollars. Different pathways that cause food shortage and climate change are described in Fig. 1.

Agriculture is a significant factor in reducing poverty and developing the economy of the world. Agriculture and food production mainly depend on two natural resources, water and land. Pakistan's population rises every day and food demand increases every year (Arif and Khalid 2007). The infrastructure, food security, and income of Pakistan rely heavily on agriculture, but these aspects are more vulnerable to the bad impacts of weather events, that is, natural disasters due to climatic changes. Mango crops of Pakistan are extremely impacted by climate change and are resulting in the loss of millions of dollars. Pakistan must secure the agriculture sector, as it has a direct effect on food production (Chaudhry and Chaudhry 2008).

2 Agricultural Challenges and Issues in Pakistan

The economy of Pakistan depends on three sectors that are commercial, industrial, agricultural; these sectors are integrated with each other. Pakistan is an agrarian country but now this trend is going to change because of the change in the rainfall pattern due to the climatic changes. In summer there is 91–97% of light rain and very few rains in the winter season due to its geographical location. But now this pattern has been changed, monsoon rainfall shifted; therefore, farmer feels very difficult to cultivate crops especially wheat and rice crop (Salma et al. 2012; Rizwan et al. 2017a). In the history of Pakistan, there are many events of floods and droughts due to inadequate water use and waterlogging, and salinity becomes a major issue. There are many other issues associated with the agriculture sector in Pakistan. These problems or challenges are characterized into four parts or segments to clarify the major issues (Azam and Shafique 2017).

2.1 Techno-Managerial Problems

2.1.1 Waterlogging and Salinity

Salinity and water pollution are minor issues in agriculture. Salt that appears on the outside of the land changes its physical properties and affects the crops automatically (Farid et al. 2017). In Pakistan, 25.5% of the country is projected to be impaired by waterlogging and salinity. Mostly, areas near the Indus Delta face this problem due to salty water. It is not only land loss but also productivity reduction (Rehman et al. 2015).

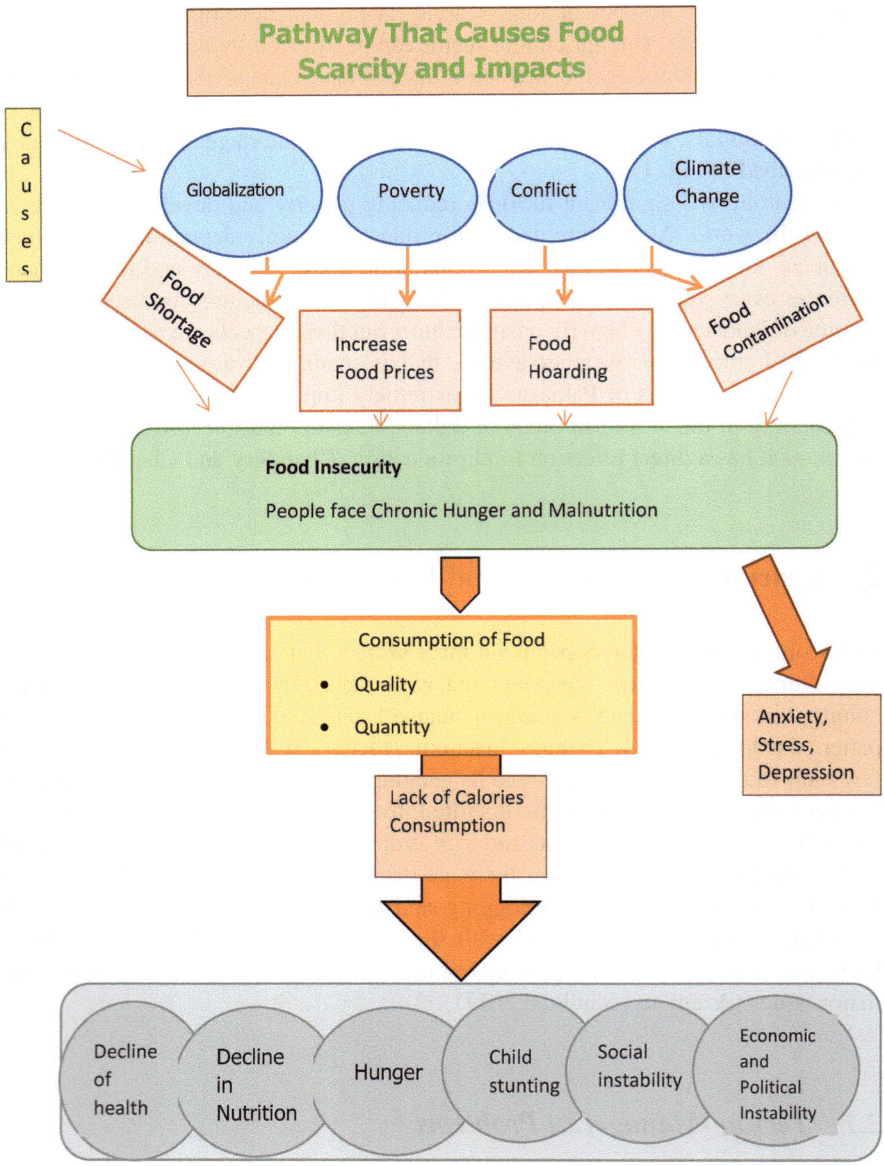

Fig. 1 Pathway that causes food scarcity and impacts

2.1.2 Lack of Irrigation Facilities

Lack of irrigation facilities leads to restraint in the expansion of Pakistan's crop field. The reduced supply of water and damages to water bodies are the major problems faced by the farm industry (Rizwan et al. 2017a; Bakt et al. 2020). Recently, water canals are not used extensively for irrigation systems in Pakistan, and it is reported that water loss in fields is about 30%. Most of the farmers use tube well water because of the lack of canal water, and this water is briny which is detrimental to crops and raises the salinity of the soil (Ali 2010).

2.1.3 Limited Infrastructure

The agriculture system exists mostly in rural areas, and there is weak infrastructure in rural areas, that is, inadequate transportation, lack of power, lack of warehouses to store crops, and problems with sanitation and hygiene; these problems disrupt the productivity of agriculture (Asghar et al. 2015).

2.1.4 Limited Cultivable Land

Pakistan's total area is nearly 80.1 million hectares, and just 23.8 million hectares of land are used for agriculture. Furthermore, 7.9 million hectares of land remain unused and not used for cultivation due to property ownership.

2.1.5 Ancient Methods of Production

Pakistani farmers use conventional farming practices because they are not well educated. Automation is growing in some areas, but some areas are using old methods for growth. These traditional techniques limit demand and thus the level of competition in the foreign market is weak (Maalik et al. 2020).

2.1.6 Inappropriate Crop Alternation

Proper crop turning is important for restoring soil fertility. Constant cropping of one or two crops maintains soil fertility. Proper crop rotation is required to preserve fertility (Ahmad et al. 2013).

2.2 Socioeconomic Problems

2.2.1 Improper Consumption and Competition

Farmers improperly use the revenue they have, and no record is maintained to show progress or development in crop growth. They have joint families and have lots of crises. There are regular and ongoing direct or indirect lawsuits among the farmers (Zaheer 2013).

2.2.2 High Illiteracy Rate in Agricultural Sector

The bulk of our country's farmers, laborers, and landlords are illiterate. They are inefficient and untrained to improve farm production. On the other hand, our farmer's well-being is deficient due to backwardness, andthe literacy rate in Pakistan is just 59.2% (Saeed et al. 2013).

2.2.3 Political Instability

Instability in politics results in a lack of investment, and the market price of crops fluctuates. That makes a huge loss in income due to political conflicts (Abdullah et al. 2013).

2.3 Natural Problems

2.3.1 Natural Climatic Changes

With the change in windrows, metrological condition changes. These natural fluctuations disrupt most of our crops. Wheat, rice, and many other crops are not fully ripe on its desired time period due to incredible rainfall fluctuation (Rizwan et al. 2017a, b).

2.3.2 Plant Diseases and Insect Attacks

Several agricultural crops such as cotton, tobacco, sugar cane, rice, and wheat are frequently invaded by rodents and insects. Pests and plant parasites are reducing annual farm production. In January 2019, desert locusts attacked various regions of Karachi and have destroyed the crops. This also results in a huge loss in crop productivity because no migratory measures have been taken to cope with the situation (Sheikh et al. 2011).

2.4 Financial Problems

2.4.1 Lack of Agricultural Finances

In Pakistan, agricultural financial facilities are not sufficient. Agricultural credit interest rates are high, and loans are not issued on time. Pakistan Human Development Report 2003 reported that about 50.8% of poor landlords borrowing seeds at very higher interest rates in Pakistan.

2.4.2 Poor Financial Position of Farmers

Most of the farmers in rural regions are very poor. Although they do hard work they have no adequate income. Therefore, they use traditional techniques of farming that do not generate enough income for them to save some financial assets for crop development instability in Market prices also disturb their financial situation (Dawn 2013).

2.4.3 Highly Expensive Fertilizers

Higher cost of fertilizers due to increases in the gas prices such as natural gas is used for making fertilizers. There is a direct increase in the prices of bags on fertilizers with an increase in gas prices. High-cost fertilizers result in less productivity as farmers are unable to afford these fertilizers to fulfill the soil needs (Arif and Chaudhry 2008).

3 Solutions and Management to Agricultural Challenges

3.1 Dams Construction

Building a large dam like Kala Bagh Dam is the only option to support the agricultural sector in the future. Smaller projects such as the Rainee Canal in Sindh, Gomal Zam Dam in Khyber Pakhtunkhwa, the Satpara Dam in Skardu, the Kachhi Canal in Balochistan, further Development of the Mangla Dam in AJ&K, and some of the new projects under construction contribute to the resolution of this issue. An overview of all management strategies and practices is described in Fig. 2.

3.2 Commercialization at Subsidized Prices of Certified Seed

Pakistan's government should take a robust initiative to market and market certified seeds in accordance with various areas and regional cropping systems that prevail in that specific area. The authority to market seed at grass root level should be granted to the federal seed certification board (World Bank 2017).

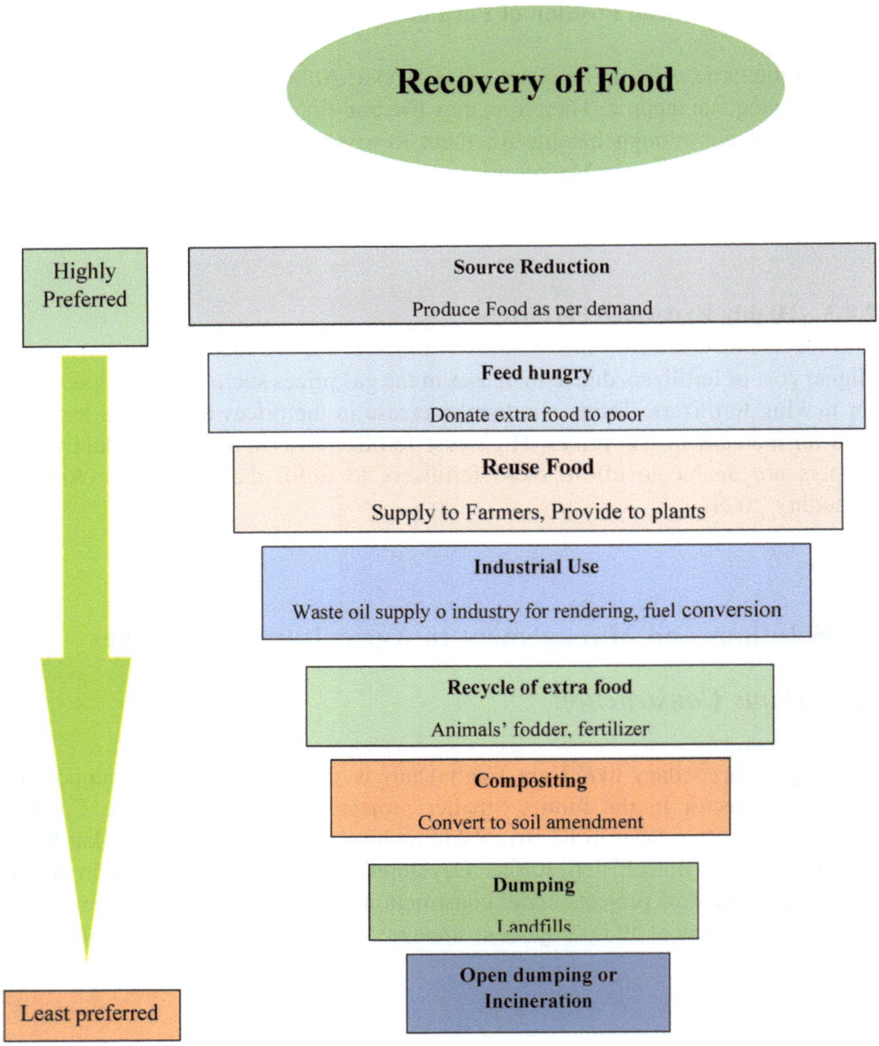

Fig. 2 Source: Sustainable management of food (US EPA)

3.3 *Integrated Development of Land Reforms*

Land reforms tend to be the main factor in empowering the poor farmers who have a small piece of land and reducing poverty. Landless farmers will get a chance to better their social standing. To counter the medieval system, great political and judicial actions are required that prevail throughout the country (Khushk 2017).

3.4 *Maintenance of Cost of Fertilizers Through Natural Gas Availability*

Natural gas seems to be the only material used in the fertilizer industry, and the availability of subsidized gas is needed for the fertilizer business. The fertilizer firms demand to provide a continuous supply of gas to lower costs per packet of fertilizers.

3.5 *Promotion of Cropping System*

The areas where monoculture cropping is common will be restricted and guided to use a multi-cropping system where different crops are grown on the same land at same time in rows (Sheikh et al. 2011).

3.6 *Controlling Agricultural Trafficking*

Pakistan's government also takes some significant steps toward the cross-border transportation of agricultural products, particularly wheat, to increase availability at reliable prices.

3.7 *Insurance of Crops*

Small landholders insured their crops to avoid financial loss. Ensuring their crops will lead to poverty reduction, and the social status of farmers will improve. Farmers are in a safe zone after Insurance, and it Will give security to farmer finance (Salam 2009)

3.8 *Improved Research Practices*

Mitigation of events that effects crop productivity is necessary to promote and develop our agriculture sector. In cases like the insect attack such as the 2019 locust attack in Karachi, the government has no migratory measures to control the situation. There should be proper research conducted in this field to cope with these events (Samejo and Sultana 2016).

4 Food Scarcity

When enough food is not produced like when crops do not grow properly due to drought, much moisture, or pests, the phenomenon of food scarcity occurs. This can also be caused by the uneven distribution of some natural resources (Conley 2012). Unfortunately, we belong to the planet where half or more people belong to countries that are facing troubles of aquifer shortages and poverty. If we will fail to control overpopulation, then the hunger of the world will increase (Vidal 2012). Food scarcity is a prediction of the production of food in the upcoming decades and the setup of the issue of food security at the top of the international agenda (Ezaz et al. 2020; Fatima et al. 2020). Agriculture production is directly proportional to the nourishment of population growth. In other words, if the agriculture production is high, more food is available for the growing population (Dexia 2010). The world hunger has increased since 2015. According to an estimate, almost 821 million people met food scarcity and were affected by hunger in 2018. Due to this extreme rate, the challenge of zero hunger in 2030 will not be achieved (Global Issues 2012). According to the estimate of US Department of Agriculture (USDA), from the 12.4 million US children, 1/5 child under the age of 5 are facing the problem of hunger and 1/6 is facing the problem of food insecurity (Feeding America 2009). Food insecurity is another term used in parallel to food scarcity. Generally, food insecurity means having no access to proper health and nutrients for the nourishment from the food. There are four terms used in food insecurity which are having, not enough, good, and healthy and culturally appropriate food. Food security is an opposite term to food insecurity and plays a major role in food justice (Monteiro et al. 2010). Food Justice is a term that determines the measuring of the right way to grow, sell, and eat proper and healthy food. Healthy food has all the qualities that are required for food security (Ruel et al. 2010). Nearly in 2018, it was estimated that 1 of 9 Americans was facing food insecurity. According to this, almost 11 million children along with 37 million people were food insecure in America in 2018. Although likely to each other, hunger and food insecurity are different concepts. Hunger means physical, sensational, and personal discomfort while food insecurity means lack of economic resources for food at the household level. Many people do not have enough and enough resources to meet their financial requirements, so they face the problem of food insecurity (Hartline-Grafton and Dean 2017).

5 Is There Enough Food?

In the year 2018, it was estimated that almost 795 million people are hungry due to the lack of food all over the globe. It is expected that about 2 billion more people will suffer from hunger along with these 795 million people till 2050. But the main thing is that enough food is being produced all over the world. According to an estimate, our farmers are producing food that can feed the population at 1.5× level. This food is so enough to feed the people in a population of 10 billion, but we are now in 7.6 billion. But the main concern is that after such an amount, how hunger can still exist here. The main reason behind the hunger after a huge amount of food production is the waste of food. It is estimated that about 30% to 40% of food is wasted all over the globe annually (Hartline-Grafton and Dean 2017). In most of the developing or underdeveloped countries, food is wasted due to the lack of knowledge and infrastructure. For example, in India the annual wastage of food is about 30% to 40% due to the shortage of cold storages. In many of the developed countries, food is wasted due to the low cost and growing portion size. Hence more food is thrown and wasted. Also, we have a very insufficient distribution system of food. We cannot rescue 2 billion people who fall into hunger by 2050 without any proper distribution. Climate change is also a factor that can lead the world into hunger. Climate change will cause colder areas to become more favorable for agriculture practices rather than current land areas of agriculture will no more able to produce food. According to the estimates, the corn production of the United States will be declined by climate change by 20%. Similarly, a clear of 16% will be seen by Brazil in corn production. Moreover, like the United States, Indonesia will also face 20% decline in its corn yield. To overcome the impacts of climate change on our food sources, it is very necessary to improve agriculture practices and adopt modern ways as well (Erdman 2018). Although we have enough food to feed the population of 7 billion people, still hunger and malnourishment occur. The main reason is probably the unequal distribution of food and poverty. We have enough food for the people who are suffering the hunger and for additional 2 billion more people can be feed by this till 2050. The main projection is to increase the production of crops by 60–100%. Increasing food production or crop production will not be necessary till we do not reduce the wastage of food and make more and more food storage. The main concern of hunger directly relates to the economic conditions (Vidal 2012). The rich people have direct demand to consume rich food which generally includes meat and other dairy products as well. According to the estimate of the United Nations, there will be an additional of 2 billion more people in the world population which will increase the food supply demand by 28%. In recent decades, the production of food is generally increased. The studies have shown that in the era of 1985 to 2005, the production of food has been increased up to 47%. The global food of all the crops and agricultural foods have been increased up to 28%. The increase in agricultural lands and the production of more and more food will not the solution to overcome hunger in the world. We are already using all or agricultural lands to produce the food so it will not be the solution. But the problem can be overcome if we reduce the wastage of food by increasing the production of crops and increase food insecurity all over the world (Colombo et al. 2016). Although the world is producing enough food for the people, but still there are cases of hunger. For the population of the world, the availability of caloric per capita and the diversity of food has been increased between the years of 1960 and 2011. The increase in the total availability of food and food growth with all the access toward the food has been increased up to 30% from 1990 to 1992 and increased up to 13% from 1998 to 2018 in the lower- and middle-income countries. The main concerned problem of hunger is that people do not have enough money or financial resources to purchase food. So, they do not meet their needs of food and lack access toward nutritional requirements (Hunger Notes 2018).

According to an estimate, the world population will become almost 9.7 billion by 2050 which means that there will be a lot more people who will have to feed. So probably, the world will face a huge hunger problem. Although we have produced enough food to feed the world till 2050, the main problem is unequal food distribution and food insecurity. It is very necessary to overcome the unequal food distribution because in this era, from 7.6 billion people almost one-quarter are facing the problem of malnutrition and almost 1 billion are hungry. So, if the world population will increase then how can we nourish billion more people with the current circumstances (Somerville 2018). A report was published by the United Kingdom and the Institute of Mechanical engineering in 2013, which estimated that about 4 billion metric tons of food is produced each year in the world and about 1.5–2 billion metric tons of that food is sent into the waste which almost becomes 30% to 50%. Similarly, the supermarkets of the United Kingdom account for the waste of food by 1.3%, the European Union accounts for the wastage of food by 5%. The food waste at household level in the EU is 42% (SGS 2015).

Now let's see the broader view of this stuff. In the scenario of 2018, the world was producing food in the range of 2,750 kcal per capita per day. And 20% of this amount gets wasted in households. So, these calories become up to 2,200 kcal per capita. The overall food production in the world is divided into three food groups. The global agriculture in 2018 was producing 12 grains serving, 5 fruits, 5 vegetables, 3 oils and fats, 1 milk, and 4 sugar servings per capita per day. But according to the HHEP (Health Hazard Evaluation Program), it was estimated that the global production of food should contain 8 grain servings, 15 fruits, 15 vegetables, 1 oil serving, 5 protein serving, and 1 milk serving per capita per day to meet the nutritional needs of a balanced diet. This shows that the world is producing grains, fats, and sugars at over rates but producing fruits, vegetables, and proteins in very small amounts. Similarly, according to the HHEP, the recommended ratio of fruits and vegetables includes 40% fruits and 60% vegetables, but the world is producing fruits and vegetables at a rate of 28%. This also contributes to and is the major cause of malnutrition in the globe although excessive food is produced (Sajjad et al. 2020).

6 Reasons for Food Shortage

The shortage of food is one of the biggest problems faced by the world. The prevalence of food shortage is much high in sub-Saharan Africa. The shortage of food is generally caused by social, economic, and environmental factors that generally include overpopulation, crop failure, and the poor policies made by the government (UNU 1998; Butt et al. 2005).

6.1 Climatic Factors

Different crops are dependent on the specific type of climate and temperature. Some Asian countries are fully able to grow their crops more than one time in a year (specifically 3 times). The production of crops is only reducing in the dry seasons or during cold temperatures. The change in climate brings floods and droughts, which are the major cause of food shortage in a country or region (Perna et al. 2020). The climate change can enhance the temperature up to 1.5°C above from the pre-industrial level. Climate change has the potential to increase the danger of a shortage of food around the globe. As the rate of droughts, floods, heatwaves, weather patterns, and wildfires are increasing, it boosts up the rate of loss of soil nutrients and causing land degradation as well. The increase in the concentration of carbon dioxide in the atmosphere which is majorly caused by the burning of fossil fuel leads to the reduction in the quality of nutrition of food. The increase in the temperature also causes harm to livestock and reduces the crop yield (Flavelle 2019). About 10% of the irrigation of food in most of the regions of the world is due to underground water. The depletion of underground water is affecting the production of food badly. Moreover, according to an estimate, if the temperature increases up to 4°C, the corn production of the US will be reduced up to half. But if the temperature will increase up to 2°C, the corn reduction in the United States will be 18%. Food production in Asia is generally subjected with the help of underground water. But due to the lack of rainfall in the regions of Asia due to climate change, almost 23 million hectares of rice cropland are lacking the production of rice. According to the National Academy of Sciences, it is estimated that crop reduction can be observed by 5% to 15% in every increase in temperature by Celsius (Cho 2018).

6.2 Economic Factors

The economic factors are directly related to poverty. Poverty is one of the major factors which can cause hunger in a country due to lack of resources to purchase food. The people who have enough resources or money do not face any kind of problem of food shortage but people especially those who are living in developing countries are facing the problems of food shortage because of a lack of resources to meet their desperate needs. The output of food is also greatly affected by poverty. For example, in Africa, due to poverty and lack of resources, many of the farmers are still unable to produce food by the modern or proper methods of irrigation. This causes less production of food and causes food shortage for the population. The population of sub-Saharan region of Africa is expected to grow up to 2.5 billion by 2050 which means more poverty will increase and more problems for the non-excess toward food will increase (Hawthorne 2017). This is the simple rule that the increase in money will increase the demand of goods and food. The inflation in the prices of food can also lead to shortage of food for those people who cannot bear such amounts (Kelly 2008).

6.3 Social Factors

Many of the social factors responsible for food shortage are overpopulation and war and conflicts between nations. Almost in 18 different countries, around 74 million people are unable to get access to food due to armed conflicts. When this type of scenario was observed from country to country, it was estimated that in Africa, almost 11 countries were facing the severe problem of chronic food shortage. Within these 11 countries, almost 37 million people are affected. The situation of the Middle East is the same as Africa. In Yemen, the quantity of people starving from the food crisis has been exceeded up to 17 million. The war and conflicts present in the countries like Palestine, Iraq, and Syria, the number of people affecting by food insecurity is 10 million (Bakır 2018). Since 2010, it was observed that state-based conflicts increased from 60% up to 125%. In all, 489 million people out of 815 million people are facing problems of food insecurity which is half of the total population due to armed conflicts. Out of 155 million, around 122 million nourished children in the world are thoroughly affected by armed conflicts. Wars impact badly on the farmers which are the producers of the food. The conflicts let them leave their lands, and it can destroy their crops and livestock. In other words, the wars and conflicts resist the output of irrigation (Samberg 2018).

7 Impacts of Food Scarcity

Due to the shortage of food, humans are facing several problems. Such problems are related to financial or economic issues, livelihood issues, and moreover health issues. Food shortage and food insecurity, both are factors that cause several issues in the lifestyle of human beings. Food shortage and food insecurity are threatening the world (FAO 2019). During childhood, many children face the issues of hunger and food shortage which lead to weight loss and malnutrition factors. Moreover, it causes personal problems to the individual like stress and hypertension, especially when a person is unable to feed his children. The shortage of food leads toward hunger, and it continuously reduces the expectancy of life and the quality of life as well (Hunger Health 2019).

7.1 Food Price Inflation

As the twenty-first century began, the rise in the prices of food had been observed as a rate of 2.6% per year as an average rate. According to the US Department of Agriculture, it was predicted that in 2020s, the rise in prices of food will be increased from 1.5% to 2.5%. Similarly, the prices of dairy products were also expected to raise in the same values. The prices of vegetables will increase up to 1%, fruit prices will increase up to 1% to 25, prices of cereals and bakery products will increase up to 2–3%, and beef and poultry values up to 1%, respectively. As the availability of food decreases, the demand for food increases which usually gives rise to the prices of food as well. According to the recent trends of the inclination of the food prices, in 2008, 6.4% rise in prices of food was observed. In 2011, the prices of food were risen by 4.8% due to wildfires, droughts, and other natural disasters all over the globe. In 2012, the prince inflation of food all over the world was observed as 2.5%. In 2013, the price inflation of food was 2%; similarly, in 2014, the price increase was 2.4%, in 2015, it was 1.9%, in 2016 it was 2%, in 2017, it was 8.2%, and in 2018 it was 1.6%. All the rise in price of food in recent years was due to the unavailability of food (Amadeo 2019). When the prices of food are increased, the poor or lower class group of a country spends a huge part of their income on purchasing food, which increases the insecurity of food at the household level. Instead of shortage or unavailability of food, the prices of food are also depending on the agricultural worth. The World Bank estimates that since 2005, almost 83% of the rise in food prices has been observed in the world (Johnson et al. 2016).

7.2 Impact on Economy

Although there are no direct impacts of hunger on economy, it has several indirect impacts. According to an estimate, the average cost of hospitalized people due to insufficient nutrients is 12,000 Dollars. The economic cost is totally dependent on public health as the hunger leads to many diseases. Due to hunger and food insecurity, the annual cost due to illness is 130.5 billion dollars. The lower earnings and the poor educational outcomes which are linked to hunger and food insecurity have a cost of 19.2 billion dollars. A huge amount of cost is donated as a charity to overcome hunger and food insecurity and this cost is almost 17.8 billion dollars. The cost of hunger for every citizen is 542 dollars. Hunger and food insecurity lead to many diseases, so in 2010, due to absentees and low grades in high school, almost 19.2 billion dollars' loss was observed in the lifetime earnings. The cost of child hunger in the United States is itself 28 billion dollars per year (IFBA 2018). According to the report of Malawi which shows that the social and economic impacts of undernutrition child caused the loss of a huge amount of money. In all, 10.3% GDP costs annually due to the undernutrition child. The economic loss of Malawi is almost 600 million dollars because of child undernutrition. In Malawi, people are engaged with different manual activities and due to lack of nutrition, their activities affect. It is estimated that due to the reduction in productivity of work, Malawi faces a loss of 67 million dollars in its economy (WFP 2015). According to an estimate, in the United States, each person pays a bill of 300 US dollars for hunger annually. By the addition of other household needs, this value increases up to 800 US dollars per year individually. It means that the total cost a person spends in his total lifetime on hunger in the United States is almost 22,000 US dollars. Studies have shown that the United States pays almost 90 billion US dollars as a direct or indirect cost for hunger or hunger-related issues (Brandeis University 2007).

8 Malnutrition

If a person does not feed due to some reason and by not feeding, he does not fulfill his need of the body, he feels hungry. Hunger is a temporary thing but when a person does not feed over a long period of time, he lacks much nutrition required for the body which leads to malnutrition. The intensity of malnutrition depends on the body requirements of a person. According to the report on Food and nutrition by the United States, it is estimated that about 815 million people have been affected by malnutrition due to issues of climate change and war and conflicts. The most vulnerable group who affects badly with malnutrition or having possible chances to become its victim are under 5 aged children, pregnant women, poor people, people living in the under developing or developing countries, and the people who are living in those areas where wars and conflicts are more often (Hunger Notes 2019). People who have enough food to eat can also become the victim of malnutrition if they do not take those nutrients that are required for their bodies. Malnutrition does not need plenty of food to eat but requires enough nutrients to meet the needs of the body. People who have a disease like intestinal problems, digestive issues, or their body does not absorb food properly are also affected with malnutrition. Malnutrition causes Fatigue, loss of energy, Dizziness, poor functioning of immune system, dry skin, underweight, poor body growth, weakness in muscles, breaking bones easily, organ functioning problems, and problems in learning as well (Perna et al. 2020). According to the report of WHO in 2017, 815 million people are facing the problem of hunger and malnutrition while 150 million children are stunted and facing the issues of malnutrition as well. Very poor progress has been shown in addressing hunger and due to this almost 151 million children under the age of 5 are facing the problems of malnutrition since 2017. This number was almost 165 million in 2012. This shows that in 5 years we have seen a reduction of only 14 million (Keulertz et al. 2020). The number of people affecting with malnutrition is 515 million in Asia, 256 million in Africa, and 39 million in the Caribbean and Latin America. Due to this malnutrition, the children under 5-year age group are facing stunting by 22.2%, wasting by 7.5%, overweight by 5.6%, and pregnant women affected with anemia are 32.8% in all over the world (WHO 2018). According to an estimate, poor nutrition causes the death of children under the of 5 by 45%. This means that 3.1 million children are dying due to poor nutrition in the world every year (Mercy Crops 2018).

9 Actions

In the current world, different types of solutions are figuring out to achieve zero hunger in the world, although different types of strategies have also been applied to achieve zero hunger by 2030 and 2050 rather countries have started working on this and making action on it (BBC 2019).

- The social protection of the poor is being raised by the countries and making them able to buy food at relevantly low costs as well.
- The infrastructure has been improved so that the transportation access of food can be easy. This helps buyers to buy food easily and sellers to sell food easily, this is making this marketing process faster. So, more food can easily be shifted from one place to another place in no time.
- The main issue is food waste all over the world which also affects the economy of a country so countries are trying to reduce the food waste as possible as they can.
- The wider capacity of crops is growing so that different types of nutrients can be consumed by the people which will reduce malnutrition.
- The focus is on child nutrition by the countries so that no child can die of hunger or be affected by any diseases caused by malnutrition (United Nations 2019).

10 Implementing Strategies to Ensure World Food Sustainability

The first thing which we can do is to improve the food security so no more damage will occur to people. There are many ways to improve the security of food.

- We can make the yield gap closer. More and more areas can be used for farming and agriculture practices. Closing the yield gap can nourish 850 million people.
- We can use the fertilizers more efficiently. We can synthesize such fertilizers that do not harm the nutrients of food or either not harm the food itself and land also.
- We can adopt methods by which low water is used to irrigate more crops so that the sustainability of water is maintained. Rice and sugar cane crops take enough water to irrigate so we can modify them in a way that they do not over-consume water.
- Food is directly used for consumption. It is estimated that a large amount of food crop is waste due to the grazing of the animals. So, we should grow crops only for the consumption of human beings.

- Food waste must be reduced. Around the world, a large amount of food is wasted which increases the scarcity of food. According to an estimate, about 30% to 50% of food is going to waste in the world. The reduction of food wastage especially in China, India, and the United States can feed 413 million people per year (Verchot 2014).

We know that food waste is one of the biggest sources of shortage of food all over the world. According to the estimate of the Food and Agriculture Organization (FAO), about one-third of food production in the world is going into waste which is almost 1300 tonnes. Due to this food waste, almost 800 million people suffer from hunger every year. So different actions have been taken by the countries to avoid waste of food. The main thing is to raise awareness among the people so that they come to know that by wasting food a large number of people in the world lack food and are dying from hunger (Joan 2017).

Further steps are taken to reduce the food shortage and overcome the crisis of food scarcity. Following steps are taken in the world to meet the solutions of food scarcity:

- Modern technologies have been developed to produce more food so that more people can be feed besides food waste as well. Modern technologies also bring enough nutrients to the food so that the human body can nourish well, and no problems of malnutrition will occur.
- A proper education is given in the institutes so that the proper awareness can be raised to tackle the problems of hunger. Moreover, this will also inform people about the nutrition requirement for the bodies of different human beings.
- The other thing is subsidies given by the government. The government gives allowance and charities to the people which have low incomes. Such people who deserve and do not purchase food. By this hunger can be controlled (Dumitrescu 2016).

11 COVID-19 (Big Risk for Food Scarcity)

The pandemic is a huge risk for both survival and likelihood. We know that the sock is huge and unusual for the food, supply, and demand as it is affecting them.

1. Supply is disturbed as the impact of the disease is affecting the life's well-being and the restricted business and its huge costs with the limited supply chain.
2. Demand is also decreasing as the change in the behavior, safety measures, and financial costs which is also affecting the lives of people, containment efforts, and the high uncertainty (Galanakis 2020).

12 Conclusions and Future Prospect

There are 113 million populations who are coping with acute insecurity. The hunger is so severe that the peoples their life's is at hug risk as it threatens it and they are relying on the external assistant for their health. The people who are facing the pandemic are willing to any further disruption or the available food they might receive.

The quarantine and panic at the time of the Ebola virus 2014–2016 showed that the food scarcity and malnutrition were faces as the restrictions were in place and the moment was limited farmers and the labor was limited for that reason farmers were unable to bring the crop to the market.

The COVID-19 is already present in more than 100 countries wherein 44 countries is proliferated and the 53 have 113 million people facing severe hunger, lot of them are facing public health issues and capacity constraints and for them, the consequences could be severe.

The FAO is observing and has concerns for those who are facing hunger and the desert locust outbreak attract on Africa, countries that import the food, and those who depend on the export of their oil.

The vulnerable group also includes the small-scale farmers as they cannot afford the working, land and seeds, fertilizers, and they must struggle due to high rates of food prices also the students who are not going to school and facing one another lacking nutrition.

As we all know these pandemics and historical severe events have one thing in common and that is the food shortage and scarcity and countries face severe drastic time managing and providing them.

References

Abdullah M, Cuixia LI, Ghazanfar S, Rehman A, Ghazanfar B, Saud S (2013) Problems faced by rice growing farmers and their behaviour to the government policies: a case from Pakistan. J Biol Agric Healthcare 3(16):1–9

Ahmad KFZ, Muhammad S, Ul HM, Tahira GH, Feehan H, Amir MS, Atif W (2013) Agricultural dynamics in Pakistan: current issues and solutions. Russ J Agric Soc-Econ Sci 20(8):20–26

Ali MH (2010) Fundamentals of irrigation and on-farm water management. Springer, New York. https://doi.org/10.1007/978-1-4419-6335-2_2

Amadeo K (2019) Why food prices are rising, recent trends, and 2021 forecast. https://www.thebalance.com/why-are-food-prices-rising-causes-of-food-price-inflation-3306099

Arif GM, Chaudhry N (2008) Demographic transition and youth employment in Pakistan. Pak Dev Rev 47(1):27–70

Arif M, Khalid N (2007) Agriculture and food security in Pakistan, South Asia Partnership, Pakistan

Asghar M, Arshad M, Afzal M, Qadeer MM, Sabir A (2015) Farmers' perceptions and beliefs regarding future of Rice crop in the Kallar tract of the Punjab, Pakistan. Int J Agric Innov Res 3(6):2319–1473

Ashfaq H, Abubakar M, Ghulzar H, Farid M, Yaqoob S, Komal N, Azam Z, Hamza A, Ali S, Adrees M (2020) Phytoremediation potential of oilseed crops for lead- and nickel-contaminated soil. In: Plant ecophysiology and adaptation under climate change: mechanisms and perspectives II, mechanisms of adaptation and stress amelioration. Springer, Singapore, pp 801–820

Azam A, Shafique M (2017) Agriculture in Pakistan and its impact on economy: a review. Int J Adv Sci Technol 103(1):47–60

Bakht S, Safdar K, Khair KU, Fatima A, Fayyaz A, Ali SM, Munir H, Farid M (2020) Response of major food crops under drought stress; physiological and biochemical response. In: Agronomic crops—volume 3: stress responses and tolerance. Springer, Singapore, pp 94–116

Bakır ZZ (2018) The global food crisis. *Insamer*. https://insamer.com/en/the-global-food-crisis_1558.html

BBC (2019) Five ways we can feed the world in 2050. https://www.bbc.com/future/bespoke/follow-the-food/five-ways-we-can-feed-the-world-in-2050.html

Brandeis University (2007) Economic impact of hunger affects all Americans: cost to nation is $90 billion. *Science Daily*, sciencedaily.com: https://www.sciencedaily.com/releases/2007/06/070605120859.htm

Bratspies RM (2012) Food, technology and hunger. Law Cult Human 10(2):212–224

Butt TA, McCarl BA, Angerer J, Dyke PT, Stuth JW (2005) The economic and food security implications of climate change in Mali. Clim Change 68:355–378

Chaudhry A, Chaudhry TT (2008) The effects of rising food and fuel costs in Pakistan. Lahore J Econ 13:117–138

Cho R (2018) How climate change will alter our food. State of the planet: Columbia School of Climate. https://blogs.ei.columbia.edu/2018/07/25/climate-change-food-agriculture/

Colombo B, West P, Smith P, Tubiello FN, Gerber J, Engstrom P, Urevig A (2016) How does agriculture change our climate? Environment Rep: Food Matters

Conley D (2012) Global food scarcity: definition, distribution, roadblocks.

Dawn News Paper (2013) Fertilizer sector raise in gas prices to hit farmers. http://beta.dawn.com/news/812724/fertiliser-sector-raise-in-gas-prices-to-hit-farmers

Dexia AM (2010) Food scarcity—trends, challenges, solutions. Doxia AM

Dumitrescu S (2016) Food security 102: What is being done to reduce global food insecurity? https://foodinsight.org/food-security-102-what-is-being-done-to-reduce-global-food-insecurity/

Erdman J (2018) We produce enough food to feed 10 billion people. So why does hunger still exist?

Ezaz Z, Azhar R, Rana A, Ashraf S, Farid M, Mansha A, Naqvi SAR, Zahoor FA, Rasool N (2020) Current trends of phytoremediation in wetlands: mechanisms and applications. In: Plant ecophysiology and adaptation under climate change: mechanisms and perspectives II, mechanisms of adaptation and stress amelioration. Springer, Singapore

FAO (2019) The state of food security and nutrition in the world 2019. http://www.fao.org/state-of-food-security-nutrition

Farid M, Ali S, Rizwan M, Ali Q, Abbas F, Bukhari SAH, Saeed R, Wu L (2017) Citric acid assisted phytoextraction of chromium by sunflower; morpho-physiological and biochemical alterations in plants. Ecotoxicol Environ Saf 145:90–102

Fatima A, Farid M, Alharby HF, Bamagoos AA, Rizwan M, Ali S (2020) Efficacy of fenugreek plant for ascorbic acid assisted phytoextraction of copper (cu); a detailed study of cu induced morpho-physiological and biochemical alterations. Chemosphere 251:126424

Feeding America (2009) New report focuses on economic toll of child hunger. feedingamerica.org: https://www.feedingamerica.org/about-us/press-room/new-report-focuses-on-economic-toll-of-child-hunger

Flavelle C (2019) Climate change threatens the world's food supply united nations warns. https://www.nytimes.com/2019/08/08/climate/climate-change-food-supply.html

Galanakis CM (2020) The food systems in the era of the coronavirus (COVID-19) pandemic crisis. Foods 9(4):523

Global Issues (2012) Food. https://www.un.org/en/sections/issues-depth/food/index.html

Hartline-Grafton H, Dean O (2017) The impact of poverty, food insecurity, and poor nutrition on health and Well-being. FRAC, pp. 1–14. https://frac.org/wp-content/uploads/hunger-health-impact-poverty-food-insecurity-health-well-being.pdf. Accessed 27 Mar 2020

Hawthorne J (2017) 8 critical factors behind every food crisis. Social Sciene in Humanitarian Action Platform. https://www.socialscienceinaction.org/resources/8-critical-factors-behind-every-food-crisis/

Hunger Health (2019) What is food insecurity? Feeding America. hungerandhealth.feedingamerica.org. https://hungerandhealth.feedingamerica.org/understand-food-insecurity

Hunger Notes (2018) World hunger and poverty facts and statistics. worldhunger.org: https://www.worldhunger.org/world-hunger-and-poverty-facts-and-statistics

Hunger Notes (2019) Hunger and nutrition facts. worldhunger.org: https://www.worldhunger.org/hunger-and-nutrition-facts

Iowa Food Bank Association IFBA (2018) Impacts of hunger on the economy. https://www.iowafba.org/impacts-hunger-economy

Joan JJ (2017) Preventing waste, a recipe for food security. Sustainable Development Goals Fund. https://www.sdgfund.org/preventing-food-waste-recipe-sdgs

Johnson EJ, Welch D, Maynard JA, Bell JD, Pecl G, Robins J, Saunders T (2016) Assessing and reducing vulnerability to climate change: moving from theory to practical decision-support. Mar Policy 74:220–229

Kelly K (2008) The cause of the food shortage. The free market. The Free Market. https://mises.org/library/cause-food-shortage

Keulertz M, Mulligan M, Allan JA (2020) The impact of COVID-19 on water and food systems: flattening the much bigger curve ahead. Water Int 45(5):430–434

Khushk GM (2017). Participation in crop maximization project, empowerment and Well-being among small farmers of Sindh Province, Pakistan. Doctoral thesis, Universiti Putra Malaysia

Maalik U, Farid M, Zubair M, Ali S, Rizwan M, Shafqat M, Ishaq HK (2020) Rice production, augmentation, escalation and yield under water stress. In: Agronomic crops - volume 3: stress responses and tolerance. Springer, Singapore

Manap NMA, Ismail NW (2019) Food security and economic growth. Int J Mod Trends Sci Technol 2(8):108–118

Mercy Corps (2018) Growing up hungry: quick facts about malnutrition. mercycorps.org: https://europe.mercycorps.org/en-gb/blog/quick-facts-malnutrition

Monteiro C, Levy R, Claro R, de Castro I, Cannon G (2010) Increasing consumption of ultra-processed foods and likely impact on human Health: evidence from Brazil. Public Health Nutr 14:5–13

Perna L, Zhang Y, Wild B, Kliegel M, Ihle A, Schöttker B, Mons U, Brenner H (2020) Childhood exposure to hunger: associations with health outcomes in later life and epigenetic markers. Epigenomics 12(21):1861–1870

Rehman A, Jingdong L, Shahzad B, Chandio AA, Hussian I, Nabi G, Iqbal MS (2015) Economic perspectives of major field crops of Pakistan: an empirical study. Pacific Sci Rev 1(3):145–158

Rizwan M, Ali S, Hussain A, Ali Q, Shakoor MB, Zia-ur-Rehman M, Farid M, Asma M (2017a) Effect of zinc-lysine on growth, yield and cadmium uptake in wheat (Triticum aestivum L.) and health risk assessment. Chemosphere 187:35–42

Rizwan M, Ali S, Qayyum MF, Ok YS, Adress M, Ibrahim M, Zia-ur-Reham M, Farid M, Abbas F (2017b) Effect of metal and metal oxide nanoparticles on growth and physiology of globally important food crops: a critical review. J Hazard Mater 322:2–16. https://doi.org/10.1016/j.jhazmat.2016.05.061

Ruel MT, Garrett JL, Hawkes C, Cohen MJ (2010) The food, fuel, and financial crises affect the urban and rural poor disproportionately: a review of the evidence. J Nutr 140:170S–176S

Saeed R, Lodhi R, Saeed Z (2013) Effect of micro finance on poverty reduction of small scale farmers of Pakistan. Sci Papers Series Manage Econ Eng Agric Rural Develop 13(2):363–368

Sajjad A, Jabeen F, Farid M, Fatima Q, Akbar A, Ali Q, Hussain I, Iftikhar U, Farid S, Ishaq HK (2020) Biochar: a sustainable product for remediation of contaminated soils. Current trends of phytoremediation in wetlands: mechanisms and applications. In: Plant ecophysiology and adaptation under climate change: mechanisms and perspectives II, mechanisms of adaptation and stress amelioration. Springer, Singapore, pp 787–800

Salam A (2009) Distortions in incentives to production of major crops in Pakistan: 1991-2008. J Int Trade Econ Dev 5(2):185–208

Salma S, Rehman S, Shah MA (2012) Rainfall trends in different climate zones of Pakistan. Pak J Meteorol 9(17)

Samberg L (2018) World hunger is increasing thanks to wars and climate change. The Conversation. http://theconversation.com/: http://theconversation.com/world-hunger-is-increasing-thanks-to-wars-and-climate-change-84506

Samejo AA, Sultana R (2016) Comparative study on the various species of locusts with special reference to its population fluctuation from Thar Desert, Sindh. J Entomol Zool Stud 4(6):38–45

SGS (2015) Food waste: a global epidemic? sgs.com https://www.sgs.com/en/news/2015/11/food-waste-a-global-epidemic

Sheikh SA, Nizamani SM, Jamali AA, Kumbhar MI (2011) Pesticides and associated impact on human health: a case of small farmers in southern Sindh, Pakistan. J Pharm Nutrition Sci 1:82–86

United nation (2019) Can we feed the world and ensure no one goes hungry? Global Perspective Human Stories. https://news.un.org/en/story/2019/10/1048452

UNU (1998) Food shortage. http://archive.unu.edu. Accessed 27 Aug 2012

Verchot M (2014) 5 ways to improve global food security. https://www.treehugger.com/sustainable-agriculture/5-ways-improve-global-food-security.html

Vidal J (2012) Food scarcity: the Timebomb setting nation against nation. Our World. https://ourworld.unu.edu/en/food-scarcity-the-timebomb-setting-nation-against-nation. Accessed 19 Feb 2020

World Bank (2017) Pakistan - strengthening Markets for Agriculture and Rural Transformation in Punjab program project (English). World Bank Group, Washington

World Food Program (2015) New study reveals huge impact of hunger on economy of Malawi. https://www.wfp.org/news/new-study-reveals-huge-impact-hunger-economy-malawi

World Health Organization (WHO) (2018) Global hunger continues to rise, new UN report says. https://www.who.int/news/item/11-09-2018-global-hunger-continues-to-rise%2D%2D-new-un-report-says

Zaheer R (2013) Analyzing the performance of agriculture sector in Pakistan. Int J Human Soc Sci Invent 2(5):1–10

Increasing the Heavy Metals Accumulation Ability of Transgenic Plants by Expressing Bacterial Enzymes

Pooja Saraswat, Dipinte Gupta, and Rajiv Ranjan

Abstract Heavy metal toxicity is one of the most critical and persistent problems of global concern which poses a serious threat to human life. Advancements in science and technology have opened up the new opportunity of using plants for the reduction and extraction of heavy metal contamination from the soil with the help of phytoremediation. Many plants itself do not possess the ability to extract heavy metals, and the plant abilities to uptake, translocate, and transform heavy metals, as well as to limit their toxicity, may be significantly enhanced by overexpression of specialized bacterial metal-binding protein with the help of genetic engineering. The overexpression of metal-binding proteins has been widely exploited to increase the metal-binding capacity, tolerance, or accumulation of plants. Metallothioneins and metalloregulatory proteins found in bacteria play an important role in heavy metal accumulation, sequestration, and translocation. In this chapter, we provide a comprehensive overview of present bacterial enzymes responsible for providing heavy metal tolerance and recent strategies highly useful for increasing the phytoremediation abilities of transgenic plants.

Keywords Abiotic stress · Metal toxicity · Phytochelatins · Transgenic

1 Introduction

Heavy metals are considered an important pollutant, highly toxic, and the main abiotic stress for plants and animals due to their heavy use in industries and agriculture techniques. The potent metal ions like arsenic (As), lead (Pb), Zinc (Zn), cadmium (Cd), nickel (Ni), and chromium (Cr) are highly poisonous which

---

Pooja Saraswat and Dipinte Gupta contributed equally with all other contributors.

P. Saraswat · D. Gupta · R. Ranjan (✉)
Plant Biotechnology Lab, Department of Botany, Faculty of Science, Dayalbagh Educational Institute (Deemed University), Agra, Uttar Pradesh, India

M. Hasanuzzaman et al. (eds.), *Managing Plant Production Under Changing Environment*, https://doi.org/10.1007/978-981-16-5059-8_19

retards the growth of plant by decreasing its photosynthetic efficiency (Krupa et al. 1993), affects the quality of life, and public health (Gupta et al. 2016). Some heavy metals are toxic even at very small concentrations. Due to heavy metal toxicity, biomolecules get inactivated by binding to the functional groups or displacing the essential metal ions (Goyer 1997). Some organisms being susceptible to heavy metal toxicity respond accordingly to avoid the harmful effects. Being a part of the defense system, bacteria synthesize several metal-binding proteins, metal sensors that form a complex with the metal ion and reduces its toxicity.

Microbes possess various mechanisms including oxidation, reduction, precipitation, intracellular bioaccumulation, extracellular sequestration, compartmentalization, and synthesis of metal-binding protein to withstand the toxicity of heavy metals (Roane 1999; Nies 1999; Borremans et al. 2001; Naik and Dubey 2011; De et al. 2008). Microorganism having these unique abilities of metal absorption, accumulation, and resistance has been isolated from nature. These microorganisms are often engineered for remediation of the polluted environment by overexpression of bacterial metal-binding protein is widely done for increasing metal-binding capability of plants (Gupta et al. 2019). Bacterial metallothionein has been identified as SmtA (*Synechoccous)* (Olafson et al. 1988) and MymT (*Mycobacterium tuberculosis*) (Gold et al. 2008), BmtA from *Pseudomonas putida* is able to maintain cytosolic metal homeostasis. Bacterial metallothioneins have been recognized in *Pseudomonas aeruginosa, Anabaena* PCC7120, *E. coli.*, and *P. putida.* In cyanobacteria, SmtA (cysteine-rich small protein) is found with a high affinity for Zn(II) in comparison to Hg(II), Cd(II), and Cu(II). Thus, bacteria having metallothioneins are considered to be an ideal tool for developing transgenics that may accumulate heavy metals. Other than metallothioneins, some proteins present in bacteria are responsible for controlling the gene expression of membrane transporters and other related proteins that manage metal homeostasis and play an important role in regulating metal bioavailability (Ma et al. 2009; Waldron et al. 2009; Mishra et al. 2019). The expression of such genes is controlled by transcriptional regulators termed as “metal sensor” proteins or metalloregulatory proteins. Overexpression of these metal-binding proteins in plants has resulted in heavy metal tolerance or/and accumulation which can be beneficial for phytoremediation.

In the last few years, information has been gathered on several key steps involved in metal uptake, transportation, and sequestration, which may be utilized for the development of transgenic plants for phytoremediation. Genetic modification in plants opens up new avenues, which are useful to develop the improved quality of plants in terms of high yield, tolerance to biotic and abiotic stress, and disease resistance. With the development in plant molecular biology, there is progress in many aspects including regulation of genes, transformation, and so on. The advantage of developing transgenic plants over conventional method is the cost-effectiveness, specificity, and less time-consuming. In terms of heavy metal toxicity, plant biotechnology is employed to develop transgenics with overexpression of metallothioneins to accumulate heavy metals. The very first report related to stable transgenics with expressed metallothioneins was reported in 1989 (Misra and Gedamu 1989), in which the human MT-II gene was incorporated to enhance the cadmium tolerance in oilseed and tobacco. Later, many of the studies were done using metallothioneins, metalloregulatory proteins, and other bacterial enzymes to generate metal accumulating transgenic plants which are discussed later in this chapter. From laboratory to field, many of the tests are to be done to assess their ability to withstand the heavy metal stressed environment.

2 Types of Metal-Binding Enzymes in Microorganisms

Microorganisms have a well-established mechanism to develop resistance to toxic metals to maintain the hemostasis of biologically required metal ions without sacrificing the viability of the cell. Different defense systems have been adopted which include exclusion, compartmentalization, synthesis of metal-binding proteins such as metallothioneins (MTs), and phytochelatins (PCs).

2.1 Metallothioneins

Metallothioneins (MTs) is a heterogeneous family of low molecular mass (4–8 kDa) (Koszucka and Dąbrowska 2006), ubiquitous, cysteine-rich, metal-binding proteins that have been isolated from a wide range of organisms like bacteria, fungi, plants, vertebrates, insects, and mammals (Binz and Kägi 1999). The presence of sulfhydryl groups enables their bonding with heavy metals by forming metal ion coordination. MTs are essential for metal detoxification and maintain tolerance to increased concentrations of metal ions (Kumar et al. 2012). Plant MTs are involved in response to oxidative stress and detoxification of heavy metals (Thirumoorthy et al. 2011). In animals, MTs function in metal homeostasis, ROS scavenging, metal detoxification, and so on. (Gu et al. 2014). Bacterial metalloprotein possesses a distinguishable function, but all contribute to ensuring metal homeostasis in the cell (Waldron and Robinson 2009). The detailed physiological function of MTs has not yet been fully expounded, its anticipatory roles include maintaining the homeostasis between essential and toxic heavy metals, sequestrating and detoxification, and protection of cell against intracellular damage (Gasic and Korban 2007). Based on the cysteine content and structure, MTs are of three different types such as Cys-Cys, Cys-X-Cys and Cys-X-X-Cys motifs (in which X denotes any amino acid). Bacterial MTs are found to occur only in some cyanobacteria and some *Pseudomonas* (Olafson et al. 1988). A cysteine-rich small 56 amino acid metal-binding protein was isolated and sequenced from the cyanobacterium *Synechococcus* TX-20 (Olafson et al. 1988) and has been defined as a class I metallothionein (Kojima 1991). Due to the heavy metal tolerance, accumulation, and high metal-binding MTs are widely exploited for bio and phytoremediation (Hamer 1986). Bacterial gene 1-aminocyclopropane-1-carboxylic acid (ACC) deaminase under the control 35*S* cauliflower mosaic virus promoters (constitutive expression), the *rolD* promoter from *Agrobacterium* *rhizogenes* (root-specific expression) or the pathogenesis-related PRB-1*b* promoter was expressed in transgenic tomato to analyze its efficiency of accumulating heavy metals (Grichko et al. 2000). Transgenics lines of *Brassica juncea* expressing ECS (glutamylcysteine synthetase) and GS (glutathione synthetase) possess significantly higher capacity to tolerate and accumulate a variety of metals such as As, Cd, Cr, Zn, and Pb (Reisinger et al. 2008). γ-glutamylcysteine synthetase-glutathione synthetase of *Streptococcus thermophiles* was overexpressed in three transgenic lines of sugar beet (*Beta vulgaris* L.) (s2, s4, and s5) had shown enhanced tolerance to different concentrations of cadmium, zinc, and copper (Liu et al. 2015).

2.2 *Phytochelatins*

Many of the eukaryotes possess glutathione (GSH) and its derivative phytochelatins as important binding factors for the homeostasis of transition metal. GSH is also involved in chromate, Zn(II), Cd(II), and Cu(II) homeostasis and resistance in *Escherichia coli.* The precursor of GSH, γ-glutamylcysteine (γEC), a cysteine sulfhydryl residue–rich class of small peptides, plays important roles in detoxification and sequestration of thiol-reactive heavy metals and metalloids in plants (Xiang et al. 2001). In contrast, not much is known about the interplay of heavy metals and GSH in bacteria, except that GSH influences resistance to arsenite and mercury in *E. coli* (Latinwo et al. 1998) and cadmium tolerance in *Rhizobium leguminosarum bv. viciae* (Lima et al. 2006). On the other hand, GSH is heavily involved in transition-metal homeostasis in eukaryotes (Gharieb and Gadd 2004; Cobbett and Goldsbrough 2002).

3 Mechanism Involved in Transgenic Plant Development

Various microorganisms are known to have different systems for metal resistance and accumulation (Silver 1996). Such genes from bacteria are transferred to plants but this is complicated because of the fact that resistance or tolerance is controlled by several genes of an operon. In general, there is a specific molecular mechanism(s) for the accumulation of heavy metals. With fundamental knowledge of all the physiological processes (uptake and translocation) of plant and microorganism, regulatory controllers, and specific promoters (Gupta and Ranjan 2017) and with the use of molecular biology tools scientists are now able to develop effective and economically beneficial transgenic plants for phytoremediation (Fig. 1).

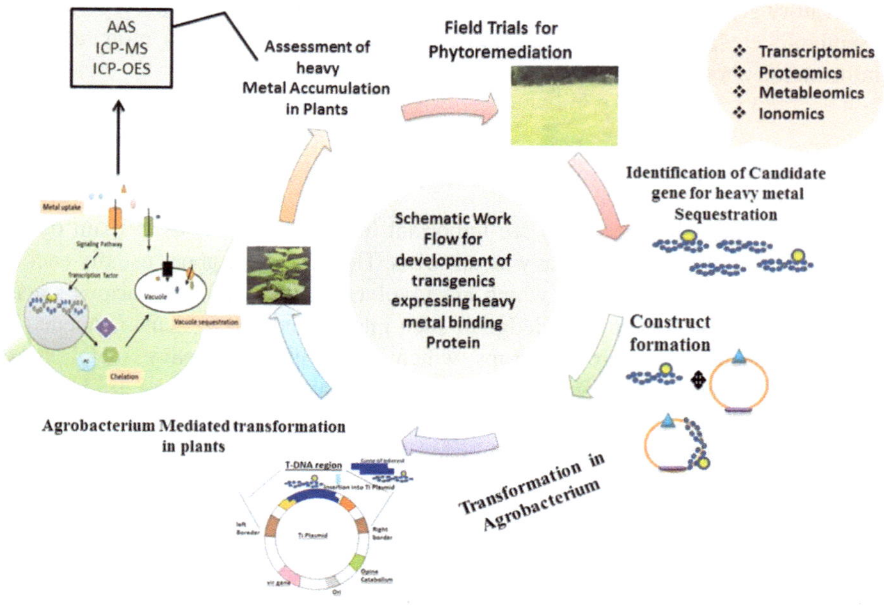

Fig. 1 Sequential events for generation of transgenic plants suitable for heavy metals remediation

3.1 *Identification of Candidate Gene Involved in Heavy Metal Tolerance*

Transgenic plant development involves various processes including gene identification, isolation, cloning, and metal assessment. Genetic and molecular techniques are useful in identification of gene families that are involved in metal transport and can be identified in bacteria, fungi, and plants. Bacteria are known to have heterologous genes for coding of metallothioneins, metal regulatory proteins, and metal transporters. Development of omics technologies has made identification and analysis of microbial community relevant to phytoremediation. The different techniques including genomics, transcriptomics, proteomics, metagenomics, metatranscriptomics, metaproteomics, ionomics, and so on are useful in searching the candidate gene for metal tolerance or accumulation.

3.1.1 Metagenomics/Metatranscriptomics/Metaproteomics

Metagenomics deals with the identification of plasmids that may allow the distribution of genes to a wider range of microbes, so as to examine the transformation efficiency and when combine with metatranscriptomics it determines the effects of introducing plasmid in the microbial community. As plasmids are a source of innovatory genome which allow microorganism to adapt in various stress conditions (Sentchilo et al. 2013). The metatranscriptome analysis provides more accurate information about the microbial activities.

3.1.2 Transcriptomics

Gene expression and regulation are important biological processes in plant development and its adaptation to heavy metal stress. The regulatory genes usually encode for transcription factors (TFs) which are involved in regulation of various stress-responsive genes. A deep knowledge of such interrelated mechanisms is needed to develop genetically improved crops which can withstand heavy metal stress (Nakashima et al. 2009). For this, transcriptomics has been employed in identification of active genes involved in phytoremediation as in *Brassica juncea* growing in sulfur (S) and chromium (Cr) contaminated environments (Schiavon et al. 2012).

3.1.3 Proteomics

Proteomics deals with the expression studies of proteins encoded by genes in an organism (Anderson and Anderson 1998). Therefore, proteomics help in identifying protein involved in biological processes and interaction between pathways that are useful in metal (Ahsan et al. 2009). It contributes to a better understanding of the mechanism related to heavy metal stress and about the signaling cascade that contributes to the expression of genes at their transcriptional stages (Kosová et al. 2011).

3.1.4 Inome and Inomics

Inome consists of studying the role of mineral ions, namely, Nitrogen (N), Potassium (K), phosphorous (P), calcium(Ca), magnesium (Mn), molybdenum (Mo), Zinc (Zn), and cobalt (Co) in metal toxicity. Furthermore, the study of elemental composition in response to external stimuli in living organisms is called ionomics.

3.2 Construction of the Gene for Plant Expression

For the development of ideal transgenic plants, a proper gene construct is to be made. The gene of interest after the identification is isolated by constructing genome libraries and use of specific probes to recognize the desired gene and amplification of gene using appropriate primers. The transgene construct generally has a gene of interest, its promoters, terminator regions, and a selectable marker to track the transformed plants. When the heterologous genes are introduced, its level of expression in new cellular and genetic context is an issue. To solve this, desired genes are incorporated in an expression cassette. This will allow proper transcription and translation of the gene. The optimal expression of bacterial genes has been found in case of MerA (Ruiz and Daniell 2009). The selection of promoters is done among consecutive promoters which shows a high level of expression in plant, cauliflower mosaic virus 35S (CaMV 35S) (Thomine et al. 2000) actin and ubiquitin promoters (Dhankher et al. 2002) are largely utilized in studies related to phytoremediation. In *Agrobacterium*-mediated transformation, the gene is transferred through Ti plasmid or in a binary vector.

3.3 Transformation of the Construct in Plant System

Several methods have been developed to introduce the desired genes in plants. Transformation can be done by direct or indirect methods. The most common method is by *Agrobacterium tumefaciens* (natural genetic engineer), a plant pathogen that is now used as a vector to transfer genes (Gelvin 2003). T-DNA complex along with the vir genes is transferred into the plant nucleus where it is expressed. It is believed that integration of T-DNA complex is random, but some studies have proven that integration occurs at specific transcriptionally active regions (Gelvin 2003). *A. tumefaciens*-mediated transformation has been utilized for phytoremediation in plants like tobacco (Nagata et al. 2010), *A. thaliana* (Rugh et al. 1996), and Indian mustard (Reisinger et al. 2008). The direct method of transformation like the biolistic method is used in plants where transformation by *Agrobacterium* is not possible. In this method, DNA is coated on tungsten or gold and transferred to the target cell by acceleration. The microprojectiles or particle gun or biolistics is widely applicable in plant genetic engineering (Rugh et al. 1998). Chemical gene transfer methods involve chemical treatments on the cell wall that enable the entry of foreign DNA. This method is quite not preferably used in plant systems because it can only be applicable for protoplasts.

3.4 Confirmation of Transgene Expression

After transformation, transgenes are to be confirmed based on the transgene construct, reported genes, and selectable markers used. Some of the genes are fluorometric or colometric, namely, GUS, GFP, and Luc, and their expressions can be observed under a microscope or visually (Naylor 1999). To identify the transgenes inserted in the host, transgene rearrangement and integrity southern blotting is often performed (Dai et al. 2001; Gheysen et al. 1987). The polymerase chain reaction (PCR) method is among the easiest techniques used for the verification of transgene by using primers specific to the gene of interest. Amplified DNA fragments with desired gene sequence indicate the presence of transgene, and obtained sequence is further confirmed by sequencing. The emergence of next-generation sequencing (NGS) technologies allows massive parallel generation of sequences from the whole genome in a relatively short time with a lower cost. NGS data analysis allows identification of transgene insertion in genomic location of the organism having transposable elements and repetitive sequences which could previously not be done by PCR-based method (Elbaidouri et al. 2013).

3.5 Establishing an Efficient Assay System

Many analytical methods can be applied to measure the concentration level of heavy metals in various samples. The most common techniques are atomic absorption spectrometry (AAS); inductively coupled plasma optical emission spectrometry (ICP-OES); atomic emission/ fluorescence spectrometry (AES/AFS); inductively coupled plasma mass spectrometry (ICP-MS); neutron activation analysis (NAA); X-ray fluorescence (XRF); and so on. Generally, metals are quantified by using ICP-MS, ICP-OES, and AAS due to their high sensitivity. These techniques have been utilized by various researchers in the detection of heavy metal concentration, the mercury concentration in merC-transgenic plant sample of *Arabidopsis* was analyzed by ICP-MS (HP 4500, Hewlett-Packard). On the other hand, X-ray fluorescence spectrometry (XRF) is a powerful technique in the analysis of multi-elements. Atomic absorption spectroscopy has been used to measure the total nickel concentration in transgenic canola (Stearns et al. 2005). Hsieh et al. (2009) used Inductively Coupled Plasma-Mass Spectrometry (ICP-MS) to measure the heavy metal concentration in both wild and transgenic plant samples of Arabidopsis.

4 Heavy Metal Tolerance in Transgenic Plants by Expression of Bacterial Genes

Metal ions are considered to be essential trace elements at higher concentrations they cause toxicity for flora as well as fauna. As heavy metals are indestructible, they are non-degradable by chemical or biological means, hence were difficult to remove from the environment in comparison to other persistent pollutants. Many micro-organisms have developed a mechanism with the help of metal-binding enzymes, which innocuous the toxic effect of these heavy metals. Genes that encode proteins involved in heavy metal uptake will play an important role in the multigene engineering of plants that are capable of accumulating heavy metals (Eapen and D'souza 2005). These bacteria are often exploited for their enzymes and metal transporters generating heavy metal tolerance in plants (Table 1). In the following section, we have discussed the transgenic plants raised for the tolerance of several heavy metals.

Table 1 List of transgenic plants expressing bacterial metal-binding protein for heavy metal tolerance

Gene	Transgenic Plant	Reference
Mercury accumulation		
merA	*Arabidopsis*	Rugh et al. (1996)
merA and mer B	*N. tabacum* and *Liriodendron tulipifera; Arabidopsis*	Rugh et al. (1998), Bizily et al. (2000)
mer A9 and mer A18	*Populus*	Che et al. (2003)
merC	*Arabidopsis* and tobacco	Sasaki et al. (2006)
Bacterial polyphosphate (polyP)	Tobacco	Nagata et al. (2006a)
Polyphosphate kinase (PPK)	Tobacco	Nagata et al. (2006b)
γ-Gultaylcystein synthetase	*Arabidopsis*	Li et al. (2005)
merP	*Arabidopsis*	Hsieh et al. (2009)
merT and ppk	Tobacco	Nagata et al. (2009)
merB	Tobacco	Nagata et al. (2010)
MerT	*Arabidopsis*	Xu et al. (2017)
MerC-SYP121	*Arabidopsis*	Uraguchi et al. (2019)
Cadmium tolerance		
Wt cysK	Tobacco	Liszewska and Sirko (2003)
Znt A-heavy metal transporters	*Arabidopsis*	Lee et al. (2003)
ACC deaminase	Canola	
CUP1	Sunflower	Watanabe et al. (2005)
CcMT1	*Arabidopsis*	Sekhar et al. (2011)
merC	*Arabidopsis*	Kiyono et al. (2012)
ECS and GS	*Brassica juncea*	Reisinger et al. (2008)
γ-Glutamylcysteine synthetase	*Populus tremula*	He et al. (2015)
ACC deaminase	*Arabidopsis*	Grichko et al. (2000)
BjMT2	*Arabidopsis*	Zhigang et al. (2006)
g-Glutamylcysteine synthetase	Indian mustard	Zhu et al. (1999)
StGCS-GS	*Beta vulgaris* (sugar beet)	Liu et al. (2015)
Zinc tolerance		
ACC deaminase	*Arabidopsis*	Grichko et al. (2000)
ECS and GS	*Brassica juncea*	Reisinger et al. (2008)
smtA	*Arabidopsis*	Xu et al. (2009)
StGCS-GS	*Beta vulgaris* (sugar beet)	Liu et al. (2015)
Glutathione synthetase 1	*Brassica juncea*	Bennett et al. (2003)
Copper tolerance		
StGCS-GS	*Beta vulgaris* (sugar beet)	Liu et al. (2015)
ACC deaminase	*Arabidopsis*	Grichko et al. (2000)

(continued)

Table 1 (continued)

Gene	Transgenic Plant	Reference
BjMT2	*Arabidopsis*	Zhigang et al. (2006)
copC	*Arabidopsis*	Rodríguez-Llorente et al. (2012)
Glutathione synthetase 1	*Brassica juncea*	Bennett et al. (2003)
Arsenic tolerance		
SRS1p/ArsC and ACT2p/ gamma-ECS	*Arabidopsis*	Dhankher et al. (2002)
Arsenate reductase	Indian mustard	Dhankher et al. (2002)
γ-Gultaylcystein synthetase	*Arabidopsis*	Li et al. (2005)
ECS and GS	*Brassica juncea*	Reisinger et al. (2008)
Ni/Co/Pb/Se tolerance		
ACC deaminase	*Arabidopsis* (Co/Ni/Pb)	Grichko et al. (2000)
ZntA	*Arabidopsis* (Pb)	Lee et al. (2003)
ACC deaminase gene	Canola (Ni)	Stearns et al. (2005)
ECS and GS	*Brassica juncea* (Pb)	Reisinger et al. (2008)
iaaM and ACC	Tobacco	Zhang et al. (2008)
Citrate synthase	*Arabidopsis* (P)	Koyama et al. (2000)

4.1 Arsenic Tolerance

Arsenic (As) is one of the extremely toxic metalloids that persists in nature (Jackson et al. 2006; Williams et al. 2009). As mostly gets accumulated in the edible parts of plants which makes its way toward the food chain of ecosystems and makes it highly toxic and a serious threat for human health (Abedin et al. 2002). Arsenite *Sadenosylmethionine methyltransferase* gene (arsM) of *Rhodopseudomonas palustris,* metalloregulatory protein (ArsR), offers immense potential of phytoremediation against arsenic (Yuan et al. 2008). Overexpression of ArsR genes in *E. coli* was found to bioaccumulate As ions. Overexpression of arsC gene of *E. coli* encoding arsenate reductase in the regulation of light-induced soybean rubisco promoter (SRS1p) in leaves were found consequently hypersensitive to arsenate. A moderate tolerance to arsenic was observed in *Arabidopsis* plants expressing the *E. coli* gene encoding gamma-glutamylcysteine synthetase (gamma-ECS) from a strong constitutive actin promoter (ACT2p) (Dhankher et al. 2002).

4.2 Cadmium Tolerance

Globally, cadmium is one of the most noteworthy environmental pollutants (Vido et al. 2001), and due to anthropogenic activities and natural activities it has been released into the environment in substantial quantities. merC, a bacterial heavy metal transporter, is found to transport both mercury and cadmium (Sasaki et al. 2005), and overexpression of this gene in *Arabidopsis* had shown increased cadmium accumulation and tolerance (Kiyono et al. 2012). Serine acetyltransferase (SAT) and cysteine synthase, also called O-acetyl-serine (thiol) lyase (OASTL), are the other important enzymes that catalyzed the last step of cysteine biosynthesis. The constitutive overexpression of *Escherichia coli* cysE gene, which encodes SAT, had been studied in tobacco and potato, and they had not shown any effect on the rate of Cadmium accumulation (Sirko et al. 2004) while overproduction of OASTL in tobacco had shown increase tolerance to cadmium (Sirko et al. 2004). Producing higher glutathione (GSH) concentrations in leaves by overexpressing bacterial γ-glutamylcysteine synthetase in the cytosol of *Populus tremula* × *P. alba* produces indicates its potentiality for cadmium phytoremediation. He et al. (2015) had reported that transgenic plants exhibit higher cadmium ion accumulation in the aerial parts in comparison to wild plants on exposure to cadmium. Similarly, Zhu et al. (1999) had shown that overexpression of *E. coli* gshI in the plastids of transgenic plants was grown better in hydroponic system Cd concentrations were 40–90% higher in comparison to wild type plants (Zhu et al. 1999).

4.3 Copper Tolerance

Copper-binding metallothionein was discovered in the *Mycobacterium tuberculosis* (MymT) (Gold et al. 2008), which is found to be induced by copper, cadmium, zinc, cobalt, and nickel. Festa et al. (2011) had reported that MymT is an important component of a five-loci regulon for copper resistance (Festa et al. 2011). On the expression of *Pseudomonas sp.* Az13 copC gene, which encodes periplasmic Cu-binding protein in *Arabidopsis thaliana* in the regulation of strong constitutive CaMV35S promoter, up to fivefold increase in Cu accumulation was observed in roots which are up to 2000 μg Cu. g(−1), while up to 400 μg Cu. g(−1) was observed in shoots (Rodríguez-Llorente et al. 2012).

4.4 Lead Tolerance

Lead is the most important environmental pollutant (Salt et al. 1998) and poses a serious threat to human health. Several bacterial enzymes were found to be involved in accumulation and transporters, which include ZntA, bmtA, and other metallotheonin. *E. Coli* ZntA is an ATPase involved in transporting Pb(II)/Cd(II)/Zn(II) and hence plays an essential role in resistance in bacterium by flushing out these metals during toxic concentration (Rensing et al. 1997). In a study, a successful expression of ZntA was carried out in the plasma membrane of *Arabidopsis*, which confers resistance Pb(II) and Cd(II) in transgenic plants (Lee et al. 2003).

Pseudomonas aeruginosa strain WI-1 accumulates a significantly high amount of lead (26.5 mg/g), which is often exploited for bioremediation of lead-polluted environmental sites. In a study it was found that these strains exhibit the presence of bmtA gene encoding metallothionein (11 kDa), which makes it a potential candidate for lead phytoremediation (Naik et al. 2012).

4.5 Mercury Tolerance

Gram-negative bacteria have mercury resistance, which is regulated by an operon that possesses six genes (Summers 1986). In an intensive study, it was detected that this mercury resistance operon consists of linked genes, which are involved in the detection and regulation (merR), recognition and mobilization (merP, merT, merC), and a mercuric ion reductase gene involved in enzymatic detoxification (merA) (Barkay et al. 2003). MerP is a periplasmic protein of gram-negative bacteria, which recognizes and binds the mercury ion through highly conserved domains have two cysteine metal-binding residue (Lund and Brown 1987) and hence suggests the potentiality of MerP in improving the efficiency of metal absorption. Strong constitutive expression of MerP in the regulation of actin A2 promoter in *A. thaliana*, localized at the cell membrane and vesicles of plant cells, had significantly improved mercury tolerance and accumulation (Hsieh et al. 2009). MerA is a soluble NADPH-dependent, FAD-containing disulfide oxidoreductase which converts toxic Hg^{2+} to the less toxic metallic mercury (Hg0) state. A. thaliana (Bizily et al. 2000), *N. tabacum*, and *Liriodendron tulipifera* L. (Rugh et al. 1998) transgenics developed with overexpressed mer A and mer B genes were able to grow in the toxic concentration of mercury which suggests the accumulation of mercury in these plants (Bizily et al. 2000). In a similar study, transgenic yellow poplar was developed using merA gene for mercury phytoremediation (Rugh et al. 1998). On overexpressing mer A9 and mer A18 gene in transgenic *Populus deltoids*, two- to fourfold Hg (0) were evolved in comparison to wild plants (Che et al. 2003). In a study, merC gene from *Acidithiobacillus ferrooxidans* that functions as a mercury uptake pump was overexpressed in *Arabidopsis thaliana* and tobacco plants under the regulation of *Cauliflower mosaic virus* 35S promoter. The resulting transgenic plants are more hypersensitive to mercury ions and found to accumulate twice ion in comparison to wild-type plants.

Transgenic tobacco developed with Mercury transporter (MerT)-integrated polyphosphate kinase gene was found to have accelerated and enhanced mercury uptake (Nagata et al. 2009). Similarly, transgenic *Arabidopsis* ectopically expressing merT from *Pseudomonas alcaligenes* had enhanced the tolerance to $HgCl_2$ (Xu et al. 2017). Mercuric ion transporters such as MerC is another potential candidate gene for producing transgenic plants that hyperaccumulate Hg^{2+} ion and may act as an ideal phenotype for phytoremediation in mercury-contaminated environment. MerC-SYP121 fusion proteins were expressed in the endodermis of *Arabidopsis*. These transgenic *Arabidopsis* roots have significant uptake of mercury ions and accumulation in shoots (Uraguchi et al. 2019).

4.6 Nickel Tolerance

Nickel is the recalcitrant pollutant in comparison to other heavy metals (Deng et al. 2003, 2005). It has been demonstrated that overexpressed metallothionein (MT) and glutathione S-transferase fusion protein (GSM-MT) (Chen and Wilson 1997) as well as nixA gene, encoding a 37-kDa integral membrane protein consisting of eight transmembrane domains, exhibit very high affinity for Ni^{2+}. BacterialACC deaminase in the phytoremediation of metals from the environment two transgenic canola lines with the gene for this enzyme were generated and tested. In these transgenic canola plants, expression of the ACC deaminase gene is driven by either tandem constitutive cauliflower mosaic virus (CaMV) 35S promoters or the root-specific rolD promoter from Agrobacterium rhizogenes. Following the growth of transgenic and nontransformed canola in nickel-contaminated soil, it was observed that the rolD plants demonstrate significantly increased tolerance to nickel compared to the nontransformed control plants (Stearns et al. 2005).

4.7 Selenium Tolerance

Larger doses of selenium have toxic and detrimental effects (Wilber 1983), and to suppress the effect of this toxic heavy metal ATP sulfurylase gene is oven exploited in plants. Transgenic plants developed by overexpressing ATP sulfurylase gene (APS) have four times higher enzymatic activity and threefold accumulation of selenium in comparison to wild plants (Pilon-Smits et al. 1999). Transgenic Indian mustard developed with a bacterial glutathione reductase by chloroplast and cytoplast transformation. Both types of transgenic had better growth in comparison to wild type in agar medium having toxic selenate or selenite concentrations (De Souza et al. 2000).

4.8 Zinc Tolerance

Few *Pseudomonas* and *Cyanobacteria* species expressed metallothioneins under the regulation of zinc sensors such as SmtB. The elevated level of Zinc leads to the conformational changes in SmtA which increase its zinc-binding efficiency (Osman and Cavet 2010). Shi et al. (1992) had reported that glutathione S-transferase (GST)-fusion protein of SmtA is capable of binding with zinc, cadmium, copper, and mercury. Overexpression of SmtA in transgenic plants had shown high survival rate than wild type in high zinc stress conditions (Xu et al. 2009).

5 Conclusion and Future Prospects

Several studies were reported for increased metal tolerance and accumulation using engineered metallothioneins and small peptides; however, most of these studies were conducted only in model plants. The heterologous expression of native and/or genetically modified peptides and proteins appears to be an attractive solution to improve the metal-binding abilities of these organisms. The development of transgenic in crop plants and their field trials is still in its infancy years (van Huysen et al. 2004). Extreme information is available about the expression of different proteins on induction of metals which can be interlinked with proteome and DNA array technology for identification of candidate gene for metal phytoremediation subsequently cleaning up the environment. More research is required to develop a clear understanding of the transgenic metabolic pathway and the use of associated microbes in an integrated manner to ascertain the effectiveness and possible side effects for phytoremediation. Environmental safety, techno-economic perspective, and industrially acceptable concepts are other major concerns that need to be closely monitored. Considering the importance of transgenic microbes in greatly enhancing detoxification and degradation of xenobiotic and heavy metal contaminants, more studies should be carried out to enhance their survival when released into the environment for bioremediation because their survivability is currently poor. The other possible risks involved in the development of transgenics are their heightened exposure of heavy metals to wildlife and subsequently to humans, which can be somehow prevented by the use of suitable fencing and nonpalatable plants. So far no transgenic for phytoremediation has been commercialized, and this opens new avenues of research in the field of transgenics development to accumulate heavy metals from the environment.

References

Abedin MJ, Feldmann J, Meharg AA (2002) Uptake kinetics of arsenic species in rice plants. Plant Physiol 128:1120–1128

Ahsan N, Renaut J, Komatsu S (2009) Recent developments in the application of proteomics to the analysis of plant responses to heavy metals. Proteomics 9:2602–2621

Anderson NL, Anderson NG (1998) Proteome and proteomics: new technologies, new concepts, and new words. Electrophoresis 19:1853–1861

Barkay T, Miller SM, Summers AO (2003) Bacterial mercury resistance from atoms to ecosystems. FEMS Micro Rev 27:355–384

Bennett LE, Burkhead JL, Hale KL, Terry N, Pilon M, Pilon-Smits EA (2003) Analysis of transgenic Indian mustard plants for phytoremediation of metal-contaminated mine tailings. J Environ Qual 32:432–440

Binz P-A, Kägi JHR (1999) Metallothionein: molecular evolution and classification. Metallothio IV:7–13

Bizily SP, Rugh CL, Meagher RB (2000) Phytodetoxification of hazardous organomercurials by genetically engineered plants. Nat Biotechnol 18:213–217

Borremans B, Hobman JL, Provoost A, Brown NL, van Der Lelie D (2001) Cloning and functional analysis of thepbr lead resistance determinant of *Ralstonia metallidurans* CH34. J Bacteriol 183: 5651–5658

Che D, Meagher RB, Heaton AC, Lima A, Rugh CL, Merkle SA (2003) Expression of mercuric ion reductase in eastern cottonwood (*Populus deltoides*) confers mercuric ion reduction and resistance. Plant Biotechnol J 1:311–319

Chen S, Wilson DB (1997) Genetic engineering of bacteria and their potential for hg 2+ bioremediation. Biodegradation 8:97–103

Cobbett C, Goldsbrough P (2002) Phytochelatins and metallothioneins: roles in heavy metal detoxification and homeostasis. Annu Rev Plant Biol 53:159–182

Dai S, Zheng P, Marmey P, Zhang S, Tian W, Chen S et al (2001) Comparative analysis of transgenic rice plants obtained by agrobacterium-mediated transformation and particle bombardment. Mol Breeding 7:25–33

De J, Ramaiah N, Vardanyan L (2008) Detoxification of toxic heavy metals by marine bacteria highly resistant to mercury. Marine Biotechnol 10:471–477

De Souza MP, Pilon-Smits EAH, Terry N (2000) The physiology and biochemistry of selenium volatilization by plants. In: Phytoremediation of toxic metals: using plants to clean-up the environment. Wiley, New York, pp 171–190

Deng X, Li QB, Lu YH, He N, Jiang J (2005) Genetic engineering of E. coli SE5000 and its potential for Ni2+ bioremediation. Process Biochem 40:425–430

Deng X, Li QB, Lu YH, Sun DH, Huang YL, Chen XR (2003) Bioaccumulation of nickel from aqueous solutions by genetically engineered Escherichia coli. Water Res 37:2505–2511

Dhankher OP, Li Y, Rosen BP, Shi J, Salt D, Senecoff JF et al (2002) Engineering tolerance and hyperaccumulation of arsenic in plants by combining arsenate reductase and γ-glutamylcysteine synthetase expression. Nat Biotechnol 20:1140

Eapen S, D'souza SF (2005) Prospects of genetic engineering of plants for phytoremediation of toxic metals. Biotechnol Adv 23:97–114

Elbaidouri M, Chaparro C, Panaud O (2013) Use of next generation sequencing (NGS) technologies for the genome-wide detection of transposition. In: Plant transposable element. Humana Press, Totowa, NJ, pp 265–274

Festa RA, Jones MB, Butler-Wu S, Sinsimer D, Gerads R, Bishai WR et al (2011) A novel copper-responsive regulon in Mycobacterium tuberculosis. Mol Microbiol 79:133–148

Gasic K, Korban SS (2007) Expression of Arabidopsis phytochelatin synthase in Indian mustard (Brassica juncea) plants enhances tolerance for Cd and Zn. Planta 225:1277–1285

Gelvin SB (2003) Agrobacterium-mediated plant transformation: the biology behind the "gene-jockeying" tool. Microbiol Mol Biol Rev 67:16–37

Gharieb MM, Gadd GM (2004) Role of glutathione in detoxification of metal (loid) s by Saccharomyces cerevisiae. Biometals 17:183–188

Gheysen G, Van Montagu M, Zambryski P (1987) Integration of agrobacterium tumefaciens transfer DNA (T-DNA) involves rearrangements of target plant DNA sequences. Proc National Acad Sci 84:6169–6173

Gold B, Deng H, Bryk R, Vargas D, Eliezer D, Roberts J et al (2008) Identification of a copper-binding metallothionein in pathogenic mycobacteria. Nat Chem Biol 4:609

Goyer RA (1997) Toxic and essential metal interactions. Annu Rev Nutr 17:37–50

Grichko VP, Filby B, Glick BR (2000) Increased ability of transgenic plants expressing the bacterial enzyme ACC deaminase to accumulate Cd, Co, Cu, Ni, Pb, and Zn. J Biotechnol 81:45–53

Gu CS, Liu LQ, Zhao YH, Deng YM, Zhu XD, Huang SZ (2014) Overexpression of Iris. Lactea var. chinensis metallothionein llMT2a enhances cadmium tolerance in Arabidopsis thaliana. Ecotoxicol Environ Safety 105:22–28

Gupta A, Joia J, Sood A, Sood R, Sidhu C, Kaur G (2016) Microbes as potential tool for remediation of heavy metals: a review. J Microb Biochem Technol 8:364–372

Gupta D, Ranjan R (2017) In silico comparative analysis of promoters derived from plant pararetroviruses. Virus Disease 28:416–421

Gupta D, Satpati S, Dixit A, Ranjan R (2019) Fabrication of biobeads expressing heavy metal-binding protein for removal of heavy metal from wastewater. Appl Microbiol Biotechnol 103: 5411–5420

Hamer DH (1986) Metallothionein. Annu Rev Biochem 55:913–951

He J, Li H, Ma C, Zhang Y, Polle A, Rennenberg H et al (2015) Overexpression of bacterial γ-glutamylcysteine synthetase mediates changes in cadmium influx, allocation and detoxification in poplar. New Phytol 205:240–254

Hsieh JL, Chen CY, Chiu MH, Chein MF, Chang JS, Endo G, Huang CC (2009) Expressing a bacterial mercuric ion binding protein in plant for phytoremediation of heavy metals. J Hazard Mater 161:920–925

Jackson BP, Seaman JC, Bertsch PM (2006) Fate of arsenic compounds in poultry litter upon land application. Chemosphere 65:2028–2034

Kiyono M, Oka Y, Sone Y, Tanaka M, Nakamura R, Sato MH et al (2012) Expression of the bacterial heavy metal transporter MerC fused with a plant SNARE, SYP121, in Arabidopsis thaliana increases cadmium accumulation and tolerance. Planta 235:841–850

Kojima Y (1991) Definitions and nomenclature of metallothioneins. Methods Enzymol 205:8–10

Kosová K, Vítámvás P, Prášil IT, Renaut J (2011) Plant proteome changes under abiotic stress-contribution of proteomics studies to understanding plant stress response. J Proteomics 74: 1301–1322

Koszucka AM, Dąbrowska G (2006) Plant metallothioneins. Adv Cell Biol 33:285–302

Koyama H, Kawamura A, Kihara T, Hara T, Takita E, Shibata D (2000) Overexpression of mitochondrial citrate synthase in Arabidopsis thaliana improved growth on a phosphorus-limited soil. Plant Cell Physiol 41:1030–1037

Krupa Z, Siedlecka A, Maksymiec W, Baszyński T (1993) In vivo response of photosynthetic apparatus of Phaseolus vulgaris L. to nickel toxicity. J Plant Physiol 142:664–668

Kumar G, Kushwaha HR, Panjabi-Sabharwal V, Kumari S, Joshi R, Karan R et al (2012) Clustered metallothionein genes are co-regulated in rice and ectopic expression of OsMT1e-Pconfers multiple abiotic stress tolerance in tobacco via ROS scavenging. BMC Plant Biol 12:107

Latinwo LM, Donald C, Ikediobi C, Silver S (1998) Effects of intracellular glutathione on sensitivity ofEscherichia colito mercury and arsenite. Biochem Biophys Res Commun 242: 67–70

Lee J, Bae H, Jeong J, Lee JY, Yang YY, Hwang I et al (2003) Functional expression of a bacterial heavy metal transporter in Arabidopsis enhances resistance to and decreases uptake of heavy metals. Plant Physiol 133:589–596

Li Y, Dhankher OP, Carreira L, Balish RS, Meagher RB (2005) Arsenic and mercury tolerance and cadmium sensitivity in Arabidopsis plants expressing bacterial γ-glutamylcysteine synthetase. Environ Toxicol Chem 24:1376–1386

Lima AIG, Corticeiro SC, Figueira EMDAP (2006) Glutathione-mediated cadmium sequestration in Rhizobium leguminosarum. Enzyme Microb Technol 39:763–769

Liszewska F, Sirko A (2003) Analysis of transgenic tobacco lines expressing bacterial cysK gene encoding O-acetylserine (thiol) lyase A. In: Sulphur transport and assimilation in plants. Regulation, interaction and signaling. Backhuys, Leiden, pp 269–271

Liu D, An Z, Mao Z, Ma L, Lu Z (2015) Enhanced heavy metal tolerance and accumulation by transgenic sugar beets expressing Streptococcus thermophilus StGCS-GS in the presence of Cd, Zn and Cu alone or in combination. PLoS One 10:e0128824

Lund PA, Brown NL (1987) Role of the merT and merP gene products of transposon Tn501 in the induction and expression of resistance to mercuric ions. Gene 52:207–214
Ma Z, Jacobsen FE, Giedroc DP (2009) Coordination chemistry of bacterial metal transport and sensing. Chem Rev 109:4644–4681
Mishra S, Gupta D, Ranjan R (2019) Molecular approaches for enhancing abiotic stress tolerance in plants: profiling and counter action. https://doi.org/10.1201/9781351104722-23
Misra S, Gedamu L (1989) Heavy metal tolerant transgenic Brassica napusL. and Nicotiana tabacum L. plants. Theor Appl Genet 78:161–168
Nagata T, Ishikawa C, Kiyono M, Pan-Hou H (2006a) Accumulation of mercury in transgenic tobacco expressing bacterial polyphosphate. Biol Pharm Bull 29:2350–2353
Nagata T, Kiyono M, Pan-Hou H (2006b) Engineering expression of bacterial polyphosphate kinase in tobacco for mercury remediation. Appl Microbiol Biotechnol 72:777–782
Nagata T, Morita H, Akizawa T, Pan-Hou H (2010) Development of a transgenic tobacco plant for phytoremediation of methylmercury pollution. Appl Microbiol Biotechnol 87:781–786
Nagata T, Nakamura A, Akizawa T, Pan-Hou H (2009) Genetic engineering of transgenic tobacco for enhanced uptake and bioaccumulation of mercury. Biol Pharm Bull 32:1491–1495
Naik M, Dubey SK (2011) Lead-enhanced siderophore production and alteration in cell morphology in a Pb-resistant Pseudomonas aeruginosa strain 4EA. Curr Microbiol 62:409–414
Naik MM, Pandey A, Dubey SK (2012) Pseudomonas aeruginosa strain WI-1 from Mandovi estuary possesses metallothionein to alleviate lead toxicity and promotes plant growth. Ecotoxicol Environ Safety 79:129–133
Nakashima K, Ito Y, Yamaguchi-Shinozaki K (2009) Transcriptional regulatory networks in response to abiotic stresses in Arabidopsis and grasses. Plant Physiol 149:88–95
Naylor LH (1999) Reporter gene technology: the future looks bright. Biochem Pharmacol 58:749–757
Nies DH (1999) Microbial heavy-metal resistance. Appl Microbiol Biotechnol 51:730–750
Olafson RW, McCubbin WD, Kay CM (1988) Primary- and secondary structural analysis of a unique prokaryotic metallothionein from a Svnechococctls sp. cyanobactcrimn. Biochem J 251: 691–699
Osman D, Cavet JS (2010) Bacterial metal-sensing proteins exemplified by ArsR–SmtB family repressors. Nat Prod Rep 27:668–680
Pilon-Smits EA, Hwang S, Lytle CM, Zhu Y, Tai JC, Bravo RC et al (1999) Overexpression of ATP sulfurylase in Indian mustard leads to increased selenate uptake, reduction, and tolerance. Plant Physiol 119:123–132
Reisinger S, Schiavon M, Terry N, Pilon-Smits EAH (2008) Heavy metal tolerance and accumulation in Indian mustard (Brassica JunceaL.) expressing bacterial γ-glutamylcysteine synthetase or glutathione synthetase. Int J Phytoremediation 10:440–454
Rensing C, Mitra B, Rosen BP (1997) The zntA gene of Escherichia coli encodes a Zn (II)-translocating P-type ATPase. Proc National Acad Sci 94:14326–14331
Roane TM (1999) Lead resistance in two bacterial isolates from heavy metal- contaminated soils. Microb Ecol 37:218–224
Rodríguez-Llorente ID, Lafuente A, Doukkali B, Caviedes MA, Pajuelo E (2012) Engineering copper hyperaccumulation in plants by expressing a prokaryotic copC gene. Environ Sci Technol 46:12088–12097
Rugh CL, Seneco JF, Meagher RB, Merkle SA (1998) Development of transgenic yellow poplar for mercury phytoremediation. Nat Biotechnol 16:925–928
Rugh CL, Wilde HD, Stack NM, Thompson DM, Summers AO, Meagher RB (1996) Mercuric ion reduction and resistance in transgenic Arabidopsis thaliana plants expressing a modified bacterial merA gene. Proc Natl Acad Sci 93:3182–3187
Ruiz ON, Daniell H (2009) Genetic engineering to enhance mercury phytoremediation. Curr Opin Biotechnol 20:213–219
Salt DE, Smith RD, Raskin I (1998) Phytoremediation. Annu Rev Plant Physiol Plant Mol Biol 49: 643–668

Sasaki Y, Hayakawa T, Inoue C, Miyazaki A, Silver S, Kusano T (2006) Generation of mercury-hyperaccumulating plants through transgenic expression of the bacterial mercury membrane transport protein MerC. Transgenic Res 15:615

Sasaki Y, Minakawa T, Miyazaki A, Silver S, Kusano T (2005) Functional dissection of a mercuric ion transporter, MerC, from Acidithiobacillus ferrooxidans. Biosci Biotechnol Biochem 69: 1394–1402

Schiavon M, Galla G, Wirtz M, Pilon-Smits EA, Telatin V, Quaggiotti S et al (2012) Transcriptome profiling of genes differentially modulated by sulfur and chromium identifies potential targets for phytoremediation and reveals a complex S–Cr interplay on sulfate transport regulation in B. juncea. J Hazard Mater 239:192–205

Sekhar K, Priyanka B, Reddy VD, Rao KV (2011) Metallothionein 1 (CcMT1) of pigeonpea (Cajanus cajan, L.) confers enhanced tolerance to copper and cadmium in Escherichia coli and Arabidopsis thaliana. Environ Exp Bot 72:131–139

Sentchilo V, Mayer AP, Guy L, Miyazaki R, Tringe SG, Barry K et al (2013) Community-wide plasmid gene mobilization and selection. The ISME J 7:1173–1186

Shi J, Lindsay WP, Huckle JW, Morby AP, Robinson NJ (1992) Cyanobacterial metallothionein gene expressed in Escherichia coli metal-binding properties of the expressed protein. FEBS Lett 303:159–163

Silver S (1996) Bacterial resistances to toxic metal ions—a review. Gene 179:9–19

Sirko A, Błaszczyk A, Liszewska F (2004) Overproduction of SAT and/or OASTL in transgenic plants: a survey of effects. J Exp Bot 55(404):1881–1888

Stearns JC, Shah S, Greenberg BM, Dixon DG, Glick BR (2005) Tolerance of transgenic canola expressing 1-aminocyclopropane-1-carboxylic acid deaminase to growth inhibition by nickel. Plant Physiol Biochem 43:701–708

Summers AO (1986) Organization, expression, and evolution of genes for mercury resistance. Annu Rev Microbiol 40:607–634

Thirumoorthy N, Sunder AS, Kumar KM, Ganesh GNK, Chatterjee M (2011) A review of metallothionein isoforms and their role in pathophysiology. World J Surg Oncol 9:54

Thomine S, Wang R, Ward JM, Crawford NM, Schroeder JI (2000) Cadmium and iron transport by members of a plant metal transporter family in Arabidopsis with homology to Nramp genes. Proc Nation Acad Sci 97:4991–4996

Uraguchi S, Sone Y, Yoshikawa A, Tanabe M, Sato H, Otsuka Y et al (2019) SCARECROW promoter-driven expression of a bacterial mercury transporter MerC in root endodermal cells enhances mercury accumulation in Arabidopsis shoots. Planta 250:667–674

Van Huysen T, Terry N, Pilon-Smits EA (2004) Exploring the selenium phytoremediation potential of transgenic Indian mustard overexpressing ATP sulfurylase or cystathionine-gamma-synthase. Int J Phytoremediation 6:111-8

Vido K, Spector D, Lagniel G, Lopez S, Toledano MB, Labarre J (2001) A proteome analysis of the cadmium response in Saccharomyces cerevisiae. J Biol Chem 276:8469–8474

Waldron KJ, Robinson NJ (2009) How do bacterial cells ensure that metalloproteins get the correct metal? Natl Rev 6:25–35

Waldron KJ, Rutherford JC, Ford D, Robinson NJ (2009) Metalloproteins and metal sensing. Nature 460:823–830

Watanabe M, Shinmachi F, Noguchi A, Hasegawa I (2005) Introduction of yeast metallothionein gene (CUP1) into plant and evaluation of heavy metal tolerance of transgenic plant at the callus stage. Soil Sci Plant Nutr 51:129–133

Wilber CG (1983) Selenium: a potential environmental poison and a necessary food constituent. Thomas, Springfield

Williams PN, Lei M, Sun GX, Huang Q, Lu Y, Deacon C, Meharg AA, Zhu YG (2009) Occurrence and partitioning of cadmium, arsenic and lead in mine impacted paddy rice: hunan. China Environ Sci Technol 43:637–642

Xiang C, Werner BL, E'Lise MC, Oliver DJ (2001) The biological functions of glutathione revisited in Arabidopsis transgenic plants with altered glutathione levels. Plant Physiol 126:564–574

Xu J, Tian Y-S, Peng R-H, Xiong A-S, Zhu B, Hou X-L, Yao Q-H (2009) Cyanobacteria MT gene SmtA enhance zinc tolerance in Arabidopsis. Mol Biol Rep 37:1105–1110

Xu S, Sun B, Wang R, He J, Xia B, Xue Y, Wang R (2017) Overexpression of a bacterial mercury transporter MerT in Arabidopsis enhances mercury tolerance. Biochem Biophysi Res Comm 490:528–534

Yuan CG, Lu XF, Qin J, Rosen BP, Le XC (2008) Volatile arsenic species released from Escherichia coli expressing the AsIII S-adenosylmethionine methyltransferase gene. Environ Sci Technol 42:3201–3206

Zhang Y, Zhao L, Wang Y, Yang B, Chen S (2008) Enhancement of heavy metal accumulation by tissue specific co-expression of iaaM and ACC deaminase genes in plants. Chemosphere 72: 564–571

Zhigang A, Cuijie L, Yuangang Z, Yejie D, Wachter A, Gromes R, Rausch T (2006) Expression of BjMT2, a metallothionein 2 from Brassica juncea, increases copper and cadmium tolerance in Escherichia coli and Arabidopsis thaliana, but inhibits root elongation in Arabidopsis thaliana seedlings. J Exp Bot 57:3575–3582

Zhu YL et al (1999) Cadmium tolerance and accumulation in Indian mustard is enhanced by overexpressing γ-glutamylcysteine synthetase. Plant Physiol 121:1169–1177

MIX
Papier aus verantwortungsvollen Quellen
Paper from responsible sources
FSC® C105338

If you have any concerns about our products,
you can contact us on
ProductSafety@springernature.com

In case Publisher is established outside the EU,
the EU authorized representative is:
Springer Nature Customer Service Center GmbH
Europaplatz 3, 69115 Heidelberg, Germany

Printed by Libri Plureos GmbH
in Hamburg, Germany